Photosynthesis Bibliography

volume 7 1976

References no. 25162-28495 / ABD-ZVE

Editors Z. Šesták & J. Čatský

Springer-Science+Business Media, B.V. 1980

Contributors:
Z. Šesták
J. Čatský
I. Tichá
J. Pospíšilová
J. Solárová
D. Hodáňová

ISBN 978-90-6193-046-4   ISBN 978-94-009-9166-8 (eBook)
DOI 10.1007/978-94-009-9166-8

PREFACE

The bibliography includes papers in all fields of photosynthesis research - from studies of model biochemical and biophysical systems of the photosynthesis mechanism to primary production studied by the so-called growth analysis. In addition to papers devoted entirely to photosynthesis, papers on other topics are included if they contain data on photosynthetic activity, photorespiration, chloroplast structure, chlorophyll and carotenoid synthesis and destruction, *etc.*, or if they contain valuable methodological information (measurement of selected environmental factors, leaf area, *etc.*). In many branches it has been very difficult to define the limits of interest for photosynthesis researchers. This problem has arisen *e.g.* in topics dealing with the transport of gases, where - in addition to the papers on CO_2 transfer - some papers on water vapour transfer are included, these being of general application. On the other hand, many papers dealing with the anatomy and physiology of stomata have been omitted, if the aspect of carbon dioxide or water vapour exchange has not been discussed.

This volume contains references to papers published in the year 1976, and, similarly to Vol. 6, also addenda including references published in the preceding period (*i.e.* 1966 - 1975). The numbers of these additional references are labelled with an asterisk in the list of references.

To maximize the value of the bibliography the references are arranged alphabetically by authors' names, and each volume is provided with three indexes. The Authors' Index contains all names of authors, co-authors and editors. The Subject Index covers only primary items chosen according to their interest for photosynthesis researchers. Starting with Volume 6, the Subject Index has been newly arranged and enlarged. It contains more details on the electron transport chain, carbon fixation pathways, gas exchange on leaf and canopy level, *etc.* In the Plant Index, only important crop plants and selected plant types and groups are indexed.

Cumulative indexes accompany Volumes 1, 5, and then every fifth volume, *i.e.* Volumes 10, 15, *etc.*

We have tried to cover fully the relevant papers which have appeared in the most important scientific periodicals and books. Articles published in local journals, mimeographed booklets, *etc.*, were chosen mostly from reprints and lists of publications received directly from the authors. Only abstracts published in regular journals were included.

Since some 3000 relevant papers are currently published every year and included in this bibliography, and since the majority of citations have been checked with the originals, collecting and preparing for publication of such a large amount of material would have been impossible without the collaboration of the authors of the relevant publications. The courtesy of those authors who have already supplied us with reprints is highly appreciated.

We acknowledge with thanks the cooperation of our colleagues from the Institute of Experimental Botany of the Czechoslovak Academy of Sciences in Prague, especially Mrs. DRAHOMÍRA TĚŽKÁ, Mrs. VLASTA FLORIÁNOVÁ, Mrs. LENKA KOLČABOVÁ, Mrs. MARIE MANDLOVÁ, and Mrs. MARTA ŠMÍDOVÁ who helped in preparing the card material. Mr. PETR ZÁZVORKA and the librarian of the Institute, Mrs. ZORA ZAWOYSKA helped us with checking the references.

Dr. Z. ŠESTÁK and Dr. J. ČATSKÝ

Institute of Experimental Botany
Czechoslovak Academy of Sciences

Flemingovo n. 2
CS-160 00 PRAHA 6
Czechoslovakia

INSTRUCTIONS FOR USE

All references are arranged alphabetically according to the authors' names and the year of publication. They are numbered and these numbers are used in the indexes. In case of a book title, the number is preceded by B. An asterisk preceding the number denotes the reference published in the preceding period (1966 - 1975).

The references contain the original unshortened title of the paper (book). English, French and German titles are cited in the original language. Titles in other languages are supplemented with a translation in English (sometimes using the title of the respective English abstract or a shortened title with omitted deadweight words). Titles of Japanese, Chinese *etc.* papers are given in English translation only. The journals' names are abbreviated mainly according to the "Style Manual for Biological Journals" (Second Edition, Amer. Institute of Biological Sciences, Washington, D.C. 1964), *e.g.* :

Abhandlungen	chinese	Industry	Publishers
Abstract	Chromatography	inorganic	quantitative
Abteilung	Commission	Institute	Quarterly
Academy	Communication	international	Radiation
Acta	comparative	Investigation	Radiobiology
Africa	Comptes rendus	italian	Rastenií
agricultural	Conference	Izvestiya	Recherche
Agriculture	Congress	Jahrbuch	Report
Agronomy	Contribution	japanese	Research
Akademie (-emiya)	Cytochemistry	Japan	Review
Algology	Cytology	Journal	royal
allgemeine	czechoslovak	Klasse	russian
american	Dendrology	Laboratory	russkií
America	Department	Landwirtschaft	scandinavicus
analytical	Deutschland	Letters	Science
Anatomy	Disease	Limnology	Section
angewandte	Dissertation	Magazin	Series (-iya)
Annals	Doklady	marine	Society
annual	Dopovidi	Mathematics	sovetskií
anorganisch (-nic)	Ecology	Microbiology	soviet
applied	Education	miscellaneous	special
Arbeit	Embryology	molecular	SSSR
Archiv	Encyclopedia	Monograph	Station
Atmosphere	Engineer	moskovskií	Supplement
atomic	Enzymology	Mycology	Survey
Australia	european	national	Symposium
Beiheft	experimental	natural	technical
Belgique	Experiment	Naturforschung	Technology
Bericht	Faculty	neerlandicus	Tijdschrift
biochemical	Federation	Netherland	Transaction
Biochemistry	Fizika	New Zealand	Travail (-aux)
biokhimicheskií	Fiziologiya	nuclear	tropical
Biokhimiya	Forestry	Oceanography	Trudy
biological(-ogicheskií)	Forschung	Optics	ukrainian
Biology (-ogiya)	Foundation	organic	UK
biophysical	France	original	US, USA
Biophysics	Gazette	Otdelenie	USSR
Bodenkunde	general	Pathology	University
bolgarskií	genetical	Pflanzen-	végétal
botanical (-anicheskií)	Genetics	Philosophy	Virology
Botany	Gesellschaft	physical	Virusforschung
british	Giornale	Physics	Volume
Bulletin	helveticus	physiological	Weekblad
Canada	Histochemistry	Physiology	Wetenschappen
cellular (-ulaire)	Histology	Phytopathology	Wissenschaft
central	Horticulture	Plant (-arum)	Zeitschrift
chemical	hungaricus	polish	Zeitung
Chemistry	Husbandry	Proceedings	Zentralblatt
chimicus	imperial	Publication	Zhurnal

The numbers at the end of each reference of a journal article denote : volume (issue) : first page - last page, year of publication. The number of issue is given only in the journal where each issue is paginated separately.

Book titles are cited according to the title page, not to the book jacket or cover.(if the names of the editors are not given on the title page, they are not cited in the reference). The publishing house, place and year of publication are included.

Brackets at the end of the reference give bibliographic details and explanations to the contents, not given in the original. The following abbreviations are used most often :

ab	abstract	Jap.	Japanese
Arm.	Armenian	Latv.	Latvian
Belorus.	Belorussian	Lithu.	Lithuanian
Bulg.	Bulgarian	Norweg.	Norwegian
Car	carotenoids	PC	paper chromatography
CC	column chromatography	PhAR	photosynthetically active radiation
Chin	Chinese		
Chl	chlorophyll	Pol.	Polish
Croat.	Croatian	Ps	photosynthesis
E	English	R	Russian
F	French	Roum.	Roumanian
G	German	Span.	Spanish
GC	gas chromatography	Swed.	Swedish
Georg.	Georgian	TLC	thin-layer chromatography
Hung.	Hungarian	Tr	transpiration
IRGA	infra-red gas analyser	Ukr.	Ukrainian
Ital.	Italian	Uz.	Uzbeg

The transliteration of Cyrillic characters is in accordance with the BSI-ASA//SC-Z39 draft table, *i.e.* :

Translit.	Cyrill.	Translit.	Cyrill.
a	а	p	п
b	б	r	р
ch	ч	s	с
d	д	sh	ш
e	е	shch	щ
ê	э	t	т
f	ф	ts	ц
g	г	u	у
i	и	v	в
ï	й	y	ы
k	к	ya	я
kh	х	yu	ю
l	л	z	з
m	м	zh	ж
n	н	"	ъ
o	о	'	ь

Several exceptions apply for Ukrainian, Belorussian and Serbian :

Translit.	Cyrill.		Translit.	Cyrill.	Translit.	Cyrill.
Ukr. y	и	Serbian :	ć	ħ	č	ч
і	і		dj	ђ	š	ш
ï	ї		dž	џ	c	ц
Beloruss.			h	х		
ŭ	ў		j	j		
			lj	љ		
			nj	њ		

Authors' names are presented in spelling used in the original paper. If this spelling does not correspond to the original spelling used by the author (*e.g.* Russian papers of English authors), one spelling is referred to the other in the Authors' Index.

Printers' errors in the original papers are marked by underlining the respective words (letters).

E R R A T A

Ref. no.	For	Read
Volume 1, Part 1		
361	Respitation	Respiration
Volume 1, Part 2		
7217	vieillisement	vieillissement
Volume 2		
9955	229	299
10311	resistance	resistances
Volume 4		
17212 - MIZRAKH, S.A.	1721	17212
Volume 5, Part 2		
p. 281 Age ... chlorophyll	18431	19431
Volume 6		
21537	pigmnet	pigment
24780	evapotranspirazione	evapotraspirazione
24827	dela	de la
p. 225		GOVÏNDJEE
		GOVINDJEE, R.
p. 282 Leaf chamber ...	.chember	chamber
p. 285 Model of canopy ...	23471	23871
p. 286 O_2 and gas ...	22442	22542

25162 - **ABDEL-RAHMAN, M.** : Effects of fungicide programs on tree growth, vigor, leaf development, chlorophyll contents and fruit set of apples. - HortScience *11* (3, Sect. 2) : 318, 1976.

25163 - **ABELIOVICH, A., AZOV, Y.** : Toxicity of ammonia to algae in sewage oxidation ponds. - Appl. environ. Microbiol. *31* : 801 - 806, 1976. [Ps.]

*25164 - **ABERNETHY, R.H., WRIGHT, L.N.** : Inflorescence gas exchange and seed weight of blue panicgrass. - Crop Sci. *15* : 383 - 384, 1975.

*25165 - **ACKEFORS, H., HERNROTH, L., LINDAHL, O.** : Studies on the primary and secondary production of plankton in the Baltic. - Medd. Havsfiskelab. Lysekil *186* : 1 - 3, 1975.

25166 - **ACKER, S., PICAUD, A., DURANTON, J.** : Des activités photosynthétiques sans les complexes pigmentaires CP1 et CP2. - Biochim. biophys. Acta *440* : 269 - - 277, 1976.

25167 - **ACOCK, B., CHARLES-EDWARDS, D.A., HAND, D.W.** : An analysis of some effects of humidity on photosynthesis by a tomato canopy under winter light conditions and a range of carbon dioxide concentrations. - J. exp. Bot. *27* : 933 - - 941, 1976.

25168 - **ACOCK, B., HAND, D.W., THORNLEY, J.H.M., WARREN WILSON, J.** : Photosynthesis in stands of green peppers. An application of empirical and mechanistic models to controlled-environment data. - Ann. Bot. *40* : 1293 - 1307, 1976.

25169 - **ADAMS, J.E., ARKIN, G.F., RITCHIE, J.T.** : Influence of row spacing and straw mulch on first stage drying. - Soil Sci. Soc. Amer. J. *40* : 436 - 442, 1976. [Growth analysis.]

25170 - **ADEDIPE, N.O.** : Comparative sensitivity of ^{14}C-sucrose and ^{32}P translocation to water stress pulse in *Hibiscus esculentus* (okra). - Biochem. Physiol. Pflanzen *169* : 347 - 353, 1976.

25171 - **ADEDIPE, N.O., FLETCHER, R.A., ORMROD, D.P.** : Distribution of ^{14}C assimilates in the cowpea (*Vigna unguiculata* L.) in relation to fruit abscission and treatment with benzyladenine. - Ann. Bot. *40* : 731 - 737, 1976.

25172 - **ADELANA, B.O.** : Effects of plant density on tomato yields in Western Nigeria. - Exp. Agr. *12* : 43 - 47, 1976. [Growth analysis.]

25173 - **ADJEI-TWUM, D.C., SPLITTSTOESSER, W.E.** : The effect of soil water regimes on leaf water potential, growth and development of soybeans. - Physiol. Plant. *38* : 131 - 137, 1976. [Ps.]

25174 - **ADLER, K.** : Sequential synthesis of membrane polypeptides in chloroplast thylakoids of synchronized *Chlorella pyrenoidosa* cells. - Plant Sci. Lett. *6* : 261 - 266, 1976.

25175 - **ADYGEZALOV, V.F., GRODZINSKIĬ, D.M.** : Izuchenie metabolicheskoĭ prirody fotoindutsiruemykh biopotentsialov lista. [Metabolic nature of photoinduced biopotentials in the leaf.] - Fiziol. Biokhim. kul't. Rast. *8* : 601 - 606, 1976. [In R, ab : E.]

25176 - **AĔROV, I.L.** : Vliyanie rasseyaniya sveta mezhkletnikami na opticheskie parametry list'ev. [Effect of light scattering by intercellular spaces on optical parameters of leaves.] - Fiziol. Biokhim. kul't. Rast. *8* : 94 - 98, 1976. [In R, ab : E.]

25177 - **AFLAFO, C., SHAVIT, N.** : Phosphorylation of nucleotides bound to chloroplast membranes and their role in photophosphorylation. - Biochim. biophys. Acta *440* : 522 - 530, 1976.

25178 - **AKANOV, Ĕ.N.** : Metod avtomaticheskogo izmereniya i registratsii gazoobmena v fitotrone. [Automatic measuring and recording of gas exchange in a phytotron.] - In : Gazometricheskoe Issledovanie Fotosinteza i Dykhaniya Rasteniĭ. Pp. 7 - 8. Akad. Nauk SSSR, Tartu 1976. [In R.]

25179 - **AKAZAWA, T., OSMOND, C.B.** : Structural properties and ribulosebisphosphate carboxylase and oxygenase activity of fraction-1 protein from the marine alga *Halimeda cylindracea (Chlorophyta)*. - Aust. J. Plant Physiol. *3* : 93 - 103, 1976.

25180 - ÅKERLUND, H.-E., ANDERSSON, B., ALBERTSSON, P.-Å. : Isolation of Photosystem II enriched membrane vesicles from spinach chloroplasts by phase partition. - Biochim. biophys. Acta *449* : 525 - 535, 1976.

25181 - AKERS, C.P. : Tobacco leaf trichomes : ultrastructure of glandular trichomes. - Plant Physiol. *57* (Suppl.) : 44, 1976. [Chloroplast.]

25182 - AKULOVA, E.A. : Dva puti transporta èlektronov v khloroplastakh : fosforili-ruyushchiĭ i nefosforiliruyushchiĭ. [Two ways of electron transport in chloroplasts : phosphorylating and non-phosphorylating.] - In : Itogi Issledovaniya Mekhanizma Fotosinteza. Pp. 29 - 34. Pushchino 1976. [In R.]

25183 - AKULOVA, E.A., ROSHCHINA, V.V. : Svetoindutsirovannye okislitel'no-vosstano-vitel'nye prevrashcheniya tsitokhroma f v list'yakh gorokha. [Light-induced redox conversions of cytochrome f in pea leaves.] - Biokhimiya *41* : 1742 - - 1747, 1976. [In R, ab : E.]

25184 - ALBERTE, R.S., HESKETH, J.D., KIRBY, J.S. : Comparisons of photosynthetic activity and lamellar characteristics of virescent and normal green peanut leaves. - Z. Pflanzenphysiol. *77* : 152 - 159, 1976.

25185 - ALBERTE, R.S., HESKETH, J.D., THORNBER, J.P., KLEINHOFS, A. : Comparisons of photosynthetic activity and chloroplast lamellar characteristics of chlorophyll mutants of barley. - Plant Physiol. *57* (Suppl.) : 72, 1976.

25186 - ALBERTE, R.S., McCLURE, P.R., THORNBER, J.P. : Photosynthesis in trees. Organization of chlorophyll and photosynthetic unit size in isolated gymnosperm chloroplasts. - Plant Physiol. *58* : 341 - 344, 1976.

25187 - ALBERTINE, K.H., MARAVOLO, N.C., KAUSTINEN, H. : Effects of gibberellin and several growth retardants on plastid ultrastructure in the hepatic *Marchantia polymorpha*. - Bryologist *79* : 22 - 34, 1976.

25188 - ALBERTSSON, P.-Å., LARSSON, C. : Properties of chloroplasts isolated by phase partition. - Mol. cell. Biochem. *11* : 183 - 189, 1976.

25189 - ALDERFER, R.G., EAGLES, C.F. : The effects of partial defoliation on the growth and photosynthetic efficiency of bean leaves. - Bot. Gaz. *137* : 351 - - 355, 1976.

25190 - ALEXANDER, A.G., BIDDULPH, O. : Effects of Paraquat on carbon assimilation and transport by sugarcane leaves in white light and darkness. - Plant Cell Physiol. *17* : 601 - 609, 1976.

25191 - ALFANO, R.R., YU, W., GOVINDJEE, R., BECHER, B., EBREY, T.G. : Picosecond kinetics of the fluorescence from the chromophore of the purple membrane protein of *Halobacterium halobium*. - Biophys. J. *16* : 541 - 545, 1976.

25192 - ALIEV, K.A., VASIL'EVA, V.N. : Deĭstvie khloramfenikola i tsiklogeksimida na sintez i karboksilaznuyu aktivnost' ribulezo-1,5-difosfat-karboksilazy. [Effect of chloramphenicol and cycloheximide on the synthesis and carboxylating activity of ribulose-1,5-diphosphate carboxylase.] - Fiziol. Rast. *23*:786 - - 791, 1976. [In R, ab : E.]

25193 - ALLAWAY, W.G. : Influence of stomatal behaviour on long distance transport. - In : WARDLAW, I.F., PASSIOURA, J.B. (ed.) : Transport and Transfer Processes in Plants. Pp. 295 - 311. Academic Press, New York - San Francisco - London 1976. [Photosynthates.]

25194 - ALLAWAY, W.G., MILTHORPE, F.L. : Structure and functioning of stomata. - In : KOZLOWSKI, T.T. (ed.) : Water Deficits and Plant Growth. Vol. IV. Pp. 57 - 102. Academic Press, New York - San Francisco - London 1976. [Ps.]

25195 - ALLEN, L.H., Jr., LEMON, E.R. : Carbon dioxide exchange and turbulence in a Costa Rican tropical rain forest. - In : MONTEITH, J.L. (ed.) : Vegetation and the Atmosphere. Vol. 2. Case Studies. Pp. 265 - 308. Academic Press, London - New York - San Francisco 1976.

25196 - ALLEN, M.J., CRANE, A.E. : Null potential voltammetry - an approach to the study of plant photosystems. - Bioelectrochem. Bioenerg. *3* : 84 - 91, 1976.

25197 - ALLEN, M.M., HUTCHISON, F. : Effect of some environmental factors on cyanophage AS-1 development in *Anacystis nidulans*. - Arch. Microbiol. *110* : 55 - 60, 1976. [Ps.]

25198 - **ALLEN, R.J., Jr.** :   A five-year comparison of solar radiation and sugarcane production in the Everglades agricultural area. - Soil Crop Sci. Soc. Florida Proc. *36* : 197 - 200, 1976. [Prediction.]

25199 - **ALSCHER, R., SMITH, M.A., PETERSEN, L.W., HUFFAKER, R.C., CRIDDLE, R.S.** : *In vitro* synthesis of the large subunit of ribulose diphosphate carboxylase on 70 S ribosomes. - Arch. Biochem. Biophys. *174* : 216 - 225, 1976.

25200 - **ALSCHER, R.G., HAWKES, S.P., SAUER, K.** : The association of protein synthesis with protochlorophyllide holochrome regeneration in dark-grown barley leaves. - Biochem. biophys. Res. Commun. *73* : 240 - 247, 1976.

25201 - **ALVAGER, T., GROSS, J.A., MAGYAR, R.** : Nanosecond fluorescence study of chloroplast systems. - Proc. indian Acad. Sci. *85* : 343 - 349, 1976.

25202 - **AMESZ, J., DE GROOTH, B.G.** : Photosynthetic electron transport and electrochromic effects at sub-zero temperatures. - Biochim. biophys. Acta *440* : 301- - 313, 1976.

25203 - **AMIRDZHANOV, A.G., KIRPICHEV, I.V., PRIVALOVA, E.A.** : Matematicheskoe opisanie zavisimosti radiatsionnogo rezhima krony vinogradnogo kusta ot fitometricheskikh kharakteristik. [Mathematical description of the effect of phytometric characteristics on the radiation regime of the crown of grapevine.] - Fiziol. Rast. *23* : 58 - 66, 1976. [In R, ab : E.]

*25204 - **AMIRI, Z.** : Contribution à l'etude des pigments chlorophylliens de la feuille de tabac. - Ann. Dir. Etud. Equip. SEITA, Sect. 2, *8* : 73 - 121, 1971. [Chl determination.]

25205 - **AMPOFO, S.T., MOORE, K.G., LOVELL, P.H.** : The role of the cotyledons in four *Acer* species and in *Fagus sylvatica* during early seedling development. - New Phytol. *76* : 31 - 39, 1976. [Ps, Chl.]

25206 - **AMPOFO, S.T., MOORE, K.G., LOVELL, P.H.** : Cotyledon photosynthesis during seedling development in *Acer*. - New Phytol. *76* : 41 - 52, 1976.

25207 - **AMPOFO, S.T., MOORE, K.G., LOVELL, P.H.** : The influence of leaves on cotyledon photosynthesis and export during seedling development in *Acer*. - New Phytol. *76* : 247 - 255, 1976.

25208 - **ANDERSAG, R., PIRSON, A.** :   Verwertung von Glukose in Chlorellakulturen bei Blau- und Rotlichtbestrahlung. - Biochem. Physiol. Pflanzen *169* : 71 - 85, 1976. [Chl.]

25209 - **ANDERSEN, A.S.** :   Regulation of apical dominance by ethephon, irradiance and CO_2. - Physiol. Plant. *37* : 303 - 308, 1976. [Growth analysis.]

25210 - **ANDERSEN, F.Ø.** :   Primary production in a shallow water lake with special reference to a reed swamp. - Oikos *27* : 243 - 250, 1976. [Chl.]

25211 - **ANDERSON, L.E., AVRON, M.** : Light modulation of enzyme activity in chloroplasts. Generation of membrane-bound vicinal-dithiol groups by photosynthetic electron transport. - Plant Physiol. *57* : 209 - 213, 1976.

25212 - **ANDERSON, L.E., AVRON, M.** : Reductive generation of vicinal-dithiols by photosynthetic electron transport system is involved in light-regulation of chloroplast enzyme activity. - In : SHALTIEL, S. (ed.) : Metabolic Interconversion of Enzymes 1975. Pp. 220 - 226. Springer-Verlag, Berlin - Heidelberg - New York 1976.

25213 - **ANDERSON, L.E., DUGGAN, J.X.** : Light modulation of glucose-6-phosphate dehydrogenase. Partial characterization of the light inactivation system and its effects on the properties of the chloroplastic and cytoplasmic forms of the enzyme. - Plant Physiol. *58* : 135 - 139, 1976.

25214 - **ANDERSON, L.E., NEHRLICH, S.C.** : Are the dark forms of the light-modulated carbon metabolism enzymes Hill reagents ? - Plant Physiol. *57* (Suppl.) : 5, 1976.

25215 - **ANDERSON, O.R.** : Respiration and photosynthesis during resting cell formation in *Amphora coffeaeformis* (AG.)KÜTZ. - Limnol. Oceanogr. *21* : 452 - 456, 1976.

25216 - **ANDERSON, W.R., JOHNSON, M., LAYCOCK, R., SMITH, B.N.** : $^{13}C/^{12}C$ fractionation by ribulose bisphosphate carboxylase and phosphoenolpyruvate carboxylase *in vitro* as influenced by environmental factors. - Plant Physiol. *57* (Suppl.) : 5, 1976.

25217 - **ANDERSSON, B., ÅKERLUND, H.-E., ALBERTSSON, P.-Å.** : Separation of subchloroplast membrane particles by counter-current distribution. - Biochim. biophys. Acta *423* : 122 - 132, 1976.

25218 - **ANDRE, C., VERCRUYSSE, A.** : Seasonal fluctuations of the carotene (α,β) content in freshwater phytoplankton. - Acta bot. neerl. *25* : 495 - 501, 1976.

25219 - **ANDREEVA, N.E., CHIBISOV, A.K.** : Vozbuzhdennye kompleksy v okislitel'no-vosstanovitel'nykh fotoreaktsiyakh pigmentov. I. Spektral'noe obnaruzhenie éksipleksa v tripletnom sostoyanii v reaktsii okisleniya khlorofilla. [Excited complexes in oxidative-reductive photoreactions of pigments. I. Spectral observation of exciplex in the triplet state of chlorophyll photooxidation.]-Biofizika *21* : 24 - 28, 1976. [In R, ab : E.]

25220 - **ANDREEVA, N.E., PESHKIN, A.F., CHIBISOV, A.K.** : Vliyanie spektral'nogo sostava sveta na mekhanizm reaktsii okisleniya khlorofilla. [Effect of wavelength on the mechanism of chlorophyll photooxidation.] - Biofizika *21* : 29 - - 34, 1976. [In R, ab : E.]

25221 - **ANDREEVA, T.F., AVDEEVA, T.A.** : Adaptatsiya fotosinteza C_3- i C_4-rasteniĭ k usloviyam vneshneĭ sredy. [Adaptation of C_3- and C_4-plants photosynthesis to environment.] - Fiziol. Biokhim. kul't. Rast. *8* : 236 - 241, 1976. [In R, ab : E.]

25222 - **ANDREO, C.S., VALLEJOS, R.H.** : Sulphydryl groups in photosynthetic energy conservation. I. Light-dependent inhibition of photophosphorylation by the sulphydryl reagent 2-2'dithio bis-(5-nitropyridine). - Biochim. biophys. Acta *423* : 590 - 601, 1976.

25223 - **ANDREWS, A.K., SVEC, L.V.** : Pod and leaf photosynthesis and disease incidence in soybean (*Glycine max* (L.) MERR.) with potassium fertilization. - Commun. Soil Sci. Plant Anal. *7* : 345 - 363, 1976.

25224 - **ANGUS, J.F., WILSON, J.H.** : Photosynthesis of barley and wheat leaves in relation to canopy models. - Photosynthetica *10* : 367 - 377, 1976.

25225 - **ANIKEENKO, A.P.** : K opredeleniyu krasyashchikh veshchestv v krasnom pertse. [Determination of colouring agents in red pepper.] - Fiziol. Biokhim. kul't. Rast. *8* : 437 - 438, 1976. [Car; in R, ab : E.]

25226 - **ANISIMOV, A.A., LEONT'EVA, A.N., KON'KOVA, E.A., OLYUNINA, L.N., GANICHEVA, O.P., NEIMAN, V.A.** : Metabolicheskaya regulyatsiya transporta assimilyatov. [Metabolic regulation of photosynthate transport.] - Fiziol. Biokhim. kul't. Rast. *8* : 367 - 371, 1976. [In R, ab : E.]

25227 - **ANTON, J.A., KWONG, J., LOACH, P.A.** : Synthesis of covalently linked porphyrin dimers and trimers (1). - J. heterocyclic Chem. *13* : 717 - 725, 1976. [Ps.]

25228 - **APEL, K., MILLER, K.R., BOGORAD, L., MILLER, G.J.** : Chloroplast membranes of the green alga *Acetabularia mediterranea*. II. Topography of the chloroplast membrane. - J. Cell Biol. *71* : 876 - 893, 1976.

25229 - **APEL, P.** : Beziehungen zwischen Photosyntheseleistung und Ertragsbildung bei Getreide. - Wiss. Z. Humboldt-Univ. Berlin, math.-naturwiss. Reihe *25* : 747 - - 751, 1976.

25230 - **APEL, P.** : Grain growth and carbohydrate content in spring wheat at different CO_2-concentrations. - Biochem. Physiol. Pflanzen *169* : 355 - 362, 1976.

25231 - **APEL, P.** : Untersuchungen über die Beeinflußbarkeit des Einzelährenertrags von Sommerweizen. - Experimente in Pflanzenwachstumskammern. - Kulturpflanze *24* : 143 - 157, 1976. [Yield formation.]

25232 - **APEL, P., NÄTR, L.** : Carbohydrate content and grain growth in wheat and barley. - Biochem. Physiol. Pflanzen *169* : 437 - 446, 1976. [Production.]

25233 - **APELBAUM, A., GOLDSCHMIDT, E.E., BEN-YEHOSHUA, S.** : Involvement of endogenous ethylene in the induction of color change in Shamouti oranges. - Plant Physiol. *57* : 836 - 838, 1976. [Chl.]

25234 - **APONASENKO, A.D., FRANK, N.A., SID'KO, F.Ya.** : Differentsial'nyĭ spektrofoto-
metr dlya gidroopticheskikh issledovaniĭ. [Differential spectrophotometer for
hydrooptical studies.] - Okeanologiya *16* : 924 - 928, 1976. [Chl; in R, ab :
E.]

*25235 - **ap REES, T.** : Changes in the activity of fructose-1,6 diphosphate during the
differentiation of photosynthetic cells. - In : HALL, D.O., HAWKINS, S.E.
(ed.) : Laboratory Manual of Cell Biology. Pp. 174 - 176. The English Univer-
sities Press, London 1975. [Enzyme determination.]

25236 - **ARAD (MALIS), S., RICHMOND, A.E.** : Leaf cell water and enzyme activity. -
Plant Physiol. *57* : 656 - 658, 1976. [Chl.]

25237 - **ARGYROUDI-AKOYUNOGLOU, J.H., KONDYLAKI, S., AKOYUNOGLOU, G.** : Growth of grana
from "primary" thylakoids in *Phaseolus vulgaris*. - Plant Cell Physiol. *17* :
939 - 954, 1976.

25238 - **ARKIN, G.F., VANDERLIP, R.L., RITCHIE, J.T.** : A dynamic grain sorghum growth
model. - Trans. ASAE *19* : 622 - 626, 1976. [Ps.]

25239 - **ARMENTANO, T.V., WOODWELL, G.M.** : The production and standing crop of litter
and humus in a forest exposed to chronic *gamma* irradiation for twelve years.
- Ecology *57* : 360 - 366, 1976.

25240 - **ARMOND, P.A., ARNTZEN, C.J., BRIANTAIS, J.-M., VERNOTTE, C.** : Differentiation of chloroplast lamellae. Light harvesting efficiency and grana development. - Arch. Biochem. Biophys. *175* : 54 - 63, 1976.

25241 - **ARMOND, P.A., STAEHELIN, L.A., ARNTZEN, C.J.** : A model for the spatial relationships of Photosystem I, Photosystem II and the light harvesting complex
in chloroplast membranes. - Biophys. J. *16* (2, Part 2) : 160 a, 1976.

25242 - **ARNASON, T., SINCLAIR, J.** : Studies on the rate-limiting reaction of photosynthetic oxygen evolution in spinach chloroplasts. - Biochim. biophys. Acta
430 : 517 - 523, 1976.

25243 - **ARNASON, T., SINCLAIR, J.** : An investigation of water splitting in the Kok
scheme of photosynthetic oxygen evolution. - Biochim. biophys. Acta *449* :
581 - 586, 1976.

25244 - **ARNOLD, W.** : Path of electrons in photosynthesis. - Proc. nat. Acad. Sci. USA
73 : 4502 - 4505, 1976.

25245 - **ARNTZEN, C.J., ARMOND, P.A., DITTO, C.H.** : Dynamic interactions between pigment-protein complexes in chloroplasts. - Biophys. J. *16* (2, Part 2) : 160 a,
1976.

25246 - **ARNTZEN, C.J., DITTO, C.L.** : Effects of cations upon chloroplast membrane subunit interactions and excitation energy distribution. - Biochim. biophys. Acta
449 : 259 - 274, 1976.

25247 - **ARPIN, N., SVEC, W.A., LIAAEN-JENSEN, S.** : New fucoxanthin-related carotenoids from *Coccolithus huxleyi*. - Phytochemistry *15* : 529 - 532, 1976.

25248 - **ASADA, K.** : [Photosynthesis and photooxidative damage.] - J. agr. chem. Soc.
Jap. *50* : R115 - R120, 1976. [In Jap.]

25249 - **ASAMI, S., AKAZAWA, T.** : Biosynthetic mechanism of glycolate in *Chromatium*
III. Effects of α-hydroxy-2-pyridinemethanesulfonate, glycidate and cyanide
on glycolate excretion. - Plant Cell Physiol. *17* : 1119 - 1129, 1976.

25250 - **ASCENSO, J.C., SOOST, R.K.** : Relationships between leaf surface area and linear dimensions in seedling *Citrus* populations. - J. amer. Soc. hort. Sci.
101 : 696 - 698, 1976.

25251 - **ASTON, M.J.** : Variation of stomatal diffusive resistance with ambient humidity
in sunflower (*Helianthus annuus*). - Aust. J. Plant Physiol. *3* : 489 - 501,
1976.

*25252 - **ATANASIU, L.** : Photosynthesis and respiration of three mosses at winter low
temperatures. - Bryologist *74* : 23 - 27, 1971.

25253 - **AUCLAIR, A.N.D., BOUCHARD, A., PAJACZKOWSKI, J.** : Plant standing crop and productivity relations in a *Scirpus-Equisetum* wetland. - Ecology *57* : 941 - 952,
1976.

25254 - AUCLAIR, A.N.D., BOUCHARD, A., PAJACZKOWSKI, J. : Productivity relations in
a *Carex*-dominated ecosystem. - Oecologia *26* : 9 - 31, 1976.

25255 - AUCLAIR, D. : Effets des poussières sur la photosynthèse. I. Effets des poussières de ciment et de charbon sur la photosynthèse de l'Épicéa. - Ann. Sci.
forest. *33* : 247 - 255, 1976.

25256 - AUFHAMMER, W., SOLANSKY, S. : Beeinflussung der Assimilatspeicherungsprozesse
in der Sommergerstenähre durch Kinetinbehandlungen. - Z. Pflanzenernähr. Bodenkunde *131* : 503 - 515, 1976.

25257 - AUGUSTIN, P. : Messung der Photosyntheserate an Blattscheiben. I. Aufbau einer
kleinen 4-Kammerküvette zur gleichzeitigen Messung des CO_2-Gaswechsels mehre-
rer Varianten. - Arch. Gartenbau *24* : 121 - 123, 1976.

25258 - AUGUSTIN, P. : Messung der Photosyntheserate an Blattscheiben. II. Physiolo-
gisches Verhalten der Blattscheiben der Gewächshausgurke (*Cucumis sativus*
L.) während der Messung der Photosyntheserate. - Arch. Gartenbau *24* : 125 -
- 133, 1976.

25259 - AUGUSTINE, J.J., STEVENS, M.A., BREIDENBACH, R.W. : Inheritance of carboxylation efficiency in the tomato. - J. amer. Soc. hort. Sci. *101* : 456 - 460,
1976.

25260 - AUGUSTINE, J.J., STEVENS, M.A., BREIDENBACH, R.W. : Inheritance of carboxylation efficiency in tomatoes. - HortScience *11* (3, Sect. 2) : 305, 1976.

25261 - AUGUSTINE, J.J., STEVENS, M.A., BREIDENBACH, R.W., PAIGE, D.F. : Genotypic
variation in carboxylation of tomatoes. - Plant Physiol. *57* : 325 - 333,
1976.

25262 - AUSTENFELD, F.-A. : The effect of various alkaline salts on the glycolate
oxidase of *Salicornia europaea* and *Pisum sativum in vitro*. - Physiol. Plant.
36 : 82 - 87, 1976. [Ps.]

25263 - AUSTIN, L.A. : Characterization of light-harvesting bacteriochlorophyll-protein complexes from photosynthetic bacteria. - Report *LBL-5512* : 1 - 4, II -
- X, 1 - 137, 1976.

25264 - AUSTIN, R.B., FORD, M.A., EDRICH, J.A., HOOPER, B.E. : Some effects of leaf
posture on photosynthesis and yield in wheat. - Ann. appl. Biol. *83* : 425 -
- 446, 1976.

25265 - AVETISOV, V.A., DEMCHENKO, S.I., BUTENKO, R.G. : Kallusoobrazovanie i morfogenez v kul'ture normal'nykh i defektnykh po khlorofillu tkanei *Arabidopsis
thaliana*. [Callus formation and morphogenesis in the culture of normal and
chlorophyll deficient tissues of *Arabidopsis thaliana*.] - Fiziol. Rast. *23* :
353 - 360, 1976. [In R, ab : E.]

25266 - AVIRAM, I., PETTIGREW, G.W., SCHEJTER, A. : A spectrophotometric and fluori-
metric study of alkaline transitions of *Euglena* cytochrome *c* 552. - Biochemistry *15* : 635 - 638, 1976.

25267 - AVRON, M. : Energy transduction in isolated chloroplast membranes. - In :
HATEFI, Y., DJAVADI-OHANIANCE, L. (ed.) : The Structural Basis of Membrane
Function. Pp. 227 - 238. Academic Press, New York - San Francisco - London
1976.

25268 - AXELSSON, L. : The photostability of different chlorophyll forms in dark
grown leaves of wheat. I. Stability to high intensity red light of forms
appearing after photoreduction of protochlorophyllide. - Physiol. Plant. *38* :
327 - 332, 1976.

25269 - AXELSSON, L. : The photostability of different chlorophyll forms in dark
grown leaves of wheat. II. Reaction kinetics for the photodecomposition of
the 684-form and 673-form. - Physiol. Plant. *38* : 333 - 336, 1976.

25270 - AYLES, G.B., LARK, J.G.I., BARICA, J., KLING, H. : Seasonal mortality of rain-
bow trout (*Salmo gairdneri*) planted in small eutrophic lakes of central Canada. - J. Fish. Res. Board Can. *33* : 647 - 655, 1976. [Chl.]

25271 - AYRES, P.G. : Patterns of stomatal behaviour, transpiration, and CO_2 exchange in pea following infection by powdery mildew (*Erysiphe pisi*). - J. exp. Bot. *27* : 1196 - 1205, 1976.

25272 - **AZAM, F., CHISHOLM, S.W.** :   Silicic acid uptake and incorporation by natural marine phytoplankton populations. - Limnol. Oceanogr. *21* : 427 - 435, 1976. [Ps.]

25273 - **AZZI, A., MONTECUCCO, C.** : Probes of energy transduction in membranes. - J. Bioenerg. Biomembr. *8* : 257 - 269, 1976. [Chloroplast.]

25274 - **BAARS, J.A.** : Seasonal distribution of pasture production in New Zealand. VIII. Dargaville. - New Zeal. J. exp. Agr. *4* : 151 - 156, 1976.

25275 - **BAARS, J.A.** : Seasonal distribution of pasture production in New Zealand. IX. Hamilton. - New Zeal. J. exp. Agr. *4* : 157 - 161, 1976.

25276 - **BABCOCK, G.T., BLANKENSHIP, R.E., SAUER, K.** : Reaction kinetics for positive charge accumulation on the water side of chloroplast photosystem II. - FEBS Lett. *61* : 286 - 289, 1976.

25277 - **BABENKO, V.I., NARIICHUK, F.D.** : Izmenenie aktivnosti nekotorykh fermentov v khloroplastakh pshenitsy pri zakalivanii. [Changes in activities of some enzymes in wheat chloroplasts.] - Fiziol. Biokhim. kul't. Rast. *8* : 590 - 594, 1976. [In R, ab : E.]

25278 - **BABENKO, V.I., NARIICHUK, F.D.** : Vliyanie ponizhennoĭ temperatury na malat-i glyutamatdegidrogenazu khloroplastov ozimoĭ pshenitsy. [The effects of lowered temperature on malate- and glutamate dehydrogenase of winter wheat chloroplasts.] - Dokl. Akad. Nauk SSSR *228* : 1252 - 1255, 1976. [In R.]

25279 - **BACHE, D.H.** :   Modèle numérique de dispersion des polluants atmosphériques en presence de couverts végétaux. - Atmos. Environ. *10* : 675, 1976. [Micro-climatological methods.]

25280 - **BACHMANN, P., KORNMANN, P., ZETSCHE, K.** :   Regulation der Entwicklung und des Stoffwechsels der Grünalge *Urospora* durch die Temperatur. - Planta *128* : 241 - 245, 1976. [Ps.]

25281 - **BACHOFEN, R., HANSELMANN, K., SNOZZI, M., ZÜRRER, H.** : Reactioncenter complexes (RC) from chromatophore membranes of *Rhodospirillum rubrum*. - Experientia *32* : 785, 1976.

25282 - **BACKMAN, T.W., BARILOTTI, D.C.** : Irradiance reduction : Effects on standing crops of the eelgrass *Zostera marina* in a coastal lagoon. - Mar. Biol. *34* : 33 - 40, 1976.

25283 - **BACONE, J., BAZZAZ, F.A., BOGGESS, W.R.** : Correlated photosynthetic responses and habitat factors of two successional tree species. - Oecologia *23* : 63 - - 74, 1976.

25284 - **BADGER, M.R., LORIMER, G.H.** : Activation of ribulose-1,5-bisphosphate oxygenase. The role of Mg^{2+}, CO_2, and pH. - Arch. Biochem. Biophys. *175* : 723 - 729, 1976.

25285 - **BADGER, M.R., LORIMER, G.H.** : Activation of RuDP oxygenase by CO_2 and Mg^{2+}. - Plant Physiol. *57* (Suppl.) : 4, 1976.

*25286 - **BAGGE, P., NIEMI, Å.** :   Dynamics of phytoplankton primary production and bio-mass in Loviisa archipelago (Gulf of Finland). - Merentutkimuslait. Julk./ /Havsforskningsinst. Skr. *233* : 19 - 41, 1971.

25287 - **BAHL, J., FRANCKE, B., MONÉGER, R.** : Lipid composition of envelopes, prola-mellar bodies and other plastid membranes in etiolated, green and greening wheat leaves. - Planta *129* : 193 - 201, 1976.

25288 - **BAHL, J., MONÉGER, R.** : Sur les lipides des corps lamellaires isolés de feu-illes de Blé étiolées et verdissantes. - Compt. rend. Acad. Sci. Paris, Sér. D *282* : 177 - 180, 1976.

25289 - **BAHL, J., MONÉGER, R.** : Sur les lipides plastidiaux de feuilles étiolées de Blé exposées à des lumières oligochromatiques. - Compt. rend. Acad. Sci. Pa-ris, Sér. D *282* : 1001 - 1004, 1976.

25290 - BAHR, J.T., JENSEN, R.G. : Activity of RuBP carboxylase in intact spinach chloroplasts. - Plant Physiol. 57 : 4, 1976.

25291 - BAIER, W., DAVIDSON, H., DESJARDINS, R.L., OUELLET, C.E., WILLIAMS, G.D.V. : Recent biometeorological applications to crops. - Int. J. Biometeorol. 20 : 108 - 127, 1976. [Ps.]

25292 - BAIRD, B.A., HAMMES, G.G. : Chemical cross-linking studies of chloroplast coupling factor 1. - J. biol. Chem. 251 : 6953 - 6962, 1976.

25293 - BAKER, N.R., BUTLER, W.L. : Development of the primary photochemical apparatus of photosynthesis during greening of etiolated bean leaves. - Plant Physiol. 58 : 526 - 529, 1976.

25294 - BAKER, N.R., HARDWICK, K. : Development of the photosynthetic apparatus in cocoa leaves. - Photosynthetica 10: 361 - 366, 1976.

25295 - BAKER, N.R., STRASSER, R.J., BUTLER, W.L. : Development of primary photochemical activity in greening etiolated bean leaves. - Plant Physiol. 57 (Suppl.) : 95, 1976.

25296 - BAKKER, E.P., ROTTENBERG, H., CAPLAN, S.R. : An estimation of the light-induced electrochemical potential difference of protons across the membrane Halobacterium halobium. - Biochim. biophys. Acta 440 : 557 - 572, 1976.

25297 - BAKR AHMED, M., ASHOUR, N.I., EL-BASYOUNI, S.Z., SAYED, A.M. : Response of photosynthetic apparatus of corn (Zea mays) to presowing seed treatment with gamma rays and ammonium molybdate. - Environ. exp. Bot. 16 : 217 - 222, 1976.

25298 - BALASHOV, S.P., LITVIN, F.F. : Pervichnye fotokhimicheskie reaktsii bakteriorodopsina v purpurnykh membranakh galobakteriĭ pri 4 °K. [Initial photochemical transformations of bacteriorhodopsin in the Halobacterium purple membranes at 4 K.] - Bioorg. Khim. 2 : 565 - 566, 1976. [In R.]

25299 - BALASUBRAMANIAN, V., SINHA, S.K. : Effects of salt stress on growth, nodulation and nitrogen fixation in cowpea and mung beans. - Physiol. Plant. 36 : 197 - 200, 1976. [Photosynthates.]

25300 - BALDING, F.R., CUNNINGHAM, G.L. : A comparison of heat transfer characteristics of simple and pinnate leaf models. - Bot. Gaz. 137 : 65 - 74, 1976.

25301 - BALNOKIN, Yu.V., NIKIFOROVA, T.A. : Metod vydeleniya intaktnykh khloroplastov, ne zagryaznennykh drugimi kletochnymi komponentami. [Procedure for isolating intact chloroplasts free of other cell components contamination.] - Biokhimiya 41 : 1573 - 1575, 1976. [In R, ab : E.]

25302 - BALTSCHEFFSKY, M. : The carotenoid absorbance change and energy transduction in chromatophores. - Biophys. J. 16 (2, Part 2) : 223 a, 1976.

25303 - BAMBERG, S.A., VOLLMER, A.T., KLEINKOPF, G.E., ACKERMAN, T.L. : A comparison of seasonal primary production of Mojave desert shrubs during wet and dry years. - Amer. Midland Naturalist 95 : 398 - 405, 1976.

25304 - BANASZAK, J., BARR, R., CRANE, F.L. : Evidence for multiple sites of ferricyanide reduction in chloroplasts. - J. Bioenerg. 8 : 83 - 92, 1976.

25305 - BANCROFT, K., PAUL, E.A., WIEBE, W.J. : The extraction and measurement of adenosine triphosphate from marine sediments. - Limnol. Oceanogr. 21 : 473 - - 480, 1976.

25306 - BANKS, M.S., BONNETT, H.T. : The uptake of chloroplasts by higher plant protoplasts. - Plant Physiol. 57 (Suppl.) : 51, 1976.

25307 - BANSE, K. : Rates of growth, respiration and photosynthesis of unicellular algae as related to cell size - A review. - J. Phycol. 12 : 135 - 140, 1976.

25308 - BARADAS, M.W., BLAD, B.L., ROSENBERG, N.J. : Reflectant induced modification of soybean canopy radiation balance. IV. Leaf and canopy temperature. - Agron. J. 68 : 843 - 848, 1976.

25309 - BARADAS, M.W., BLAD, B.L., ROSENBERG, N.J. : Reflectant induced modification of soybean canopy radiation balance. V. Longwave radiation balance. - Agron. J. 68 : 848 - 852, 1976.

25310 - **BARANOV, A.A., SAAKOV, V.S., BARANOVA, L.S., BOYARSHINOVA, G.A., RUTMAN, G.I.:**
Analiz spektrov pogloshcheniya plastid pri issledovanii reaktsiĭ ustoĭchivosti rasteniĭ k ėkstremal'nym vozdeĭstviyam. [Analysis of absorption spectra
of plastids in studies of resistance of plants to extreme influences.] -
Byull. vses. nauch. issled. Inst. Rastenievod. Im. N.I. Vavilova *63* : 3 - 14,
1976. [In R.]

25311 - **BARASHKOVA, Ė.A., BADINA, G.V., SMIRNOVA, V.S. :** Vliyanie vozdeĭstviya niz-
kikh temperatur pri prorastanii semyan raznykh kul'tur na izmenenie nativno-
go sostoyaniya khlorofill-belkovogo kompleksa list'ev. [Effect of low temperatures during germination of seeds of various crops on changes in native
state of chlorophyll-protein complex of leaves.] - Byull. vses. nauch. issled.
Inst. Rastenievod. Im. N.I. Vavilova *63* : 55 - 59, 1976. [In R.]

25312 - **BĂRBAT, I., SUCIU, T., SALONTAI, A., MUNTEAN, L. :** Studiul unor procese fi-
ziologice la principalele soiuri de hamei cultivate în țara noastră. [Photo-
synthesis and respiration in relation to the yield of five productive hop varieties.] - Bul. Inst. agron. Cluj-Napoca *30* : 37 - 40, 1976. [In Roum., ab :
E.]

*25313 - **BARBER, J. :** Delayed light emission from chloroplasts as an indicator of electrical gradients across the thylakoid membranes. - J. Physiol. *223* (1) : 23P -
- 24P, 1972.

25314 - **BARBER, J. :** Cation control in photosynthesis. - Trends biochem. Sci. *1* (2) :
33 - 36, 1976.

25315 - **BARBER, J. :** Ionic regulation in intact chloroplasts and its effect on primary photosynthetic processes. - In : **BARBER, J.** (ed.) : The Intact Chloroplast.
Pp. 89 - 134. Elsevier, Amsterdam - New York - Oxford 1976.

25316 - **BARBER, J., MILLS, J. :** Control of chlorophyll fluorescence by the diffuse
double layer. - FEBS Lett. *68* : 288 - 292, 1976.

25317 - **BARINOV, G.V. :** Kinetyka fotosyntezu ta okyslyuval'no-vidnovnyĭ mekhanizm
perenesennya ėnergiĭ nukleotydtryfosfatamy. [Kinetics of photosynthesis and
redox mechanism of energy transfer by nucleotidetriphosphates.] - Dopovidi
Akad. Nauk ukr. RSR, Ser. B *1976* : 740 - 742, 1976. [In Ukr., ab : E.]

25318 - **BARLOW, E.W.R., BOERSMA, L. :** Interaction between leaf elongation, photosynthesis, and carbohydrate levels of water-stressed corn seedlings. - Agron. J.
68 : 923 - 926, 1976.

25319 - **BARLOW, E.W.R., BOERSMA, L., YOUNG, J.L. :** Root temperature and soil water potential effects on growth and soluble carbohydrate concentration of corn seedlings. - Crop Sci. *16* : 59 - 62, 1976.

25320 - **BARR, R., CRANE, F.L. :** Control of photosynthesis by CO_2 : Evidence for a bi-
carbonate-inhibited redox feedback in Photosystem II. - Proc. Indiana Acad.
Sci. *85* : 120 - 128, 1976.

25321 - **BARR, R., CRANE, F.L. :** Inhibition of spinach chloroplast electron transport
by radicals and partial reversal by α-tocopherol and analogs. - Plant Physiol.
57 (Suppl.) : 24, 1976.

25322 - **BARR, R., CRANE, F.L. :** Organization of electron transport in photosystem II
of spinach chloroplasts according to chelator inhibition sites. - Plant Physiol. *57* : 450 - 453, 1976.

25323 - **BARSKIĬ, E.L., KONDRASHIN, A.A., SAMUILOV, V.D., SKULACHEV, V.P. :** Rekonstruk-
tsiya funktsii obrazovaniya membrannogo potentsiala izolirovannymi pigment-
-belkovymi kompleksami *Rhodospirillum rubrum*. [Reconstitution of function of
membrane potential generation by isolated pigment-protein complexes from *Rhodospirillum rubrum*.] - Biokhimiya *41* : 513 - 519, 1976. [In R, ab : E.]

25324 - **BARTA, A.L. :** Transport and distribution of $^{14}CO_2$ assimilate in *Lolium perenne* in response to varying nitrogen supply to halves of a divided root system.
- Physiol. Plant. *38* : 48 - 52, 1976.

25325 - **BARTAKKE, S.P., JOSHI, G.V. :** Crassulacean acid metabolism and photosynthesis
in *Aloe barbadensis* MILL. - Proc. Indian nat. Sci. Acad. *42* : 227 - 233, 1976.

25326 - **BARTHAKUR, N.** : Stomatal response to microwave induced thermal stresses. - J. Microwave Power *11* : 247 - 254, 1976. [Stomatal resistance.]

25327 - **BARTHOLOMEW, D.P., KADZIMIN, S.B.** : Porometer cup to measure leaf resistance of pineapple. - Crop Sci. *16* : 565 - 568, 1976.

25328 - **BASAK, A., JANKIEWICZ, L.S.** : The effect of defoliants on photosynthesis on the level of chemical constituents in the leaves, and on the shoot anatomy of apple trees. - Fruit Sci. Rep. (Skierniewice) *3* : 19 - 33, 1976.

25329 - **BASIOUNY, F.M., BIGGS, R.H.** : Photosynthesis and carbonic anhydrase activity in Zn-deficient peaches irradiated with ultraviolet light. - HortScience *11* : 408 - 410, 1976.

25330 - **BASIOUNY, F.M., BIGGS, R.H.** : Rates of photosynthesis and the Hill reaction in *Citrus* seedlings affected by Fe, Mn, and Zn nutrition. - J. amer. Soc. hort. Sci. *101* : 193 - 196, 1976.

25331 - **BASSHAM, J.A.** : Photosynthesis. - Proc. Robert A. Welch Found. Conf. chem. Res. *19* (Photon Chem.) : 279 - 316, 1976.

25332 - **BASSI, P.K., TREGUNNA, E.B., JOLLIFFE, P.A.** : Carbon dioxide exchange and phytochrome control of flowering in *Xanthium pennsylvanicum*. - Can. J. Bot. *54* : 2881 - 2887, 1976.

25333 - **BATES, S.S.** : Effects of light and ammonium on nitrate uptake by two species of estuarine phytoplankton. - Limnol. Oceanogr. *21* : 212 - 218, 1976. [Chl.]

25334 - **BATTERSBY, A.R., McDONALD, E.** : Biosynthesis of porphyrins and corrins. - Phil. Trans. roy. Soc. London *B273* : 161 - 180, 1976. [Chl.]

25335 - **BAUER, K., WILD, A.** : Die Wirkung von Blaulicht auf den photosynthetischen Elektronentransport bei gelbgrünen Mutanten von *Chlorella fusca*. - Z. Pflanzenphysiol. *80* : 443 - 454, 1976.

25336 - **BAUER, M.E., PENDLETON, J.W., BEUERLEIN, J.E., GHORASHY, S.R.** : Influence of terminal bud removal on the growth and seed yield of soybeans. - Agron. J. *68* : 709 - 711, 1976. [Growth analysis.]

25337 - **BAUER, P.-J., DENCHER, N.A., HEYN, M.P.** : Evidence from chromophore-chromophore interactions in the purple membrane from reconstitution experiments of the chromophore-free membrane. - Biophys. Struct. Mech. *2* : 79 - 92, 1976.

25338 - **BAUER, R., HUBER, W., SANKHLA, N.** : Effect of abscisic acid on photosynthesis in *Lemna minor* L. - Z. Pflanzenphysiol. *77* : 237 - 246, 1976.

*25339 - **BAUER, R., WIJNANDS, M.J.G.** : The inhibition of photosynthetic electron transport by DBMIB and its restoration by *p*-phenylenediamines; studied by means of prompt and delayed chlorophyll fluorescence of green algae. - Z. Naturforsch. *29C* : 725 - 732, 1974.

25340 - **BAUMANN, G., GÜNTHER, G.** : The metabolism of glycolate in green and amitrole-chlorotic oat leaves. - Biochem. Physiol. Pflanzen *169* : 337 - 345, 1976.

25341 - **BAUMERT, H.** : Mathematisches Modell zur Deutung der durch intermittierende Belichtung von Phytoplanktern hervorgerufenen Mehrleistung der Photosynthese. - Int. Rev. ges. Hydrobiol. *61* : 517 - 527, 1976.

25342 - **BAUMERT, H.** : Abschätzung von Turbulenzkorrekturen für die phytoplanktische O_2-Produktion bei schwacher Turbulenz. - Int. Rev. ges. Hydrobiol. *61* : 627 - 637, 1976.

25343 - **BAUMERT, H.** : Über Probleme der mathematischen Modellierung aquatischer Ökosysteme. - Acta hydrochim. hydrobiol. *4* : 565 - 579, 1976.

25344 - **BAZZAZ, M.B., REBEIZ, C.A.** : Repair, maintenance and degradation of mature chloroplasts incubated in a simple medium. - Plant Physiol. *57* (Suppl.) : 72, 1976.

25345 - **BEALE, S.I.** : The biosynthesis of δ-aminolaevulinic acid in plants. - Phil. Trans. roy. Soc. London B *273* : 99 - 108, 1976.

25346 - **BEARDALL, J., MUKERJI, D., GLOVER, H.E., MORRIS, I.** : The path of carbon in photosynthesis by marine phytoplankton. - J. Phycol. *12* : 409 - 417, 1976.

25347 - **BEARDEN, A.J., MALKIN, R.** : Correlation of reaction-center chlorophyll (P-700) oxidation and bound iron-sulfur protein photoreduction in chloroplast photosystem I at low temperatures. - Biochim. biophys. Acta *430* : 538 - 547, 1976.

25348 - **BEARDEN, A.J., MALKIN, R.** : Photon-rate dependent photochemistry of chloroplast Photosystem I at low temperatures. - Biophys. J. *16* (2, Part 2) : 160a, 1976.

25349 - **BEAUMONT, G., BASTIN, R., THERRIEN, H.P.** : Effets physiologiques de l'atrazine à doses sublétales sur *Lemna minor* L. II. Influence sur la photosynthèse et sur la respiration. - Nat. can. (Quebec) *103* : 535 - 541, 1976.

25350 - **BEAVER, J.E., FINN, R.D., HUPF, H.B.** : A new method for the production of high concentration oxygen-15 labeled carbon dioxide with protons. - Int. J. appl. Rad. Isotop. *27* : 195 - 197, 1976.

25351 - **BECHER, B., CASSIM, J.Y.** : Effects of light adaptation on the purple membrane structure of *Halobacterium halobium*. - Biophys. J. *16* : 1183 - 1200, 1976.

25352 - **BECHER, B., EBREY, T.G.** : Evidence for chromophore-chromophore (exciton) interaction in the purple membrane of *Halobacterium halobium*. - Biochem. biophys. Res. Commun. *69* : 1 - 6, 1976.

25353 - **BECKER, J.F., BRETON, J., GEACINTOV, N.E., TRENTACOSTI, F.** : Anisotropy of photosynthetic membranes and the degree of fluorescence polarization. - Biochim. biophys. Acta *440* : 531 - 544, 1976.

25354 - **BEDBROOK, J.R., BOGORAD, L.** : Endonuclease recognition sites mapped on *Zea mays* chloroplast DNA. - Proc. nat. Acad. Sci. USA *73* : 4309 - 4313, 1976.

25355 - **BEDDARD, G.** : Chlorophyll dimer and triplet states - key roles in photosynthesis ? - Nature *263* : 459 - 460, 1976.

25356 - **BEDDARD, G.S., PORTER, G.** : Concentration quenching in chlorophyll. - Nature *260* : 366 - 367, 1976.

25357 - **BEDENKO, V.P., FEDYUSHIN, A.A., USHAROVA, G.P., INTYKBAEVA, B.B., MALEVA, M. N.** : Fotosinteticheskaya deyatel'nost' i produktivnost' ozimoĭ pshenitsy v posevakh na yugo-vostoke Kazakhstana. [Photosynthetic activity and productivity of winter wheat in crops of the south-eastern Kazakhstan.] - In : Fotosintez i Produktivnost' Ozimoĭ Pshenitsy na Yugo-Vostoke Kazakhstana. Pp. 3- - 20, 129. Nauka kazakh. SSR, Alma-Ata 1976. [In R.]

25358 - **BEDENKO, V.P., USHAROVA, G.P.** : Teplotvornaya sposobnosť ozimoĭ pshenitsy. [Energy accumulating ability of winter wheat.] - In : Fotosintez i Produktivnosť Ozimoĭ Pshenitsy na Yugo-Vostoke Kazakhstana. Pp. 50 - 57, 130 - 131. Nauka kazakh. SSR, Alma-Ata 1976. [In R.]

25359 - **BEEVERS, H.** : Crassulacean acid metabolism : an assessment. - In : BURRIS, R.H., BLACK, C.C. (ed.) : CO_2 Metabolism and Plant Productivity. Pp. 405 - - 406. Univ. Park Press, Baltimore - London - Tokyo 1976.

25360 - **BEHNKE, H.-D.** : Ultrastructure of sieve-element plastids in *Caryophyllales (Centrospermae)*, evidence for the delimitation and classification of the order. - Plant Syst. Evol. *126* : 31 - 54, 1976.

25361 - **BEKASOVA, O.D., EVSTIGNEEV, V.B.** : Ob obratimom fotookislenii allofikotsianina i fikotsianina. [Reversible photooxidation of allophycocyanin and phycocyanin.] - Izv. Akad. Nauk SSSR, Ser. biol. *1976* (1) : 149 - 154, 1976. [In R, ab : E.]

25362 - **BEKINA, R.M., LEBEDEVA, A.F., RUBIN, B.A.** : O lokalizatsii reaktsii Melera s ètanolkatalaznoĭ lovushkoĭ v tsepi fotosinteticheskogo transporta èlektronov. [Localization of Mehler reaction with ethanolcatalase trap in the photosynthetic electron transport chain.] - Biokhimiya *41* : 815 - 821, 1976. [In R, ab : E.]

25363 - **BEKINA, R.M., LEBEDEVA, A.F., SHUVALOV, V.A.** : Nechuvstvitel'noe k diuronu vosstanovlenie kisloroda v fotosisteme II v prisutstvii kremnievomolibdenovoĭ kisloty. [Diuron-insensitive reduction of oxygen in Photosystem 2 in the presence of molybdosilicic acid.] - Dokl. Akad. Nauk SSSR *231* : 739 - 742, 1976. [In R.]

25364 - **BELL, J.N.B., MUDD, C.H.** : Sulphur dioxide resistance in plants : a case study of *Lolium perenne*. - In : **MANSFIELD, T.A.** (ed.) : Effects of Air Pollutants on Plants. Pp. 87 - 103. Cambridge Univ. Press, Cambridge - London - New York - Melbourne 1976. [Chl.]

25365 - **BELL, L.N.** : Nekotorye osobennosti énergetiki fotosinteziruyushchikh rasteniĭ. [Some peculiarities of energetics of photosynthesizing plants.] - In : Itogi Issledovaniya Mekhanizma Fotosinteza. Pp. 62 - 65. Pushchino 1976. [In R.]

25366 - **BELL, L.N., SHUVALOVA, N.P.** : Vozmozhnoe fiziologicheskoe znachenie fotodykhaniya. [Possible physiological role of photorespiration.] - In : Gazometricheskoe Issledovanie Fotosinteza i Dykhaniya Rasteniĭ. Pp. 9 - 10. Akad. Nauk SSSR, Tartu 1976. [In R.]

25367 - **BELL, L.N., SHUVALOVA, N.P., VOLKOVA, T.V., KRUPENKO, A.N.** : Discrepancy between gas exchange and energy storage rates in *Chlorella*. - Photosynthetica *10* : 147 - 154, 1976.

25368 - **BELYAEVA, M.G., LUKPANOV, Zh.L.** : Vliyanie bordoskoĭ zhidkosti na obmen veshchestv v tkanyakh ogurtsov. [Effect of Bordeaux mixture on the metabolism in cucumber tissues.] - Khim. sel'. Khoz. *1976* (7) : 57 - 60, 1976. [Chl; in R.]

*25369 - **BELYAKOVA, T.N., KADZYAUSKAS, Yu.P., SKULACHEV, V.P., SMIRNOVA, I.A., CHEKULAEVA, L.N., YASAITIS, A.A.** : Generatsiya élektrokhimicheskogo potentsiala ionov H^+ i fotofosforilirovanie v kletkakh *Halobacterium halobium*. [Generation of electrochemical potential of H^+-ions and photophosphorylation in *Halobacterium halobium* cells.] - Dokl. Akad. Nauk SSSR *223* : 483 - 486, 1975. [In R.]

*25370 - **BEN-AMOTZ, A., GIBBS, M.** : H_2 metabolism in photosynthetic organisms. II. Light-dependent H_2 evolution by preparations from *Chlamydomonas, Scenedesmus* and spinach. - Biochem. biophys. Res. Commun. *64* : 355 - 359, 1975.

25371 - **BENECKE, U., HAVRANEK, W., McCRACKEN, I.** : Comparative study of water use by tree species in a mountain environment. - In : Proceedings of Soil and Plant Water Symposium. Pp. 191 - 199. Palmerston North 1976. [Ps.]

25372 - **BENEDETTI, E., DE, FORTI, G., GARLASCHI, F.M., ROSA, L.** : On the mechanism of ammonium stimulation of photosynthesis in isolated chloroplasts. - Plant Sci. Lett. *7* : 85 - 90, 1976.

25373 - **BENEDICT, C.R.** : RuDPCase in non-photosynthetic endosperms of germinating castor beans. - Plant Physiol. *57* (Suppl.) : 5, 1976.

25374 - **BENEDICT, C.R., SCOTT, J.R.** : Photosynthetic carbon metabolism of a marine grass. - Plant Physiol. *57* : 876 - 880, 1976.

25375 - **BEN-HAYYIM, G., DRECHSLER, Z., NEUMANN, J.** : Photosystem 2 mediated electron transport and phosphorylation with ferricyanide and dibromothymoquinone. The uncoupling activity of dibromothymoquinone. - Plant Sci. Lett. *7* : 171 - 178, 1976.

25376 - **BENK, E., TREIBER, H., BERGMANN, R.** : Zum Nachweis von *Tagetes*-Extrakt in Orangensaftkonzentraten, entsprechenden Grundstoffen und Getränken. - Riechstoffe, Aromen, Körperpflegemittel *26* (10) : 216, 218, 220 - 221, 1976. [Car high pressure liquid chromatography.]

25377 - **BENNETT, J.** : Inhibition of chloroplast development by tentoxin. - Phytochemistry *15* : 263 - 265, 1976.

25378 - **BENNOUN, P., CHUA, N.H.** : Methods for the detection and characterization of photosynthetic mutants in *Chlamydomonas reinhardi*. - In : **BÜCHER, T., NEUPERT, W., SEBALD, W., WERNER, S.** (ed.) : Genetics and Biogenesis of Chloroplasts and Mitochondria. Pp. 33 - 39. North-Holland Publ. Co., Amsterdam - New York - Oxford 1976.

25379 - **BENNOUN, P., JUPIN, H.** : Spectral properties of system I-deficient mutants of *Chlamydomonas reinhardi*. Possible occurrence of uphill energy transfer. - Biochim. biophys. Acta *440* : 122 - 130, 1976.

25380 - **BENSASSON, R., LAND, E.J., MAUDINAS, B.** : Triplet states of carotenoids from photosynthetic bacteria studied by nanosecond ultraviolet and electron pulse irradiation. - Photochem. Photobiol. *23* : 189 - 193, 1976.

25381 - BERDYKULOV, Kh.A., AGZAMOV, A., BAKHRAMDZHANOVA, N.A. : Usvoenie uglekislogo gaza suspenzieĭ khlorelly, kul'tiviruemoĭ barbatazhnym sposobom. [CO$_2$ assimilation by *Chlorella* suspension cultivated by a bubbling method.] - In : Al'goflora i Mikoflora Sredneĭ Azii. Pp. 199 - 202. Fan, Tashkent 1976. [In R.]

25382 - BERDYKULOV, Kh.A., AGZAMOV, A., NURIEVA, D. : Usvoenie rastvorennykh form uglekisloty v zavisimosti ot vozrasta i plotnosti *Chlorella pyrenoidosa* 123. [Assimilation of dissolved forms of CO$_2$ in dependence on growth and density of *Chlorella pyrenoidosa* 123.] - In : Al'goflora i Mikoflora Sredneĭ Azii. Pp. 203 - 207. Fan, Tashkent 1976. [In R.]

25383 - BERESNEV, G.F., SID'KO, F.Ya. : Éffektivnost' fotosinteza mikrovodorosli khlorelly i ee pigmentnyĭ sostav. [Effectivity of photosynthesis of the microscopic alga *Chlorella* and its pigment composition.] - In : Gazometricheskoe Issledovanie Fotosinteza i Dykhaniya Rasteniĭ. P. 11. Akad. Nauk SSSR, Tartu 1976. [In R.]

25384 - BEREZIN, I.V., VARFOLOMEEV, S.D., ZAITSEV, S.V. : Biofotoliz vody. [Biophotolysis of water.] - Dokl. Akad. Nauk SSSR *229* : 94 - 97, 1976. [Ps; in R.]

25385 - BERG, S., IZAWA, S. : Inhibition of electron transport and uncoupling of photophosphorylation in chloroplasts by salicylaldoxime. - Plant Physiol. *57* (Suppl.) : 24, 1976.

25386 - BERG, S.P., IZAWA, S. : Concentration-dependent effects of salicylaldoxime on chloroplast reactions. - Biochim. biophys. Acta *440* : 483 - 494, 1976.

25387 - BERING, C.L., DILLEY, R.A., CRANE, F.L. : Inhibition of energy-transducing functions of chloroplast membranes by lipophilic iron chelators. - Biochim. biophys. Acta *430* : 327 - 335, 1976.

25388 - BERKOVICH, Yu.A., KORBUT, V.L., NORKIN, K.B., PRUT, V.M., SUSLOVA, O.B., TIME, I.V. : Izmerenie bystrykh perekhodnykh protsessov fotosinteza v zakrytykh sistemakh. [Measurement of rapid transient processes of photosynthesis in closed systems.] - In : Gazometricheskoe Issledovanie Fotosinteza i Dykhaniya Rasteniĭ. Pp. 12 - 14. Akad. Nauk SSSR, Tartu 1976. [In R.]

25389 - BERLAND, B.R., BONIN, D.J., MAESTRINI, S.Y. : L'emploi concomitant d'enceintes dialysantes et de tests biologiques pour la détermination des facteurs nutritionnels limitant la production primaire des eaux marines. - Ann. Inst. Océanogr. *52* : 45 - 55, 1976.

25390 - BERMAN, T. : Light penetrance in Lake Kinneret. - Hydrobiologia *49* : 41 - 48, 1976. [Chl.]

25391 - BERMAN, T. : Release of dissolved organic matter by photosynthesizing algae in Lake Kinneret, Israel. - Freshwater Biol. *6* : 13 - 18, 1976.

*25392 - BERNDORFERNÉ KRASZNER, É., TELEGDY KOVÁTS, L. : Néhány újabb adat a plasztokinonok ismeretéhez. [New data on plastoquinones.] - Élelmezési Ipar *27* (3) : 65 - 69, 1973. [In Hung., ab : E, G, R.]

25393 - BERNS, D.S. : Photosensitive bilayer membranes as model systems for photobiological processes. - Photochem. Photobiol. *24* : 117 - 139, 1976.

25394 - BERNS, D.S., CHEN, C.-H., ILANI, A. : The modification by biliproteins of intensity and direction of electron flow across chlorophyll-containing membranes. - In : PULLMAN, B. (ed.) : Environmental Effects on Molecular Structure and Properties. Pp. 547 - 560. D. Reidel Publ. Comp., Dordrecht 1976.

25395 - BERRY, J., BOYNTON, J., KAPLAN, A., BADGER, M. : Growth and photosynthesis of *Chlamydomonas reinhardtii* as a function of CO$_2$ concentration. - Carnegie Inst. Year Book *75* : 423 - 432, 1976.

25396 - BERRY, J.A., BJÖRKMAN, O.E., FORK, D.C. : Damage to the photosynthetic capacity of leaves by high temperature. - Plant Physiol. *57* (Suppl.) : 24, 1976.

25397 - BERTIN, C., PANOUILLÉ, A., RAUTOU, S. : Obtention de variétés de maïs "prolifiques en épis" productives en grain et à large adaption écologique. - Ann. Amélior. Plantes *26* : 387 - 418, 1976.

25398 - **BERTRAMS, M., HEINZ, E.** : Experiments on enzymatic acylation of *sn*-glycerol 3-phosphate with enzyme preparations from pea and spinach leaves. - Planta *132* : 161 - 168, 1976. [Chl.]

25399 - **BERZBORN, R.J., SCHRÖER, P.** : Photophosphorylation : mechanism of reconstitution by coupling factor 1 (CF_1). - FEBS Lett. *70* : 271 - 275, 1976.

25400 - **BESECKE, S., EVANS, B., BARNETT, G.H., SMITH, K.M., FUHRHOP, J.-H.** : Reaktionen von Magnesiumporphyrinaten in imidazolhaltigen Polymeren. - Angew. Chem. *88* (18) : 616, 1976. [Chl.]

25401 - **BETHLENFALVAY, G.J., CASTELFRANCO, P.A.** : Energy conservation in washed chloroplast fragments. - Plant Physiol. *57* (Suppl.) : 23, 1976.

25402 - **BEUHLER, R.J., PIERCE, R.C., FRIEDMAN, L., SIEGELMAN, H.W.** : Cleavage of phycocyanobilin from C-phycocyanon. Separation and mass spectral identification of the products. - J. biol. Chem. *251* : 2405 - 2411, 1976.

25403 - **BEYER, G., CRANE, F.L.** : Thiol inhibition of photosystem I electron transport in spinach chloroplasts. - Plant Physiol. *57* (Suppl.) : 24, 1976.

25404 - **BHAGSARI, A.S., BROWN, R.H.** : Photosynthesis in peanut (*Arachis*) genotypes. - Peanut Sci. *3* : 1 - 5, 1976.

25405 - **BHAGSARI, A.S., BROWN, R.H.** : Translocation of photosynthetically assimilated ^{14}C in peanut (*Arachis*) genotypes. - Peanut Sci. *3* : 5 - 9, 1976.

25406 - **BHAGSARI, A.S., BROWN, R.H.** : Relationship of net photosynthesis to carbon dioxide concentration and leaf characteristics in selected peanut (*Arachis*) genotypes. - Peanut Sci. *3* : 10 - 14, 1976.

25407 - **BHAGSARI, A.S., BROWN, R.H., SCHEPERS, J.S.** : Effect of moisture stress on photosynthesis and some related physiological characteristics in peanut. - Crop Sci. *16* : 712 - 715, 1976.

25408 - **BHAGWAT, A.S., SANE, P.V.** : Studies on enzymes of C_4 pathway : partial purification & kinetic properties of maize phosphoenol pyruvate carboxylase. - Indian J. exp. Biol. *14* : 155 - 158, 1976.

25409 - **BHAGWAT, A.S., SANE, P.V.** : Studies on enzymes of C-4 pathway : Part II - Regulation of CO_2 fixation by malic enzyme in C-4 plants. - Indian J. exp. Biol. *14* : 691 - 693, 1976.

25410 - **BHAN, A.K., KAUL, M.L.H.** : Frequency and spectrum of chlorophyll-deficient mutations in rice after treatment with radiation and alkylating agents. - Mutat. Res. *36* : 311 - 318, 1976.

25411 - **BHATT, J.G.** : Translocation of labelled assimilate in morphologically contrasting cotton plants. - New Phytol. *76* : 53 - 57, 1976.

25412 - **BHATT, J.G., SHAH, R.C., SHARMA, A.N.** : Net assimilation rate of cotton in relation to spacing. - J. agr. Sci. *86* : 281 - 285, 1976.

*25413 - **BHATTY, R.S., WU, K.K.** : Determination of gross energy of cereals and legumes with a ballistic bomb calorimeter. - Can. J. Plant Sci. *54* : 439 - 441, 1974.

25414 - **BIAMONTE, R.L., LANGHANS, R.W.** : Greenhouse measurements of net photosynthesis of the rose "Forever Yours". - HortScience *11* (3, Sect. 2) : 325, 1976.

25415 - **BIAMONTE, R.L., LANGHANS, R.W.** : Laboratory measurements of net photosynthesis of the rose "Forever Yours". - HortScience *11* (3, Sect. 2) : 325, 1976.

25416 - **BICKEL, H., SCHULTZ, G.** : Biosynthesis of plastoquinone and β-carotene in isolated chloroplasts. - Phytochemistry *15* : 1253 - 1255, 1976.

25417 - **BICKEL-SANDKÖTTER, S., STROTMANN, H.** : Effects of external factors on photophosphorylation and exchange of CF_1-bound adenine nucleotides. - FEBS Lett. *65* : 102 - 106, 1976.

25418 - **BIELECKI, K., SKRABKA, H.** : Wpływ niektórych herbicydów na fotosyntezę *Spirodela polyrrhiza (Lemnaceae)*. [Effect of some herbicides on photosynthesis of *Spirodela polyrrhiza (Lemnaceae)*.] - Acta agrobot. *29* : 59 - 68, 1976. [In Pol., ab : E.]

25419 - BIERHUIZEN, J.F. : Irrigation and water use efficiency. - In : LANGE, O.L.,
        KAPPEN, L., SCHULZE, E.-D. (ed.) : Water and Plant Life. Pp. 421 - 431. Sprin-
        ger-Verlag, Berlin - Heidelberg - New York 1976. [Ps.]

25420 - BIL', K.Ya., BELOBRODSKAYA, L.K., KARPILOV, Yu.S. : Lokalizatsiya ATFaz v
        kletochnykh strukturakh assimilyatsionnykh tkaneĭ list'ev shchiritsy. [Loca-
        lization of ATPases in cell structures of assimilation tissues of leaves of
        Amaranthus retroflexus.] - Dokl. Akad. Nauk SSSR 226 : 1229 - 1231, 1976.
        [In R.]

25421 - BILLECOCQ, A. : Structures of membranes : localisation of glycolipids in chlo-
        roplast membranes by immunoenzymatic techniques to electron microscopy. - In :
        FELDMANN, G., DRUET, P., BIGNON, J., AVRAMEAS, S. (ed.) : Immunoenzymatic
        Techniques. INSERM Symposium 2. Pp. 487 - 490. North-Holland Publ. Comp.,
        Amsterdam 1976.

25422 - BILLORE, S.K., MALL, L.P. : Seasonal variation in chlorophyll content of a
        grassland community. - Trop. Ecol. 17 : 39 - 44, 1976.

25423 - BILLORE, S.K., MEHTA, S.C., MALL, L.P. : Changes in chlorophylls a, b and ca-
        rotenoid in summer leaves of tree species in a dry deciduous forest. - J. in-
        dian bot. Soc. 55 : 56 - 59, 1976.

25424 - BILLORE, S.K., SINGH, V.P. : Chlorophyll and carotenoid contents of the tree
        bark in a dry deciduous forest. - Geobios 3 : 94 - 95, 1976.

25425 - BINDER, A., TEL-OR, E., AVRON, M. : Photosynthetic activities of membrane
        preparations of the blue-green alga Phormidium luridum. - Europe. J. Biochem.
        67 : 187 - 196, 1976.

25426 - BINDLOSS, M.E. : The light-climate of Loch Leven, a shallow Scottish lake, in
        relation to primary production by phytoplankton. - Freshwater Biol. 6 : 501 -
        - 518, 1976.

25427 - BINGHAM, S., SCHIFF, J.A. : Cellular origins of plastid membrane polypeptides
        in Euglena. - In : BÜCHER, T., NEUPERT, W., SEBALD, W., WERNER, S. (ed.) :
        Genetics and Biogenesis of Chloroplasts and Mitochondria. Pp. 79 - 86. North-
        -Holland Publ. Co., Amsterdam - New York - Oxford 1976.

25428 - BINGHAM, S., SCHIFF, J.A. : Plastid membrane polypeptides from Euglena graci-
        lis var. bacillaris. - Plant Physiol. 57 (Suppl.) : 73, 1976.

25429 - BIRKY, C.W., Jr. : The inheritance of genes in mitochondria and chloroplasts.
        - BioScience 26 : 26 - 33, 1976.

25430 - BISCOE, P.V., COHEN, Y., WALLACE, J.S. : Daily and seasonal changes of water
        potential in cereals. - Phil. Trans. roy. Soc. London B 273 : 565 - 580, 1976.
        [Ps.]

25431 - BISHOP, D.G., REED, M.L. : The C₄ pathway of photosynthesis : ein Kranz-Typ
        Wirtschaftswunder ? - In : SMITH, K.C. (ed.) : Photochemical and Photobiolo-
        gical Reviews. Vol. 1. Pp. 1 - 69. Plenum Press, New York - London 1976.

25432 - BISWAL, U.C., MOHANTY, P. : Aging induced changes in photosynthetic electron
        transport of detached barley leaves. - Plant Cell Physiol. 17 : 323 - 331,
        1976.

25433 - BISWAL, U.C., MOHANTY, P. : Dark stress induced senescence of detached barley
        leaves : II. Alteration in the absorption characteristic and photochemical
        activity of the chloroplasts isolated from senescing leaves. - Plant Sci.
        Lett. 7 : 371 - 379, 1976.

25434 - BISWAL, U.C., SHARMA, R. : Phytochrome regulation of senescence in detached
        barley leaves. - Z. Pflanzenphysiol. 80 : 71 - 73, 1976. [Chl.]

25435 - BISWAS, A.K., CHOUDHURI, M.A. : Control of senescence of rice by nutrient
        sprays at different developmental stages and its impact on yield. - Indian J.
        agr. Sci. 46 : 225 - 228, 1976. [Chl.]

25436 - BITTAKER, H.F., IVERSON, R.L. : Thalassia testudinum productivity : A field
        comparison of measurement methods. - Mar. Biol. 37 : 39 - 46, 1976.

25437 - BJÖRKMAN, O. : Adaptive and genetic aspects of C₄ photosynthesis. - In : BUR-

RIS, R.H., BLACK, C.C. (ed.) : CO_2 Metabolism and Plant Productivity. Pp. 287 - 309. Univ. Park Press, Baltimore - London - Tokyo 1976.

25438 - BJÖRKMAN, O., BOYNTON, J., BERRY, J. : Comparison of the heat stability of photosynthesis, chloroplast membrane reactions, photosynthetic enzymes, and soluble protein in leaves of heat-adapted and cold-adapted C_4 species. - Carnegie Inst. Year Book 75 : 400 - 407, 1976.

25439 - BJÖRN, G.S., BJÖRN, L.O. : Photochromic pigments from blue-green algae : Phycochromes a, b and c. - Physiol. Plant. 36 : 297 - 304, 1976.

25440 - BJÖRN, L.O. : The state of protochlorophyll and chlorophyll in corn roots. - Physiol. Plant. 37 : 183 - 184, 1976.

25441 - BJÖRN, L.O. : Why are plants green ? Relationships between pigment absorption and photosynthetic efficiency. - Photosynthetica 10 : 121 - 129, 1976.

25442 - BJØRNLAND, T., AGUILAR-MARTINEZ, M. : Carotenoids in red algae. - Phytochemistry 15 : 291 - 296, 1976.

25443 - BLACK, C.C. : An assessment of C_4 photosynthesis and productivity. - In : BURRIS, R.H., BLACK, C.C. (ed.) : CO_2 Metabolism and Plant Productivity. Pp. 397 - 403. Univ. Park Press, Baltimore - London - Tokyo 1976.

25444 - BLACK, C.C. : Checklists of C_4-dicarboxylic acid photosynthesis plant species. - In : BURRIS, R.H., BLACK, C.C. (ed.) : CO_2 Metabolism and Plant Productivity. Pp. 425 - 426. Univ. Park Press, Baltimore - London - Tokyo 1976.

25445 - BLACK, C.C., GOLDSTEIN, L.D., RAY, T.B., KESTLER, D.P., MAYNE, B.C. : The relationship of plant metabolism to internal leaf and cell morphology and to the efficiency of CO_2 assimilation. - In : BURRIS, R.H., BLACK, C.C. (ed.) : CO_2 Metabolism and Plant Productivity. Pp. 113 - 139. Univ. Park Press, Baltimore - London - Tokyo 1976.

25446 - BLACK, C.C., WILLIAMS, S. : Plant exhibiting characteristics common to Crassulacean acid metabolism. - In : BURRIS, R.H., BLACK, C.C. (ed.) : CO_2 Metabolism and Plant Productivity. Pp. 407 - 424. Univ. Park Press, Baltimore - London - Tokyo 1976.

25447 - BLACK, C.C., Jr., BENDER, M.M. : $\delta^{13}C$ values in marine organisms from the Great Barrier Reef. - Aust. J. Plant Physiol. 3 : 25 - 32, 1976.

25448 - BLACK, C.C., Jr., BURRIS, J.E., EVERSON, R.G. : Influence of oxygen concentration on photosynthesis in marine plants. - Aust. J. Plant Physiol. 3 : 81 - 86, 1976.

25449 - BLACKWOOD, G.C., MIFLIN, B.J. : The effect of nitrate and ammonium feeding on carbon dioxide assimilation in maize. - J. exp. Bot. 27 : 735 - 747, 1976.

25450 - BLAD, B.L., ROSENBERG, N.J. : Measurement of crop temperature by leaf thermocouple, infrared thermometry and remotely sensed thermal imagery. - Agron. J. 68 : 635 - 641, 1976.

25451 - BLAD, B.L., ROSENBERG, N.J. : Evaluation of resistance and mass transport evapotranspiration models requiring canopy temperature data. - Agron. J. 68 : 764 - 769, 1976.

25452 - BLAKE, N.J., JOHNSON, D.L. : Oxygen production-consumption of the pelagic Sargassum community in a flow-through system with arsenic additions. - Deep Sea Res. 23 : 773 - 778, 1976. [Ps.]

25453 - BLONDON, F., CLABAULT, G., DERONNE, M. : Activité photosynthétique de la feuille adulte et induction florale chez le Perilla ocymoides L. - Compt. rend. Acad. Sci. Paris, Sér. D 283 : 1493 - 1495, 1976.

25454 - BLUM, A., NAVEH, M. : Improved water-use efficiency in dryland grain sorghum by promoted plant competition. - Agron. J. 68 : 111 - 116, 1976. [Growth analysis.]

25455 - BNIŃSKA, M., HILLBRICHT-ILKOWSKA, A., KAJAK, Z., WĘGLEŃSKA, T., ZDANOWSKI, B.: Influence of mineral fertilization of lake ecosystem functioning. - Limnologica (Berlin) 10 : 255 - 267, 1976. [Productivity.]

25456 - **BOARDMAN, N.K., THORNE, S.W.** : Cation effects on light-induced chlorophyll a fluorescence in chloroplasts lacking both chlorophyll b and chlorophyll-protein complex-II. - Plant Sci. Lett. 7 : 219 - 224, 1976.

25457 - **BÖGER, P.** : Crop productivity in the light of basic photosynthesis research. - In : Plant Research and Development. Vol. 3. Pp. 60 - 76. Inst. sci. Co--operation, Tübingen 1976.

25458 - **BÖGER, P.** : Ist der Sauerstoff der Luft in Gefahr ? - Naturwiss. Rundschau 29 : 221 - 223, 1976. [Ps.]

25459 - **BÖGER, P., SCHLUE, U.** : Long-term effects of herbicides on the photosynthetic apparatus. I. Influence of diuron, triazines and pyridazinones. - Weed Res. 16 : 149 - 154, 1976.

25460 - **BOGGESS, S.F., ASPINALL, D., PALEG, L.G.** : Stress metabolism. IX. The significance of end-product inhibition of proline biosynthesis and of compartmentation in relation to stress-induced proline accumulation. - Aust. J. Plant Physiol. 3 : 513 - 525, 1976. [Chl.]

25461 - **BOGOMOLNI, R.A., BAKER, R.A., LOZIER, R.H., STOECKENIUS, W.** : Light-driven proton translocations in Halobacterium halobium. - Biochim. biophys. Acta 440 : 68 - 88, 1976.

25462 - **BOGORAD, L.** : Chlorophyll biosynthesis. - In : GOODWIN, T.W. (ed.) : Chemistry and Biochemistry of Plant Pigments. 2nd Ed. Vol. 1. Pp. 64 - 148. Academic Press, London - New York - San Francisco 1976.

25463 - **BOGORAD, L., DAVIDSON, J.N., HANSON, M.R.** : Genes affecting erythromycin resistance and sensitivity of Chlamydomonas reinhardi chloroplast ribosomes. - In : BÜCHER, T., NEUPERT, W., SEBALD, W., WERNER, S. (ed.) : Genetics and Biogenesis of Chloroplasts and Mitochondria. Pp. 61 - 67. North-Holland Publ. Co., Amsterdam - New York - Oxford 1976.

*25464 - **BOGUSLAVSKIĬ, L.I., VOLKOV, A.G.** : Fotoindutsirovannyĭ perenos protonov cherez granitsu razdela faz dekan/voda v prisutstvii khlorofilla. [Photoinduced transfer of protons through the decane-water interface in the presence of chlorophyll.] - Dokl. Akad. Nauk SSSR 224 : 1201 - 1204, 1975. [In R.]

25465 - **BOGUSLAVSKIĬ, L.I., VOLKOV, A.G., KANDELAKI, M.D.** : Perenos élektronov cherez granitsu razdela oktan/voda v prisutstvii khlorofilla. [Electron transfer from water into octane catalyzed with chlorophyll.] - Biofizika 21 : 808 - 811, 1976. [In R, ab : E.]

25466 - **BOGUSLAVSKIĬ, L.I., VOLKOV, A.G., KANDELAKI, M.D., NIZHNIKOVSKIĬ, E.A.** : Fotookislenie vody i transport protonov cherez granitsu razdela faz dvukh nesmeshivayushchikhsya zhidkosteĭ v prisutstvii khlorofilla. [Photooxidation of water and proton transport through the interface of two immiscible liquids in the presence of chlorophyll.] - Dokl. Akad. Nauk SSSR 227 : 727 - 730, 1976. [In R.]

25467 - **BOGUSLAVSKIĬ, L.I., VOLKOV, A.G., KOZLOV, I.A., MAL'YAN, A.N.** : Perenos protonov iz vody v oktan, sopryazhennyĭ s reaktsieĭ gidroliza ATF rastvorimoĭ ATFazoĭ khloroplastov. [ATP-hydrolysis coupled proton transfer from water to octane by soluble ATPase from chloroplasts.] - Biofizika 21 : 286 - 288, 1976. [In R, ab : E.]

25468 - **BOGUSLAVSKY, L.I., VOLKOV, A.G., KANDELAKI, M.D.** : Transfers of electrons and protons at the decane/water interface in the presence of chlorophyll. - FEBS Lett. 65 : 155 - 188, 1976.

25469 - **BOHÁČ, J.** : Detekçia produktívnosti rastlín. [Detection of plant productivity.] - Genetika Šlechtění (Praha) 12 : 259 - 265, 1976. [In Slovak.]

25470 - **BÖHME, H.** : Photoreactions of cytochrome b_6 and cytochrome f in chloroplast photosystem I fragments. - Z. Naturforsch. 31C : 68 - 77, 1976.

25471 - **BOĬCHENKO, E.A.** : Zavisimost' aktivnosti reduktazy uglekisloty ot soderzhaniya lipidov. [Effect of lipid content on carbon dioxide reductase activity.] - Fiziol. Rast. 23 : 720 - 725, 1976. [In R, ab : E.]

25472 - **BOĬKO, A.V., SCHASTNYI, A.K., ARABEI, N.M.** : Korrelatsionnaya zavisimost'

biomassy lesnykh fitotsenozov ot summarnogo raskhoda vlagi iz zony aëratsii
pochvy. [A correlation between the forest phytocoenose biomass and the total
moisture consumption from the soil aeration zone.] - Dokl. Akad. Nauk belorus.
SSR *20* : 552 - 554, 575, 1976. [In R.]

25473 - BOKHARI, U.G. : Influence of temperatures, water stress, and nitrogen treatments on chlorophyll and dry matter of western wheatgrass. - J. Range Manage.
29 : 127. - 131, 1976.

25474 - BOKHARI, U.G. : The influence of stress conditions on chlorophyll content of
two range grasses with contrasting photosynthetic pathways. - Ann. Bot. *40* :
969 - 979, 1976.

25475 - BOLGHARI, H.A. : Estimation de la production de jeunes plantations de pin
rouge et de pin gris du sud du Québec. - Can. J. Forest Res. *6* : 478 - 486,
1976.

25476 - BOLHÀR-NORDENKAMPF, H.R. : Messungen und Modellvorstellungen zum CO_2-Gasstoff-
wechsel von *Phaseolus vulgaris* var. *nanus* L. mit besonderer Berücksichtigung
der Photorespiration sowie der Atrazinwirkung. - Biochem. Physiol. Pflanzen
169 : 121 - 161, 1976.

25477 - BOLTON, J.R., WARDEN, J.T. : Paramagnetic intermediates in photosynthesis. -
Annu. Rev. Plant Physiol. *27* : 375 - 383, 1976.

25478 - BONDAR', V.A., GOGOTOVA, G.I., ZYAKUN, A.M. : O fraktsionirovanii izotopov
ugleroda fotoavtotrofnymi mikroorganizmami s razlichnymi putyami assimilyatsii uglekisloty. [Fractionation of carbon isotopes by photoautotrophic microorganisms with different ways of carbon dioxide assimilation.] - Dokl. Akad.
Nauk SSSR *228* : 720 - 722, 1976. [In R.]

25479 - BONDARENKO, G.L., LYADSKIĬ, P.A. : Osobennosti nakopleniya pigmentov v list'-
yakh rasteniĭ morkovi pod deĭstviem temperatury pochvy i mineral'nykh udobreniĭ. [Characteristics of pigment accumulation in leaves of carrot plants
under the effect of soil temperature and mineral fertilizers.] - Dokl. vse-
soyuz. Akad. sel'skokhoz. Nauk *1976* (12) : 15 - 17, 1976. [In R.]

25480 - BONDARENKO, V.I., ARTYUKH, A.D. : Snizhenie produktivnosti rasteniĭ ozimoĭ
pshenitsy, povrezhdennykh otritsatel'nymi temperaturami. [Decrease in productivity of winter wheat plants injured by negative temperatures.] - Fiziol.
Biokhim. kul't. Rast. *8* : 584 - 589, 1976. [In R, ab : E.]

25481 - BONDARENKO, V.I., POVZIK, M.M., KLIMOV, A.N. : Posledeĭstvie zimnikh povrezh-
deniĭ na vyzhivaemost', produktivnost' i urozhaĭ ozimoĭ pshenitsy. [Post-ef-
fect of winter damage on survival, productivity and yield of winter wheat.]
- In : Biologicheskie Osnovy Povysheniya Urozhaev Kukuruzy i Drugikh Polevykh Kul'tur v Severnoĭ Stepi USSR. Pp. 26 - 32. Sinel'nikov. selek.-op. Sta.,
Dnepropetrovsk 1976. [In R.]

25482 - BONHOMME, R. : Détermination des profils d'indice foliaire et de rayonnement
dans un couvert végétal à l'aide de photographies hémisphériques faites *in
situ.* - Ann. agron. *27* : 33 - 59, 1976.

25483 - BONHOMME, R., GANRY, J. : Mesure de l'indice foliaire du bananier par photographies hémisphériques faites "*in situ*". - Fruits *31* : 421 - 425, 1976.

25484 - BONHOMME, R., VARLET GRANCHER, C. : Assimilation nette, utilisation de l'eau
et microclimatologie d'un champ de Maïs. VII. - Variations du rapport des
énergies rouge sombre-rouge clair. - Ann. agron. *27* : 327 - 332, 1976.

*25485 - BONILLA RUIZ, J. : Notas sobre la estimacion de la productividad primaria en
la laguna de las Maritas. [Notes on the determination of primary productivity
in Las Maritas lagune.] - Lagena *32* : 3 - 12, 1973. [In Span.]

25486 - BONNETT, H.T. : On the mechanism of the uptake of *Vaucheria* chloroplasts by
carrot protoplasts treated with polyethylene glycol. - Planta *131* : 229 - 233,
1976.

25487 - BOOS, K.-S., LÜSTORFF, J., SCHLIMME, E., HESSLE, H., STROTMANN, H. : Properties of ribose modified ADP analogues in photophosphorylation of spinach chloroplasts. - FEBS Lett. *71* : 124 - 129, 1976.

25488 - **BORG, D.C., FORMAN, A., FAJER, J.** : ESR and ENDOR studies of the Ⅱ cation radical of bacteriochlorophyll. - J. amer. chem. Soc. *98* : 6889 - 6893, 1976.

25489 - **BORICHENKO, N., MANOLOV, P., RANGELOV, B.** : Vliyanie na 2,4,5-T v"rkhu fotosintezata i transporta na ^{14}C asimilati. [Effect of 2,4,5-T na photosynthesis and the transport of ^{14}C-labeled photosynthates.] - Fiziol. Rast. (Sofia) *2* (4) : 55 - 64, 1976. [In Bulg., ab : E, R.]

25490 - **BORISOV, A.Yu.** : Printsip mgnovennogo deĭstviya fotosinteza. [The principle of instant action in photosynthesis.] - Mol. Biol. (Moskva) *10* : 460 - 465, 1976. [In R, ab : E.]

25491 - **BORISOV, A.Yu., PETKOVA, R.A., IL'INA, M.D.** : Svyaz' fotoindutsirovannykh izmeneniĭ okislitel'no-vosstanovitel'nogo potentsiala suspenzii khloroplastov s aktivnost'yu vtoroĭ fotosistemy. [Connection between photoinduced changes in the oxidation-reduction potential of the chloroplast suspension and the Photosystem 2 activity.] - Dokl. Akad. Nauk SSSR *227* : 1465 - 1468, 1976. [In R.]

25492 - **BORNANCIN, M., de RENZIS, G.** : A sensitive automated method for adenosine triphosphatase kinetics. - Anal. Biochem. *75* : 374 - 381, 1976.

25493 - **BORNEFELD, T.** : The rates of photophosphorylation of the blue-green alga *Anacystis nidulans* at transition from dark to light. 1. Rates under conditions of cyclic, pseudo-cyclic, and non-cyclic electron flow and in presence and absence of DCMU [3(3,4-dichlorophenyl)-1,1-dimethylurea] and desaspidin. - Biochem. Physiol. Pflanzen *170* : 333 - 344, 1976.

25494 - **BORNEFELD, T.** : The rates of photophosphorylation of blue-green alga *Anacystis nidulans* at transition from dark to light. 2. Dependence of ATP formation on pH and phosphate concentration of the medium, temperature, and light intensity. - Biochem. Physiol. Pflanzen *170* : 345 - 353, 1976.

25495 - **BÖRNER, T., HAGEMANN, R.** : Die Evolution der eukaryotischen Zelle. Hypothesen zur Entwicklung von Chloroplasten, Mitochondrien und Zellkern der Eukaryoten-Zellen. - Biol. Rundschau *14* : 249 - 267, 1976.

25496 - **BÖRNER, T., JAHN, G., HAGEMANN, R.** : Electrophoretic mobility of subunits of fraction I protein of higher plant species. - Biochem. Physiol. Pflanzen *169*: 179 - 181, 1976.

25497 - **BÖRNER, T., SCHUMANN, B., HAGEMANN, R.** : Biochemical studies on a plastid ribosome-deficient mutant of *Hordeum vulgare*. - In : BÜCHER, T., NEUPERT, W., SEBALD, W., WERNER, S. (ed.) : Genetics and Biogenesis of Chloroplasts and Mitochondria. Pp. 41 - 48. North-Holland Publ. Co., Amsterdam - New York - Oxford 1976.

25498 - **BORNMAN, C.H.** : *Welwitschia mirabilis* callus studies. III. Some effects of morphactin. - Z. Pflanzenphysiol. *78* : 266 - 270, 1976. [Chl.]

25499 - **BORNMAN, C.H., FANSHAWE, N.C.** : *Welwitschia mirabilis* callus studies. II. Some effects of sucrose. - Z. Pflanzenphysiol. *78* : 217 - 221, 1976. [Chl.]

25500 - **BOROWITZKA, M.A.** : Some unusual features of ultrastructure of chloroplasts of the green algal order *Caulerpales* and their development. - Protoplasma *89*: 129 - 147, 1976.

25501 - **BOROWITZKA, M.A., LARKUM, A.W.D.** : Calcification in the green alga *Halimeda*. II. Exchange of Ca^{2+} and the occurrence of age gradients in calcification and photosynthesis. - J. exp. Bot. *27* : 864 - 878, 1976.

25502 - **BOROWITZKA, M.A., LARKUM, A.W.D.** : Calcification in the green alga *Halimeda*. III. The sources of inorganic carbon for photosynthesis and calcification and a model of the mechanism of calcification. - J. exp. Bot. *27* : 879 - 893, 1976.

25503 - **BOROWITZKA, M.A., LARKUM, A.W.D.** : Calcification in the green alga *Halimeda*. IV. The action of metabolic inhibitors on photosynthesis and calcification. - J. exp. Bot. *27* : 894 - 907, 1976.

25504 - **BOSCHETTI, A., SCHAFFNER, J.C., LEUENBERGER, S.** : Bildung der Thylakoidproteine während der Ergrünung einer vergilbten Mutante von *Chlamydomonas reinhardi*. - Chimia *30* (2) : 79 - 82, 1976.

25505 - **BOTHE, H.** : Die biologische Stickstoffixierung. - Naturwiss. Rundschau *29* : 316 - 324, 1976. [Ferredoxin.]

25506 - **BOTHE, H., YATES, M.G.** : The electron transport to nitrogenase in *Mycobacterium flavum*. - Arch. Microbiol. *107* : 25 - 31, 1976. [Ps.]

25507 - **BOTMAN, K.S., TELLYAEV, A.Yu., MIKHAĬLOVA, L.M.** : Raskhod vody na transpiratsiyu i fiziologicheskie protsessy v drevostoyakh archi zeravshanskoĭ razlichnoĭ somknutosti pologa. [Water expenditure in transpiration and physiological processes in zeravshan juniper timber stands differing in closed cover.] - Uzb. biol. Zh. *1976* (6) : 25 - 29, 1976. [Chl; in R, ab : Uzb.]

25508 - **BOTTOMLEY, P.J., STEWART, W.D.P.** : ATP pools and transients in the blue-green alga, *Anabaena cylindrica*. - Arch. Microbiol. *108* : 249 - 258, 1976.

25509 - **BOTTOMLEY, P.J., STEWART, W.D.P.** : The measurement and significance of ATP pools in filamentous blue-green algae. - Brit. phycol. J. *11* : 69 - 82, 1976.

25510 - **BOUCAUD, J., UNGAR, I.A.** : Influence of hormonal treatments on the growth of two halophytic species of *Suaeda*. - Amer. J. Bot. *63* : 694 - 699, 1976. [Chl.]

25511 - **BOULTER, D.** : The evolution of plant proteins with special reference to higher plant cytochromes *c*. - In : **SMITH, H.** (ed.) : Commentaries in Plant Science. Pp. 77 - 91. Pergamon Press, Oxford - New York - Toronto - Sydney - Paris - Frankfurt 1976.

25512 - **BOURDU, R.** : Discussion sur les caractéristiques structurales et infra-structurales des feuilles en fonction de l'appartenance aux trois types métaboliques. - Physiol. vég. *14* : 551 - 561, 1976. [Ps.]

25513 - **BOURQUE, D.P., McMILLAN, P.N., CLINGENPEEL, W.J., NAYLOR, A.W.** : Comparative effects of several inhibitors of chloroplast thylakoid membrane synthesis in greening jack bean. - Bot. Gaz. *137* : 279 - 284, 1976.

25514 - **BOUSQUET, J.-F., SKAJENNIKOFF, M., BETHENOD, O., CHARTIER, P.** : Effect de l' ochracine (ou melléine), metabolite isolé des filtrats de culture de *Septoria nodorum*, sur la résistance stomatique de plantules de Blé. - Compt. rend. Acad. Sci. Paris, Sér. D *283* : 1053 - 1055, 1976.

25515 - **BOUTRY, J.L., BARBIER, M., RICARD, M.** : La diatomée *Chaetoceros simplex calcitrans* PAULSEN et son environnement. II. Effets de la lumière et des irradiations ultra-violettes sur la production primaire et la biosynthèse des stérols. - J. exp. mar. Biol. Ecol. *21* : 69 - 74, 1976.

25516 - **BOWN, A., DYMOCK, I., HILL, B.** : Auxin and dark carbon dioxide fixation in *Avena sativa* coleoptile sections. - Plant Physiol. *57* (Suppl.) : 19, 1976.

25517 - **BOXER, S.G., CLOSS, G.L.** : A covalently bound dimeric derivative of pyrochlorophyllide *a*. A possible model for reaction center chlorophyll. - J. amer. chem. Soc. *98* : 5406 - 5408, 1976.

25518 - **BOYD, C.E.** : Accumulation of dry matter, nitrogen and phosphorus by cultivated water hyacinths. - Econ. Bot. *30* : 51 - 56, 1976. [Chl.]

25519 - **BOYER, J., McPHERSON, H.G.** : Physiology of water deficits in cereal grains. - In : Proceedings of the Symposium on Climate & Rice. Pp. 321 - 343. Int. Rice Res. Inst., Los Baños 1976. [Ps.]

25520 - **BOYER, J.S.** : Photosynthesis at low water potentials. - Phil. Trans. roy. Soc. London B *273* : 501 - 512, 1976.

25521 - **BOYER, J.S.** : Water deficits and photosynthesis. - In : **KOZLOWSKI, T.T.** (ed.): Water Deficits and Plant Growth. Vol. IV. Pp. 153 - 190. Academic Press, New York - San Francisco - London 1976.

*25522 - **BOYER, P.D.** : A model for conformational coupling of membrane potential and proton translocation to ATP synthesis and to active transport. - FEBS Lett. *58* : 1 - 6, 1975.

25523 - **BOYLEN, C.W., SHELDON, R.B.** : Submergent macrophytes : growth under winter ice cover. - Science *194* : 841 - 842, 1976. [Ps.]

25524 - **BOYNTON, J.E., GILLHAM, N.W., HARRIS, E.H., TINGLE, C.L., Van WINKLE-SWIFT, K., ADAMS, G.M.W.** : Transmission, segregation and recombination of chloroplast

genes in *Chlamydomonas*. - In : BÜCHER, T., NEUPERT, W., SEBALD, W., WERNER,
S. (ed.) : Genetics and Biogenesis of Chloroplasts and Mitochondria. Pp. 313-
- 322. North-Holland Publ. Co., Amsterdam - New York - Oxford 1976.

25525 - BØYUM, A. : Limnology and paleolimnology of Lake Nordvann, south-eastern Norway. - Arch. Hydrobiol. *77* : 277 - 329, 1976. [Chl.]

25526 - BOZCUK, S. : Effect of sodium chloride upon growth and transpiration in *Sta-
tice* sp. and *Pisum sativum* L. - In : VARDAR, Y., SHEIKH, K.H., ÖZTÜRK, M.A.
(ed.) : Proceedings of the Third MPP Meeting. Pp. 37 - 44. Ege Univ., Izmir
1976. [Growth analysis.]

25527 - BOZHKOVA, M.D., GLAGOLEVA, T.A., FILIPPOVA, L.A. : Posledeĭstvie ponizhennoĭ
nochnoĭ temperatury na rost i fotosinteticheskiĭ metabolizm ugleroda u *Chlorella pyrenoidosa* CHICK. [After-effects of low night temperature on the
growth and photosynthetic metabolism of carbon in *Chlorella pyrenoidosa*
CHICK.] - Bot. Zh. *61* : 1297 - 1303, 1976. [In R.]

25528 - BRADBEER, J.W. : Chloroplast development in greening leaves. - In : SUNDER-
LAND, N. (ed.) : Perspectives in Experimental Biology. Vol. 2. Botany. Pp.
131 - 143. Pergamon Press, Oxford - New York 1976.

25529 - BRADBEER, J.W., ARRON, G.P., KEMBLE, R., WARA-ASWAPATI, O. : The synthesis
of some enzymes of the photosynthetic carbon cycle in bean leaves. - In :
Acides Nucléiques et Synthèse des Protéines chez les Végétaux. Colloq. Int.
C.N.R.S. No. 261. Pp. 453 - 456. C.N.R.S., Paris 1976.

25530 - BRADBEER, J.W., MONTES, G. : The photocontrol of chloroplast development -
ultrastructural aspects and photosynthetic activity. - In : SMITH, H. (ed.):
Light and Plant Development. Pp. 213 - 227. Butterworths, London - Boston -
Sydney - Wellington - Durban - Toronto 1976.

25531 - BRADBURY, I.K., HOFSTRA, G. : Vegetation death and its importance in primary
production measurements. - Ecology *57* : 209 - 211, 1976.

25532 - BRAND, J.J., KIRCHANSKI, S. : Effects of chilling on the ultrastructure of
Anacystis nidulans. - Plant Physiol. *57* (Suppl.) : 73, 1976. [Chloroplast.]

25533 - BRANDT, A.B., FINAKOV, G.Z. : Energy yield of *Chlorella* photosynthesis in
different spectral regions. - Stud. biophys. *54* : 233 - 238, 1976.

25534 - BRANDT, A.B., KISELEVA, M.I. : Kvantovaya produktivnost' sinkhronnoĭ kul'tury khlorelly v raznykh oblastyakh spektra. [Quantum productivity of synchronous culture of *Chlorella* in different spectral regions.] - Biofizika *21* :
180, 1976. [In R, ab : E.]

25535 - BRANDT, A.B., KISELEVA, M.I. : O korrelyatsii stroeniya kletok khlorelly s
opticheskoĭ plotnost'yu ee suspenzii. [Correlation between the structure of
Chlorella cells and optical density of its suspensions.] - Fiziol. Rast. *23*:
1068 - 1070, 1976. [In R.]

25536 - BRANDT, A.B., KISELEVA, M.I., FINAKOV, G.Z. : Zavisimost' vydeleniya i pogloshcheniya kisloroda khlorelloĭ ot kharaktera raspredeleniya svetovogo polya v suspenzii. [Oxygen evolution and absorption by *Chlorella* depending on
the character of distribution of light field in suspension.] - Fiziol. Rast.
23 : 710 - 714, 1976. [In R, ab : E.]

25537 - BRANSON, F.A. : Water use on rangelands. - In : HEADY, H.F., FALKENBORG, D.
H., RILEY, J.P. (ed.) : Watershed Management on Range and Forest Lands. Pp.
193 - 209. Utah State Univ., Logan 1976. [Stomatal resistance.]

25538 - BRANSON, F.A., MILLER, R.F., McQUEEN, I.S. : Moisture relationships in twelve
northern desert shrub communities near Grand Junction, Colorado. - Ecology
57 : 1104 - 1124, 1976. [Growth analysis.]

25539 - BRASSEUR, F., DE SLOOVER, J.R. : L'extinction du rayonnement dans les gammes
spectrales bleu, rouge et rouge lointain. Comparaison de deux peuplements
forestièrs de Haute Ardenne. - Bull. Soc. roy. Bot. Belg. *109* : 319 - 334,
1976. [Radiation in canopy.]

25540 - BRAUN, H.J. : Rhythmus und Grösse von Wachstum, Wasserverbrauch und Produktivität des Wasserverbrauchs bei Holzpflanzen. II. *Acer platanoides* L.,

Acer pseudoplatanus L. und *Fraxinus excelsior* L. mit einem Vergleich aller untersuchter Baumarten einschließlich einiger *Populus*-Klone. - Allgem. Forst-Jagdzeitung *147* : 163 - 168, 1976. [Growth analysis.]

25541 - **BRAUN, J.G., ESCANEZ, J.E., DE LEON, A.R.** : Observaciones químicas y biológicas en el NW de Africa, entre cabo Juby y cabo Ghir. (Campaña CINECA - "Cornide de Saavedra", Febrero 1973). [Chemical and biological observations in NW of Africa, between Cape Juby and Cape Ghir. (CINECA programme - "Cornide de Saavedra", February 1973).] - Bol. Inst. esp. Oceanogr. *209* : 1 - 11, 1976. [Chl; in Span., ab : E.]

25542 - **BREIDENBACH, R.W.** : Microbodies. - In : **BONNER, J., VARNER, J.E.**(ed.) : Plant Biochemistry. Third Edition. Pp. 91 - 114. Academic Press, New York - San Francisco - London 1976.

25543 - **BRENNER, D., VALIELA, I., VAN RAALTE, C.D., CARPENTER, E.J.** : Grazing by *Talorchestia longicornis* on an algal mat in a New England salt marsh. - J. exp. mar. Biol. Ecol. *22* : 161 - 169, 1976. [Ps, Chl.]

25544 - **BRETON, J., GEACINTOV, N.E.** : Quenching of fluorescence of chlorophyll *in vivo* by long-lived excited states. - FEBS Lett. *69* : 86 - 89, 1976.

25545 - **BREZEANU, A.G., DAVIS, D.G., SHIMABUKURO, R.H.** : Ultrastructural effects and translocation of methyl-2-(4-(2,4-dichlorophenoxy)-phenoxy)propanoate in wheat (*Triticum aestivum*) and wild oat (*Avena fatua*). - Can. J. Bot. *54* : 2038 - - 2048, 1976. [Chloroplast.]

25546 - **BRICK, M.A., DOBRENZ, A.K., SCHONHORST, M.H.** : Transmittance of the multifoliolate leaf characteristic into non-dormant alfalfa. - Agron. J. *68* : 134 - - 136, 1976. [Growth analysis.]

25547 - **BRIDGEN, J., WALKER, I.D.** : Photoreceptor protein from the purple membrane of *Halobacterium halobium*. Molecular weight and retinal binding site. - Biochemistry *15* : 792 - 798, 1976.

25548 - **BRIDGES, S., WARD, B.** : Effect of hydrogen ion buffers on photosynthetic oxygen evolution in the blue-green alga, *Agmenellum quadruplicatum*. - Microbios *15* : 49 - 56, 1976.

25549 - **BRINKHUIS, B.H., TEMPEL, N.R., JONES, R.F.** : Photosynthesis and respiration of exposed salt-marsh fucoids. - Mar. Biol. *34* : 349 - 359, 1976.

25550 - **BRINKMANN, K.** : Circadian rhythm in the kinetics of acid denaturation of cell membranes of *Euglena gracilis*. - Planta *129* : 221 - 227, 1976. [Chl.]

25551 - **BRISKER, H.E., GOLDSCHMIDT, E.E., GOREN, R.** : Ethylene-induced formation of ABA in citrus peel as related to chloroplast transformations. - Plant Physiol. *58* : 377 - 379, 1976.

25552 - **BRITTON, C.M., DODD, J.D.** : Relationships of photosynthetically active radiation and shortwave irradiance. - Agr. Meteorol. *17* : 1 - 7, 1976.

25553 - **BRITTON, G.** : Biosynthesis of carotenoids. - In : **GOODWIN, T.W.** (ed.) : Chemistry and Biochemistry of Plant Pigments. 2nd Ed. Vol. 1. Pp. 262 - 327. Academic Press, London - New York - San Francisco 1976.

25554 - **BRITTON, G.** : Later reactions of carotenoid biosynthesis. - Pure appl. Chem. *47* : 223 - 236, 1976.

25555 - **BRITTON, G., MALHOTRA, H.C., SINGH, R.K., GOODWIN, T.W., BEN-AZIZ, A.** : Cross--conjugated carotenoid aldehydes in *Rhodopseudomonas viridis*. - Phytochemistry *15* : 1749 - 1751, 1976.

25556 - **BRITTON, G., MALHOTRA, H.C., SINGH, R.K., TAYLOR, S., GOODWIN, T.W., BEN-AZIZ, A.** : Methoxyspheroidene and methoxyspheroidenone, two carotenoids from *Rhodopseudomonas spheroides*. - Phytochemistry *15* : 1971 - 1972, 1976.

25557 - **BRITZ, S.J.** : A multi-sample automatic monitoring device for the circadian rhythm of transmittance change in *Ulva*. - Carnegie Inst. Year Book *75* : 383 - - 392, 1976.

25558 - **BRITZ, S.J.** : Two applications of a multi-sample automatic monitor for rhythmic transmittance changes in *Ulva*. - Carnegie Inst. Year Book *75* : 392 - 394, 1976.

25559 - BRITZ, S.J., BRIGGS, W.R. : Circadian rhythms of chloroplast orientation and photosynthetic capacity in *Ulva*. - Plant Physiol. *58* : 22 - 27, 1976.

25560 - BRITZ, S.J., PFAU, J., NULTSCH, W., BRIGGS, W.R. : Automatic monitoring of a circadian rhythm of change in light transmittance in *Ulva*. - Plant Physiol. *58* : 17 - 21, 1976. [Chloroplast movement.]

25561 - BROCK, T.D., PETERSEN, S. : Some effects of light on the viability of rhodopsin-containing halobacteria. - Arch. Microbiol. *109* : 199 - 200, 1976.

25562 - BROCKMANN, H. Jr. : Bacteriochlorophyll *e* : structure and stereochemistry of a new type of chlorophyll from *Chlorobiaceae*. - Phil. Trans. roy. Soc. London B *273* : 277 - 285, 1976.

25563 - BROCKMANN, H. Jr., GLOE, A., RISCH, N., TROWITZSCH, W. : Zur absoluten Konfiguration der Chlorophylle, VII. Bacteriochlorophyll *e*, ein neues Chlorophyll aus braunen Arten von *Chlorobiaceae*. - Liebigs Ann. Chem. *1976* : 566 - 577, 1976.

25564 - BRODERSEN, P. : Factors affecting the photoconversion of protochlorophyllide to chlorophyllide in etioplast membranes isolated from barley. - Photosynthetica *10* : 33 - 39, 1976.

25565 - BRODY, S.S., OWENS, N.F. : Photosynthetic electron carriers at a heptane-water interface. - Z. Naturforsch. *31C* : 569 - 574, 1976.

25566 - BROUÉ, P., MARSHALL, D.R., MUNDAY, J. : The response of lupins to waterlogging. - Aust. J. exp. Agr. anim. Husb. *16* : 549 - 554, 1976. [Growth analysis.]

*25567 - BROUERS, M. : Agregation de la protochlorophyllide dans des solvants non polaires. - In : BRODA, E., LOCKER, A., SPRINGER-LEDERER, H. (ed.) : Proceedings of the First European Biophysics Congress. Vol. 4. Pp. 23 - 26. Verlag wiener med. Akad., Wien 1971.

25568 - BROVCHENKO, M.I., SLOBODSKAYA, G.A., CHMORA, S.N., LIPATOVA, T.F. : Vliyanie CO_2 i O_2 na fotosintez i sopryazhennyi s nim vykhod assimilyatov v svobodnoe prostranstvo lista sakharnoi svekly. [Effect of CO_2 and O_2 on photosynthesis and coupled with it transport of assimilates into the free space of sugar beet leaves.] - Fiziol. Rast. *23* : 1232 - 1240, 1976. [In R, ab : E.]

25569 - BROWN, J.K. : Estimating shrub biomass from basal stem diameters. - Can. J. Forest Res. *6* : 153 - 158, 1976.

25570 - BROWN, J.S. : *P*700-chlorophyll *a*-protein complexes. - Carnegie Inst. Year Book *75* : 460 - 465, 1976.

25571 - BROWN, K.W. : Sugar beet and potatoes. - In : MONTEITH, J.L. (ed.) : Vegetation and the Atmosphere. Vol. 2. Case Studies. Pp. 65 - 86. Academic Press, London - New York - San Francisco 1976.

25572 - BROWN, K.W., JORDAN, W.R., THOMAS, J.C. : Water stress induced alterations of the stomatal response to decreases in leaf water potential. - Physiol. Plant. *37* : 1 - 5, 1976. [Stomatal resistance.]

25573 - BROWN, R.H. : Characteristics related to photosynthesis and photorespiration of *Panicum milioides*. - In : BURRIS, R.H., BLACK, C.C. (ed.) : CO_2 Metabolism and Plant Productivity. Pp. 311 - 325. Univ. Park Press, Baltimore 1976.

25574 - BROWN, R.H., ARMITAGE, T.L., MERRETT, M.J. : Ribulose diphosphate carboxylase synthesis in *Euglena*. III. Serological relationships of the intact enzyme and its subunits. - Plant Physiol. *58* : 773 - 776, 1976.

25575 - BRUCHWALD, A., GROCHOWSKI, J. : Comparison of the accuracy in pine stand volume determination by way of sample trees measurements and by means of form factor and volume tables. - Bull. Acad. pol. Sci., Sér. Sci. biol. *24* : 43 - - 51, 1976.

25576 - BRUN, W.A. : The relation of N_2 fixation to photosynthesis. - In : HILL, L.D. (ed.) : World Soybean Research. Pp. 135 - 143. Interstate Printers Publ., Inc., Danville 1976.

25577 - BRUNOLD, C., SCHIFF, J.A. : Studies of sulfate utilization by algae. 15. Enzymes of assimilatory sulfate reduction in *Euglena* and their cellular localization. - Plant Physiol. *57* : 430 - 436, 1976. [Chl.]

25578 - **BRYAN, W., WRIGHT, R.** : The effect of enhanced CO_2 levels and variable light intensities on net photosynthesis in competing mountain trees. - Amer. Midland Naturalist *95* : 446 - 450, 1976.

25579 - **BRYANT, D.A., GLAZER, A.N., EISERLING, F.A.** : Characterization and structural properties of the major biliproteins of *Anabaena* sp. - Arch. Microbiol. *110* : 61 - 75, 1976.

*25580 - **BRYDGES, T.G.** : Chlorophyll *a* - total phosphorus relationships in Lake Erie. - Proc. Conf. Great Lakes Res. *14* : 185 - 190, 1971.

25581 - **BRZOSKA, W.** : Produktivität und Energiegehalte von Gefäßpflanzen im Adventdalen (Spitzbergen). - Oecologia *22* : 387 - 398, 1976.

*25582 - **BUCHANAN, B.B.** : Orthophosphate requirement for the formation of phosphoenolpyruvate from pyruvate by enzyme preparations from photosynthetic bacteria. - J. Bacteriol. *119* : 1066 - 1068, 1974.

25583 - **BUCHANAN, B.B., SCHÜRMANN, P., WOLOSIUK, R.A.** : Appearance of sedoheptulose 1,7-diphosphatase activity on conversion of chloroplast fructose 1,6-diphosphatase from dimer form to monomer form. - Biochem. biophys. Res. Commun. *69* : 970 - 978, 1976.

25584 - **BUCHANAN, B.B., SIREVÅG, R.** :Ribulose 1,5-diphosphate carboxylase and *Chlorobium thiosulfatophilum*. - Arch. Microbiol. *109* : 15 - 19, 1976.

25585 - **BUCHANAN, B.B., WOLOSIUK, R.A.** : Photosynthetic regulatory protein found in animal and bacterial cells. - Nature *264* : 669 - 670, 1976.

25586 - **BUCHECKER, R., LIAAEN-JENSEN, S., BORCH, G., SIEGELMAN, H.W.** : Carotenoids of *Anacystis nidulans*, structures of caloxanthin and nostoxanthin. - Phytochemistry *15* : 1015 - 1018, 1976.

25587 - **BUETOW, D.E.** : Phylogenetic origin of the chloroplast. - J. Protozool. *23* : 41 - 47, 1976.

25588 - **BUL'ON, V.V.** : Pervichnaya produktsiya planktona v ozere Baĭkal. [Primary plankton production in Lake Baikal.] - Zh. obshch. Biol. *37* : 517 - 524, 1976. [In R, ab : E.]

25589 - **BULYCHEV, A.A., ANDRIANOV, V.K., KURELLA, G.A., LITVIN, F.F.** : Photoinduction kinetics of electrical potential in a single chloroplast as studied with microelectrode technique. - Biochim. biophys. Acta *430* : 336 - 351, 1976.

25590 - **BULYCHEV, A.A., VREDENBERG, W.J.** : The effect of cations and membrane permeability modifying agents on the dark kinetics of the photoelectric response in isolated chloroplasts. - Biochim. biophys. Acta *423* : 548 - 556, 1976.

25591 - **BULYCHEV, A.A., VREDENBERG, W.J.** : Effect of ionophores A23187 and nigericin on the light-induced redistribution of Mg^{2+}, K^+ and H^+ across the thylakoid membrane. - Biochim. biophys. Acta *449* : 48 - 58, 1976.

25592 - **BUNCE, J.A., MILLER, L.N.** : Differential effects of water stress on respiration in the light in woody plants from wet and dry habitats. - Can. J. Bot. *54* : 2457 - 2464, 1976. [Ps.]

25593 - **BUNTING, A.H.** : Maximizing the product, or how to have it both ways. - In : CANNELL, M.G.R., LAST, F.T. (ed.) : Tree Physiology and Yield Improvement. Pp. 1 - 20. Academic Press, London - New York - San Francisco 1976. [Ps.]

25594 - **BUNTING, E.S.** : Effects of grain formation on dry matter distribution and forage quality in maize. - Exp. Agr. *12* : 417 - 428, 1976.

25595 - **BUNUS, F.T.** : Plate efficiency measurement in operational conditions of a CO_2, liquid absorption column with radioisotopes. - Int. J. appl. Rad. Isotop. *27* : 260 - 262, 1976.

25596 - **BURIAN, K., WINTER, C.** : Die Wirkung verschieden langer Lichtperioden auf die Photosynthese einiger Gräser. - Photosynthetica *10* : 25 - 32, 1976.

25597 - **BURIEL, J.F., KRAMER, P.J.** : Response of gas exchange in loblolly pine to short term changes in temperature. - Plant Physiol. *57* (Suppl.) : 104, 1976.

*25598 - BURKHOLDER, P.R., ALMODÓVAR, L.R. : Studies on mangrove algal communities in
Puerto Rico. - Florida Scientist *38* : 66 - 74, 1973. [Ps, Chl.]

25599 - BURLEIGH, R., HEWSON, A. : Evidence for short term atmospheric ^{14}C variations
about 4,000 yr BP. - Nature *262* : 128 - 130, 1976.

25600 - BURNETT, J.H. : Functions of carotenoids other than in photosynthesis. - In :
GOODWIN, T.W. (ed.) : Chemistry and Biochemistry of Plant Pigments. 2nd Ed.
Vol. 1. Pp. 655 - 679. Academic Press, London - New York - San Francisco 1976.

25601 - BURNHAM, J.C., STETAK, T., LOCHER, G. : Extracellular lysis of the bluegreen
alga *Phormidium luridum* by *Bdellovibrio bacteriovorus*. - J. Phycol. *12* : 306-
- 313, 1976. [Ps, Chl.]

25602 - BURRIS, J.E., HOLM-HANSEN, O., BLACK, C.C., Jr. : Glycine and serine production in marine plants as a measure of photorespiration. - Aust. J. Plant Physiol. *3* : 87 - 92, 1976.

B25603 - BURRIS, R.H., BLACK, C.C. (ed.) : CO_2 Metabolism and Plant Productivity. -
Univ. Park Press, Baltimore - London - Tokyo 1976.

25604 - BURROWS, F.J., MILTHORPE, F.L. : Stomatal conductance in the control of gas
exchange. - In : KOZLOWSKI, T.T. (ed.) : Water Deficits and Plant Growth. Vol.
IV. Pp. 103 - 152. Academic Press, New York - San Francisco - London 1976.

25605 - BURSTRÖM, H.G. : Growth and transpiration of *Pisum* stems under water stress.
- Z. Pflanzenphysiol. *79* : 419 - 427, 1976. [Growth analysis.]

25606 - BUSCH, G.E., RENTZEPIS, P.M. : Picosecond chemistry. A variety of ultrafast
molecular processes have been measured by picosecond spectroscopic techniques.
- Science *194* : 276 - 283, 1976. [Ps, Chl.]

25607 - BUTCHER, T.B., HAVEL, J.J. : Influence of moisture relationships on thinning
practice. - New Zeal. J. Forest Sci. *6* : 158 - 170, 1976. [Growth analysis.]

25608 - BÜTTNER, R. : Untersuchungen der Primärproduktion an Obstgehölzen und ihre
mögliche Bedeutung für die Züchtung neuer Sorten. - Wiss. Z. tech. Univ. Dresden *25* : 245 - 246, 1976.

25609 - BYKOV, O.D. : CO_2-gazoobmen i metabolicheskie puly ugleroda. [CO_2 exchange and
metabolic pools of carbon.] - In : Gazometricheskoe Issledovanie Fotosinteza
i Dykhaniya Rastenii. Pp. 15 - 18. Akad. Nauk SSSR, Tartu 1976. [In R.]

25610 - BYKOV, O.D., LEVIN, E.S. : K ob"yasneniyu perekhodnykh protsessov CO_2-gazoob-
mena. [Elucidation of transient processes in CO_2 exchange.] - In : Gazometricheskoe Issledovanie Fotosinteza i Dykhaniya Rastenii. Pp. 19 - 22. Akad. Nauk
SSSR, Tartu 1976. [In R.]

25611 - BYKOV, O.D., LEVIN, E.S. : Vnutrenie i vneshnie protsessy fotosinteticheskogo
gazoobmena v svyazi s diffuziei CO_2 v liste. [Internal and external processes
of photosynthetic gas exchange in relation to CO_2 diffusion in leaves.] - Fiziol. Rast. *23* : 238 - 246, 1976. [In R, ab : E.]

25612 - BYOTT, G.S. : Leaf air space systems in C_3 and C_4 species. - New Phytol. *76* :
295 - 299, 1976.

25613 - BYRNE, G.F., TORSSELL, B.W.R., SASTRY, P.S.N. : Plant growth curves in mixtures and climatological response. - Agr. Meteorol. *16* : 37 - 44, 1976. [Ps.]

25614 - BYSTROVA, M.I., MAL'GOSHEVA, I.N., KRASNOVSKII, A.A. : Izuchenie molekulyarnoi
organizatsii agregirovannykh form khlorofilla i ego analogov. [Molecular arrangement of aggregated forms of chlorophyll and its analogues.] - Mol. Biol.
(Moskva) *10* : 193 - 205, 1976. [In R, ab : E.]

25615 - CABANGBANG, R.P., GOMEZ, A.A. : Photosynthesis and respiration measurements
in some grain sorghum cultivars. - Philippine J. Crop Sci. *1* : 156 - 160,
1976.

25616 - CAHEN, D., MALKIN, S., SHOCHAT, S., OHAD, I. : Development of photosystem II
complex during greening of *Chlamydomonas reinhardi* y-1. - Plant Physiol. *58* :
257 - 267, 1976.

25617 - **CAHET, G., GADEL, F.** : Bilan du carbone dans des sédiments lagunaires et marins méditerranéens : effets des processus biologiques saisonniers et diagénétiques. - Arch. Hydrobiol. *77* : 109 - 138, 1976. [Ps.]

25618 - **CALÈ, M.T., FIGLIOLIA, A., IZZA, C., DE ROSSI, C.** : Studi sulla nutrizione idrica e minerale di *Vicia faba*. Nota II. [Studies on water relations and mineral nutrition of *Vicia faba*. Note II.] - Ann. Ist. sperim. Nutr. Piante *7* (3) : 1 - 14, 1976. [Ps; in Ital., ab : E.]

25619 - **CALMÉS, J., VIALA, G., CAVALIÉ, G.** : Métabolisme du glycolate dans la feuille de Vigne vierge *in situ*. - Physiol. vég. *14* : 487 - 497, 1976.

25620 - **CALMÉS, J., VIGNES, D.** : Modifications des échanges gazeux du metabolisme glucidique de la feuille isolée par rapport à la feuille sur pied. - Photosynthetica *10* : 255 - 263, 1976.

25621 - **CALVAYRAC, R., LEDOIGT, G.** : Croissance des euglènes en presence de DCMU : Évolution du plastidome en fonction de la tension d'oxygène. - Plant Sci. Lett. *7* : 249 - 263, 1976.

25622 - **CALVAYRAC, R., LEFORT-TRAN, M.** : Organisation spatiale des chloroplastes chez *Euglena* à l'aide de coupes sériées semi-fines. - Protoplasma *89* : 353 - 358, 1976.

25623 - **CALVERT, A., SLACK, G.** : Effect of carbon dioxide enrichment on growth, development and yield of glasshouse tomatoes. II. The duration of daily periods of enrichment. - J. hort. Sci. *51* : 401 - 409, 1976.

25624 - **CALVERT, H.E., DAWES, C.J.** : Ontogenetic membrane transitions in plastids of the coenocytic alga *Caulerpa (Chlorophyceae)*. - Phycologia *15* : 37 - 40, 1976.

25625 - **CALVERT, H.E., DAWES, C.J., BOROWITZKA, M.A.** : Phylogenetic relationships of *Caulerpa (Chlorophyta)* based on comparative chloroplast ultrastructure. - J. Phycol. *12* : 149 - 162, 1976.

25626 - **CALVIN, M.** : Photosynthesis as a resource for energy and materials. - Amer. Sci. *64* : 270 - 278, 1976.

25627 - **CALVIN, M.** : Photosynthesis as a resource for energy and materials. - Photochem. Photobiol. *23* : 425 - 444, 1976.

*25628 - **CAMMACK, R., HALL, D.O., RAO, K.K.** : Ferredoxine erhellen die Entwicklungsgeschichte der Pflanzen. - Naturwiss. Rundschau *25* : 443 - 444, 1972.

25629 - **CAMMACK, R., TEL-OR, E., STEWART, W.D.P.** : EPR spectra of photosystem I constituents in heterocyst preparations from *Anabaena cylindrica*. - FEBS Lett. *70* : 241 - 244, 1976.

25630 - **CAMPILLO, A.J., KOLLMAN, V.H., SHAPIRO, S.L.** : Intensity dependence of the fluorescence lifetime of *in vivo* chlorophyll excited by a picosecond light pulse. - Science *193* : 227 - 229, 1976.

25631 - **CAMPILLO, A.J., SHAPIRO, S.L.** : Use of picosecond lasers for studying photosynthesis. - Proc. Soc. photo-opt. Instrum. Eng. *94* (High Speed Optical Techniques) : 89 - 99, 1976.

25632 - **CAMPILLO, A.J., SHAPIRO, S.L., KOLLMAN , V.H., WINN, K.R., HYER, R.C.** : Picosecond exciton annihilation in photosynthetic systems. - Biophys. J. *16* : 93- - 97, 1976.

25633 - **CAMURRI, L., FERRARI, I., VILLANI, M.** : Biomassa e produzione del fitoplancton nel Lago Santo Parmense nella stagione delle acque aperte. [Biomass and production of phytoplankton in a mountain lake (Lago Santo Parmense) during the open water season.] - Arch. Oceanogr. Limnol. *18* : 237 - 253, 1976. [Chl; in Ital., ab : E.]

*25634 - **CANCEL, L.E., DE MONTALVO, M.R., RIVERA, J.M.** : Chlorophyll in citron (*Citrus medica* L.). - J. Agr. Univ. Puerto Rico *53* : 244 - 252, 1972.

25635 - **CANDAU, P., MANZANO, C., LOSADA, M.** : Bioconversion of light energy into chemical energy through reduction with water of nitrate to ammonia. - Nature *262* : 715 - 717, 1976.

25636 - **CANELLAKIS, E.S., AKOYUNOGLOU, G.** : Iodination of chloroplasts. I. Properties of iodinated chloroplasts. - Biochim. biophys. Acta *440* : 163 - 175, 1976.

B25637 - **CANNELL, M.G.R., LAST, F.T.** (ed.) : Tree Physiology and Yield Improvement. - Academic Press, London - New York - San Francisco 1976. [Ps.]

25638 - **CANTLEY, L.C., Jr., HAMMES, G.G.** : Investigation of quercetin binding sites on chloroplast coupling factor 1. - Biochemistry *15* : 1 - 8, 1976.

25639 - **CANTLEY, L.C., Jr., HAMMES, G.G.** : Characterization of sulfhydryl groups on chloroplast coupling factor 1 exposed by heat activation. - Biochemistry *15* : 9 - 14, 1976.

25640 - **CANVIN, D.T., LLOYD, N.D.H., FOCK, H., PRZYBYLLA, K.** : Glycine and serine metabolism and photorespiration. - In : BURRIS, R.H., BLACK, C.C. (ed.) : CO_2 Metabolism and Plant Productivity. Pp. 161 - 176. Univ. Park Press, Baltimore - London - Tokyo 1976.

25641 - **CAPRON, T.M., MANSFIELD, T.A.** : Inhibition of net photosynthesis in tomato in air polluted with NO and NO_2. - J. exp. Bot. *27* : 1181 - 1186, 1976.

25642 - **CARBONNEAU, A.** : Mise au point bibliographique sur la photosynthèse chez la Vigne. - Connaiss. Vigne Vin *10* : 249 - 267, 1976.

25643 - **CARBONNEAU, A.** : Principes et méthodes de mesure de la surface foliaire. Essai de caractérisation des types de feuilles dans le genre *Vitis*. - Ann. Amélior. Plant. *26* : 327 - 343, 1976.

25644 - **CARDOZO, G.,H., SCHNETTER, M.-L.** : Estudios ecologicos en el paramo de Cruz Verde, Colombia. III. La biomasa de tres asociaciones vegetales y la productividad de *Calamagrostis effusa* (H.B.K.) STEUD. y *Paepalanthus columbiensis* RUHL. en comparacion con la concentration de clorofila. [Ecological studies in the Cruz Verde paramo, Colombia. III. Biomass of three plant associations and productivity of *Calamagrostis effusa* (H.B.K.) STEUD. and *Paepalanthus columbiensis* RUHL. in comparison with chlorophyll concentration.] - Caldasia *11* (54) : 69 - 83, 1976. [In Span., ab : G.]

25645 - **CARITHERS, R.P., PARSON, W.W.** : Delayed fluorescence from *Rhodopseudomonas sphaeroides* following single flashes. - Biochim. biophys. Acta *440* : 215 - - 232, 1976.

25646 - **CARLSON, R.W., BAZZAZ, F.A., STUKEL, J.J., WEDDING, J.B.** : Physiological effects, wind reentrainment, and rainwash of Pb aerosol particulate deposited on plant leaves. - Environ. Sci. Technol. *10* : 1139 - 1142, 1976. [Ps.]

25647 - **CARPENTER, D.J., CARPENTER, S.M.** : Numerical analysis of primary production and associated data from waters off the east coast of Australia. - Aust. J. mar. Freshwater Res. *27* : 431 - 439, 1976.

25648 - **CARRETO, J.I., CATOGGIO, J.A.** : Variations in pigment contents of diatom *Phaeodactylum tricornutum* during growth. - Mar. Biol. *36* : 105 - 112, 1976.

25649 - **CARSON, S.D.** : Ammonium molybdate-stannous chloride determination of orthophosphate in the presence of *Triton X-100*. - Anal. Biochem. *75* : 472 - 477, 1976.

25650 - **CASHMORE, A.R.** : Protein synthesis in plant leaf tissue. The sites of synthesis of the major proteins. - J. biol. Chem. *251* : 2848 - 2853, 1976. [Ps enzymes.]

25651 - **CASTENHOLZ, R.W.** : The effect of sulfide on the blue-green algae of hot springs. I. New Zealand and Iceland. - J. Phycol. *12* : 54 - 68, 1976. [Ps.]

25652 - **CASTRO, P.R.C., LOURENÇO, R.S., FILHO, W.J., CARELLI, M.L.C., TURKIEWICZ, L., SOBRAL, L.F.** : Efeito de GA, CEPA e confinamento em polietileno na maturaçao do maracujá (*Passiflora edulis* SIMS). [Effects of GA, CEPA and polyethylene bags on the ripening of yellow passion fruit (*Passiflora edulis* SIMS).] - Solo *68* (1) : 7 - 11, 1976. [Chi; in Port., ab : E.]

25653 - **CATARINO, F.M., BENTO-PEREIRA, F.** : Ecological characteristics and CO_2 fixation in a xerophytic plant (*Ceratonia siliqua*). - In : VARDAR, Y., SHEIKH, K. H., ÖZTÜRK, M.A. (ed.) : Proceedings of the Third MPP Meeting. Pp. 68 - 79. Ege Univ., Izmir 1976.

25654 - CATSKÝ, J., TICHÁ, I., SOLÁROVÁ, J. : Ontogenetic changes in the internal limitations to bean-leaf photosynthesis. 1. Carbon dioxide exchange and conduc-
tances for carbon dioxide transfer. - Photosynthetica 10 : 394 - 402, 1976.

25655 - CATTOLICO, R.A., BOOTHROYD, J.C., GIBBS, S.P. : Synchronous growth and plastid replication in the naturally wall-less alga Olisthodiscus luteus. - Plant
Physiol. 57 : 497 - 503, 1976.

25656 - CAVELL, S., SCOPES, R.K. : Isolation and characterization of the "photosynthetic" phosphoglycerate kinase from Beta vulgaris. - Europe. J. Biochem. 63 :
483 - 690, 1976.

25657 - CEDEÑO-MALDONADO, A., ASENCIO, C.I. : Carbonic anhydrase : an alternate site
for cadmium inhibition of photosynthesis. - Plant Physiol. 57 (Suppl.) : 7,
1976.

25658 - CEDEÑO-MALDONADO, A., LIU, L.C. : Effect of two substituted urea and two s-
-triazine type herbicides on the photosynthesis of Lemna perpusilla TORR. -
J. Agr. Univ. Puerto Rico 60 : 369 - 374, 1976.

25659 - CERNUSCA, A. : Bestandesstruktur, Bioklima und Energiehaushalt von alpinen
Zwergstrauchbeständen. - Ecol. Plant. 11 : 71 - 102, 1976.

25660 - CERNUSCA, A. : Energy exchange within individual layers of a meadow. - Oecologia 23 : 141 - 149, 1976.

25661 - CERNUSCA, A. : Energie- und Wasserhaushalt eines alpinen Zwergstrauchbestan-
des während einer Föhnperiode. - Arch. Meteorol. Geophys. Bioklimatol., Ser.
B 24 : 219 - 241, 1976.

25662 - CESARENI, G., IANNUCCI, C. : Diurnal variation of corn crop albedo. - Oecol.
Plant. 11 : 257 - 265, 1976.

25663 - CEULEMANS, R., LEMEUR, R., MOERMANS, R., SAMSUDDIN, Z., IMPENS, I. : Studie
van bladanatomie en -morfologie en hun samenhang met de diffusiekarakteris-
tieken van verschillende Populus-klonen. [Study of leaf anatomy and -morpho-
logy and their relation with the diffusion characteristics of several Populus
clones.] - In : Rapport Nr. 2 - 1976. Pp. 1 - 41. Univ. Instelling, Antwerpen
1976. [Resistances; in Flem.]

25664 - CHABOT, B.F., LEWIS, A.R. : Thermal acclimation of photosynthesis in northern
red oak. - Photosynthetica 10 : 130 - 135, 1976.

25665 - CHADEFAUD, M. : Sur l'origine des plastes, les plastes "cyanelloïdes" et la
"classe?" des Glaucophycées. - Compt. rend. Acad. Sci. Paris, Sér. D 283 :
1029 - 1032, 1976.

25666 - CHALLA, H. : An analysis of the diurnal course of growth, carbon dioxide exchange and carbohydrate reserve content of cucumber. - Agr. Res. Rep. (Wageningen) 861 : VIII + 1 - 88, 1976.

25667 - CHAMBROY, Y., BURET, M., FLANZY, C. : Appareil pour le dosage conductimétri-
que du gaz carbonique dégagé par des organes végétaux. - Ann. Technol. agr.
25 : 191 - 202, 1976.

25668 - CHAMOROVSKIĬ, S.K., NOKS, P.P., REMENNIKOV, S.M., KONONENKO, A.A., RUBIN, A.
B. : Issledovanie temperaturnoĭ zavisimosti kinetiki temnovogo vosstanovleniya bakteriokhlorofilla P870, okislennogo impul'snym lazernym i postoyannym
svetom v preparatakh fotosinteticheskikh reaktsionnykh tsentrov iz Rhodopseudomonas spheroides, shtamm 1760-1. [Study of temperature dependence of kinetics of dark reduction of bacteriochlorophyll P870 oxidized by pulsed laser
and continuous light in photosynthetic reaction center preparations from
Rhodopseudomonas spheroides, strain 1760-1.] - Biofizika 21 : 300 - 306, 1976.
[In R, ab : E.]

25669 - CHAMOROVSKY, S.K., REMENNIKOV, S.M., KONONENKO, A.A., VENEDIKTOV, P.S., RUBIN,
A.B. : New experimental approach to the estimation of rate of electron transfer from the primary to secondary acceptors in the photosynthetic electron
transport chain of purple bacteria. - Biochim. biophys. Acta 430 : 62 - 71,
1976.

25670 - CHAMPAGNOL, F. : Chromatographie d'affinité de l'anhydrase carbonique végétale. - J. Chromatogr. 120 : 489 - 490, 1976.

25671 - **CHAMPIGNY, M.-L.** : La régulation du cycle de Calvin. - Physiol. vég. *14* :
607 - 628, 1976.

25672 - **CHAMPIGNY, M.-L., BISMUTH, E.** : Role of photosynthetic electron transfer in
light activation of Calvin cycle enzymes. - Physiol. Plant. *36* : 95 - 100,
1976.

25673 - **CHAND, P., SHARMA, N.N.** : Constant for determining leaf area index in maize.
- Indian J. Agron. *21* : 171 - 172, 1976.

25674 - **CHANG, N.K., LIM, M.K., YUN, I.S.** : Studies on the gross metabolism of carbon in a grassland of *Miscanthus sinensis*. - Korean J. ani. Sci. *18* : 231 -
- 236, 1976. [Ps; in Korean, ab : E.]

25675 - **CHAPARRO, A., MALDONADO, J.M., DIEZ, J., RELIMPIO, A.M., LOSADA, M.** : Nitrate
reductase inactivation and reducing power and energy charge in *Chlorella*
cells. - Plant Sci. Lett. *6* : 335 - 342, 1976. [Ps.]

*25676 - **CHAPMAN, A.R.O., CAMPBELL, C.C.M.** : Quanta vs. watts. - Limnol. Oceanogr.
20 : 496, 1975. [Ps.]

25677 - **CHAPMAN, A.R.O., HOYT, B., PATON, B.E.** : A self-contained instrument for lo-
garithmic recording of submarine quantum irradiance. - Mar. Biol. *38* : 91 -
- 94, 1976.

25678 - **CHAPMAN, D.J., LEECH, R.M.** : Phosphoserine as an early product of photosynthesis in isolated chloroplasts and in leaves of *Zea mays* seedlings. - FEBS
Lett. *68* : 160 - 164, 1976.

25679 - **CHAPMAN, G.W., Jr., ROBERTSON, J.A., BURDICK, D.** : Fatty acid and chlorophyll
levels of coastal bermudagrass during the day and during maturation. - Physiol. Plant. *36* : 66 - 70, 1976.

25680 - **CHAPRA, S.C., TARAPCHAK, S.J.** : A chlorophyll *a* model and its relationship
to phosphorus loading plots for lakes. - Water Resources Res. *12* : 1260 -
- 1264, 1976.

25681 - **CHARLES-EDWARDS, D.A.** : Shoot and root activities during steady-state plant
growth. - Ann. Bot. *40* : 767 - 772, 1976. [Growth analysis.]

25682 - **CHARLES-EDWARDS, D.A., HO, L.C.** : Translocation and carbon metabolism in tomato leaves. - Ann. Bot. *40* : 387 - 389, 1976.

25683 - **CHARLES-EDWARDS, D.A., THORPE, M.R.** : Interception of diffuse and direct-beam
radiation by a hedgerow apple orchard. - Ann. Bot. *40* : 603 - 613, 1976.

25684 - **CHARTIER, P., PRIOUL, J.L.** : The effects of irradiance, carbon dioxide and
oxygen on the net photosynthetic rate of the leaf : A mechanistic model. -
Photosynthetica *10* : 20 - 24, 1976.

25685 - **CHATTAR, M.S., MARÓTI, I.** : Seasonal changes in total content of soluble carbohydrates in ever-green plants. - Acta biol. szeged. *22* (1 - 4) : 3 - 5,
1976.

25686 - **CHATTERJEE, A., MANDAL, R.K., SIRCAR, S.M.** : Effects of growth substances on
productivity, photosynthesis and translocation of rice varieties. - Indian J.
Plant Physiol. *19* : 131 - 138, 1976.

25687 - **CHATTERTON, N.J.** : Photosynthesis of 22 alfalfa populations differing in resistance to diseases, insect pests, and nematodes. - Crop Sci. *16* : 833 - 834,
1976.

25688 - **CHECCUCCI, A., COLOMBETTI, G., FERRARA, R., LENCI, F.** : Action spectra for
photoaccumulation of green and colorless *Euglena* : Evidence for identification of receptor pigments. - Photochem. Photobiol. *23* : 51 - 54, 1976.

25689 - **CHELM, B.K., HALLICK, R.B.** : Changes in the expression of the chloroplast genome of *Euglena gracilis* during chloroplast development. - Biochemistry *15* :
593 - 599, 1976.

25690 - **CHEMERILOVA, V.I., KVITKO, K.V.** : Izuchenie modifitsiruyushchikh pigmentatsiyu
mutatsiĭ u shtammov *Chlamydomonas reinhardi* raznoĭ ploidnosti. Soobshchenie I.
Diploidy, geterozigotnye po mutatsii Its-31. [Pigmentation-modifying mutations
in *Chlamydomonas reinhardi* strains of different ploidy. I. Diploids heterozy-

gous for Its-31 mutation.] - Genetika *12* (9) : 44 - 49, 1976. [In R, ab : E.]

25691 - **CHEN, C.-H., BERNS, D.S.** : Photosensitivity of artificial bilayer membranes : lipid-chlorophyll interaction. - Photochem. Photobiol. *24* : 255 - 260, 1976.

25692 - **CHEN, K., JOHAL, S., WILDMAN, S.G.** : Role of chloroplast and nuclear DNA genes during evolution of Fraction I protein. - In : BÜCHER, T., NEUPERT, W., SEBALD, W., WERNER, S. (ed.) : Genetics and Biogenesis of Chloroplasts and Mitochondria. Pp. 3 - 11. North-Holland Publ. Co., Amsterdam - New York - Oxford 1976.

25693 - **CHEN, K., KUNG, S.D., GRAY, J.C., WILDMAN, S.G.** : Subunit polypeptide composition of fraction I protein from various plant species. - Plant Sci. Lett. *7* : 429 - 434, 1976.

25694 - **CHENG, D.M.H., TYLER, P.A.** : Primary productivity and trophic status of lakes Sorell and Crescent, Tasmania. - Hydrobiologia *48* : 59 - 64, 1976.

25695 - **CHENIAE, G., MARTIN, I.F.** : Mechanism of Tris inactivation of O_2 evolution. - Plant Physiol. *57* (Suppl.) : 25, 1976.

25696 - **CHERMNYKH, L.N., CHUGUNOVA, N.G.** : Formirovanie i funktsionirovanie fotosinteticheskogo apparata rastenii pri izmenenii svetovogo i temperaturnogo faktorov v kontroliruemykh usloviyakh. [Formation and functioning of the photosynthetic apparatus of plants at changing the light and temperature factors in controlled conditions.] - In : Itogi Issledovaniya Mekhanizma Fotosinteza. Pp. 75 - 78. Pushchino 1976. [In R.]

*25697 - **CHERNAVINA, I.A., ZHOGOVA, E.P., MATUS, V.K.** : Tsitokhromnye komponenty khloroplastov ovsa, vyrashchennogo v usloviyakh izbytka, margantsa. [Cytochrome components of chloroplasts of oats grown under manganese surplus.] - Nauch. Dokl. vyssh. Shkoly, biol. Nauki *17* (2) : 76 - 80, 1974. [In R.]

25698 - **CHERNOMORSKII, S.A., KURNYGINA, V.T., FRAGINA, A.I.** : Poluchenie feofitina na ionoobmennoi smole KU-1. [Production of pheophytin on ion-exchange resin KU-1.] - Prikl. Biokhim. Mikrobiol. *12* : 421 - 422, 1976. [In R, ab : E.]

*25699 - **CHERNYAD'EV, I.I., TEREKHOVA, I.V., KOMAROVA, Yu.M., DOMAN, N.G., GORONKOVA, O.I., AL'BITSKAYA, O.N.** : Ob uchastii karboangidrazy v fotosinteze. [Participation of carbonic anhydrase in photosynthesis.] - Dokl. Akad. Nauk SSSR *223* : 501 - 503, 1975. [In R.]

25700 - **CHEVALLIER, D.** : Effets d'une carence en manganèse sur la structuration et le fonctionnement de l'appareil photosynthétique des spores *Funaria hygrometrica*. - Biol. Plant. *18* : 132 - 139, 1976.

25701 - **CHEVALLIER, D., DOUCE, R.** : Interactions between mitochondria and chloroplasts in cells. I. Action of cyanide and of 3-(3,4-dichlorophenyl)-1,1-dimethylurea on the spore of *Funaria hygrometrica*. - Plant Physiol. *57* : 400 - 402, 1976.

25702 - **CHEVROU, R.B.** : Précisions des mesures de superficie estimée par grille de points ou intersections de parallèles. - Ann. Sci. forest. *33* : 257 - 269, 1976. [Dot grid method, transects.]

25703 - **CHIBISOV, A.K., ZAKHAROVA, N.I., PESHKIN, A.F., SLAVNOVA, T.D.** : Issledovanie promezhutochnykh sostoyanii v fotoreaktsiyakh khlorofill-belkovykh kompleksov. [Transient states in photoreactions of chlorophyll-protein complexes.] - Mol. Biol. (Moskva) *10* : 1002 - 1010, 1976. [In R, ab : E.]

25704 - **CHIN, P., BRODY, S.S.** : Mixed monomolecular films of chlorophyll and cytochromes. - Z. Naturforsch. *31C* : 44 - 47, 1976.

*25705 - **CHIRANJEEVI, V., TRIPATHI, R.K.** : Changes in chlorophyll and carotenoid contents in sorghum leaves due to zonate leaf spot and anthracnose. - Indian J. Mycol. Plant Pathol. *5* : 98 - 99, 1975.

25706 - **CHISHOLM, S.W., STROSS, R.G.** : Phosphate uptake kinetics in *Euglena gracilis* (Z) (*Euglenophyceae*) grown in light/dark cycles. II. Phased PO_4-limited cultures. - J. Phycol. *12* : 217 - 222, 1976.

25707 - **CHLOUPEK, O.** : Die Bewertung des Wurzelsystems von Senfpflanzen auf Grund der dielektrischen Eigenschaften und mit Rücksicht auf den Endertrag. - Biol. Plant. *18* : 44 - 49, 1976.

25708 - **CHMORA, S.N., SLOBODSKAYA, G.A., NICHIPOROVICH, A.A.** : Uglekislotnyĭ gazoobmen i fotodykhanie rasteniĭ s razlichnoĭ aktivnost'yu fotosinteticheskogo
apparata. [CO_2 exchange and photorespiration in plants with different activity of photosynthetic apparatus.] - In : Gazometricheskoe Issledovanie Fotosinteza i Dykhaniya Rasteniĭ. Pp. 141 - 144. Akad. Nauk SSSR, Tartu 1976.
[In R.]

25709 - **CHMORA, S.N., SLOBODSKAYA, G.A., NICHIPOROVICH, A.A.** : Ingibirovanie uglekislotnogo gazoobmena C-3 rasteniĭ kislorodom v usloviyakh vysokoĭ kontsentratsii CO_2. [Inhibition of CO_2 exchange in C_3 plants by oxygen in high CO_2
concentration.] - In : Gazometricheskoe Issledovanie Fotosinteza i Dykhaniya
Rasteniĭ. Pp. 145 - 147. Akad. Nauk SSSR, Tartu 1976. [In R.]

25710 - **CHMORA, S.N., SLOBODSKAYA, G.A., NICHIPOROVICH, A.A.** : Ingibirovanie uglekislotnogo gazoobmena C-3 rasteniĭ kislorodom v usloviyakh vysokoĭ kontsentratsii CO_2. [Inhibition of carbon dioxide gas exchange of C_3 plants by oxygen in conditions of high CO_2 concentration.] - Fiziol. Rast. *23* : 885 - 892,
1976. [In R, ab : E.]

25711 - **CHOLLET, R.** : C_4 control of photorespiration : Studies with isolated mesophyll cells and bundle sheath strands. - In : BURRIS, R.H., BLACK, C.C. (ed.):
CO_2 Metabolism and Plant Productivity. Pp. 327 - 341. Univ. Park Press, Baltimore - London - Tokyo 1976.

25712 - **CHOLLET, R.** : Effect of glycidate on glycolate formation and photosynthesis
in isolated spinach chloroplasts. - Plant Physiol. *57* : 237 - 240, 1976.

25713 - **CHOLLET, R., ANDERSON, L.L.** : Regulation of ribulose 1,5-bisphosphate carboxylase-oxygenase activities by temperature pretreatment and chloroplast metabolites. - Plant Physiol. *57* (Suppl.) : 4, 1976.

25714 - **CHOPRA, N.M.** : Investigations into the fate of plant pigments in some Canadian soils. - Soil Sci. *121* : 103 - 113, 1976.

25715 - **CHOUDHARY, D.K., KAUL, B.L.** : Effect of sodium azide on X-rays induced seedling injury, chromosome aberrations & frequency of chlorophyll deficient mutations in barley *Hordeum vulgare* L. - Indian J. exp. Biol. *14* : 607 - 609,
1976.

25716 - **CHOUDHURY, N.K., GURU, B.C., DASH, M.C.** : Observations on the relationship
between leaf pigment concentration, moisture and dry weight before leaf fall
in *Dalbergia sissoo* ROXB. - Comp. Physiol. Ecol. *1* : 147 - 149, 1976.

25717 - **CHOW, C.T.** : Cell-free, protein-synthesizing system of photosynthetic and
heterotrophic *Rhodospirillum rubrum* . - Can. J. Microbiol. *22* : 304 - 308,
1976.

25718 - **CHOW, C.T.** : Functional and structural differences between photosynthetic
and heterotrophic *Rhodospirillum rubrum* ribosomes and S-100 fractions. - Can.
J. Microbiol. *22* : 1522 - 1539, 1976.

25719 - **CHOW, W.S., HOPE, A.B.** : Light-induced pH gradients in isolated spinach chloroplasts. - Aust. J. Plant Physiol. *3* : 141 - 152, 1976.

25720 - **CHOW, W.S., WAGNER, G., HOPE, A.B.** :Light-dependent redistribution of ions in
isolated spinach chloroplasts. - Aust. J. Plant Physiol. *3* : 853 - 861, 1976.

25721 - **CHRISTELLER, J.T., LAING, W.A., TROUGHTON, J.H.** : Isotope discrimination by
ribulose 1,5-diphosphate carboxylase. No effect of temperature or HCO_2^- concentration. - Plant Physiol. *57* : 580 - 582, 1976.

25722 - **CHRISTY, A.L.** : Mathematical models of Münch pressure flow : basic concepts
and assumptions. - In : WARDLAW, I.F., PASSIOURA, J.B. (ed.) : Transport and
Transfer Processes in Plants. Pp. 363 - 368. Academic Press, New York - San
Francisco - London 1976. [Photosynthate transport.]

25723 - **CHRISTY, A.L., SWANSON, C.A.** : Control of translocation by photosynthesis and
carbohydrate concentrations of the source leaf. - In : WARDLAW, I.F., PASSI-
OURA, J.B. (ed.) : Transport and Transfer Processes in Plants. Pp. 329 - 338.
Academic Press, New York - San Francisco - London 1976.

25724 - **CHRÓST, R.J., SIKORSKA, U.** : The effect of pollution on photosynthetic activity of algae and physiological activity of bacteria in lake. - Pol. Arch. Hydrobiol. *23* : 357 - 364, 1976.

25725 - **CHU, A.C.P., TILLMAN, R.F.** : Growth of a forage sorghum hybrid under two soil moisture regimes in the Manawatu. - New Zeal. J. exp. Agr. *4* : 351 - 355, 1976. [Growth analysis.]

25726 - **CHUA, N.-H.** : A uniparental mutant of *Chlamydomonas reinhardtii* with a variant thylakoid membrane polypeptide. - In : BÜCHER, T., NEUPERT, W., SEBALD, W., WERNER, S. (ed.)· : Genetics and Biogenesis of Chloroplasts and Mitochondria. Pp. 323 - 330. North-Holland Publ. Co., Amsterdam - New York - Oxford 1976.

25727 - **CHUB, A.I., BURTSEVA, R.A.** : Sakhara fotosinteziruyushchikh i provodyashchikh tkaneĭ u soi. [Sugars of photosynthesizing and conducting tissues in soybean.] - Fiziol. Biokhim. kul't. Rast. *8* : 519 - 523, 1976. [In R, ab : E.]

25728 - **CHURCH, J.M.F., REES, W.A.** : A computer programme for primary productivity studies. - East afr. Wildlife J. *14* : 169 - 170, 1976.

25729 - **CHUROVÁ, K.** : Analýza produkčného procesu ovsa na zelenú hmotu v závislosti od termínu aplikácie dusíkatého hnojenia. [Analysis of production process of oats for green matter dependent on time of nitrogen application.] - Ved. Práce výsk. Ustavu rastl. Výroby Piešťanoch *13* : 101 - 112, 1976. [In Slovak, ab : E, R.]

25730 - **CIFERRI, G.O., TIBONI, O.** : Evidence for the synthesis in the chloroplast of elongation factor G. - Plant Sci. Lett. *7* : 455 - 466, 1976.

*25731 - **ČIRKOVA-GEORGIEVA, M., PEŠEVSKA, V., PETROVSKA, V., VESOVA, N.** : Sodržina na karotin kaj nekoi populacii spanak' (*S. oleracea* L.) vo Makedonija. [Carotene levels in some spinach (*Spinacia oleracea* L.) crops in Macedonia.] - God. Zb. zemjodel.-šumarskiot Fak. Univ. Skopje *24* : 65 - 70, 1972. [In Macedonian, ab : E.]

*25732 - **ČIRKOVA-GEORGIEVSKA, M., PEŠEVSKA-SOTIROVA, V., VESOVA, N., PETROVSKA, V.** : Zastapenost na askorbinskata kiselina, na karotinot i na oksalatite vo nekoi populacii loboda (*Atriplex hortensis* L.) vo Makedonija. [Contents of ascorbic acid, carotene and oxalates in some populations of *Atriplex hortensis* L. in Macedonia.] - Maked. Akad. Nauk. Umetnost. Prilozi *4* (1 - Odd. prirod.-mat. Nauki) : 11 - 18, 1972. [In Macedonian, ab : E.]

25733 - **CLARKE, R.H., CONNORS, R.E.** : Optically detected zero-field triplet state magnetic resonance in photosynthetic bacteria. - Chem. phys. Lett. *42* : 69 - - 72, 1976.

25734 - **CLARKE, R.H., CONNORS, R.E., FRANK, H.A.** : Investigation of the structure of the reaction center in photosynthetic bacteria by optical detection of triplet stage magnetic resonance. - Biochem. biophys. Res. Commun. *71* : 671 - - 675, 1976.

25735 - **CLARKE, R.H., CONNORS, R.E., SCHAAFSMA, T.J., KLEIBEUKER, J.F., PLATENKAMP, R.J.** : The triplet state of chlorophylls. - J. amer. chem. Soc. *98* : 3674 - - 3677, 1976.

25736 - **CLARKE, R.H., HOFELDT, R.H.** : Optical detection of zero field magnetic resonance in the triplet state of chlorophyll. - In : BIRKS, J.B. (ed.) : Excited States of Biological Molecules. Pp. 309 - 313. Wiley, Chichester 1976.

25737 - **CLARKSON, N.M., RUSSELL, J.S.** : Effect of water stress on the phasic development of annual *Medicago* species. - Aust. J. agr. Res. *27* : 227 - 234, 1976.

*25738 - **CLASBY, R., HORNER, R., ALEXANDER, V.** : Arctic ice algae studies. - In : Science in Alaska. Proc. Twenty-third Alaska Sci. Conf. *1972* : 54, 1972. [Ps.]

25739 - **CLASBY, R.C., ALEXANDER, V., HORNER, R.** : Primary productivity of sea-ice algae. - In : Assessment of the Arctic Marine Environment : Selected Topics. Pp. 289 - 304. Univ. Alaska, Fairbanks 1976.

25740 - **CLAUSSEN, W.** : Einfluß der Frucht auf die Trockensubstanzverteilung in der Aubergine (*Solanum melongena* L.). - Gartenbauwissenschaft *41* : 236 - 239, 1976.

25741 - **CLAYTON, R.K., YAMAMOTO, T.** : Photochemical quantum efficiency and absorption spectra of reaction centers from *Rhodopseudomonas sphaeroides* at low temperature. - Photochem. Photobiol. *24* : 67 - 70, 1976.

25742 - CLINE, R.G., CAMPBELL, G.S. : Seasonal and diurnal water relations of selected forest species. - Ecology *57* : 367 - 373, 1976. [Stomatal resistance.]

25743 - CLOERN, J.E. : Recent limnological changes in southern Kootenay Lake, British
Columbia. - Can. J. Zool. *54* : 1571 - 1578, 1976. [Chl.]

*25744 - COATS, G.E., FOY, C.L. : Effects of atrazine-phytobland oil combinations on
$^{14}CO_2$-fixation and transpiration. - Weed Sci. *22* : 215 - 220, 1974.

25745 - COBB, A.H., WELLBURN, A.R. : Polypeptide binding to plastid envelopes during
chloroplast development. - Planta *129* : 127 - 131, 1976.

25746 - COGDELL, R.J., PARSON, W.W., KERR, M.A. : The type, amount, location, and
energy transfer properties of the carotenoid in reaction centers from *Rhodopseudomonas sphaeroides*. - Biochim. biophys. Acta *430* : 83 - 93, 1976.

25747 - COHEN, C.E., REBEIZ, C.A. : Contribution of various protochlorophyll holochro-
mes to the greening process of higher plants. - Plant Physiol. *57* (Suppl.) :
72, 1976.

25748 - COHEN, C.E., SCHIFF, J.A. : Events surrounding the early development of *Euglena* chloroplasts - XI. Protochlorophyll(ide) and its photoconversion. - Photochem. Photobiol. *24* : 555 - 566, 1976.

25749 - COHEN, D., SCHIFF, J.A. : Events surrounding the early development of *Euglena*
chloroplasts. Photoregulation of the transcription of chloroplastic and cytoplasmic ribosomal RNAs. - Arch. Biochem. Biophys. *177* : 201 - 216, 1976.

25750 - COHEN, W.S., McCARTY, R.E. : Reversibility of the cyanide inhibition of electron transport in spinach chloroplast thylakoids. - Biochem. biophys. Res.
Commun. *73* : 679 - 685, 1976.

25751 - COLBOW, K., DANYLUK, R.P. : Energy transfer in photosynthesis. - Biochim.
biophys. Acta *440* : 107 - 121, 1976.

*25752 - COLE, D.R., PLAPP, F.W., Jr. : Inhibition of growth and photosynthesis in
Chlorella pyrenoidosa by a polychlorinated biphenyl and several insecticides.
- Environ. Entomol. *3* : 217 - 220, 1974.

25753 - COLEMAN, D.C., ANDREWS, R., ELLIS, J.E., SINGH, J.S. : Energy flow and partitioning in selected man-managed and natural ecosystems. - Agro-Ecosystems *3* :
45 - 54, 1976.

25754 - COLEMAN, W.K., GREYSON, R.I. : The growth and development of the leaf in tomato (*Lycopersicon esculentum*) I. The plastochron index, a suitable basis for
description. - Can. J. Bot. *54* : 2421 - 2428, 1976.

25755 - COLLATZ, J., FERRAR, P.J., SLATYER, R.O. : Effects of water stress and differential hardening treatments on photosynthetic characteristics of a xeromor-
phic shrub, *Eucalyptus socialis*, F. MUELL. - Oecologia *23* : 95 - 105, 1976.

25756 - COLLINS, M.L.P., NIEDERMAN, R.A. : Membranes of *Rhodospirillum rubrum* : isolation and physicochemical properties of membranes from aerobically grown
cells. - J. Bacteriol. *126* : 1316 - 1325, 1976. [Chl.]

25757 - COLLINS, M.L.P., NIEDERMAN, R.A. : Membranes of *Rhodospirillum rubrum* : physicochemical properties of chromatophore fractions isolated from osmotically
and mechanically disrupted cells. - J. Bacteriol. *126* : 1326 - 1338, 1976.

25758 - COLLINS, W.B. : Effect of carbon dioxide enrichment on growth of the potato
plant. - HortScience *11* : 467 - 469, 1976. [Growth analysis.]

25759 - COLMAN, B., CHENG, K.H., INGLE, R.K. : The relative activities of PEP carboxylase and RuDP carboxylase in blue-green algae. - Plant Sci. Lett. *6* : 123 -
- 127, 1976.

25760 - COLWELL, G.L., WICKSTROM, C.E. : Cell, cyanelle, and chlorophyll *a* relationships in *Glaucocystis nostochinearum* ITZ. - J. Phycol. *12* (Suppl.) : 11, 1976.

25761 - CONJEAUD, H., MICHEL-VILLAZ, M. : Photoinduced electric field effect on the
optical properties of photosynthetic membranes. - J. theor. Biol. *62* : 1 - 16,
1976.

25762 - CONJEAUD, H., MICHEL-VILLAZ, M., VERMEGLIO, A., MATHIS, P. : Location of field-sensitive carotenoid molecules in the chloroplast membrane. Arguments from low-temperature studies. - FEBS Lett. *71* : 138 - 141, 1976.

25763 - CONKLIN, P.J. : Size fractions of phytoplankton from a Florida estuary : their relative composition, numbers, pigmentation and primary productivity. - J. Phycol. *12* (Suppl.) : 21, 1976.

25764 - CONOVER, C.A., POOLE, R.T. : Production of acclimatized *Ficus benjamina.* - HortScience *11* (3, Sect. 2) : 302, 1976. [Chl.]

25765 - COOK, M.G., EVANS, L.T. : Effect of sink size, geometry and distance from source on the distribution of assimilates in wheat. - In : WARDLAW, I.F., PASSIOURA, J.B. (ed.) : Transport and Transfer Processes in Plants. Pp. 393 - - 400. Academic Press, New York - San Francisco - London 1976.

*25766 - COOKSEY, K.E., COOKSEY, B. : Turnover of photosynthetically fixed carbon in reef corals. - Mar. Biol. *15* : 289 - 292, 1972.

25767 - COOMBS, J. : β-carboxylation, photorespiration and photosynthetic carbon as-similation in C4 plants. - In : SMITH, H. (ed.) : Commentaries in Plant Scien-ce. Pp. 1 - 12. Pergamon Press, Oxford - New York - Toronto - Sydney - Paris - Frankfurt 1976.

25768 - COOMBS, J. : Interactions between chloroplasts and cytoplasm in C4 plants. - In : BARBER, J. (ed.) : The Intact Chloroplast. Pp. 279 - 313. Elsevier, Am-sterdam - New York - Oxford 1976.

*25769 - COOMBS, J., BALDRY, C.W., BUCKE, C. : C4 photosynthesis. - In : SUNDERLAND, N. (ed.) : Perspectives in Experimental Biology. Vol. 2. Botany. Pp. 177 - - 188. Pergamon Press, Oxford - New York - Toronto - Sydney - Paris - Braun-schweig 1975.

25770 - COOMBS, J., GREENWOOD, A.D. : Compartmentation of the photosynthetic appara-tus. - In : BARBER, J. (ed.) : The Intact Chloroplast. Pp. 1 - 51. Elsevier, Amsterdam - New York - Oxford 1976.

25771 - COOPER, A.J., THORNLEY, J.H.M. : Response of dry matter partitioning, growth, and carbon and nitrogen levels in the tomato plant to changes in root tempe-rature : Experiment and theory. - Ann. Bot. *40* : 1139 - 1152, 1976.

25772 - COOPER, J.P. : Photosynthetic efficiency of the whole plant. - In : DUCKHAM, A.N., JONES, J.G.W., ROBERTS, E.H. (ed.) : Food Production and Consumption : The Efficiency of Human Food Chains and Nutrient Cycles. Pp. 107 - 126. North--Holland Publ. Co., Amsterdam 1976.

25773 - CORKER, G.A. : A survey of EPR investigations of bacterial photosynthesis. - Photochem. Photobiol. *24* : 617 - 628, 1976.

25774 - CORLEY, R.H.V. : Photosynthesis and Productivity. - In : CORLEY, R.H.V., HARD-ON, J.J., WOOD, B.J. (ed.) : Developments in Crop Science (1).Oil Palm Re-search. Pp. 55 - 76. Elsevier Sci. Publ. Co., Amsterdam - New York 1976.

B25775 - CORLEY, R.H.V., HARDON, J.J., WOOD, B.J. (ed.) : Developments in Crop Science (1). Oil Palm Research. - Elsevier Sci. Publ. Co., Amsterdam - New York 1976. [Ps.]

25776 - CORNIC, G. : Effet exercé sur l'activité photosynthétique du *Sinapis alba* L. par une inhibition temporaire de la photorespiration se déroulant dans un air sans CO_2. - Compt. rend. Acad. Sci. Paris, Sér. D *282* : 1955 - 1958, 1976.

25777 - CORREDOR, J.E., CAPONE, D.G., COOKSEY, K.E. : On the use of liquid scintilla-tion vials in ATP-photometry. - Anal. Biochem. *70* : 624 - 627, 1976.

25778 - COSSON, J., GAYRAL, P., JACQUES, R. : Action de la composition spectrale de la lumière sur la croissance et la reproduction des gamétophytes de la *Lami-naria digitata* (L.) LAM (Phéophycée, Laminariale). - Compt. rend. Acad. Sci. Paris, Sér. D *283* : 1293 - 1296, 1976. [Ps.]

25779 - COSTES, C., BURGHOFFER, C., CARRAYOL, E., DUCET, G., DIANO, M. : Occurrence of carotenoids in non-plastidial materials from potato tuber cells. - Plant Sci. Lett. *6* : 253 - 259, 1976.

25780 - COUTTS, M.P., ARMSTRONG, W. : Role of oxygen transport in the tolerance of trees to waterlogging. - In : CANNELL, M.G.R., LAST, F.T. (ed.) : Tree Physiology and Yield Improvement. Pp. 361 - 385. Academic Press, London - New York - San Francisco 1976. [Ps.]

25781 - **COX, E.F., McKEE, J.M.T., DEARMAN, A.S.** : The effect of growth rate on tipburn occurrence in lettuce. - J. hort. Sci. *51* : 297 - 309, 1976. [Growth analysis.]

25782 - **COX, R., DELOSME, R.** : Variations rapides d'absorption après un éclair chez *Chlorella*. - Compt. rend. Acad. Sci. Paris, Sér. D *282* : 775 - 778, 1976.

*25783 - **CRAMER, J., MYERS, A.L.** : Rate of increase of atmospheric carbon dioxide. - Atmosph. Environ. *6* : 563 - 573, 1972. [Ps.]

25784 - **CRAWFORD, R.M.M.** : Tolerance of anoxia and the regulation of glycolysis in the roots. - In : CANNELL, M.G.R., LAST, F.T. (ed.) : Tree Physiology and Yield Improvement. Pp. 387 - 401. Academic Press, London - New York - San Francisco 1976. [Ps.]

25785 - **CREMER, K.W.** : Daily patterns of shoot elongation in *Pinus radiata* and *Eucalyptus regnans*. - New Phytol. *76* : 459 - 468, 1976.

25786 - **CREWS, C.E., SEARS, B.B., KOHEL, R.J., BENEDICT, C.R.** : RuDPCase in genetically produced white sectors of cotton leaves. - Plant Physiol. *57* (Suppl.) : 5, 1976.

25787 - **CREWS, C.E., WILLIAMS, S.L., VINES, H.M., BLACK, C.C.** : Changes in the metabolism and physiology of Crassulacean acid metabolism plants grown in controlled environments. - In : BURRIS, R.H., BLACK, C.C. (ed.) : CO_2 Metabolism in Plant Production. Pp. 235 - 250. Univ. Park Press, Baltimore - London - Tokyo 1976.

25788 - **CRISP, D.J.** : The British contribution to the I.B.P. programme on marine productivity. - Phil. Trans. roy. Soc. London B *274* : 393 - 399, 1976.

*25789 - **CSEH, E.** : Fotofoszforiláció és iontranszport. [Photophosphorylation and ion transport.] - MTA Biol. Oszt. Közl. *17* : 203 - 224, 1974. [In Hung.]

25790 - **CSORBA, I., ERDEI, L., FAJSZI, Cs.** : Arrangement of crystalline chlorophyll in lipid membranes. - Acta biochim. biophys. Acad. Sci. hung. *11* : 184, 1976.

25791 - **CUENDET, P.A., ZUBER, H.** : A bacteriochlorophyll-containing protein (light harvesting-complex) from *Rhodospirillum rubrum*. - Experientia *32* : 790, 1976.

25792 - **CULLER, R.C., HANSON, R.L., JONES, J.E.** : Relation of consumptive use coefficient to description of vegetation. - Water Resources Res. *12* : 40 - 46, 1976. [Growth analysis.]

25793 - **CUNNINGHAM, C.C., CONRAD, D.D.W., MOFFATT, J.D., LEVY, E.M.** : Improved reagent dispenser for the determination of dissolved oxygen in sea water. - J. Fish. Res. Board Can. *33* : 2076 - 2078, 1976.

25794 - **CZECZUGA, B.** : Carotenoid pigments in some phytobenthos species of different systematic position in coastal area of Ofotfjord/Norway. - Nova Hedwigia *27* : 223 - 229, 1976.

25795 - **CZYGAN, F.-C.** : Carotinoid-Garnitur und -Stoffwechsel der Flechte *Haematomma ventosum* (L.) MASSAL. s. str. und ihres Phycobionten. - Z. Pflanzenphysiol. *79* : 438 - 445, 1976.

25796 - **DALEY, L.S., BIDWELL, R.G.S.** : Phosphoserine and phosphohydroxypyruvate : evidence for their role as early intermediates in photosynthesis. - Plant Physiol. *57* (Suppl.) : 5, 1976.

25797 - **DALEY, L.S., RAY, T., VINES, H.M., BLACK, C.C.** : Pineapple PEP carboxykinase: partial purification and properties. - Plant Physiol. *57* (Suppl.) : 32, 1976.

25798 - **DALTON, C.C., STREET, H.E.** : The role of the gas phase in the greening and growth of illuminated cell suspension cultures of spinach (*Spinacia oleracea, L.*). - In Vitro *12* : 485 - 494, 1976. [Chl, Car.]

25799 - **DALY, J.M.** : 5.3 The carbon balance of diseased plants : changes in respiration, photosynthesis and translocation. - In : HEITEFUSS, R., WILLIAMS, P.H.

(ed.) : Physiological Plant Pathology. (Encycl. Plant Physiol. N.S. Vol. 4.)
Pp. 450 - 479. Springer Verlag, Berlin - Heidelberg - New York 1976.

25800 - **DAM, R.J., KONGSLIE, K.F., GRIFFITH, O.H.** : Photoelectron quantum yields and
photoelectron microscopy of chlorophyll and chlorophyllin. - Biophys. J. *16*
(2, Part 2) :158a, 1976.

25801 - **DAMISCH, W.** : Über die Beziehungen zwischen Ertragsbildung und CO_2-Gaswechsel
bei Weizen (*Triticum aestivum*). - Wiss. Z. Humboldt-Univ. Berlin, math.-natur-
wiss. Reihe *25* : 752 - 758, 1976.

25802 - **DAMSZ, B., MIKULSKA, E.** : Ontogenesis of photosynthetic membranes in the plastids of *Cattleya* sp. leaves grown in the light. - Biochem. Physiol. Pflanzen
169 : 257 - 263, 1976.

25803 - **DANCSHÁZY, Z., KARVALY, B.** : Incorporation of bacteriorhodopsin into a bilayer
lipid membrane; a photoelectric-spectroscopic study. - FEBS Lett. *72* : 136 -
- 138, 1976.

25804 - **DANON, A., CAPLAN, S.R.** : Stimulation of ATP synthesis in *Halobacterium halobium* R_1 by light-induced or artificially created proton electrochemical potential gradients across the cell membrane. - Biochim. biophys. Acta *423* :
133 - 140, 1976.

25805 - **DARLEY, W.M., DUNN, E.L., HOLMES, K.S., LAREW, H.G. III.** : A ^{14}C method for
measuring epibenthic microalgal productivity in air. - J. exp. mar. Biol.
Ecol. *25* : 207 - 217, 1976.

25806 - **DARLEY, W.M., WHITNEY, D.E.** : Benthic algal productivity in a Georgia salt
marsh. - J. Phycol. *12* (Suppl.) : 15, 1976.

25807 - **DARLING, M.S.** : Interpretation of global differences in plant calorific values. The significance of desert and arid woodland vegetation. - Oecologia
23 : 127 - 139, 1976.

25808 - **DAS GUPTA, D.K., BARBAREZ, M.K.** : Steady state electrical conduction in β-carotene and Rose-Bengal under high electric field conditions. - In : BIRKS, J.
B. (ed.) : Excited States of Biological Molecules. Pp. 353 - 360. Wiley, Chichester 1976.

25809 - **DAS GUPTA, D.K., BASUCHAUDHURI, P., SEN GUPTA, P.** : Effect of molybdenum on
carbon assimilation by the rice plant. - Indian Agr. *20* : 51 - 54, 1976.
[Assimilation chamber.]

25810 - **DASH, M.C., CHOUDHURY, N.K., BEHERA, N., MOHAPATRA, P.K.** : Primary production
in four species of *Ipomoea*. - Geoblos *3* : 13 - 15, 1976.

25811 - **DAUBENMIRE, R.** : The use of vegetation in assessing the productivity of forest lands. - Bot. Rev. *42* : 115 - 143, 1976.

25812 - **DAUN, J.K.** : A rapid procedure for the determination of chlorophyll in rapeseed by reflectance spectroscopy. - J. amer. Oil Chem. Soc. *53* : 767 - 770,
1976.

25813 - **DAVENPORT, D.C., MARTIN, P.E., ROBERTS, E.B., HAGAN, R.M.** : Conserving water
by antitranspirant treatment of phreatophytes. - Water Resources Res. *12* :
985 - 990, 1976. [Stomatal resistance.]

25814 - **DAVEY, M.R., FREARSON, E.M., POWER, J.B.** : Polyethylene glycol-induced trans-
plantation of chloroplasts into protoplasts : an ultrastructural assessment.
- Plant Sci. Lett. *7* : 7 - 16, 1976.

25815 - **DAVIES, B.H.** : Carotenoids. - In : GOODWIN, T.W. (ed.) : Chemistry and Biochemistry of Plant Pigments. 2nd Ed. Vol. 2. Pp. 38 - 165. Academic Press,
London - New York - San Francisco 1976.

25816 - **DAVIES, B.H., TAYLOR, R.F.** : Carotenoid biosynthesis - the early steps. - Pure
appl. Chem. *47* : 211 - 221, 1976.

25817 - **DAVIES, I.** : Developmental characteristics of grass varieties in relation to
their herbage production. 1. An analysis of high-digestibility varieties of
Dactylis glomerata at three stages of development. - J. agr. Sci. *87* : 25 -
- 32, 1976. [Carbohydrate content.]

37 25818 - 25834 / DAV - DEL

25818 - **DAVIES, I.** : Developmental characteristics of grass varieties in relation to
their herbage production. 2. Spring defoliation of *Dactylis glomerata* : the
fate of reproductive tillers which are cut, but whose stem apex is retained.
- J. agr. Sci. *87* : 33 - 38, 1976.

25819 - **DAVIS, D.J., ARMOND, P.A., GROSS, E.L., ARNTZEN, C.J.** : Differentiation of
chloroplast lamellae. Onset of cation regulation of excitation energy distribution. - Arch. Biochem. Biophys. *175* : 64 - 70, 1976.

25820 - **DAVIS, D.J., GROSS, E.L.** : Protein-protein interactions of the light-harvesting chlorophyll *a/b* protein. II. Evidence for two stages of cation indepen-
dent association. - Biochim. biophys. Acta *449* : 554 - 564, 1976.

25821 - **DAVIS, J.T., SPARKS, D.** : Assimilation of carbon-14 by pistillate inflorescen-
ces and fruits of "Stuart" pecan. - HortScience *11* : 262 - 263, 1976.

25822 - **DAVTYAN, V.A., KAZARYAN, V.V., MOVSESYAN, G.M.** : Ob izmenenii soderzhaniya
khlorofilla i prochnosti ego svyazi s lipoproteidnym kompleksom u nekotorykh
listopadnykh porod. [Change in the content of chlorophyll and its binding
strength with lipoprotein complex in some deciduous species.] - Biol. Zh.
Arm. *29* (11) : 57 - 61, 1976. [In R, ab : Arm.]

25823 - **DAVYDOV, N.A., PAVLENKO, V.A., SHEĬNIN, D.M., SHUTOV, M.D.** : Ispol'zovanie
termomagnitnykh gazoanalizatorov dlya izmereniya malykh izmeneniĭ kontsentratsii kisloroda v gazovykh smesyakh. [Use of thermomagnetic gas analysers
for measuring small changes in oxygen concentration in gas mixtures.] - In :
Gazometricheskoe Issledovanie Fotosinteza i Dykhaniya Rasteniĭ. Pp. 43 - 45.
Akad. Nauk SSSR, Tartu 1976. [In R.]

25824 - **DAWES, C.J., LaCLAIRE, J.W., MOON, R.E.** : Culture studies on *Eucheuma nudum*
J. AGARDH., a carrageenan producing red alga from Florida. - Aquaculture *7* :
1 - 9, 1976.

25825 - **DAWES, C.J., MOON, R., LaCLAIRE, J.** : Photosynthetic responses of the red
alga, *Hypnea musciformis* (WULFEN) LAMOUROUX (*Gigartinales*). - Bull. mar. Sci.
26 : 467 - 473, 1976.

25826 - **DAWSON, F.H.** : The annual production of the aquatic macrophyte *Ranunculus
penicillatus* var. *calcareus* (R.W. BUTCHER) C.D.K. COOK. - Aquatic Bot. *2* :
51 - 73, 1976.

25827 - **DECAU, J., PUJOL, B.** : Influence de réductions artificielles de la féconda-
tion de l'épi sur la photosynthèse nette et le rendement du Maïs (*Zea mays*
L). Étude comparée d'un Maïs normal et d'un mutant "opaque 2" (o_2). - Compt.
rend. Acad. Sci. Paris, Sér. D *283* : 923 - 926, 1976.

25828 - **DEDIO, W., STEWART, D.W., GREEN, D.G.** : Evaluation of photosynthesis measuring methods as possible screening techniques for drought resistance in wheat.
- Can. J. Plant Sci. *56* : 243 - 247, 1976.

25829 - **De GREEF, J.A., CAUBERGS, R., VERBELEN, J.P., MOEREELS, E.** : Phytochrome-me-
diated inter-organ dependence and rapid transmission of the light stimulus.
- In : SMITH, H. (ed.) : Light and Plant Development. Pp. 295 - 316. Butterworths, London - Boston - Sydney - Wellington - Durban - Toronto 1976. [Ps,
Chl.]

25830 - **De GREEF, J.A., VERBELEN, J.P., MOEREELS, E.** : Cytochemical and biochemical
studies of an energy-regulating enzyme complex modulated by light during de-
-etiolation processes of *Phaseolus vulgaris* L. - Plant Physiol. *57* (Suppl.) :
14, 1976.

25831 - **DeGROOTE, D., KENNEDY, R.A.** : C_4 acid production in *Elodea canadensis* MICHX.
- Plant Physiol. *57* (Suppl.) : 31, 1976.

25832 - **DEINUM, B.** : Photosynthesis and sink size : An explanation for the low productivity of grass swards in autumn. - Neth. J.agr. Sci. *24* : 238 - 246, 1976.

25833 - **DELANEY, M.E., WALKER, D.A.** : A reconstituted chloroplast system from *Helianthus annuus*. - Plant Sci. Lett. *7* : 285 - 294, 1976.

25834 - **DELBRÜCK, M.** : Light and life III. - Carlsberg Res. Commun. *41* : 299 - 309,
1976. [Chl.]

25835 - **DELEENS, É.** : La discrimination du ^{13}C et les trois types de métabolisme des plantes. - Physiol. vég. *14* : 641 - 656, 1976.

25836 - **DELRIEU, M.J.** : Inhibition by ammonium chloride of the oxygen yield of photosynthesis. - Biochim. biophys. Acta *440* : 176 - 188, 1976.

25837 - **DEMETER, S., MUSTARDY, L., MACHOWICZ, E., GREGORY, R.P.F.** : The development of the intense circular-dichroic signal during granum formation in greening etiolated maize. - Biochem. J. *156* : 469 - 472, 1976.

25838 - **DEMETER, S., SAGROMSKY, H., FALUDI-DÁNIEL, Á.** : Orientation of chlorophyll *b* in thylakoids of barley chloroplasts. - Photosynthetica *10* : 193 - 197, 1976.

25839 - **DENCHER, N.A., RAFFERTY, C.N., SPERLING, W.** : 13-cis and trans bacteriorhodopsin : Photochemistry and dark equilibrium. - Ber. Kernforschungsanlage Jülich *1374* : 1 - 42, 1976.

25840 - **den HAAN, G.A., GORTER de VRIES, H., DUYSENS, L.N.M.** : Correlation between flash-induced oxygen evolution and fluorescence yield kinetics in the 0 to 16 μs range in *Chlorella pyrenoidosa* during incubation with hydroxylamine. - Biochim. biophys. Acta *430* : 265 - 281, 1976.

25841 - **DENMAN, K.L.** : Covariability of chlorophyll and temperature in the sea. - Deep-Sea Res. *23* : 539 - 550, 1976.

25842 - **DENMEAD, O.T., FRENEY, J.R., SIMPSON, J.R.** : A closed ammonia cycle within a plant canopy. - Soil Biol. Biochem. *8* : 161 - 164, 1976. [Stomatal resistance.]

25843 - **DENMEAD, O.T., MILLAR, B.D.** : Field studies of the conductance of wheat leaves and transpiration. - Agron. J. *68* : 307 - 311, 1976.

25844 - **DENNISS, I.S., SANDERS, J.K.M., WATERTON, J.C.** : Assignment of metalloporphyrin and chlorophyll nuclear magnetic resonance spectra *via* spin-lattice relaxation times. - J. chem. Soc., chem. Commun. *1976* : 1049 - 1050, 1976.

25845 - **DESHMUKH, P.S., TOMAR, D.P.S., SINHA, S.K.** : Contribution of sepals to seed weight per boll in linseed *Linum usitatissimum* L. - Photosynthetica *10* : 136- - 139, 1976.

*25846 - **DESMET, G., DONDEYNE, P.** : Kritisch overzicht van de literatuurgegevens omtrent de lichtreacties in het fotosynthetisch apparaat. [Critical review of literature on light reactions of the photosynthetical apparatus.] - Agricultura *18* (2) : 3 - 43, 1970. [In Flemish, ab : E.]

25847 - **DESORTOVÁ, B.** : Productivity of individual algal species in natural phytoplankton assemblage determined by means of autoradiography. - Arch. Hydrobiol. Supplem. *49* (Algol. Studies 17) : 415 - 449, 1976.

25848 - **DÉVAI, G., KOLLÁR, Gy., ÖLLÖS, G.** : Tavak eutrofizálódási folyamatairól. [Eutrophication processes of lakes.] - Acta Biol. debrecina *13* : 163 - 191, 1976. [Ps; in Hung., G, ab : E.]

25849 - **DEVAUX, J.** : Dynamique des populations phytoplanctoniques dans deux lacs du massif central français. - Ann. Sta. biol. Besse-en-Chandesse *1975-1976* (10): 1 - 185, A 1 - A 8, 1976.

25850 - **DEVAUX, J., LAIR, N.** : Production primaire et biomasse des populations algales dans un petit lac oligotrophe du massif central français. - Hydrobiologia *50* : 209 - 220, 1976.

25851 - **DEVAUX, J., MILLERIOUX, G.** : Méthode d'estimation de la biomasse totale du phytoplancton à partir des nombres de cellules, issus d'une cotation d'abondance. - Compt. rend. Acad. Sci. Paris, Sér. D *283* : 927 - 930, 1976.

25852 - **DEVLIN, R.M., KISIEL, M.J., KARCZMARCZYK, S.J.** : Chlorophyll production and chloroplast development in nonflurazon-treated plants. - Weed Res. *16* : 125- - 129, 1976.

25853 - **DIAMOND, J.** : Inhibition of chloroplast development in *Euglena* by streptomycin : Differential inhibition of the appearance of photosynthesis in the presence of the continued synthesis of chlorophyll. - Planta *130* : 145 - 149, 1976.

25854 - **DICHT, M., KOPP, A., FELLER, U., ERISMANN, K.H.** : Einfluβ von Ammonium und

Nitrat auf den Proteingehalt von *Lemna minor* L. under Photosynthesebedingungen.
- Biochem. Physiol. Pflanzen *170* : 531 - 534, 1976.

25855 - **DICKERSON, R.E., TIMKOVICH, R., ALMASSY, R.J.** : The cytochrome fold and the
evolution of bacterial energy metabolism. - J. mol. Biol. *100* : 473 - 491,
1976. [Ps.]

23856 - **DICKSON, R.E.** : Xylem translocation of ^{14}C-labeled amino acids from roots of
Populus deltoides seedlings. - Plant Physiol. *57* (Suppl.) : 78, 1976.

25857 - **DIKAN', A.P.** : Vliyanie summarnoĭ solnechnoĭ radiatsii na formirovanie gene-
rativnykh organov vinograda. [Effect of total solar radiation on formation of
grapevine generative organs.] - Fiziol. Biokhim. kul't. Rast. *8* : 643 - 648,
1976. [In R.]

25858 - **DILKS, T.J.K.** : Measurement of the carbon dioxide compensation point and the
rate of loss of ^{14}CO$_2$ in the light and dark in some bryophytes. - J. exp. Bot.
27 : 98 - 104, 1976.

25859 - **DILKS, T.J.K., PROCTOR, M.C.F.** : Seasonal variation in desiccation tolerance
in some British bryophytes. - J. Bryol. *9* : 239 - 247, 1976. [Ps.]

25860 - **DILKS, T.J.K., PROCTOR, M.C.F.** : Effects of intermittent desiccation on bryophytes. - J. Bryol. *9* : 249 - 264, 1976. [Ps.]

25861 - **DINER, B., JOLIOT, P.** : Effect of the transmembrane electric field on the photochemical and quenching properties of photosystem II *in vivo*. - Biochim. biophys. Acta *423* : 479 - 498, 1976.

25862 - **DITTRICH, P.** : Nicotinamide adenine dinucleotide-specific "malic" enzyme in
Kalanchoë daigremontiana and other plants exhibiting crassulacean acid metabolism. - Plant Physiol. *57* : 310 - 314, 1976.

25863 - **DITTRICH, P.** : Equilibration of label in malate during dark fixation of CO$_2$
in *Kalanchoë fedtschenkoi*. - Plant Physiol. *58* : 288 - 291, 1976.

25864 - **DJELEPOV, K.** : Chlorophyll mutations, induced by irradiation and chemical mu-
tagenes in *Tr. aestivum* L. and *Tr. durum* DEST. - Dokl. bulg. Akad. Nauk *29* :
895 - 897, 1976.

25865 - **DOHÁNYOS, M., GRAU, P., LISCHKE, P.** : Anwendung der Gelchromatographie zur
Bestimmung der Dehydrogenaseaktivität nach Bucksteeg in Anwesenheit von na-
türlichen Pflanzenfarbstoffen. - Acta hydrochim. hydrobiol. *4* : 501 - 503,
1976. [Chl.]

25866 - **DÖHLER, G.** : Aminotransferase-Aktivität in der Blaualge *Anacystis nidulans*. -
Z.Naturforsch. *31C* : 433 - 435, 1976.

25867 - **DÖHLER, G.** : Photosynthetische Carboxylierungsreaktionen verschieden pigmen-
tierter *Anacystis*-Zellen. - Planta *131* : 129 - 133, 1976.

25868 - **DÖHLER, G.** : Wirkung anaerober Bedingungen auf die photosynthetischen Carboxylierungsreaktionen von *Anacystis* und *Chlorella*. - Z. Pflanzenphysiol. *78* : 416-
- 420, 1976.

25869 - **DÖHLER, G., BÜRSTELL, H., JILG-WINTER, G.** : Pigmentzusammensetzung und photosynthetische CO$_2$-Fixierung von *Cyanidium caldarium* und *Porphyridium aerugi-
neum*. - Biochem. Physiol. Pflanzen *170* : 103 - 110, 1976.

25870 - **DÖHLER, G., KOHLENBACH, H.W.** : CO$_2$-Fixierung isolierter und sich teilender
Mesophyllzellen von *Pharbitis purpurea*. - Z. Pflanzenphysiol. *80* : 81 - 86,
1976.

25871 - **DOĬKOVA, M.** : S"d"rzhanie na plastidni pigmenti v listata na patladzhana v zavisimost ot toreneto. [Plastid pigments content of eggplant leaves as dependent on fertilization.] - Gradinar. lozar. Nauka *13* (1) : 58 - 62, 1976.
[In Bulg., ab : E, R.]

25872 - **DOKIĆ, P., STANAĆEV, S., STEFANOVIĆ, D.** : Effect of genotype and polyploidy
level on the content of chlorophyll *a* and *b* in sugar beet leaves of various
age. - Acta biol. iugoslav., Genetika (Zemun) *8* : 187 - 194, 1976.

25873 - **DOLEY, D., YATES, D.J.** : Gas exchange of Mitchell grass [*Astrebla lappacea*
(LINDL.) DOMIN] in relation to irradiance, carbon dioxide supply, leaf tempe-

rature and temperature history. - Aust. J. Plant Physiol. *3* : 471 - 487, 1976.

25874 - **DOLL, S., LÜTZ, C., RUPPEL, H.G.** : Biochemische und cytologische Untersuchungen zur Chloroplastenentwicklung. III. Superoxiddismutase in Etioplasten von *Avena sativa* L. - Z. Pflanzenphysiol. *80* : 166 - 176, 1976.

25875 - **DOMAN, N.G., CHERNYAD'EV, I.I., TEREKHOVA, I.V., AL'BITSKAYA, O.N.** : O skorosti i kharaktere gazoobmena u fotosintezi ruyushchikh odnokletochnykh vodoroslei. [Rate and nature of gas exchange in photosynthesizing unicellular algae.] - In : Gazometricheskoe Issledovanie Fotosinteza i Dykhaniya Rastenii. Pp. 46 - 47.. Akad. Nauk SSSR, Tartu 1976. [In R.]

25876 - **DONKIN, P.** : Ketocarotenoid biosynthesis by *Haematococcus lacustris*. - Phytochemistry *15* : 711 - 715, 1976.

25877 - **DONTSCHEV, T., LOSSNER, G.** : Der Einfluß von Chlorcholinchlorid (CCC) auf das Wachstum und den Blattfarbstoffgehalt von *Phaseolus vulgaris* L. und *Sinapis alba* L. - Arch. Phytopathol. Pflanzenschutz *12* : 49 - 56, 1976.

25878 - **DOOHAN, M.E., NEWCOMB, E.H.** : Leaf ultrastructure and $\delta^{13}C$ values of three seagrasses from the Great Barrier Reef. - Aust. J. Plant Physiol. *3* : 9 - 23, 1976.

25879 - **DÖRING, G.** : The effect of deoxycholate-treatment to the photoreactions of the active pigments in photosynthesis. - Z. Naturforsch. *31C* : 64 - 67, 1976.

25880 - **DÖRING, G.** : The chlorophyll a_{II} reaction in trypsin-treated spinach chloroplasts in the presence of potassium ferricyanide. - Z. Naturforsch. *31C* : 78- - 81, 1976.

25881 - **DORNHOFF, G.M., SHIBLES, R.** : Leaf morphology and anatomy in relation to CO_2- -exchange rate of soybean leaves. - Crop Sci. *16* : 377 - 381, 1976.

25882 - **DOUCHA, J., KUBÍN, Š.** : Measurement of *in vivo* absorption spectra of microscopic algae using bleached cells as a reference sample. - Arch. Hydrobiol. Suppl. *49* : 199 - 213, 1976.

25883 - **DOWNTON, W.J.S., BISHOP, D.G., LARKUM, A.W.D., OSMOND, C.B.** : Oxygen inhibition of photosynthetic oxygen evolution in marine plants. - Aust. J. Plant Physiol. *3* : 73 - 79, 1976.

25884 - **DRACHEV, L.A., FROLOV, V.N., KAULEN, A.D., KONDRASHIN, A.A., SAMUILOV, V.D., SEMENOV, A.Yu., SKULACHEV, V.P.** : Generation of electric current by chromatophores of *Rhodospirillum rubrum* and reconstitution of electrogenic function in subchromatophore pigment-protein complexes. - Biochim. biophys. Acta *440* : 637 - 660, 1976.

25885 - **DRAKE, B.G.** : Estimating water status and biomass of plant communities by remote sensing. - In : LANGE, O.L., KAPPEN, L., SCHULZE, E.-D. (ed.) : Water and Plant Life. Pp. 432 - 438. Springer-Verlag, Berlin - Heidelberg - New York 1976.

25886 - **DRAKE, B.G.** : Seasonal changes in reflectance and standing crop biomass in three salt marsh communities. - Plant Physiol. *58* : 696 - 699, 1976.

25887 - **DRASKOVITS, R.M., FEKETE, G.** : Chlorophyll concentration and its ecological significance in some species in beechwoods. - Acta bot. Acad. Sci. hung. *22* : 29 - 38, 1976.

*25888 - **DREWS, G.** : Structural aspects concerning the photosynthetic apparatus. - In : BRODA, E., LOCKER, A., SPRINGER-LEDERER, H. (ed.) : Proceedings of the First European Biophysics Congress. Vol. 4. Pp. 121 - 125. Verlag wiener med. Akad., Wien 1971.

25889 - **DREWS, G., DIERSTEIN, R., SCHUMACHER, A.** : Genetic transfer of the capacity to form bacteriochlorophyll-protein complexes in *Rhodopseudomonas capsulata*. - FEBS Lett. *68* : 132 - 136, 1976.

25890 - **DROBA, M.** : Resolution of chloroplast lamellar proteins by sodium tetradecyl sulphate gel electrophoresis. - Bull. Acad. pol. Sci., Sér. Sci. biol. *24* : 641 - 645, 1976.

25891 - **DROBA, M., STRZAŁKA, K.** : Zastosowanie metod immunologicznych do badania
struktury błon tylakoidów. [Application of immunological methods to investigation of thylakoid structure.] - Postępy Biochem. *22* : 511 - 525, 1976. [In
Pol.]

25892 - **DROZDOVA, I.S., GRISHINA, G.S., VOSKRESENSKAYA, N.P.** : Regulyatornaya rol'
sinego sveta v formirovanii fotosinteticheskoĭ tsepi perenosa élektronov u
khloroplastov gorokha. [Regulatory role of blue light in the formation of
photosynthetic electron transfer chain in pea chloroplasts.] - Dokl. Akad.
Nauk SSSR *226* : 221 - 224, 1976. [In R.]

25893 - **DROZDOVA, I.S., KRENDELEVA, T.E., VERKHOTUROV, V.N., TIMOFEEV, K.N., MATORIN,
D.N., VOSKRESENSKAYA, N.P.** : Regulyatornoe deĭstvie sinego sveta na sostoyanie tsitokhromov fotosinteticheskoĭ tsepi perenosa élektrona. [Regulatory
action of blue light on the state of cytochromes of the photosynthetic electron transport chain.] - Fiziol. Rast. *23* : 861 - 868, 1976. [In R, ab : E.]

25894 - **DUBINSKY, Z., BERMAN, T.** : Light utilization efficiencies of phytoplankton
in lake Kinneret (Sea of Galilee). - Limnol. Oceanogr. *21* : 226 - 230, 1976.
[Ps.]

25895 - **DUBINSKY, Z., POLNA, M.** : Pigment composition during a *Perdinium* bloom in Lake Kinneret (Israel). - Hydrobiologia *51* : 239 - 243, 1976.

25896 - **DUDA, M.** : Sezónne zmeny obsahu chlorofylov pri vybraných druhoch drevín les-
ného ekosystému. [Seasonal changes of chlorophyll content in selected forest
species of a forest ecosystem.] - Biológia (Bratislava) *31* : 233 - 240, 1976.
[In Slovak, ab : E, R.]

25897 - **DUDA, M., MASAROVIČOVÁ, E.** : Dynamika zmien obsahu karotenoidov v listoch
Quercus cerris L. a *Cornus mas* L. počas vegetačného obdobia. [The dynamics
of changes in carotenoid level in the leaves of *Quercus cerris* L. and *Cornus
mas* L. in the course of growing season.] - Acta Musei Silesiae, Ser. dendrol.
25 : 135 - 143, 1976. [In Slovak, ab : E.]

25898 - **DUDA, M., MASAROVIČOVÁ, E.** : Príspevok k ekofyziologickému štúdiu asimilač-
ných pigmentov dominantných drevín dubo-hrabového lesa v Bábe. [Ecophysio-
logical study of assimilation pigments of dominant woody plants of oak-hornbeam forest at Báb.] - Acta Musei Silesiae, Ser. dendrol. *25* : 35 - 44, 1976.
[In Slovak, ab : E.]

25899 - **DUJARDIN, E.** : Reversible transformation of the $P_{657-650}$ form into $P_{633-628}$
in etiolated bean leaves. - Plant Sci. Lett. *7* : 91 - 94, 1976.

25900 - **DUKE, S.O., FOX, S.B., NAYLOR, A.W.** : Photosynthetic independence of light-
-induced anthocyanin formation in *Zea* seedlings. - Plant Physiol. *57* : 192 -
- 196, 1976.

25901 - **DUNFORD, S.S., ZIMMERMANN, M.H.** : Factors affecting phloem exudation in white
ash (*Fraxinus americana* L.). - Plant Physiol. *57* (Suppl.) : 77, 1976.

25902 - **DUNN, E.L., SHROPSHIRE, F.M., SONG, L.C., MOONEY, H.A.** : The water factor and
convergent evolution in mediterranean-type vegetation. - In : **LANGE, O.L.,
KAPPEN, L., SCHULZE, E.-D.** (ed.) : Water and Plant Life. Pp. 492 - 505. Springer-Verlag, Berlin - Heidelberg - New York 1976. [Ps.]

25903 - **DUTTON, P.L.** : Pheophytins and the photosynthetic reaction center. - Photochem. Photobiol. *24* : 655 - 657, 1976.

25904 - **DUVAL, J.-C., TRÉMOLIÈRES, A., ROUSSEAU, B.** : Croissance et division du com-
partiment plastidial et synthèse de la chlorophylle et de l'acide linolénique
au cours du développement de la première feuille de Pois. - Physiol. vég. *14*:
67 - 76, 1976.

25905 - **DUYSEN, M.E., FREEMAN, T.P.** : Promotion of plastid pigment accumulation in
water stressed wheat leaf sections by hormone treatment. - Amer. J. Bot. *63* :
1134 - 1139, 1976.

25906 - **DUYSEN, M.E., FREEMAN, T.P.** : Stimulation of chloroplast development by hormone treatment in water-stressed wheat leaves. - Plant Physiol. *57* (Suppl.):
73, 1976.

25907 - **DUYSENS, L.N.M.** : Electronentransport in de fotosynthese. [Electron transport in photosynthesis.] - Chem. Weekbl. *1976* (12) : 642 - 644, 1976. [In Holl.]

25908 - **DYER, A.F.** : 2. The visible events of mitotic cell division. - In : YEOMAN, M.M. (ed.) : Cell Division in Higher Plants. Pp. 49 - 110. Academic Press, London - New York - San Francisco 1976. [Chloroplast.]

25909 - **DYKYJOVÁ, D., HRADECKÁ, D.** : Production ecology of *Phragmites communis*. 1. Relations of two ecotypes to the microclimate and nutrient conditions of habitat. - Folia geobot. phytotax. *11* : 23 - 61, 1976. [Growth analysis.]

25910 - **DYMINA, G.D.** : Metodika opredeleniya khozyaĭstvennoĭ produktivnosti travostoya v lugovykh soobshchestvakh po khozyaĭstvennym gruppam. [Method of determining economic productivity of grass stand in meadow communities according to the economic groups.] - Bot. Zh. *61* : 53 - 60, 1976. [In R.]

*25911 - **ECKARDT, F.E., SAUVEZON, R.** : Mesures de photosynthèse par la méthode de compensation : amélioration de la réponse du système automatique d'injection de CO_2 dans l'enceinte. - Oecol. Plant. *8* : 185 - 187, 1973.

25912 - **EDWARDS, G.E., HUBER, S.C., GUTIERREZ, M.** : Photosynthetic properties of plant protoplasts. - In : PEBERDY, J.F., ROSE, A.H., ROGERS, H.J., COCKING, E.C. (ed.) : Microbial and Plant Protoplasts. Pp. 299 - 322. Academic Press, London - New York - San Francisco 1976.

25913 - **EDWARDS, G.E., HUBER, S.C., KU, S.B., RATHNAM, C.K.M., GUTIERREZ, M., MAYNE, B.C.** : Variation in photochemical activities of C_4 plants in relation to CO_2 fixation. - In : BURRIS, R.H., BLACK, C.C. (ed.) : CO_2 Metabolism and Plant Productivity. Pp. 83 - 112. Univ. Park Press, Baltimore - London - Tokyo 1976.

25914 - **EDWARDS, P.A., JACKSON, J.B.** : The control of the adenosine triphosphatase of *Rhodospirillum rubrum* chromatophores by divalent cations and the membrane high energy state. - Europe. J. Biochem. *62* : 7 - 14, 1976.

25915 - **EFIMTSEV, E.I., BOĬCHENKO, V.A., ZATOLOKIN, N.E., LITVIN, F.F.** : Issledovanie pervichnykh protsessov fotoindutsirovannogo vydeleniya vodoroda khlorelloĭ pri impul'snom osveshchenii. [Primary processes of photoinduced hydrogen liberation by *Chlorella* during flash illumination.] - Dokl. Akad. Nauk SSSR *227* : 731 - 734, 1976. [In R.]

25916 - **EFIMTSEV, E.I., LITVIN, F.F.** : Issledovanie bystrykh fotoindutsirovannykh protsessov vydeleniya vodoroda. [Rapid photo-induced processes of hydrogen evolution.] - In : Itogi Issledovaniya Mekhanizma Fotosinteza. Pp. 55 - 58. Pushchino 1976. [In R.]

25917 - **EGLI, D.B., LEGGETT, J.E.** : Rate of dry matter accumulation in soybean seeds with varying source-sink ratios. - Agron. J. *68* : 371 - 374, 1976.

25918 - **EGNÉUS, H., SELLDÉN, G., ANDERSSON, L.** : Appearance and development of P700 oxidation and photosystem I activity in etio-chloroplasts prepared from greening barley leaves. - Planta *133* : 47 - 52, 1976.

25919 - **EGOROVA, L.I.** : Kinetika fotosinteza posle vozdeĭstviya povyshennoĭ temperatury. [Kinetics of photosynthesis after high temperature treatment.] - Bot. Zh. *61* : 945 - 950, 1976. [In R.]

25920 - **EGOROVA, L.I.** : Posledeĭstvie temperatury na kinetiku fotosinteza. [Aftereffect of temperature on photosynthesis kinetics.] - In : Gazometricheskoe Issledovanie Fotosinteza i Dykhaniya Rasteniĭ. Pp. 48 - 50. Akad. Nauk SSSR, Tartu 1976. [In R.]

25921 - **EGUNJOBI, J.K.** : An evaluation of five methods for estimating biomass of an even-aged plantation of *Pinus caribaea* L. - Oecol. Plant. *11* : 109 - 116, 1976.

25922 - **EHLERINGER, J.** : Leaf absorptance and photosynthesis as affected by pubescence in the genus *Encelia*. - Carnegie Inst. Year Book *75* : 413 - 418, 1976.

25923 - **EHLERINGER, J., BJÖRKMAN, O.** : Carbon dioxide and temperature dependence of the quantum yield for CO_2 uptake in C_3 and C_4 plants. - Carnegie Inst. Year Book *75* : 418 - 421, 1976.

25924 - EHLERINGER, J., BJÖRKMAN, O., MOONEY, H.A. : Leaf pubescence : Effects on absorptance and photosynthesis in a desert shrub. - Science *192* : 376 - 377, 1976.

25925 - EICHENBERGER, W. : Lipids of *Chlamydomonas reinhardi* under different growth conditions. - Phytochemistry *15* : 459 - 463, 1976. [Chl.]

25926 - EICKMEIER, W.G., BENDER, M.M. : Carbon isotope ratios of Crassulacean acid metabolism species in relation to climate and phytosociology. - Oecologia *25*: 341 - 347, 1976.

25927 - EISENBACH, M., BAKKER, E.P., KORENSTEIN, R., CAPLAN, S.R. : Bacteriorhodopsin : Biphasic kinetics of phototransients and of light-induced proton transfer by sub-bacterial *Halobacterium halobium* particles and by reconstituted liposomes. - FEBS Lett. *71* : 228 - 232, 1976.

25928 - EL AOUNI, M.H. : Action du déficit hydrique interne sur les mouvements stomatiques, la transpiration et la photosynthèse nette d'aiguilles excisées de Pin noir d'Autriche (*Pinus nigra* ARN.). Évolution avec l'âge foliaire. - Photosynthetica *10* : 403 - 410, 1976.

25929 - EL-GELANI, M., KHIER, M. : Effect of high humidity on the basipetal translocation of the photosynthates in the xylem tissue of *Phaseolus vulgaris*. - Plant Physiol. *57* (Suppl.) : 78, 1976.

25930 - EL-HABBASHA, K.M., SHAHEEN, A.M. : Growth, photosynthetic efficiency, water and nitrogen contents of onion (*Allium cepa* L.) seedlings, affected by water deficit in the soil. - Z. Acker- Pflanzenbau *142* : 256 - 263, 1976.

25931 - ELISEEV, I.P., BOGATOV, N.P. : Soderzhanie khlorofilla v list'yakh introdutsirovannykh vidov i sortov grushi v svyazi s ikh zimostoĭkost'yu. [Chlorophyll content in leaves of introduced genera and species of pear in connection with their winter hardiness.] - Tr. gor'kov. sel'skokhoz. Inst. *69* (Botanika i Fiziologiya RasteniĬ) : 19 - 24, 1976. [In R.]

25932 - ELLENSON, J.L., SAUER, K. : The electrophotoluminescence of chloroplasts. - Photochem. Photobiol. *23* : 113 - 123, 1976.

25933 - ELLIS, R. : Kinetics of chlorophyll synthesis in the green alga *Golenkinia*. - Plant Physiol. *57* (Suppl.) : 72, 1976.

25934 - ELLIS, R.J. : Fraction I protein. - In : SMITH, H. (ed.) : Commentaries in Plant Science. Pp. 31 - 43. Pergamon Press, Oxford - New York - Toronto - Sydney - Paris - Frankfurt 1976.

25935 - ELLIS, R.J. : Protein and nucleic acid synthesis by chloroplasts. - In : BARBER, J. (ed.) : The Intact Chloroplast. Pp. 335 - 364. Elsevier, Amsterdam - New York - Oxford 1976.

25936 - ELLIS, R.J., SCHIFF, J.A. : *In vivo* stimulation of chlorophyll synthesis by porphobilinogen. - Plant Sci. Lett. *7* : 143 - 147, 1976.

25937 - ELLSWORTH, R.K., ST. PIERRE, M.E. : Biosynthesis and inhibition of (-)-S-adenosyl-L-methionine : Magnesium protoporphyrin methyltransferase of wheat. - Photosynthetica *10* : 291 - 301, 1976.

25938 - ELLSWORTH, R.K., TSUK, R.M., ST. PIERRE, L.A. : Studies on chlorophyllase. IV. Attribution of hydrolytic and esterifying "chlorophyllase" activities observed *in vitro* to two enzymes. - Photosynthetica *10* : 312 - 323, 1976.

25939 - ELSTNER, E.F., HEUPEL, A. : Oxygen activation by isolated chloroplasts from *Euglena gracilis*. Isolation and properties of a fluorescent compound that catalyzes monovalent oxygen reduction. - Arch. Biochem. Biophys. *173* : 614 - - 622, 1976.

25940 - ELSTNER, E.F., KONZE, J. : Wege der Sauerstoffaktivierung in verschiedenen Kompartimenten von Pflanzenzellen. - Ber. deut. bot. Ges. *89* : 335 - 348, 1976. [Chloroplast.]

25941 - ELSTNER, E.F., KONZE, J.R., SELMAN, B.R., STOFFER, C. : Ethylene formation in sugar beet leaves. Evidence for the involvement of 3-hydroxytyramine and phenoloxidase after wounding. - Plant Physiol. *58* : 163 - 168, 1976. [Ps.]

25942 - **ELSTNER, E.F., WILDNER, G.F., HEUPEL, A.** : Oxygen activation by isolated chloroplasts from *Euglena gracilis*. Ferredoxin-dependent function of a fluorescent compound and photosynthetic electron transport close to photosystem I. - Arch. Biochem. Biophys. *173* : 623 - 630, 1976.

25943 - **ELSTON, J., KARAMANOS, A.J., KASSAM, A.H., WADSWORTH, R.M.** : The water relations of the field bean crop. - Phil. Trans. roy. Soc. London B *273* : 581 - - 591, 1976.

25944 - **ENAMA, M.** : Genetic variation affecting metabolic phenotypes : an approach to analyzing photosynthetic carbon reduction in a C_4 plant. - Carnegie Inst. Year Book *75* : 407 - 409, 1976.

25945 - **ENAMA, M.** : Molecular weight variation of phosphoenolpyruvate carboxylases from C_4 plants. - Carnegie Inst. Year Book *75* : 409 - 410, 1976.

25946 - **ENDRÉDI, L., HORVÁTH, I.** : Organic matter production and photosynthetic energy utilization of a plant association in Loess grassland. - Acta bot. Acad. Sci. hung. *22* : 39 - 49, 1976.

25947 - **ENDRÖDI, G., DÁVID, A.** : Stomatal resistance in different plants. - Acta agron. Acad. Sci. hung. *25* : 382 - 390, 1976.

25948 - **ENIKEEV, S.G.** : Nekotorye fiziologicheskie pokazateli sakharnoĭ svekly pri oroshenii v Tatarii i ee produktivnost'. [Some physiological characteristics of irrigated sugar beet in Tataria and its productivity.] - Tr. gor'kov. sel'skokhoz. Inst. *69* (Botanika i Fiziologiya Rasteniĭ) : 138 - 142, 1976. [Chl; in R.]

25949 - **ENIKEEV, S.G., MAZIL'NIKOV, G.V., USHAKOV, V.Yu., KIYAMOV, A.G.** : Izmenenie fotosinteticheskoĭ aktivnosti list'ev gorokha pod deĭstviem poverkhnostno- -aktivnykh veshchestv. [Changes in photosynthetic activity of pea leaves affected by surface-active substances.] - Tr. gor'kov. sel'skokhoz. Inst. *69* (Botanika i Fiziologiya Rasteniĭ) : 84 - 86, 1976. [In R.]

25950 - **ENIKEEV, S.G., MAZIL'NIKOV, G.V., USHAKOV, V.Yu., ZALYAEV, Z.K.** : Deĭstvie povyshennykh temperatur i ouabaina na raspredelenie kaliya i fotosintez v kletkakh list'ev gorokha. [Effect of elevated temperatures and ouabain on potassium distribution and photosynthesis in cells of pea leaves.] - Tr. gor'kov. sel'skokhoz. Inst. *69* (Botanika i Fiziologiya Rasteniĭ) : 81 - 83, 1976. [In R.]

25951 - **ENIKEEV, S.G., USHAKOV, V.Yu.. GARAFEEV, A.G.** : Vzaimosvyaz' fotosinteza i lipidoobrazovaniya v list'yakh gorokha pri vodnom defitsite. [Relationship between rate of photosynthesis and formation of lipids in pea leaves under water stress.] - Tr. gor'kov. sel'skokhoz. Inst. *69* (Botanika i Fiziologiya Rasteniĭ) : 91 - 96, 1976. [In R.]

25952 - **EPPLEY, R.W., RENGER, E.H., WILLIAMS, P.M.** : Chlorine reactions with seawater constituents and the inhibition of photosynthesis of natural marine phytoplankton. - Estuar. coast. mar. Sci. *4* : 147 - 161, 1976.

25953 - **ERABI, T., HIGUTI, T., SAKATA, K., KAKUNO, T., YAMASHITA, J., TANAKA, M., HORIO, T.** : Polarographic studies in presence of *Triton X-100* on oxidation-reduction components bound with chromatophores from *Rhodospirillum rubrum*. - J. Biochem. *79* : 497 - 503, 1976.

25954 - **ÉRGASHEV, A.** : Intensivnost' i dinamika obrazovaniya produktov fotosinteza u topinambura. [Rate and dynamics of photosynthates formation in Jerusalem artichoke.] - Fiziol. Biokhim. kul't. Rast. *8* : 299 - 303, 1976. [In R, ab : E.]

25955 - **ERICKSON, R.O.** : Modeling of plant growth. - Annu. Rev. Plant Physiol. *27* : 407 - 434, 1976.

25956 - **ERIKSEN, A.B., GJØNNES, B.** : Winter vigour in *Picea abies* (L.) KARST. II. Attainment of winter vigour in four-year-old spruce plants during the autumn 1972. - Medd. norsk Inst. Skogforskning *32* : 357 - 376, 1976. [Chl.]

25957 - **EROKHIN, Yu.E., CHUGUNOV, V.A., MOSKALENKO, A.A., MAKHNEVA, Z.K., AGRIKOVA, I.M.** : Molekulyarnaya organizatsiya pigmentnoĭ sistemy fotosinteziruyushchikh bakteriĭ. [Molecular organisation of pigment system in photosynthetic bacteria.] - In : Itogi Issledovaniya Mekhanizma Fotosinteza. Pp. 12 - 15. Pushchino 1976. [In R.]

25958 - **EROKHINA, L.G., KRASNOVSKIĬ, A.A.** : Issledovanie spektral'nykh éffektov dena-
turatsii *B-* i *C-*fikoéritrinov. [Spectral phenomena of *B-* and *C-*phycoerythrins
denaturation.] - Biokhimiya *41* : 1594 - 1602, 1976. [In R, ab : E.]

25959 - **ESAU, K.** : Hyperplastic phloem and its plastids in spinach infected with the
curly top virus. - Ann. Bot. *40* : 637 - 644, 1976.

25960 - **ESTRADA, M., VALLESPINÓS, F.** : Estudio estadístico de espectros de absorción
de extractos de pigmentos de comunidades de algas macrófitas. [Statistical
study of absorption spectra of pigment extracts from macrophytic algal communities.] - Invest. Pesq. *40* : 551 - 559, 1976. [In Span., ab : E.]

25961 - **ETHERTON, B.** : A comparison of different methods of measuring plant cell membrane electrical resistances. - Plant Physiol. *57* (Suppl.) : 2, 1976. [Chlo-
roplast.]

25962 - **ETMAN-GERVAIS, C.** : Composés quinoniques isoprénoides et tocophérols dans les
bulbes et les feuilles de l'*Iris hollandais* I. x Hollandica; mise en évidence
d'une nouvelle quinone. - Compt. rend. Acad. Sci. Paris, Sér. D *282* : 1171 -
- 1174, 1976. [Chl.]

25963 - **EVANS, E.H., CAMMACK, R., EVANS, M.C.W.** : Properties of the primary electron
acceptor complex of photosystem I in the blue green alga *Chlorogloea fritschii*. - Biochem. biophys. Res. Commun. *68* : 1212 - 1218, 1976.

25964 - **EVANS, E.H., GOODING, A.D.** : The effects of dibromothymoquinone on respiratory and photosynthetic electron transport in *Rhodopseudomonas capsulata* chromatophores. - Arch. Microbiol. *111* : 171 - 174, 1976.

25865 - **EVANS, L.T.** : Physiological adaptation to performance as crop plants. - Phil.
Trans. roy. Soc. London B *275* : 71 - 83, 1976. [Ps, photosynthates.]

25966 - **EVANS, L.T.** : Transport and distribution in plants. - In : **WARDLAW, I.F.,
PASSIOURA, J.B.** (ed.) : Transport and Transfer Processes in Plants. Pp. 1 -
- 13. Academic Press, New York - San Francisco - London 1976. [Photosynthates.]

25967 - **EVANS, L.T., WARDLAW, I.F.** : Aspects of the comparative physiology of grain
yield in cereals. - Advances Agron. *28* : 301 - 359, 1976. [Ps.]

25968 - **EVANS, M.C.W., SIHRA, C.K., CAMMACK, R.** : The properties of the primary electron acceptor in the photosystem I reaction centre of spinach chloroplasts
and its interaction with *P*700 and the bound ferredoxin in various oxidation-
-reduction states. - Biochem. J. *158* : 71 - 77, 1976.

25969 - **EVDOKIMOVA, I.V., KORNYUSHENKO, G.A., VASIL'EVA, V.E.** : Fenomenologicheskaya
kharakteristika prevrashcheniĭ ksantofillov v izolirovannykh khloroplastakh
gorokha. [Phenomenological characteristic of xanthophyll conversions in isolated chloroplasts of *Pisum sativum* L.] - Fiziol. Biokhim. kul't. Rast. *8* :
503 - 507, 1976. [In R, ab : E.]

25970 - **EVENARI, M., SCHULZE, E.-D., LANGE, O.L., KAPPEN, L., BUSCHBOM, U.** : Plant
production in arid and semi-arid areas. - In : **LANGE, O.L., KAPPEN, L., SCHUL-
ZE, E.-D.** (ed.) : Water and Plant Life. Pp. 439 - 451. Springer-Verlag, Berlin - Heidelberg - New York 1976. [Growth analysis.]

25971 - **EVENSON, J.P., ROSE, C.W.** : Seasonal variations in stomatal resistance in cotton. - Agr. Meteorol. *17* : 381 - 386, 1976.

25972 - **EVERT, F.** : Compatible systems for the estimation of tree and stand volume. -
Forestry Chron. *52* : 1 - 2, 1976.

25973 - **EVSTIGNEEV, V.B., BEKASOVA, O.D.** : O fluorestsentsii allofikotsianina v svyazi s vyyasneniem roli fikobiliproteidov v fotosinteze. [On the fluorescence
of allophycocyanin in connection with clearing up the role of phycobilipro-
teids in photosynthesis.] - Dokl. Akad. Nauk SSSR *227* : 752 - 755, 1976. [In
R.]

*25974 - **EVSTIGNEEV, V.B., NAZAROVA, I.G.** : O mekhanizme aktivirovaniya fotosensibili-
ziruyushchego deĭstviya vodorastvorimykh analogov khlorofilla pri svyazyvanii
ikh polimerom. [Mechanism of activation of photosensitizing action of chlorophyll water-soluble analogues in their binding by a polymer.] - Dokl. Akad.
Nauk SSSR *224* : 964 - 967, 1975. [In R.]

25975 - **EVSTIGNEEV, V.B., PARAMONOVA, L.I.** : O fotokhimicheskikh svoĭstvakh fukoksantina. [Photochemical properties of fucoxanthin .] - Biokhimiya *41* : 98 - 103,
1976. [In R, ab :, E.]

25976 - **EVSTIGNEEV, V.B., PARAMONOVA, L.I.** : Fotoėlektrokhimicheskiĭ ėffekt v tver-
dykh plenkakh fukoksantina. [Photoelectrochemical effect in solid films of
fucoxanthin.] - Biokhimiya *41* : 548 - 552, 1976. [In R, ab : E.]

25977 - **FALK, H.** : Chromoplasts of *Tropaeolum majus* L. : Structure and development. -
Planta *128* : 15 - 22, 1976. [Chloroplast.]

25978 - **FALK, J.H.** : Energetics of a suburban lawn ecosystem. - Ecology *57* : 141 -
- 150, 1976. [Production.]

25979 - **FALK, R.H., STOCKING, C.R.** : Plant membranes. - In : **STOCKING, C.R., HEBER, U.**
(ed.) : Transport in Plants III. Intracellular Interactions and Transport
Processes. Pp. 3 - 50. Springer-Verlag, Berlin - Heidelberg - New York 1976.
[Chloroplast.]

25980 - **FALKNER, G., HORNER, F., WERDAN, K., HELDT, H.W.** : pH changes in the cytoplasm
of the blue-green alga *Anacystis nidulans* caused by light-dependent proton
flux into the thylakoid space. - Plant Physiol. *58* : 717 - 718, 1976.

25981 - **FALKNER, G., WERDAN, K., HORNER, F., HELDT, H.W.** : Energieabhängige Phosphat-
aufnahme der Blaualge *Anacystis nidulans*. - Ber. deut. bot. Ges. *89* : 285 -
- 288, 1976. [Ps.]

25982 - **FALLER, N.** : Usporedna asimilacija sumpornog i ugljičnog dioksida u nekih
biljaka. [Simultaneous assimilation of sulfur and carbon dioxide by some
plants.] - Acta bot. croat. *35* : 87 - 95, 1976. [In Serbocroat., ab : G.]

25983 - **FAN I-JI** : [A survey of the work done during the last ten years on the electron theory of catalysis applied to photosynthesis.] - Acta bot. sin. *18* :
170 - 179, 1976. [In Chin.]

25984 - **FAN I-JI, CHIANG I-HWA, CHIEN YUE-CHIN** : Inorganic photoreduction of nicotinamide adenine dinucleotide phosphate. I. Photoreduction by certain oxide
semiconductors in the near ultraviolet. - Sci. sin. *19* : 716 - 722, 1976.
[Ps.]

25985 - **FAN I-JI, CHIEN YUE-CHIN, CHIANG I-HWA** : Inorganic photophosphorylation
of adenosine diphosphate to adenosine triphosphate. - Sci. sin. *19* : 805 -
- 810, 1976.

25986 - **FAN YI-ji** : [The electron theory of catalysis applied to photosynthesis - A
résumé of the work done during the last ten years.] - Acta bot. sin. *18* : 71-
- 84, 1976. [In Chin., ab : E.]

25987 - **FARINEAU, J.** : Photoassimilation du CO_2 par les cellules de gaines périvascu-
laires isolées de feuilles de Maïs. - Physiol. vég. *14* : 661 - 662, 1976.

25988 - **FARINEAU, J.** : Simultaneous appearance of C-550 and cytochrome b_{559} HP (high
potential form) in barley leaves greened under intermittent light. - Physiol.
Plant. *36* : 393 - 398, 1976.

25989 - **FARMER, R.E., jr.** : Relationships between genetic differences in yield of deciduous tree species and variation in canopy size, structure and duration. -
In : **CANNELL, M.G.R., LAST, F.T.** (ed.) : Tree Physiology and Yield Improvement. Pp. 119 - 137. Academic Press, London - New York - San Francisco 1976.

25990 - **FARRAHI-ASCHTIANI, S.** : The effect of 2-chlorethyl trimethyl-ammonium chloride (CCC) and ammonium sulfate in treating iron chlorosis in the early stage
of soybean and safflower seedlings grown on alkaline soil. - Plant Physiol.
57 (Suppl.) : 63, 1976.

25991 - **FARRAR, J.F.** : Ecological physiology of the lichen *Hypogymnia physodes*. I.
Some effects of constant water saturation. - New Phytol. *77* : 93 - 103, 1976.
[Ps, Chl.]

25992 - **FARRAR, J.F.** : Ecological physiology of the lichen *Hypogymnia physodes*. II.

47

25992 - 26009 / FAR - FER

Effects of wetting and drying cycles and the concept of "physiological buffer-
ing". - New Phytol. *77* : 105 - 113, 1976. [Ps, Chl.]

25993 - **FARRAR, J.F., SMITH, D.C.** : Ecological physiology of the lichen *Hypogymnia
physodes*. III. The importance of the rewetting phase. - New Phytol. *77* : 115-
- 125, 1976. [Ps.]

25994 - **FASHAM, M.J.R., PUGH, P.R.** : Observations on the horizontal coherence of chlorophyll *a* and temperature. - Deep-Sea Res. *23* : 527 - 538, 1976.

25995 - **FASULO, M.P., VANNINI, G.L., DALL'OLIO, G.** : Modificazioni ultrastrutturali
indotte dall'acido 2,3,5-triiodobenzoico in *Euglena gracilis* eziolata exposta
alla luce. [Ultrastructural changes induced by 2,3,5-triiodobenzoic acid in
etiolated *Euglena gracilis* exposed to light.] - G. bot. ital. *110* : 179 -
- 190, 1976. [In Ital., ab : E.]

25996 - **FAY, P.** : Factors influencing dark nitrogen fixation in a blue-green alga. -
Appl. environm. Microbiol. *31* : 376 - 379, 1976. [Ps.]

25997 - **FEDDES, R.A., Van WIJK, A.L.M.** : An integrated model-approach to the effect
of water management on crop yield. - Agr. Water Manage. *1* : 3 - 20, 1976.

25998 - **FEDERER, C.A.** : Differing diffusive resistance and leaf development may cause
differing transpiration among hardwoods in spring. - Forest Sci. *22* : 359 -
- 364, 1976.

25999 - **FEDERER, C.A., GEE, G.W.** : Diffusion resistance and xylem potential in stressed and unstressed northern hardwood trees. - Ecology *57* : 975 - 984, 1976.

26000 - **FEDYUSHIN, A.A.** : K voprosu o raschete prikhoda summarnoĭ fotosinteticheski
aktivnoĭ solnechnoĭ radiatsii (FAR) na yugo-vostoke Kazakhstana. [Calculation
of input sum of photosynthetically active radiation in the south-eastern Kazakhstan.] - In : Fotosintez i Produktivnost' Ozimoĭ Pshenitsy na Yugo-Vostoke Kazakhstana. Pp. 21 - 29, 129. Nauka kazakh. SSR, Alma-Ata 1976. [In R.]

26001 - **FEDYUSHIN, A.A.** : Kvantovyĭ raskhod vidimogo fotosinteza ozimykh pshenits na
yugo-vostoke Kazakhstana. [Quantum expenditure of net photosynthesis in winter wheat of south-eastern Kazakhstan.] - In : Fotosintez i Produktivnost'
Ozimoĭ Pshenitsy na Yugo-Vostoke Kazakhstana. Pp. 29 - 40, 130. Nauka kazakh.
SSR, Alma-Ata 1976. [In R.]

26002 - **FEE, E.J.** : The vertical and seasonal distribution of chlorophyll in lakes
of the Experimental Lakes Area, northwestern Ontario : Implications for primary production estimates. - Limnol. Oceanogr. *21* : 767 - 783, 1976.

26003 - **FEIERABEND, J.** : Temperature-sensitivity of chloroplast ribosome formation in
higher plants. - In : BÜCHER, T., NEUPERT, W., SEBALD, W., WERNER, S. (ed.) :
Genetics and Biogenesis of Chloroplasts and Mitochondria. Pp. 99 - 102. North-
-Holland Publ. Co., Amsterdam - New York - Oxford 1976.

26004 - **FEIERABEND, J., SCHRADER-REICHHARDT, U.** : Biochemical differentiation of plastids and other organelles in rye leaves with a high-temperature-induced deficiency of plastid ribosomes. - Planta *129* : 133 - 145, 1976.

26005 - **FEIGE, G.B.** : Untersuchungen zur Physiologie der Cephalodien der Flechte *Pel-
tigera aphthosa* (L.)WILLD. I. Die photosynthetische ^{14}C-Markierung der Lipid-
fraktion. - Z. Pflanzenphysiol. *80* : 377 - 385, 1976.

26006 - **FEIGE, G.B.** : Untersuchungen zur Physiologie der Cephalodien der Flechte *Pel-
tigera aphthosa* (L.)WILLD. II. Das photosynthetische ^{14}C-Markierungsmuster
und der Kohlenhydrattransfer zwischen Phycobiot und Mycobiot. - Z. Pflanzenphysiol. *80* : 386 - 394, 1976.

26007 - **FELLOWS, R.J., BOYER, J.S.** : Structure and activity of chloroplasts of sunflower leaves having various water potentials. - Planta *132* : 229 - 239, 1976.

26008 - **FENTON, R., MANSFIELD, T.A., WELLBURN, A.R.** : Effects of isoprenoid alcohols
on oxygen exchange of isolated chloroplasts in relation to their possible physiological effects on stomata. - J. exp. Bot. *27* : 1206 - 1214, 1976.

26009 - **FER, A.** : Photosynthèse et respiration des plantules de *Cuscuta lupuliformis*
KROCK., au cours de leur vie préparasitaire. - Physiol. vég. *14* : 357 - 365,
1976.

26010 - **FER, A.** : Utilisation des produits de la photosynthèse chez les plantules de *Cuscuta lupuliformis* KROCK., au cours de leur vie préparasitaire. - Compt. rend. Acad. Sci. Paris, Sér. D *282* : 1725 - 1727, 1976.

26011 - **FERENBAUGH, R.W.** : Effects of simulated acid rain on *Phaseolus vulgaris* L. (*Fabaceae*). - Amer. J. Bot. *63* : 283 - 288, 1976. [Ps, Chl.]

26012 - **FERRARI, I., BELLAVERE, C., CAMURRI, L., CATELLANI, M.** : Ricerche limnologiche in un lago appenninico in condizioni di copertura ghiacciata. [Limnological research in an Appennini lake supplied by a glacier.] - G. Geol. Ser. 2a *40* (2) : 131 - 133, 1976. [Chl; in Ital.]

26013 - **FERREE, D.C., HALL, F.R.** : Factors influencing efficiency of apple tree photosynthesis. - Ohio Rep. *61* (3) : 35 - 37, 1976.

26014 - **FERREE, D.C., HALL, F.R., SPOTTS, R.A.** : Influence of spray adjuvants and multiple applications of Benomyl and oil on photosynthesis of apple leaves. - HortScience *11* : 391 - 392, 1976.

26015 - **FETCHER, N.** : Patterns of leaf resistance to lodgepole pine transpiration in Wyoming. - Ecology *57* : 339 - 345, 1976.

26016 - **FEUILLADE, J., FEUILLADE, M.** : Importance des oligo-éléments pour la croissance d'*Oscillatoria rubescens* D.C. *in vitro*. - Ann. Hydrobiol. *7* : 1 - 9, 1976. [Chl, Car.]

26017 - **FIALA, K.** : Underground organs of *Phragmites communis*, their growth, biomass and net production. - Folia geobot. phytotax. (Praha) *11* : 225 - 259, 1976.

26018 - **FILIPPIS, L.F. De, PALLAGHY, C.K.** : The effect of sub-lethal concentrations of mercury and zinc on *Chlorella* I. Growth characteristics and uptake of metals. - Z. Pflanzenphysiol. *78* : 197 - 207, 1976. [Photosynthate export.]

26019 - **FILIPPIS, L.F. De, PALLAGHY, C.K.** : The effect of sub-lethal concentrations of mercury and zinc on *Chlorella*. II. Photosynthesis and pigment composition. - Z. Pflanzenphysiol. *78* : 314 - 322, 1976.

*26020 - **FILIPPOV, G.L., MAKSIMOVA, L.A.** : Vliyanie optimizatsii usloviĭ vyrashchivaniya na produktivnost' fotosinteza kukuruzy. [Effect of the optimization of growing conditions on the productivity of photosynthesis in maize.] - Byull. vses. nauch.-issled. Inst. Kukuruzy *1975* (2) : 37 - 44, 1975. [Growth analysis; in R.]

26021 - **FILIPPOVA, L.A., MAMUSHINA, N.S.** : K voprosu o vliyanii nakopleniya assimilyatov na fotosintez. [Effect of photosynthate accumulation on photosynthesis.] - In : Gazometricheskoe Issledovanie Fotosinteza i Dykhaniya Rasteniĭ. Pp. 135 - 138. Akad. Nauk SSSR, Tartu 1976. [In R.]

26022 - **FILIPPOVICH, I.I., NOZDRINA, V.N., KUPCHINENKO, V.V., OPARIN, A.I.** : Izuchenie membranosvyazannykh ribosom v svyazi s formirovaniem khloroplastov gorokha. [Membrane-bound ribosomes during pea chloroplasts formation.] - Biokhimiya *41* : 708 - 717, 1976. [In R, ab : E.]

26023 - **FINDENEGG, G.R.** : Correlation between accessibility of carbonic anhydrase for external substrate and regulation of photosynthetic use of CO_2 and HCO_3^- by *Scenedesmus obliquus*. - Z. Pflanzenphysiol. *79* : 428 - 437, 1976.

26024 - **FINDENEGG, G.R.** : Zur Frage eines HCO_3^-/OH^--Austausches im Plasmalemma von *Scenedesmus obliquus*. - Ber. deut. bot. Ges. *89* : 277 - 284, 1976.

26025 - **FINE, K.E., LOEBLICH, A.R. III** : Endosymbiosis in the marine dinoflagellate *Kryptoperidinium foliaceum*. - J. Protozool. *23* (2) : 8A, 1976. [Chl, Car.]

26026 - **FINN, J.T.** : Measures of ecosystem structure and function derived from analysis flows. - J. theor. Biol. *56* : 363 - 380, 1976. [Model.]

26027 - **FISCHER, K.S., WILSON, G.L.** : Studies of grain production in *Sorghum bicolor* (L. MOENCH). VI. Profiles of photosynthesis, illuminance and foliage arrangement. - Aust. J. agr. Res. *27* : 35 - 44, 1976.

26028 - **FISCHER, K.S., WILSON, G.L., DUTHIE, I.** : Studies of grain production in *Sorghum bicolor* (L. MOENCH.). VII Contribution of plant parts to canopy photosynthesis and grain yield in field stations. - Aust. J. agr. Res. *27* : 235- - 242, 1976.

26029 - **FISCHER, R.A., AGUILAR M., I.** : Yield potential in a dwarf spring wheat and the effect of carbon dioxide fertilization. - Agr. J. *68* : 749 - 752, 1976.

26030 - **FISCHER, R.A., AGUILAR M., I., MAURER, O.R., RIVAS, A.,S.** : Density and row spacing effects on irrigated short wheats at low latitude. - J. agr. Sci. *87*: 137 - 147, 1976. [Growth analysis.]

26031 - **FISCHER, R.A., LAING, D.R.** : Yield potential in a dwarf spring wheat and response to crop thinning. - J. agr. Sci. *87* : 113 - 122, 1976.

26032 - **FISHER, D.B.** : Measurement of sugar concentration in sieve elementa by negative staining. - Plant Physiol. *57* (Suppl.) : 77, 1976. [Photosynthate transport.]

26033 - **FLOWERS, T.J., HALL, J.L.** : Properties of membranes from the halophyte *Suaeda maritima*. II. Distribution and properties of enzymes in isolated membrane fractions. - J. exp. Bot. *27* : 673 - 689, 1976. [Chl.]

26034 - **FLOYD, B.W., NOBLE, R.D.** : Evaluation of infrared-sensitive film for determination of spectral quality within a forest canopy. - Ohio J. Sci. *76* : 119 - - 122, 1976.

26035 - **FLÜCKIGER, W.** : Der Einfluß von Simazin auf die apparente O_2-Produktion der einzelligen Bodenalge *Dictyococcus engadinensis* (KOL. et F.CHOD.) VISCHER. - Biochem. Physiol. Pflanzen *170* : 269 - 272, 1976.

26036 - **FLÜGGE, U.I., HELDT, H.W.** : Identification of a protein involved in phosphate transport of chloroplasts. - FEBS Lett. *68* : 259 - 262, 1976.

26037 - **FOCK, H., PRZYBYLLA, K.-R.** : Die Raten der CO_2-Entwicklung beleuchteter und verdunkelter C_3-Pflanzen in Abhängigkeit von der Temperatur. - Ber. deut. bot. Ges. *89* : 643 - 650, 1976.

26038 - **FOGG, G.E.** : Release of glycollate from tropical marine plants. - Aust. J. Plant Physiol. *3* : 57 - 61, 1976.

26039 - **FONDY, B.R., GEIGER, D.R.** : Kinetics of several leaf sugars during phloem loading in *Beta vulgaris* L. - Plant Physiol. *57* (Suppl.) : 77, 1976.

26040 - **FONG, F.K.** : Energy upconversion and the minimum quantum requirement in photosynthesis. - J. amer. chem. Soc. *98* : 7840 - 7843, 1976.

26041 - **FONG, F.K., KOESTER, V.J.** : *In vitro* preparation and characterization of a 700 nm absorbing chlorophyll-water adduct according to the proposed primary molecular unit in photosynthesis. - Biochim. biophys. Acta *423* : 52 - 64, 1976.

26042 - **FONG, F.K., KOESTER, V.J., POLLES, J.S.** : Optical spectroscopic study of (Chl $a \cdot H_2O)_2$ according to the proposed C_2 symmetrical molecular structure for the P700 photoactive aggregate in photosynthesis. - J. amer. chem. Soc. *98* : 6406- - 6408, 1976.

26043 - **FONG, F.K., WINOGRAD, N.** : *In vitro* solar conversion after the primary light reaction in photosynthesis. Reversible photogalvanic effects of chlorophyll- -quinhydrone half-cell reactions. - J. amer. chem. Soc. *98* : 2287 - 2289, 1976.

26044 - **FOOTE, K.C., SCHAEDLE, M.** : A stem cuvette for bark photosynthetic and respiratory studies. - Photosynthetica *10* : 307 - 311, 1976.

26045 - **FOOTE, K.C., SCHAEDLE, M.** : Physiological characteristics of photosynthesis and respiration in stems of *Populus tremuloides* MICHX. - Plant Physiol. *58* : 91 - 94, 1976.

26046 - **FOOTE, K.C., SCHAEDLE, M.** : Diurnal and seasonal patterns of photosynthesis and respiration by stems of *Populus tremuloides* MICHX. - Plant Physiol. *58* : 651 - 655, 1976.

26047 - **FORD, E.D.** : The canopy of a Scots pine forest : description of a surface of complex roughness. - Agr. Meteorol. *17* : 9 - 32, 1976.

26048 - **FORK, D.C.** : Temperature dependence of chlorophyll *a* fluorescence in algae and higher plants in relation to changes of state in the photosynthetic apparatus. - Carnegie Inst. Year Book *75* : 465 - 472, 1976.

26049 - FÖRSTEL, H., WEINER, B., SCHLESER, G. : Preparation of oxygen samples for $^{18}O/^{16}O$ measurements by a combined gas chromatography-burning technique. - Int. J. appl. Rad. Isotopes 27 : 211 - 215, 1976.

26050 - FORSTER, H. : Einfluß der Kaliumernährung auf Ausbildung und Chlorophyllgehalt des Fahnenblattes und auf die Kornertragskomponenten von Sommerweizen - Untersuchungen an fünf verschiedenen Sommerweizensorten. - Z. Acker- Pflanzenbau 143 : 169 - 178, 1976.

26051 - FORWARD, R.B., Jr. : Light and diurnal vertical migration : photobehavior and photophysiology of plankton. - In : SMITH, K.C. (ed.) : Photochemical and Photobiological Reviews. Vol. 1. Pp. 157 - 209. Plenum Press, New York - London 1976.

26052 - FOSTER, A., BLACK, C.C. : Changes in pathways of carbon flow in response to CO_2 concentration during leaf photosynthesis in *Panicum maximum*. - Plant Physiol. 57 (Suppl.) : 59, 1976.

*26053 - FOSTER, R.J., GIBBONS, G.C., GOUGH, S., HENNINGSEN, K.W., KAHN, A., NIELSEN, O.F., WETTSTEIN, D. von : Protochlorophyll holochrome. - In : BRODA, E., LOCKER, A., SPRINGER-LEDERER, H. (ed.) : Proceedings of the First European Biophysics Congress. Vol. 4. Pp. 137 - 149. Verlag wiener med. Akad., Vienna 1971.

B26054 - Fotosintez i Produktivnost' Ozimoĭ Pshenitsy na Yugo-Vostoke Kazakhstana. [Photosynthesis and Productivity of Winter Wheat in South-eastern Kazakhstan.] - Nauka kazakh. SSR, Alma-Ata 1976. [In R.]

26055 - FOUTZ, A.L., WILHELM, W.W., DOBRENZ, A.K. : Relationship between physiological and morphological characteristics and yield of nondormant alfalfa clones. - Agron. J. 68 : 587 - 591, 1976. [Ps.]

26056 - FOWLER, C.F., KOK, B. : Determination of H^+/e^- ratios in chloroplasts with flashing light. - Biochim. biophys. Acta 423 : 510 - 523, 1976.

26057 - FOX, S.B., NAYLOR, A.W. : Comparison of RUDP and PEP carboxylase and chlorophyll levels in greening *Zea mays* leaves. - Plant Physiol. 57 (Suppl.) : 4, 1976.

26058 - FOY, R.H., GIBSON, C.E., SMITH, R.V. : The influence of daylength, light intensity and temperature on the growth rates of planktonic blue-green algae. - Brit. phycol. J. 11 : 151 - 163, 1976.

26059 - FOYER, C.H., HALLIWELL, B. : The presence of glutathione and glutathione reductase in chloroplasts : A proposed role in ascorbic acid metabolism. - Planta 133 : 21 - 25, 1976.

26060 - FRACKOWIAK, D., FIKSIŃSKI, K. : Przekazywanie energii w jednostkach fotosyntetycznych. [Energy transfer in photosynthetic units.] - Postepy Biochem. 22: 439 - 465, 1976. [In Pol.]

26061 - FRACKOWIAK, D., GRABOWSKI, J., MANIKOWSKI, H. : Circular dichroism spectra of biliproteins. - Photosynthetica 10 : 204 - 207, 1976.

26062 - FRACKOWIAK, D., SKOWRON, A., SALAMON, Z. : Photopotential of biliproteins in electrochemical cell. - Photosynthetica 10 : 339 - 342, 1976.

26063 - FRADKIN, L.I., KOLYAGO, V.M., MORDACHEVA, G.S., SAĬ, P.K. : O submembrannykh chastitsakh khloroplastov. [Submembrane chloroplast particles.] - Fiziol. Rast. 23 : 666 - 674, 1976. [In R, ab : E.]

26064 - FRAISSE, D. : Automatic microanalyzers. II. Automatic analyzer for rapid microdetermination of oxygen. - Microchem. J. 21 : 178 - 197, 1976.

26065 - FRANK, A.B., BARKER, R.E. : Rates of photosynthesis and transpiration and diffusive resistance of six grasses grown under controlled conditions. - Agron. J. 68 : 487 - 490, 1976.

26066 - FREDRICKS, W.W., GEHL, J.M. : Multiple forms of ferredoxin-nicotinamide adenine dinucleotide phosphate reductase from spinach. - Arch. Biochem. Biophys. 174 : 666 - 674, 1976.

*26067 - FRENYÓ, V., NINH THAI DUY : Rézionok hatása *Elodea* levelek katalázaktivitására.

[Effect of copper ions on the catalase activity of *Elodea* leaves.] - Bot. Köz-lem. *59* : 233 - 235, 1972. [Ps; in Hung.]

26068 - **FREY, N.M., MOSS, D.N.** : Variation in RuDPCase activity in barley. - Crop Sci. *16* : 210 - 213, 1976.

26069 - **FREYSSINET, G.** : Relation between paramylum content and the length of the lag period of chlorophyll synthesis during greening of dark-grown *Euglena gracilis*. - Plant Physiol. *57* : 824 - 830, 1976.

26070 - **FREYSSINET, G.** : Influence of culture conditions on the length of the lag period of chlorophyll synthesis in preilluminated dark-grown *Euglena*. - Plant Physiol. *57* : 831 - 835, 1976.

26071 - **FRICK, H., JONES, R.F.** : Physiology and plastid fine structure of deetiolating *Lemna minor*. - Can. J. Bot. *54* : 1819 - 1826, 1976.

26072 - **FRIEDMANN, E.I., OCAMPO, R.** : Endolithic blue-green algae in the dry valleys: Primary producers in the Antarctic desert ecosystems. - Science *193* : 1247 - - 1249, 1976.

26073 - **FRIEND, D.J.C., HELSON, V.A.** : Thermoperiodic effects on the growth and photosynthesis of wheat and other crop plants. - Bot. Gaz. *137* : 75 - 84, 1976.

26074 - **FROMMHOLD, I.** : Physiologie der Gramineen-Stomata. I. Struktur, Funktion und Regulation der Öffnungsweite. - Wiss. Z. Humboldt-Univ. Berlin, math.-naturwiss. Reihe *25* : 803 - 810, 1976. [Stomatal resistance.]

26075 - **FROMMHOLD, I.** : Physiologie der Gramineen-Stomata. II. Beziehungen zwischen der Stomatafrequenz und -apertur sowie der Transpirations- bzw. Photosyntheserate. - Wiss. Z. Humboldt-Univ. Berlin, math.-naturwiss. Reihe *25* : 811 - - 817, 1976.

26076 - **FROMMHOLD, I., HOANG-THI-HÀ** : Untersuchungen über potentielle Antitranspirationsmittel. - Arch. Acker- Pflanzenbau Bodenkunde *20* : 637 - 643, 1976. [Chl.]

26077 - **FROSCH, S., BERGFELD, R., MOHR, H.** : Light control of plastogenesis and ribulosebisphosphate carboxylase levels in mustard seedling cotyledons. - Planta *133* : 53 - 56, 1976.

26078 - **FRY, D.J., PHILLIPS, I.D.J.** : Photosynthesis of conifers in relation to annual growth cycles and dry matter production. I. Some C_4 characteristics in photosynthesis of Japanese larch (*Larix leptolepis*). - Physiol. Plant. *37* : 185 - - 190, 1976.

26079 - **FRYDRYCH, J.** : Photosynthetic characteristics of cucumber seedlings grown under two levels of carbon dioxide. - Photosynthetica *10* : 335 - 338, 1976.

26080 - **FUCHS, M., STANHILL, G., MORESHET, S.** : Effect of increasing foliage and soil reflectivity on the solar radiation balance of wide-row grain sorghum. - Agron. J. *68* : 865 - 871, 1976.

26081 - **FUCHS, W.H.** : 1.1 History of physiological plant pathology. - In : **HEITEFUSS, R., WILLIAMS, P.H.** (ed.) : Physiological Plant Pathology. (Encycl. Plant Physiol. N.S. Vol.4.) Pp. 1 - 26. Springer-Verlag, Berlin - Heidelberg - New York 1976. [Ps.]

26082 - **FUGATE, R.D., SONG, P.-S.** : Excited states of biomolecules - IV. - Photochem. Photobiol. *24* : 629 - 640, 1976. [Chl, Car, biliproteins.]

26083 - **FUHRHOP, J.-H.** : Molekülkomplexe und Sauerstoffaddukte von Tetrapyrrolfarbstoffen. - Angew. Chem. *88* : 704 - 716, 1976. [Chl dimers.]

26084 - **FUJITA, Y.** : The C550 photoresponse at room temperature observed in membrane fragments of the blue-green alga *Anabaena variabilis*. - Plant Cell Physiol. *17* : 187 - 191, 1976.

26085 - **FUKAI, S., KOH, S., KUMURA, A.** : Dry matter production and photosynthesis of *Hordeum vulgare* L. in the field. - J. appl. Ecol. *13* : 877 - 887, 1976.

26086 - **FUKAI, S., LOOMIS, R.S.** : Leaf display and light environments in row-planted cotton communities. - Agr. Meteorol. *17* : 353 - 379, 1976. [Ps.]

26087 - **FUKAI, S., SILSBURY, J.H.** : Responses of subterranean clover communities to temperature. I Dry matter production and plant morphogenesis. - Aust. J. Plant. Physiol. *3* : 527 - 543, 1976. [Growth analysis.]

26088 - GABRYŚ-MIZERA, H. : Model considerations of the light conditions in noncylindrical plant cells. - Photochem. Photobiol. *24* : 453 - 461, 1976. [Chloroplast.]

26089 - GÄCHTER, R. : Untersuchungen über die Beeinflussung der planktischen Photosynthese durch anorganische Metallsalze im eutrophen Alpnachersee und der mesotrophen Horwer Bucht. - Schweiz. Z. Hydrol. *38* : 97 - 119, 1976. [Ps.]

26090 - GAFF, D.F., ZEE, S.-Y., O'BRIEN, T.P. : The fine structure of dehydrated and reviving leaves of *Borya nitida* LABILL. - a desiccation-tolerant plant. - Aust. J. Bot. *24* : 225 - 236, 1976. [Chloroplast.]

26091 - GALE, J., TAKO, T. : Response of *Zea mays* and *Phaseolus vulgaris* to supra-atmospheric concentrations of oxygen. - Photosynthetica *10* : 89 - 92, 1976.

26092 - GALLAGHER, J.N., BISCOE, P.V., SCOTT, R.K. : Barley and its environment. VI. Growth and development in relation to yield. - J. appl. Ecol. *13* : 563 - 583, 1976. [Growth analysis.]

26093 - GALLAHER, R.N., BROWN, R.H., ASHLEY, D.A., JONES, J.B., Jr. : Photosynthesis of, and $^{14}CO_2$-photosynthate translocation from, calcium-deficient leaves of crops. - Crop Sci. *16* : 116 - 119, 1976.

26094 - GALLING, G. : Synthesis of chlorophyll-free thylakoids in *Chlorella* after clindamycin-treatment and in a temperature sensitive mutant of *Chlorella*. - In : BUCHER, T., NEUPERT, W., SEBALD, W., WERNER, S. (ed.) : Genetics and Biogenesis of Chloroplasts and Mitochondria. Pp. 53 - 60. North-Holland Publ. Co., Amsterdam - New York - Oxford 1976.

26095 - GAMALEĬ, Yu.V., KULIKOV, G.V. : Struktura khloroplastov u predstaviteleĭ semeĭstva *Oleaceae*. [The structure of chloroplasts in representatives of the family *Oleaceae*.] - Bot. Zh. *61* : 3 - 11, 1976. [In R, ab : E.]

26096 - GANCHARYK, M.M., LYAGENCHANKA, B.I., MIKUL'SKAYA, S.A., TALANAVA, K.S., MATSYUSHEĬSKAYA, V.P. : Uplyŭ impul'snaga dazhdzhavannya na fiziyalogiyu i praduktsyĭnasts' aŭsa pry roznaĭ vil'gotnastsi tarfyanoĭ gleby. [The effect of impulse irrigation on physiology and productivity of oat under different moisture of peat soil.] - Vestsi Akad. Navuk belarus. SSR, Ser. biyal. Navuk *1976* (6) : 5 - 11, 137, 1976. [Ps; in Belorus., ab : R.]

26097 - GANOZA, V.G., McFEETERS, R.F. : Chlorophyllase activity during pigmentation changes in *Chlorella prototheaoides*. - Photosynthetica *10* : 1 - 6, 1976.

26098 - GANTT, E., LIPSCHULTZ, C.A., ZILINSKAS, B. : Further evidence for a phycobilisome model from selective dissociation, fluorescence emission, immunoprecipitation, and electron microscopy. - Biochim. biophys. Acta *430* : 375 - 388, 1976.

B26099 - GAPONENKO, V.I. : Vliyanie Vneshnikh Faktorov na Metabolizm Khlorofilla. [Effect of External Factors on Chlorophyll Metabolism.] - Nauka i Tekhnika, Minsk 1976. [In R.]

26100 - GARAB, G.I., BRETON, J. : Polarized light spectroscopy on oriented spinach chloroplasts. Fluorescence emission at low temperature. - Biochem. biophys. Res. Commun. *71* : 1095 - 1102, 1976.

26101 - GARAB, G.I., CHERNISHEVA, S., KISS, J.G., FALUDI-DÁNIEL, Á. : Chlorophyll forms affected by 3(3,4-dichlorophenol)-1,1-dimethylurea as shown by low temperature fluorescence spectra of chloroplasts and fragments. - FEBS Lett. *61* : 140 - 143, 1976.

26102 - GARBER, M.B., EROKHIN, Yu.E., RESHETNIKOVA, L.S., CHUGUNOV, V.A. : Vydelenie i kristallizatsiya plastotsianina iz list'ev gorokha (*Pisum sativum* L.). [Isolation and crystallization of plastocyanin from leaves of pea.] - Dokl. Akad. Nauk SSSR *230* : 469 - 471, 1976. [In R.]

26103 - GARBER, M.P. : Effects of light and chilling temperatures on chloroplast thylakoids of thermophilic and cool-season plants. - HortScience *11* (3, Sect.2): 299, 1976.

26104 - GARBER, M.P., STEPONKUS, P.L. : Alterations in chloroplast thylakoids during an *in vitro* freeze-thaw cycle. - Plant Physiol. *57* : 673 - 680, 1976.

26105 - **GARBER, M.P., STEPONKUS, P.L.** : Alterations in chloroplast thylakoids during cold acclimation. - Plant Physiol. *57* : 681 - 686, 1976.

26106 - **GARCIA, L.R., HANWAY, J.J.** : Foliar fertilization of soybeans during the seed--filling period. - Agron. J. *68* : 653 - 657, 1976. [Ps.]

26107 - **GARG, I.D., MANDAHAR, C.L.** : Physiology of powdery mildew infected leaves of *Abelmoschus esculentus*. I. Respiration and photosynthesis. - Phytopathol. Z. *85* : 298 - 307, 1976.

26108 - **GARGAS, E., NIELSEN, C.S., LØNHOLDT, J.** : An incubator method for estimating the actual daily planktonalgae primary production. - Water Res. *10* : 853 - - 860, 1976.

26109 - **GARNER, J.O., Jr., KAPLAN, S.L., LUDFORD, P.M., OZBUN, J.L.** : Partitioning of photosynthate in two dry bean cultivars. - HortScience *11* (3, Sect. 2) : 299, 1976.

26110 - **GARRELS, R.M., LERMAN, A., MACKENZIE, F.T.** : Controls of atmospheric O_2 and CO_2 : Past, present, and future. - Amer. Sci. *64* : 306 - 315, 1976. [Ps.]

26111 - **GARRETT, M.K., ALLEN, M.D.B.** : Photosynthetic purification of the liquid phase of animal slury. - Environ. Pollut. *10* : 127 - 139, 1976.

26112 - **GÄRTNER, M.** : Photosynthetische Leistung und Stoffproduktion höherer Wasserpflanzen, dargestellt am Beispiel von *Typha latifolia* und *Typha angustifolia*. - Wiss. Z. tech. Univ. Dresden *25* : 246 - 250, 1976.

26113 - **GASSMAN, M.L., CASTELFRANCO, P.A.** : Metabolic fate of 5-aminolevulinic acid in etiolated and greening barley seedlings. - Plant Physiol. *57* (Suppl.) : 46, 1976.

26114 - **GATES, D.M.** : Energy exchange and transpiration. - In : **LANGE, O.L., KAPPEN, L., SCHULZE, E.-D.** (ed.) : Water and Plant Life. Pp. 137 - 147. Springer-Verlag, Berlin - Heidelberg - New York 1976.

26115 - **GAUDILLÈRE, J.P.** : Effet de la salinité du milieu de culture sur le rendement photosynthétique de conversion de la lumière d'une plante halophyte (*Plantago maritima*) et d'une plante glycophyte (*Plantago lanceolata*). - Physiol. vég. *14* : 662, 1976.

26116 - **GAUHL, E.** : Photosynthetic response to varying light intensity in ecotypes of *Solanum dulcamara* L. from shaded and exposed habitats. - Oecologia *22* : 275 - - 286, 1976.

26117 - **GAUSMAN, H.W., RODRIGUEZ, R.R., RICHARDSON, A.J.** : Infinite reflectance of dead compared with live vegetation. - Agron. J. *68* : 295 - 296, 1976. [Chl.]

26118 - **GAVRILENKO, V.F., RUBIN, B.A., AVSIEVICH, N.A., ZHIGALOVA, T.V.** : Issledovanie sistem assimilyatsii uglekisloty u prorostkov pshenits razlichnoĭ produktivnosti.· [Carbon assimilation systems in wheat seedlings of different productivity.] - Sel'skokhoz. Biol. *11* : 347 - 353, 1976. [In R, ab : E.]

26119 - **GAYLER, K.R., MORGAN, W.R.** : An NADP-dependent glutamate dehydrogenase in chloroplasts from the marine green alga *Caulerpa simpliciuscula*. - Plant Physiol. *58* : 283 - 287, 1976.

B26120 - Gazometricheskoe Issledovanie Fotosinteza i Dykhaniya Rasteniĭ. [Gasometric Studies of Photosynthesis and Respiration of Plants.] - Akad. Nauk SSSR, Tartu 1976. [In R.]

26121 - **GEIGER, D.R.** : Effects of translocation and assimilate demand on photosynthesis. - Can. J. Bot. *54* : 2337 - 2345, 1976.

26122 - **GEIGER, D.R.** : Phloem loading in source leaves. - In : **WARDLAW, I.F., PASSIOURA, J.B.** (ed.) : Transport and Transfer Processes in Plants. Pp. 167 - 183. Academic Press, New York - San Francisco - London 1976. [Photosynthates.]

26123 - **GEJ, B.** : Charakterystyka fizjologiczna poszczególnych liści sałaty traktowanej symazyną. [Physiological characteristics of individual leaves of lettuce treated with simazine.] - Roczn. Nauk roln. A *101*(4): 7 - 19,1976. [Ps, Chl; in Pol., ab : E, R.]

26124 - **GELFI, N., BLANCHET, R., BOSC, M.** : Influence de l'alimentation hydrique sur la transpiration et la photosynthèse du Soja (*Glycine Max*. L. MERRIL); conséquences relatives a l'efficience de l'eau consommée. - Compt. rend. Acad. Sci. Paris, Sér. D *283* : 495 - 498, 1976.

26125 - **GENCHEV, S., MIKHOV, A.** : V"rkhu s"d"rzhanieto na plastidni pigmenti i aktivnostta na nyakoi enzimi v"v vr"zka s kheterozisniya efekt pri krastavitsata. [Plastid pigment content and activity of certain enzymes in relation to heterosis effect in cucumber.] - Fiziol. Rast. (Sofia) *2* (2) : 3 - 13, 1976. [In Bulg., ab : E, R.]

26126 - **GENEROZOVA, I.P.** : Zakalivanie rasteniĭ kak sposob povysheniya ustoĭchivosti membran khloroplastov k obezvozhivaniyu na primere prorostkov pshenitsy. [Hardening of plants as a means to increase stability of chloroplast membranes to dehydration studied with wheat seedlings.] - Fiziol. Rast. *23* : 921 - - 927, 1976. [In R, ab : E.]

26127 - **GENEVES, L., CHOUSSY, M., BARBIER, M., NEUVILLE, D., DASTE, P.** : Ultrastructure et composition pigmentaire comparées des chromatophores de la Diatomée *Navicula ostrearia* (GAILLON) BORY normale et bleue. - Compt. rend. Acad. Sci. Paris, Sér. D *282* : 449 - 452, 1976.

26128 - **GENKEL', P.A.** : Fiziologicheskie osnovy adaptatsii rasteniĭ k zasukhe. [Physiological principles of plants adaptation to drought.] - Fiziol. Biokhim. kul't. Rast. *8* : 132 - 137, 1976. [Chloroplast; in R, ab : E.]

26129 - **GEORGE, D.G.** : A pumping system for collecting horizontal plankton samples and recording continuously sampling depth, water temperature, turbidity and *in vivo* chlorophyll. - Freshwater Biol. *6* : 413 - 419, 1976.

26130 - **GEORGE, D.G., EDWARDS, R.W.** : The effect of wind on the distribution of chlorophyll *a* and crustacean plankton in a shallow eutrophic reservoir. - J. appl. Ecol. *13* : 667 - 690, 1976.

26131 - **GEORGIEVA, D., SALCHEVA, G., PETKOVA, M.** : Vliyanie na niskite temperatury v"rkhu s"d"rzhanieto na plastidni pigmenti pri rasteniya ot zimna pshenitsa. [Effect of low temperatures on the plastid pigment content in winter wheat plants.] - Fiziol. Rast. (Sofia) *2* (2) : 14 - 27, 1976. [In Bulg., ab : E, R.]

26132 - **GERHARDT, B., BETSCHE, T.** : The change of microbodies from glyoxysomal to peroxysomal function within fatty, greening cotyledons : Hypotheses, results, problems. - Ber. deut. bot. Ges. *89* : 321 - 334, 1976.

26133 - **GHOLZ, H.L., FITZ, F.K., WARING, R.H.** : Leaf area differences associated with old-growth forest communities in the western Oregon Cascades. - Can. J. Forest Res. *6* : 49 - 57, 1976.

26134 - **GHOSH, N., GOTOH, K., NAKASEKO, K., MORI, Y.** : Influence of soybean dwarf virus on growth and yield. - Proc. Crop Sci. Soc. Jap. *45* : 624 - 629, 1976.

26135 - **GIAQUINTA, R.** : Evidence for phloem loading from the apoplast. Chemical modification of membrane sulfhydryl groups. - Plant Physiol. *57* : 872 - 875, 1976. [Ps.]

26136 - **GIAQUINTA, R.** : Phloem loading from the apoplast : chemical modification of membrane sulfhydryl groups. - Plant Physiol. *57* (Suppl.) : 77, 1976.

26137 - **GIBBONS, L.K., KOLDENHOVEN, E.F., NETHERY, A.A., MONTGOMERY, R.E., PURCELL, W.P.** : Quantitative structure-activity relationships among selected pyrimidinones and Hill reaction inhibition. - J. agr. Food Chem. *24* : 203 - 206, 1976.

26138 - **GIBSON, R.W., WHITEHEAD, D., AUSTIN, D.J., SIMKINS, J.** : Prevention of potato top-roll by aphicide and its effect on leaf area and photosynthesis. - Ann. appl. Biol. *82* : 151 - 153, 1976.

26139 - **GIENAPP, K., KHOFFMANN, P.** : Obrazovanie lipidov v protsesse zeleneniya étiolirovannykh prorostkov pshenitsy. [Formation of lipids during greening of etiolated wheat seedlings.] - Fiziol. Rast. *23* : 938 - 944, 1976. [In R, ab : E.]

26140 - **GIFFORD, R.M.** : On overview of fuel used for crops and national agricultural systems. - Search 7 : 412 - 417, 1976. [Crop energetics.]

26141 - **GILES, K.L., COHEN, D., BEARDSELL, M.F.** : Effects of water stress on the ultrastructure of leaf cells of *Sorghum bicolor.* - Plant Physiol. 57 : 11 - 14, 1976. [Chloroplast.]

26142 - **GILLEN, L.A.F., WONG, J.H.H., BENEDICT, C.R.** : Development of ribulose 1,5- -diphosphate carboxylase in nonphotosynthetic endosperms of germinating castor beans. - Plant Physiol. 57 : 589 - 593, 1976.

26143 - **GILLER, Yu.E., SHCHERBAKOVA, I.Yu.** : O deĭstvii spetsificheskikh ingibitorov transkriptsii i translatsii na obrazovanie nativnykh form khlorofilla. [Action of specific inhibitors of transcription and translation on the formation of native forms of chlorophyll.] - Dokl. Akad. Nauk tadzh. SSR 19 (10) : 45 - - 48, 1976. [In R, ab : Tajik.]

26144 - **GILLHAM, N.W., BOYNTON, J.E., HARRIS, E.H., FOX, S.B., BOLEN, P.L.** : Genetic control of chloroplast ribosome biogenesis in *Chlamydomonas.* - In : BUCHER, T., NEUPERT, W., SEBALD, W., WERNER, S. (ed.) : Genetics and Biogenesis of Chloroplasts and Mitochondria. Pp. 69 - 76. North-Holland Publ. Co., Amsterdam - New York - Oxford 1976.

26145 - **GIL'MANOV, T.G.** : Submodel' rastitel'nosti v tselostnoĭ modeli travyanoĭ ėkosistemy (v osobennosti ee podzemnoĭ chasti). [Plant submodel in the integral model of a grassland ecosystem with special reference to its subsurface part.] - Bot. Zh. 61 : 1185 - 1197, 1976. [Biomass; in R, ab : E.]

26146 - **GIMÉNEZ-GALLEGO, G., del VALLE-TASCÓN, S., RAMÍREZ, J.M.** : A possible physiological function of the oxygen-photoreducing system of *Rhodospirillum rubrum.* - Arch. Microbiol. 109 : 119 - 125, 1976.

26147 - **GIMMLER, H.** : Different uptake and binding of benzoquinone and hydroquinone in unicellular algae and type A chloroplasts. - Z. Pflanzenphysiol. 78 : 76 - - 81, 1976.

26148 - **GINZBERG, D., PADAN, E., SHILO, M.** : Metabolic aspects of LPP cyanophage replication in the cyanobacterium *Plectonema boryanum.* - Biochim. biophys. Acta 423 : 440 - 449, 1976. [Ps.]

26149 - **GIRAULT, G., GALMICHE, J.M.** : Nucleotides effect on the decay kinetics of the 520 nm absorbance change in tightly coupled chloroplasts. - Biochem. biophys. Res. Commun. 68 : 724 - 729, 1976.

*26150 - **GIRS, G.I., KAVERZINA, L.N.** : Izmenchivost' uglevodnogo obmena khvoi listvennitsy sibirskoĭ v svyazi s vysotnoĭ zonal'nost'yu lesov. [Heterogeneity of carbohydrate metabolism of *Larix sibirica* needles in relation to altitudinal distribution of forests.] - Ėkologiya 1975 (4) : 42 - 46, 1975. [Photosynthates; in R.]

26151 - **GITELSON, I.I., TERSKOV, I.A., KOVROV, B.G., SIDKO, F.Ya., LISOVSKY, G.M., OKLADNIKOV, Yu.N., BELYANIN, V.N., TRUBACHOV, I.N., RERBERG, M.S.** : Life support system with autonomous control employing plant photosynthesis. - Acta astronaut. 3 : 633 - 650, 1976.

26152 - **GIVAN, A.C.V.** : Glutamine synthesis and its relation to photophosphorylation in *Pisum* chloroplasts. Effects of 3-(3,4-dichlorophenyl)-1,1-dimethylurea and antimycin A. - Plant Physiol. 57 : 623 - 627, 1976.

26153 - **GIVAN, C.V., HARWOOD, J.L.** : Biosynthesis of small molecules in chloroplasts of higher plants. - Biol. Rev. Cambridge phil. Soc. 51 : 365 - 406, 1976.

26154 - **GIVNISH, T.J., VERMEIJ, G.J.** : Sizes and shapes of liane leaves. - Amer. Natur. 110 : 743 - 778, 1976. [Ps.]

26155 - **GLAGOLEVA, T.A., MOKRONOSOV, A.T., ZALENSKII, O.V.** : Vliyanie kisloroda na fotosintez i metabolizm u rasteniĭ yugo-vostochnykh Karakumov. [Effect of oxygen on photosynthesis and metabolism in plants of southeastern Karakum.] - In : Gazometricheskoe Issledovanie Fotosinteza i Dykhaniya Rasteniĭ. Pp. 29 - 32. Akad. Nauk SSSR, Tartu 1976. [In R.]

26156 - **GLÄSER, M., WOLFF, C., RENGER, G.** : Indirect evidence for a very fast recovery kinetics of chlorophyll-a_{II} in spinach chloroplasts. - Z. Naturforsch. *31C* : 712 - 721, 1976.

26157 - **GLASS, R.W., SIMPSON, K.L.** : The isolation of γ-carotene and a poly-*cis*-γ-carotene from the tangerine tomato. - Phytochemistry *15* : 1077 - 1078, 1976.

26158 - **GLAZER, A.N.** : Phycocyanins : structure and function. - In : SMITH, K.C. (ed.) : Photochemical and Photobiological Reviews. Vol. 1. Pp. 71 - 115. Plenum Press, New York - London 1976.

26159 - **GLAZER, A.N., APELL, G.S., HIXSON, C.S., BRYANT, D.A., RIMON, S., BROWN, D. M.** : Biliproteins of cyanobacteria and *Rhodophyta* : Homologous family of photosynthetic accessory pigments. - Proc. nat. Acad. Sci. USA *73* : 428 - 431, 1976.

*26160 - **GLAZER, A.N., BRYANT, D.A.** : Allophycocyanin B (λ_{max} 671, 618 nm). A new cyanobacterial phycobiliprotein. - Arch. Microbiol. *104* : 15 - 22, 1975.

26161 - **GLERUM, C.** : Frost hardiness in forest trees. - In : CANNELL, M.G.R., LAST, F.T. (ed.) : Tree Physiology and Yield Improvement. Pp. 403 - 420. Academic Press, London - New York - San Francisco 1976. [Photosynthates.]

26162 - **GLIDEWELL, S.M., RAVEN, J.A.** : Photorespiration : ribulose diphosphate oxygenase of hydrogen peroxide ? - J. exp. Bot. *27* : 200 - 204, 1976.

26163 - **GLIWICZ, Z.M.** : Stratification of kinetic origin and its biological consequences in a neotropical man-made lake. - Ekol. pol. *24* : 197 - 210, 1976. [Ps, Chl.]

26164 - **GLOSER, J.** : Photosynthesis and respiration of some alluvial meadow grasses : Responses to irradiance, temperature and CO_2 concentration. - Přírodověd. Práce Ústavu českoslov. Akad. Věd Brně *10* (2) : 1 - 39, 1976.

26165 - **GNANAM, A., FRANCIS, K.** : Distribution of photosynthetic enzymes in the leaf cell types of *Sorghum Vulgare*. - Plant biochem. J. *3* : 11 - 23, 1976.

26166 - **GNAUCK, A., WEISE, G.** : Zur Stabilität der Stoffproduktion submerser Makrophyten bei Nitratbelastung gemessen am CO_2-Umsatz. - Wiss. Z. tech. Univ. Dresden *25* : 250 - 252, 1976. [Ps.]

26167 - **GÖBEL, E., RIESSNER, R., POHL, P.** : Einfluß von DCMU auf die Bildung von Lipiden und Fettsäuren und auf die Ultrastruktur von *Euglena gracilis*. - Z. Naturforsch. *31C* : 687 - 692, 1976.

26168 - **GOEDHEER, J.C.** : Spectral properties of the blue-green alga *Anacystis nidulans* grown under different environmental conditions. - Photosynthetica *10* : 411 - - 422, 1976.

26169 - **GOESCHL, J.D., NEWTON, R.J., SHIVE, J.** : Theoretical and physiological significance of unloading mechanisms and sieve tube diameter in phloem translocation. - Plant Physiol. *57* (Suppl.) : 77, 1976.

26170 - **GOFF, L.J.** : The biology of *Harveyella mirabilis (Cryptonemiales; Rhodophyceae)*. V. Host responses to parasite infection. - J. Phycol. *12* : 313 - 328, 1976. [Ps.]

26171 - **GOGOTOV, I.N.** : Gidrogenaznaya aktivnost' i metabolizm vodoroda u fotosinteziruyushchikh mikroorganizmov. [Hydrogenase activity and hydrogen metabolism in photosynthesizing bacteria.] - In : Itogi Issledovaniya Mekhanizma Fotosinteza. Pp. 51 - 54. Pushchino 1976. [In R.]

26172 - **GOLBECK, J.H., LIEN, S., SAN PIETRO, A.** : The relationship between acid-labile sulfide (ALS) content and *P*700 photochemistry in spinach photosystem I particles. - Plant Physiol. *57* (Suppl.) : 95, 1976.

26173 - **GOLBECK, J.H., LIEN, S., SAN PIETRO, A.** : Quantitation of labile sulfide content and *P*700 photochemistry in spinach photosystem 1 particles. - Biochem. biophys. Res. Commun. *71* : 452 - 458, 1976.

26174 - **GOLBECK, J.H., SAN PIETRO, A.** : Determination of acid-labile sulfide in subchloroplast particles containing Triton X-100. - Anal. Biochem. *73* : 539 - - 542, 1976.

26175 - **GOL'D, V.M., GAEVSKIĬ, N.A., GRIGOR'EV, Yu.S.** : Izuchenie sootnosheniya bystroĭ i medlennoĭ komponent tusheniya fluorestsentsii khlorofilla a pri netsiklicheskom transporte ėlektronov. [Relation between fast and slow components of quenching chlorophyll a fluorescence at non-cyclic electron transport.] - Studia biophys. *54* : 139 - 146, 1976. [In R, ab : E.]

26176 - **GOL'DFEL'D, M.G., TSAPIN, A.I., SHUTILOVA, N.I., KHANGULOV, S.V.** : Ob opticheskom pogloshchenii pri 700 nm i spektrakh ĖPR khloroplastov i subkhloroplastnykh chastits. [Optical absorption and ESR spectra of chloroplasts and subchloroplast particles.] - Biofizika *21* : 183 - 185, 1976. [In R, ab : E.]

26177 - **GOLDHABER, A.S., NIETO, M.M.** : The mass of the photon. - Sci. Amer. *234* : 86 - 96, 1976.

26178 - **GOLDKORN, T., SCHEJTER, A.** : The redox potential of cytochrome c-552 from *Euglena gracilis* : A thermodynamic study. - Arch. Biochem. Biophys. *177* : 39 - 45, 1976.

*26179 - **GOLDMAN, C.R., de AMEZAGA, E.** : Primary productivity in the littoral zone of Lake Tahoe, California-Nevada. - In : Limnology of Shallow Waters. Pp. 49 - - 62. Akad. Kiadó, Budapest 1975.

26180 - **GOLDSCHMIDT, C.R., OTTOLENGHI, M., KORENSTEIN, R.** : On the primary quantum yields in the bacteriorhodopsin photocycle. - Biophys. J. *16* : 839 - 843, 1976.

26181 - **GOLDSTEIN, L.D., RAY, T.B., KESTLER, D.P., MAYNE, B.C., BROWN, R.H., BLACK, C.C.** : Biochemical characterization of *Panicum* species which are intermediate between C_3 and C_4 photosynthesis plants. - Plant Sci. Lett. *6* : 85 - 90, 1976.

26182 - **GONCHARIK, M.N., KRUCHININA, S.S.** : Svetovoĭ rezhim kartofelya v shirokoryadnykh posadkakh. [Utilization of light in widely-spaced potato plantings.] - Sel'skokhoz. Biol. *11* : 34 - 37, 1976. [Ps; in R, ab : E.]

26183 - **GONCHAROVA, Ė.A.** : Deĭstvie zasukhi na metabolicheskie funktsii sochnykh plodov v periody rosta i sozrevaniya. [Action of drought on metabolic functions insappy fruits in the course of their growth and maturing.] - Tr. prikl. Bot. Genet. Selektsii *57* (2) : 68 - 76, 1976. [Chl; in R, ab : E.]

26184 - **GOOD, R.E., GOOD, N.F.** : Growth analysis of pitch pine seedlings under three temperature regimes. - Forest Sci. *22* : 445 - 448, 1976.

B26185 - **GOODWIN, T.W. (ed.)** : Chemistry and Biochemistry of Plant Pigments. 2nd Ed. Vol. 1, 2. - Academic Press, London - New York - San Francisco 1976.

26186 - **GOODWIN, T.W.** : Distribution of carotenoids. - In : **GOODWIN, T.W. (ed.)** : Chemistry and Biochemistry of Plant Pigments. 2nd Ed. Vol. 1. Pp. 225 - 261. Academic Press, London - New York - San Francisco 1976.

26187 - **GOPALAM, A., GOPALACHARI, N.C.** : Studies in leaf pigments of flue-cured tobacco. Pt. I - Their variation in different varieties and stalk positions. - Tobacco Res. *2* : 21 - 30, 1976.

26188 - **GORDON, J.C., PROMNITZ, L.C.** : Photosynthetic and enzymatic criteria for the early selection of fast-growing *Populus* clones. - In : **CANNELL, M.R.G., LAST, F.T. (ed.)** : Tree Physiology and Yield Improvement. Pp. 79 - 97. Academic Press, London - New York - San Francisco 1976.

26189 - **GORYSHINA, T.K., NESHATAEV, Yu.N., RASTVOROVA, O.G.** : Sezonnaya dinamika fitomassy travyanogo pokrova lesostepnoĭ dubravy v svyazi s dinamikoĭ fotosinteza i ėkologicheskikh faktorov. [Seasonal dynamics of biomass of grass cover in a forest-steppe oak stand as related to dynamics of photosynthesis and ecological factors.] - Tr. petergof. biol. Inst. *24* : 7 - 28, 1976. [In R.]

26190 - **GOSSETT, D.R., EGLI, D.B., LEGGITT, J.E.** : The effect of calcium deficiency on the translocation of photosynthetically fixed ^{14}C in intact soybeans. - Plant Physiol. *57* (Suppl.) : 83, 1976.

26191 - **GOUGH, S.P., BEALE, S.I., GRANICK, S.** : The biosynthesis of δ-aminolevulinic acid from the intact carbon skeleton of glutamic acid in greening barley. - Ann. clin. Res. *8* (Suppl. 17) : 70 - 73, 1976.

26192 - **GOUGH, S.P., KANNANGARA, C.G.** : Synthesis of δ-aminolevulinic acid by isolated plastids. - Carlsberg Res. Commun. *41* : 183 - 190, 1976.

26193 - **GOULD, J.M.** : Inhibition by triphenyltin chloride of a tightly-bound membrane component involved in photophosphorylation. - Europe. J. Biochem. *62* : 567 - 575, 1976.

26194 - **GOULD, J.M.** : The relationship between proton fluxes and the regulation of electron transport in chloroplasts. - FEBS Lett. *66* : 312 - 316, 1976.

26195 - **GOULDEN, P.D.** : Automated determination of carbon in natural waters. - Water Res. *10* : 487 - 490, 1976.

26196 - **GOVINDJEE, PULLES, M.P.J., GOVINDJEE, R., van GORKOM, H.J., DUYSENS, L.N.M.** : Inhibition of the reoxidation of the secondary electron acceptor of Photosystem II by bicarbonate depletion. - Biochim. biophys. Acta *449* : 602 - 605, 1976.

26197 - **GOWIN, T., UBYSZ, L.** : A descriptive model o̲ Scots pine (*Pinus silvestris* L. seedling's growth during the first growing season. - Acta Soc. Bot. Pol. *45* : 143 - 158, 1976.

26198 - **GRÄBER, P., WITT, H.T.** : Relations between the electrical potential, pH gradient, proton flux and phosphorylation in the photosynthetic membrane. - Biochim. biophys. Acta *423* : 141 - 163, 1976.

26199 - **GRACE, J.** : Wind damage to vegetation. - In : SMITH, H. (ed.) : Commentaries in Plant Science. Pp. 209 - 220. Pergamon Press, Oxford - New York - Toronto - Sydney - Paris - Frankfurt 1976. [Ps.]

26200 - **GRACE, J., WILSON, J.** : The boundary layer over a *Populus* leaf. - J. exp. Bot. *27* : 231 - 241, 1976.

*26201 - **GRACZA, P., FRIDVALSZKY, L., SÁRKÁNY, S., DÖMÖTÖR, Z.** : Functional organization of cotyledon plasts in the course of embryogenesis in *Pisum sativum* L. - Acta agron. Acad. Sci. hung. *22* : 429 - 433, 1973.

26202 - **GRAFIUS, J.E., BARNARD, J.** : Leaf canopy as related to yield in barley. - Agron. J. *68* : 398 - 402, 1976.

26203 - **GRAHAM, D., SMILLIE, R.M.** : Carbonate dehydratase in marine organisms of the Great Barrier Reef. - Aust. J. Plant Physiol. *3* : 113 - 119, 1976.

26204 - **GRAHAM, R.D.** : Anomalous water relations in copper-deficient wheat plants. - Aust. J. Plant Physiol. *3* : 229 - 236, 1976. [Growth analysis.]

*26205 - **GRANICK, S., SASSA, S.** : δ-aminolevulinic acid synthetase and the control of heme and chlorophyll synthesis. - In : GREENBERG, D.M. (ed.) : Metabolic Pathways. Vol. 5. Pp. 77 - 141. Academic Press, New York - London 1971.

26206 - **GRANT, B.R., HOWARD, R.J., GAYLER, K.R.** : Isolation and properties of chloroplasts from the siphonous green alga *Caulerpa simpliciuscula*. - Aust. J. Plant Physiol. *3* : 639 - 651, 1976.

26207 - **GRAY, B.H., COSNER, J., GANTT, E.** : Phycocyanins with absorption maxima at 637 nm and 623 nm from *Agmanellum quadruplicatum*. - Photochem. Photobiol. *24* : 299 - 302, 1976.

26208 - **GRAY, J.C., KUNG, S.D., WILDMAN, S.G.** : Polypeptide chains of the large and small subunits of Fraction I protein. - In : BÜCHER, T., NEUPERT, W., SEBALD, W., WERNER, S. (ed.) : Genetics and Biogenesis of Chloroplasts and Mitochondria. Pp. 13 - 16. North-Holland Publ. Co., Amsterdam - New York - Oxford 1976.

26209 - **GRAY, J.C., WILDMAN, S.G.** : A specific immunoabsorbent for the isolation of Fraction I protein. - Plant Sci. Lett. *6* : 91 - 96, 1976.

26210 - **GREBANIER, A.E., JAGENDORF, A.T.** : Sulfate inhibition of photophosphorylation by isolated spinach chloroplasts. - Plant Physiol. *57* (Suppl.) : 24, 1976.

26211 - **GREENBAUM, E., MAUZERALL, D.C.** : Oxygen yield per flash of *Chlorella* coupled to chemical oxidants under anaerobic conditions. - Photochem. Photobiol. *23* : 369 - 372, 1976.

26212 - **GREENE, J.C., MILLER, W.E., SHIROYAMA, T., SOLTERO, R.A., PUTMAN, K.** : Use of algal assays to assess the effects of municipal and smelter wastes upon phytoplankton production. - In : Proceedings of the Symposium on Terrestrial and Aquatic Ecological Studies of the Northwest. Pp. 327 - 336. EWSC Press, East. Washington State Coll., Cheney, Washington 1976. [Chl.]

26213 - **GREENWOOD, E.A.N., CARBON, B.A., ROSSITER, R.C., BERESFORD, J.D.** : The response of defoliated swards of subterranean clover to temperature. - Aust. J. agr. Res. *27* : 593 - 610, 1976. [Growth analysis.]

26214 - **GRIFFITHS, D.J.** : The photosynthetic capacity of the phytoplankton in the waters of a coral reef. - Aust. J. Plant Physiol. *3* : 53 - 56, 1976.

26215 - **GRIFFITHS, W.T., MORGAN, N.L., MAPLESTON, R.E.** : Chlorophyll synthesis and the development of photosynthetic activity. - In : BUCHER, T., NEUPERT, W., SEBALD, W., WERNER, S. (ed.) : Genetics and Biogenesis of Chloroplasts and Mitochondria. Pp. 111 - 118. North-Holland Publ. Co., Amsterdam - New York - Oxford 1976.

26216 - **GRIME, J.P., CURTIS, A.V.** : The interaction of drought and mineral nutrient stress in calcareous grassland. - J. Ecol. *64* : 976 - 998, 1976. [Growth analysis.]

26217 - **GROBBELAAR, J.U., STEGMANN, P.** : Biological assessment of the euphotic zone in a turbid man-made lake. - Hydrobiologia *48* : 262 - 266, 1976. [Ps productivity.]

26218 - **GRODZINSKI, B., BUTT, V.S.** : Hydrogen peroxide production and the release of carbon dioxide during glycollate oxidation in leaf peroxisomes. - Planta *128*: 225 - 231, 1976.

26219 - **GRODZINSKI, B., COLMAN, B.** : Intracellular localization of glycolate dehydrogenase in a blue-green alga. - Plant Physiol. *58* : 199 - 202, 1976.

B26220 - **GRODZINSKII, D.M.** : Plant Biophysics. - Israel Program for Scientific Translations. Jerusalem 1976. [Ps.]

26221 - **GRODZINSKII, D.M., ROZHKO, I.I., AVAKYAN, L.M., VESELOV, A.P.** : Kompartmentatsiya sakharov v fotosinteziruyushchikh list'yakh fasoli. [Compartmentation of sugars in photosynthesizing leaves of bean.] - Fiziol. Biokhim. kul't. Rast. *8* : 381 - 383, 1976. [In R, ab : E.]

26222 - **GROMET-ELHANAN, Z.** : Naphthylmercaptobenzoquinone, a new inhibitor of photophosphorylation in *Rhodospirillum rubrum* chromatophores at the level of ubiquinone. - Biochem. biophys. Res. Commun. *73* : 13 - 18, 1976.

26223 - **GROSS, E.L., DAVIS, D.J.** : Biochemical model of spillover. - Biophys. J. *16* (2, Part 2): 158 a, 1976. [Ps.]

26224 - **GROSS, E.L., ZIMMERMANN, R.J., HORMATS, G.F.** : The effect of mono- and divalent cations on the quantum yields for electron transport in chloroplasts. - Biochim. biophys. Acta *440* : 59 - 67, 1976.

26225 - **GROSS, J., CARMON, M., LIFSHITZ, A., COSTES, C.** : Carotenoids of banana pulp, peel and leaves. - Lebensm.-Wiss. Technol. *9* : 211 - 214, 1976.

26226 - **GROSS, J., COSTES, C.** : Nicotine effect on carotenogenesis in α-carotene containing banana leaves. - Physiol. vég. *14* : 427 - 435, 1976.

26227 - **GROSS, K.** : Die Abhängigkeit des Gaswechsels junger Fichtenpflanzen vom Wasserpotential des Wurzelmediums und von der Luftfeuchtigkeit bei unterschiedlichen CO_2-Gehalten der Luft. - Forstwiss. Centralbl. *95* : 211 - 225, 1976.

26228 - **GRUENSTEIN, E.I., POLLARD, A.L.** : Double-label autoradiography on polyacrylamide gels with ^3H and ^{14}C. - Anal. Biochem. *76* : 452 - 457, 1976.

26229 - **GUDIN, C., SYRATT, W.J., BOIZE, L.** : The mechanisms of photosynthetic inhibition and the development of scorch in tomato plants treated with spray oils. - Ann. appl. Biol. *84* : 213 - 219, 1976.

26230 - **GUIKEMA, J.A., YOCUM, C.F.** : Photophosphorylation by photosystem II-enriched subchloroplast particles. - Plant Physiol. *57* (Suppl.) : 23, 1976.

26231 - **GUIKEMA, J.A., YOCUM, C.F.** : The mechanism of quinonediimine acceptor activity in photosynthetic electron transport. - Biochemistry *15* : 362 - 367, 1976.

26232 - **GUKASYAN, L.A., BAGDASARYAN, E.G., TUMANYAN, É.R.** : Soderzhanie vitamina *C* i karotina u indutsirovannykh form pertsa. [Contents of vitamin *C* and carotene in induced forms of the pepper (*Capsicum annuum* L.).] - Biol. Zh. Arm. *29* (10) : 77 - 81, 1976. [In R, ab : Arm.]

26233 - **GULIEV, N.M., FEDENKO, E.P., DOMAN, N.G.** : Fosfodiésteraza 3',5'-AMP fototrofnoĭ bakterii *Rhodospirillum rubrum*. [cAMP phosphodiesterase from phototrophic bacteria *Rhodospirillum rubrum*.] - Biokhimiya *41* : 2043 - 2046, 1976. [In R, ab : E.]

26234 - **GULLER, L.** : Influence of powdery mildew (*Erysiphe graminis* f. sp. *hordei* MARCHAL) upon the submicroscopic structure of the chloroplasts of spring barley (*Hordeum sativum* L.). - Acta Fac. Rerum nat. Univ. Comeniae, Physiol. Plant. *11* : 27 - 32, 1976.

26235 - **GULYAEV, B.I.** : Vliyanie ust'its na skorost' diffuzii CO_2 pri fotosinteze. [Effect of stomata on the CO_2 diffusion rate in photosynthesis.] - Dokl. Akad. Nauk ukr. SSR, Ser. B *1976* : 53 - 56, 1976. [In R, ab : E.]

26236 - **GULYAEV, B.I.** : Vpliv prodikhiv na shvidkist' difuziĭ CO_2 v protsesi fotosintezu. [Effect of stomata on carbon dioxide diffusion rate in photosynthesis.] - Dopovidi Akad. Nauk ukr. SSR, Ser. B *1976* : 52 - 55, 1976. [In Ukr., ab : E.]

26237 - **GULYAEV, B.I.** : Vliyanie abstsizovoĭ kisloty na ust'ichnye dvizheniya i fotosintez. [Effect of abscisic acid on stomatal movements and photosynthesis.] - Dokl. Akad. Nauk ukr. SSR, Ser. B *1976* : 1022 - 1026, 1976. [In R, ab : E.]

26238 - **GULYAEV, B.I.** : Kinetika gazoobmena list'ev i voprosy mekhanizma ust'ichnykh dvizheniĭ. [Kinetics of leaf gas exchange and mechanism of stomatal movements.] - In : Gazometricheskoe Issledovanie Fotosinteza i Dykhaniya Rasteniĭ. Pp. 36 - 38. Akad. Nauk SSSR, Tartu 1976. [In R.]

26239 - **GULYAEV, B.I.** : Otsenka vliyaniya ust'its na diffuziyu CO_2. [Effect of stomata on CO_2 diffusion.] - In : Gazometricheskoe Issledovanie Fotosinteza i Dykhaniya Rasteniĭ. Pp. 39 - 40. Akad. Nauk SSSR, Tartu 1976. [In R.]

26240 - **GULYAEV, B.I., CHERNYSHENKO, T.I.** : Temperaturnye zavisimosti gazoobmena i fotosinteza lista. [Temperature dependences of leaf gas exchange and photosynthesis.]-In : Gazometricheskoe Issledovanie Fotosinteza i Dykhaniya Rasteniĭ. Pp. 41 - 42. Akad. Nauk SSSR, Tartu 1976. [In R.]

26241 - **GULYAEV, B.I., TKACHUK, K.S.** : Diya i pislyadiya gruntovoĭ posukhi na fotosintez i difuziĭni opori listkiv ozimoĭ pshenitsi. [Effect and after-effect of soil drought on photosynthesis and diffusion resistance of winter wheat leaves.] - Dopovidi Akad. Nauk ukr. RSR, Ser. B *1976* : 1111 - 1114, 1976. [In Ukr., ab : E.]

26242 - **GUNDERSEN, K.R., CORBIN, J.S., HANSON, C.L., HANSON, M.L., HANSON, R.B., RUSSELL, D.J., STOLLAR, A., YAMADA, O.** : Structure and biological dynamics of the oligotrophic ocean photic zone off the Hawaiian Islands. - Pacific Sci. *30* : 45 - 68, 1976. [Chl.]

26243 - **GUNNING, B.E.S.** : The role of plasmodesmata in short distance transport to and from the phloem. - In : **GUNNING, B.E.S., ROBARDS, A.W.** (ed.) : Intercellular Communications in Plants : Studies on Plasmodesmata. Pp. 203 - 227. Springer-Verlag, Berlin - Heidelberg - New York 1976. [Photosynthates.]

26244 - **GUPTA, R.K.** : Soluble constituents & photosynthetic products biosynthesized from $^{14}CO_2$ in leafy liverworts *Plagiochila asplenioides* (L.) DUM. & *Scapania undulata* (L.) DUM. - Indian J. exp. Biol. *14* : 595 - 598, 1976.

26245 - **GUPTA, S.K., DASS, A.D.** : Determination of leaf area in jute. - Indian J. Agron. *21* : 160, 1976.

26246 - **GUR, A., BRAVDO, B., HEPNER, J.** : The influence of root temperature on apple trees. III. The effect on photosynthesis and water balance. - J. hort. Sci. *51* : 203 - 210, 1976.

26247 - GÜRGÜN, V., KIRCHNER, G., PFENNIG, N. : Vergärung von Pyruvat durch sieben
Arten phototropher Purpurbakterien. - Z. allgem. Mikrobiol. *18* : 573 - 586,
1976. [Chl.]

26248 - GURNE, D., SHEMIN, D. : The synthesis of porphobilinogen by immobilized δ-ami-
nolevulinic acid dehydratase. - In : COLOWICK, S.P., KAPLAN, N.O. (ed.) :
Methods in Enzymology. Vol. 44. Pp. 844 - 849. Academic Press, New York - San
Francisco - London, 1976.

26249 - GUTIERREZ, M., EDWARDS, G.E., BROWN, W.V. : PEP carboxykinase containing species in the *Brachiaria* group of the subfamily *Panicoideae*. - Biochem. Syst.
Ecol. *4* : 47 - 49, 1976.

26250 - GYSI, J., ZUBER, H. : Allophycocyanin I - a second cyanobacterial allophycocyanin ? Isolation, characterization and comparison with allophycocyanin II
from the same alga. - FEBS Lett. *68* : 49 - 54, 1976.

26251 - HAALAND, E. : The effect of light and CO_2 on the carbohydrates in stock plants
and cuttings of *Campanula isophylla* MORETTI. - Sci. Hort. *5* : 353 - 361, 1976.

26252 - HACHTEL, W. : *In vitro* synthesis of membrane proteins by isolated chloroplasts
of *Vicia faba*. - Ber. deut. bot. Ges. *89* : 185 - 192, 1976.

26253 - HACHTEL, W. : Sites of synthesis of chloroplast membrane proteins in *Vicia
faba*. - In : BÜCHER, T., NEUPERT, W., SEBALD, W., WERNER, S. (ed.) : Genetics
and Biogenesis of Chloroplasts and Mitochondria. Pp. 49 - 52. North-Holland
Publ. Co., Amsterdam - New York - Oxford 1976.

26254 - HÄDER, D.-P. : Further evidence for the electron pool hypothesis. The effect
of KCN and DSPD on the photophobic reaction in the filamentous blue-green alga
Phormidium uncinatum. - Arch. Microbiol. *110* : 301 - 303, 1976.

26255 - HÄDER, D.-P. : Phobic reactions between two adjacent monochromatic light
fields. - Z. Pflanzenphysiol. *78* : 173 - 176, 1976. [Ps.]

26256 - HADJICHRISTODOULOU, A. : Effect of harvesting stage on cereal and legume forage production in low rainfall regions. - J. agr. Sci. *86* : 155 - 161, 1976.
[Seasonal course of productivity.]

26257 - HAEHNEL, W. : The ratio of the two light reactions and their coupling in
chloroplasts. - Biochim. biophys. Acta *423* : 499 - 509, 1976.

26258 - HAEHNEL, W. : The reduction kinetics of chlorophyll a_1 as an indicator for
proton uptake between the light reactions in chloroplasts. - Biochim. biophys.
Acta *440* : 506 - 521, 1976.

26259 - HAFNER, L. : Die "Chemische" und die "Chemiosmotische" Hypothese der Photo-
phosphorylierung. - Zum derzeitigen Stand der Diskussion. - Verhandl. bot.
Vereins Provinz Brandenburg *112* : 49 - 66, 1976.

26260 - HAGEMANN, R. : Plastid distribution and plastid competition in higher plants
and the induction of plastom mutations by nitroso-urea-compounds. - In :
BÜCHER, T., NEUPERT, W., SEBALD, W., WERNER, S. (ed.) : Genetics and Biogenesis of Chloroplasts and Mitochondria. Pp. 331 - 338. North-Holland Publ. Co.,
Amsterdam - New York - Oxford 1976.

26261 - HAINES, E.B. : Stable carbon isotope ratios in the biota, soils and tidal water of a Georgia salt marsh. - Estuarine coastal mar. Sci. *4* : 609 - 616,
1976.

26262 - HALES, B.J. : Semiquinones in photosynthetic systems. - Biophys. J. *16* (2,
Part 2) : 222 a, 1976.

26263 - HALEVY, J. : Growth rate and nutrient uptake of two cotton cultivars grown
under irrigation. - Agron. J. *68* : 701 - 705, 1976. [Growth analysis.]

26264 - HALL, A.E., HOFFMAN, G.J. : Leaf conductance response to humidity and water
transport in plants. - Agron. J. *68* : 876 - 881, 1976.

26265 - HALL, A.E., SCHULZE, E.-D., LANGE, O.L. : Current perspectives of steady-state
stomatal responses to environment. - In : LANGE, O.L., KAPPEN, L., SCHULZE,
E.-D. (ed.) : Water and Plant Life. Pp. 169 - 188. Springer-Verlag, Berlin -
Heidelberg - New York 1976. [Ps.]

26266 - **HALL, D.O.** : Electron transfer systems in microorganisms. - Nature *264* : 317-
- 318, 1976. [Ps.]

26267 - **HALL, D.O.** : Photobiological energy conversion. - FEBS Lett. *64* : 6 - 16,
1976.

26268 - **HALL, D.O.** : The coupling of photophosphorylation to electron transport in
isolated chloroplasts. - In : **BARBER, J.** (ed.) : The intact Chloroplast. Pp.
135 - 170. Elsevier, Amsterdam - New York - Oxford 1976.

26269 - **HALL, F.R., FERREE, D.C.** : Effects of insect injury simulation on photosynthesis of apple leaves. - J. econ. Entomol. *69* : 245 - 248, 1976.

26270 - **HALLEGRAEFF, G.M.** : Pigment diversity in freshwater phytoplankton. I. A comparison of spectrophotometric and paper chromatographic methods. - Int. Rev.
ges. Hydrobiol. *61* : 149 - 168, 1976.

26271 - **HALLEGRAEFF, G.M.** : Pigment diversity in fresh-water phytoplankton. III. Summer phytoplankton of eight lakes with widely different trophic characteristics. - Hydrobiol. Bull. *10* (2) : 87 - 95, 1976.

26272 - **HALLIWELL, B.** : Photorespiration. - FEBS Lett. *64* : 266 - 270, 1976.

26273 - **HAMEEDI, M.J.** : An evaluation of the effects of environmental variables on
marine plankton primary productivity by multivariate regression. - Int. Rev.
ges. Hydrobiol. *61* : 529 - 550, 1976.

26274 - **HAMMERTON, J.L.** : Effects of planting date on growth and yield of pigeon pea
(*Cajanus cajan* (L.) MILLSP.). - J. agr. Sci. *87* : 649 - 660, 1976.

26275 - **HAMPP, R., BEULICH, K., ZIEGLER, H.** : Effects of zinc and cadmium on photosynthetic CO_2-fixation and Hill activity of isolated spinach chloroplasts. -
Z. Pflanzenphysiol. *77* : 336 - 344, 1976.

26276 - **HAMPP, R., SCHMIDT, H.W.** : Changes in envelope permeability during chloroplast development. - Planta *129* : 69 - 73, 1976.

26277 - **HAMPP, R., WELLBURN, A.R.** : Changes in the permeability of the inner mitochondrial membrane associated with plastid development. - Planta *131* : 21 -
- 26, 1976.

26278 - **HAMPP, R., WELLBURN, A.R.** : Chlorophyllbiosynthese : Austausch möglicher Substrate zwischen Mitochondrien und Plastiden im Verlauf der Plastiden-Differen-
zierung. - Ber. deut. bot. Ges. *89* : 175 - 183, 1976.

26279 - **HAMPP, R., WELLBURN, A.R.** : Early changes in the envelope permeability of developing chloroplasts. - J. exp. Bot. *27* : 778 - 784, 1976.

26280 - **HAMPP, R., WELLBURN, A.R.** : Intraplastidic bicarbonate concentration and fixation during plastid development. - Z. Pflanzenphysiol. *79* : 246 - 253,
1976.

26281 - **HANKER, I., KŮDELOVÁ, A.** : Vliv pirimicarbu, dimethoatu a benomylu na fotosyn-
tézu a distribuci ^{14}C u řepy cukrovky. [Effect of pirimicarb, dimethoate and
benomyl on photosynthesis and ^{14}C distribution in sugar beet.] - Ochrana
Rostlin (Praha) *12* : 109 - 114, 1976. [In Czech, ab : E, G, R.]

26282 - **HANSON, Z. III, JOHNSON, H.B., TING, I.P.** : Simultaneous measurements of photosynthesis and transpiration in the field with a dual-label porometer. -
Plant Physiol. *57* (Suppl.) : 106, 1976.

26283 - **HAPPE, M., OVERATH, P.** : Bacteriorhodopsin depleted of purple membrane lipids.
- Biochem. biophys. Res. Commun. *72* : 1504 - 1511, 1976.

26284 - **HARAKI, T., NISHIKAWA, H.** : [Growth responses of rice seedlings to light wavelengths I. Growth of seedlings exposed to various light wavelengths and variations caused by different growing conditions.] - Proc. Crop Sci. Soc. Jap.
45 : 409 - 415, 1976. [Ps, Chl; in Jap., ab : E.]

26285 - **HARAKI, T., NISHIKAWA, H.** : [Growth responses of rice seedlings to light wavelengths II. Effect of light wavelength on chlorosis of seedlings.] - Proc.
Crop Sci. Soc. Jap. *45* : 416 - 421, 1976. [In Jap., ab : E.]

26286 - **HARAKI, T., NISHIKAWA, H.** : [Growth responses of rice seedlings to various
wavelengths III. Effects of various light treatments in the nursery stage on
the growth during the early stage after transplanting.] - Proc. Crop Sci. Soc.
Jap. *45* : 422 - 428, 1976. [Chl; in Jap., ab : E.]

26287 - **HARDING, L.W., Jr.** : Polychlorinated biphenyl inhibition of marine phytoplankton photosynthesis in the northern Adriatic sea. - Bull. environ. Contam. To-
xicol. *16* : 559 - 566, 1976.

26288 - **HARDT, H., KOK, B.** : Stabilization by glutaraldehyde of high-rate electron
transport in isolated chloroplasts. - Biochim. biophys. Acta *449* : 125 - 135,
1976.

26289 - **HARE, J.W., SOLOMON, E.I., GRAY, H.B.** : Infrared spectral studies of metal
binding effects on the secondary structure of bean plastocyanin. - J. amer.
chem. Soc. *98* : 3205 - 3209, 1976.

*26290 - **HARKAYNÉ VINKLER, M.** : A fűszerpaprika színezéktartalmának meghatározására
szolgáló Benedek-féle és a vékonyrétegkromatográfiás módszer összehasonlító
vizsgálata. [Comparative investigation of the Benedek method and the thin-
-layer chromatographic method for the determination of the pigment contents
of powdered paprika.] - Élelmiszervizsgálati Közlemények *21* : 195 - 201,
1975. [Car; in Hung., ab : E, F, G, R.]

26291 - **d'HARLINGUE, A.** : Stérols, acides gras, lipochromes et nucléotides pyridiniques et adényliques des frondes de *Spirodela* cultivées en présence d'acide
gibbérellique. - Physiol. vég. *14* : 713 - 723, 1976. [Ps, Chl.]

26292 - **d'HARLINGUE, A., LECHEVALLIER, D., MONÉGER, R.** : Nucléotides pyridiniques,
chlorophylles et stérols des frondes de *Spirodela* cultivés en présence de saccharose. - Physiol. vég. *14* : 367 - 376, 1976.

26293 - **HARNISCHFEGER, G.** : Possible distortions in the fluorescence emission spectra
at 77 degree K of photosynthetic organisms induced by the cooling process. -
Ber. deut. bot. Ges. *89* : 293 - 299, 1976.

26294 - **HARNISCHFEGER, G.** : Possible influence of the rate of specimen cooling on the
determination of energy distribution in photosynthesis by fluorescence emission at 77 K. - Biochim. biophys. Acta *449* : 593 - 596, 1976.

26295 - **HARRIS, E.H., BOYNTON, J.E., GILLHAM, N.W.** : Genetics of chloroplast ribosome
biogenesis in *Chlamydomonas reinhardtii*. - In : LEWIN, R.A. (ed.) : The Genetics of Algae. Pp. 119 - 144. Blackwell, Oxford; Univ. California Press, Ber-
keley 1976.

26296 - **HARRIS, G.P.** : Water content and productivity of lichens. - In : LANGE, O.L.,
KAPPEN, L., SCHULZE, E.-D. (ed.) : Water and Plant Life. Pp. 452 - 468.
Springer-Verlag, Berlin - Heidelberg - New York 1976. [Ps.]

26297 - **HARRIS, J.B.** : Changes in *Cyphonandra* chloroplast substructures during leaf
senescence. - Plant Physiol. *57* (Suppl.) : 73, 1976.

26298 - **HARRIS, L., PORTER, G., SYNOWIEC, J.A., TREDWELL, C.J., BARBER, J.** : Fluorescence lifetimes of *Chlorella pyrenoidosa*. - Biochim. biophys. Acta *449* : 329-
- 339, 1976.

26299 - **HARRIS, P.J.C., WILKINS, M.B.** : Light-induced changes in the period of the
circadian rhythm of carbon dioxide output in *Bryophyllum* leaves. - Planta
129 : 253 - 258, 1976.

26300 - **HARTGERINK, A.P., MAYO, J.M.** : Controlled-environment studies on net assimilation and water relations of *Dryas integrifolia*. - Can. J. Bot. *54* : 1884 -
- 1895, 1976.

26301 - **HARTIG, P.R.** : Fluorescence studies on chloroplast coupling factor. - Rep.
LBL-5366 : I - XII, 1 - 187, 1976.

26302 - **HARTMAN, A.** : Partition of cell particles in three-phase systems. - Acta chem.
scand. B *30* : 585 - 594, 1976. [Subchloroplast particles.]

26303 - **HARTSOCK, T.L., NOBEL, P.S.** : Watering converts a CAM plant to daytime CO_2
uptake. - Nature *262* : 574 - 576, 1976.

26304 - HARTWIG, E.O. : Nutrient cycling between the water column and a marine sediment. I. Organic carbon. - Mar. Biol. *34* : 285 - 295, 1976.

26305 - HASE, T., WADA, K., MATSUBARA, H. : Amino acid sequence of the major component of *Aphanothece sacrum* ferredoxin. - J. Biochem. (Tokyo) *79* : 329 - 343, 1976.

26306 - HASE, T., WADA, K., OHMIYA, M., MATSUBARA, H. : Amino acid sequence of the major component of *Nostoc muscorum* ferredoxin. - J. Biochem. (Tokyo) *80* : 993- - 999, 1976.

26307 - HASEBA, T., ITO, D. : [Water-vapor transfer across the boundary-layer on rice- -leaf. I. Evaporation from wet surface of leaf-shaped plate in laminar air flow.] - J. agr. Meteorol. *32* : 137 - 144, 1976. [In Jap., ab : E.]

26308 - HASHIZUME, K., NISHIMURA, S. : [Continuous forage production by means of rotating cool- and warm-season grasses IV. Growth response of warm-season grasses to the duration of shading.] - J.jap. Soc. Grassland Sci. *22* : 268 - 272, 1976. [Growth analysis; in Jap., ab : E.]

26309 - HASLETT, B.G., BOULTER, D. : The *N*-terminal amino acid sequence of plastocyanin from *Stellaria media* L. An exercise to establish criteria for the identification of residues from a sequenator. - Biochem. J. *153* : 33 - 38, 1976.

26310 - HASLETT, B.G., CAMMACK, R. : Changes in the activity of ferredoxin-NADP reductase during the greening of bean leaves. - New Phytol. *76* : 219 - 226, 1976.

26311 - HASPEL-HORVATOVIČ, E., HORIČKOVÁ, B. : Spektrophotometrische Bestimmung der gelben und roten Paprikafarbstoffe aus dem Gesamtextrakt. - J. Lebensmittel- -Untersuch. -Forsch. *160* : 275 - 276, 1976.

26312 - HASPELOVÁ-HORVATOVIČOVÁ, A., HORIČKOVÁ, B. : Rýchla metóda pre stanovenie obsahu červených a žltých farbív papriky. [Rapid method for determining the contents of red and yellow pigments in pepper.] - Průmysl Potravin *27* : 233 - - 234, 1976. [In Slovak.]

26313 - HATA, M., YOKOHAMA, Y. : Photosynthesis-temperature relationships in seaweeds and their seasonal changes in the colder region of Japan. - Bull. jap. Soc. Phycol. *24* : 1 - 7, 1976.

25314 - HATCH, M.D. : Photosynthesis : the path of carbon. - In : BONNER, J., VARNER, J.E.(ed.) : Plant Biochemistry. Third Edition. Pp. 797 - 844. Academic Press, New York - San Francisco - London 1976.

26315 - HATCH, M.D. : The C_4 pathway of photosynthesis: Mechanism and function. - In : BURRIS, R.H., BLACK, C.C. (ed.) : CO_2 Metabolism and Plant Productivity. Pp. 59 - 81. Univ. Park Press, Baltimore - London - Tokyo 1976.

26316 - HATCH, M.D., KAGAWA, T. : Photosynthetic activities of isolated bundle sheath cells in relation to differing mechanisms of C_4 pathway photosynthesis. - Arch. Biochem. Biophys. *175* : 39 - 53, 1976.

26317 - HATCH, M.D., OSMOND, C.B. : Compartmentation and transport in C_4 photosynthesis. - In : STOCKING, C.R., HEBER, U. (ed.) : Transport in Plants III. Intracellular Interactions and Transport Processes. Pp. 144 - 184. Springer-Verlag, Berlin - Heidelberg - New York 1976.

26318 - HATFIELD, J.L., STANLEY, C.D., CARLSON, R.E. : Evaluation of an electronic foliometer to measure leaf area in corn and soybeans. - Agron. J. *68* : 434 - - 436, 1976.

26319 - HATTERSLEY, P.W., WATSON, L. : C_4 grasses : an anatomical criterion for distinguishing between NADP-malic enzyme species and PCK or NAD-malic enzyme species. - Aust. J. Bot. *24* : 297 - 308, 1976.

26320 - HATTERSLEY, P.W., WATSON, L., OSMOND, C.B. : Metabolite transport in leaves of C_4 plants : specification and speculation. - In : WARDLAW, I.F., PASSIOURA, J.B. (ed.) : Transport and Transfer Processes in Plants. Pp. 191 - 201. Academic Press, New York - San Francisco - London 1976.

26321 - **HAUPT, W.** : Le phytochrome, principal pigment photorécepteur dans le photopériodisme des végétaux : sa localisation dans la cellule. - Bull. Groupe Étude Rythmes biol. *8* (1) : 9 - 20, 1976. [Chloroplast.]

26322 - **HAUPT, W.** : Problèmes concernant les mouvements de chloroplastes orientés par
la lumière. - Bull. Groupe Étude Rythmes biol. *8* (1) : 21 - 26, 1976.

26323 - **HAUPT, W.** : Perception de la direction de la lumière dans le mouvement d'orientation d'organites cellulaires et d'organismes mobiles. - Bull. Groupe Etude
Rythmes biol. *8* (1) : 27 - 40, 1976. [Chloroplast.]

26324 - **HAUPT, W., BRETZ, N.** : Short-term reactions of phytochrome in *Mougeotia* ? -
Planta *128* : 1 - 3, 1976. [Chloroplast.]

26325 - **HAVEMAN, J., MATHIS, P.** : Flash-induced absorption changes of the primary donor of Photosystem II at 820 nm in chloroplasts inhibited by low pH or Tris-
-treatment. - Biochim. biophys. Acta *440* : 346 - 355, 1976.

26326 - **HAXO, F.T., KYCIA, J.H., SOMERS, G.F., BENNETT, A., SIEGELMAN, H.W.** : Peridinin-chlorophyll *a* proteins of the dinoflagellate *Amphidinium carterae* (Ply-
mouth 450). - Plant Physiol. *57* : 297 - 303, 1976.

26327 - **HAYDEN, D.B., HOPKINS, W.G.** : Membrane polypeptides and chlorophyll-protein
complexes of maize mesophyll chloroplasts. - Can. J. Bot. *54* : 1684 - 1689,
1976.

26328 - **HAYMAN, E.P., YOKOYAMA, H.** : Effects of 4-[β-(diethylamino)-ethoxy]-benzophe-
none upon carotenogenesis in *Rhodospirillum rubrum*. - J. Bacteriol. *127* :
1030 - 1031, 1976.

26329 - **HAYS, R.L.** : The influence of leaf anatomy on water-use efficiency : A modeling approach. - Plant Physiol. *57* (Suppl.) : 77, 1976. [CO_2 transport.]

26330 - **HEAD, A.** : Primary production in an estuarine environment : a comparison of
in situ and simulated *in situ* ^{14}C techniques. - Estuarine coastal mar. Sci.
4 : 575 - 578, 1976.

26331 - **HEAPY, L.A., ROBERTSON, J.A., McBEATH, D.K., von MAYDELL, U.M., LOVE, H.C.,
WEBSTER, G.R.** : Development of a barley yield equation for central Alberta.
I. Effect of soil and fertilizer N and P. - Can. J. Soil Sci. *56* : 233 - 247,
1976. [Model.]

26332 - **HEBER, U.** : Energy coupling in chloroplasts. - J. Bioenerg. Biomembr. *8* : 157-
- 172, 1976.

26333 - **HEBER, U., ANDREWS, T.J., BOARDMAN, N.K.** : Effects of pH and oxygen on photosynthetic reactions of intact chloroplasts. - Plant Physiol. *57* : 277 - 283,
1976.

26334 - **HEBER, U., BOARDMAN, N.K., ANDERSON, J.M.** : Cytochrome *b*-563 redox changes
in intact CO_2-fixing spinach and in developing pea chloroplasts. - Biochim.
biophys. Acta *423* : 275 - 292, 1976.

26335 - **HEBER, U., SANTARIUS, K.A.** : Water stress during freezing. - In : **LANGE, O.L.,
KAPPEN, L., SCHULZE, E.-D.** (ed.) : Water and Plant Life. Pp. 253 - 267. Springer-Verlag, Berlin - Heidelberg - New York 1976. [Ps.]

26336 - **HEICHEL, G.H., ANAGNOSTAKIS, S.** : Modifying leaf diffusive resistance with an
epidermal graft. - Plant Physiol. *57* (Suppl.) : 43, 1976.

26337 - **HEILMAN, J.L., KANEMASU, E.T.** : An evaluation of a resistance form of the
energy balance to estimate evapotranspiration. - Agron. J. *68* : 607 - 611,
1976.

26338 - **HEINZE, M., FIEDLER, H.-J.** : Beziehungen des Chlorophyllgehaltes zu Standorts-
faktoren, Ernährungszustand und Wachstum bei Koniferen. - Flora *165* : 269 -
- 293, 1976.

26339 - **HEITEFUSS, R., WOLF, G.** : Nucleic acids in host-parasite interactions. - In :
HEITEFUSS, R., WILLIAMS, P.H. (ed.) : Physiological Plant Pathology. Pp. 480-
- 508. Springer Verlag, Berlin - Heidelberg - New York 1976. [Chloroplast.]

25340 - **HEIZMANN, P.** : Contrôle des synthèses d'ARN au cours de la morphogenèse du chloroplaste, chez *Euglena gracilis*. - Ann. biol. *15* (5/6) : 197 - 226, 1976.

26341 - **HELDT, H.W.** : Metabolite carriers of chloroplasts. - In : **STOCKING, C.R., HEBER, U.** (ed.) : Transport in Plants III. Intracellular Interactions and Transport Processes. Pp. 137 - 143. Springer-Verlag, Berlin - Heidelberg - New York 1976.

26342 - **HELDT, H.W.** : Metabolite transport in intact spinach chloroplasts. - In : **BARBER, J.** (ed.) : The Intact Chloroplast. Pp. 215 - 234. Elsevier, Amsterdam - New York - Oxford 1976.

26343 - **HELDT, H.W.** : Transfer of substrates across the chloroplast envelope. - Horizons Biochem. Biophys. *2* : 199 - 229, 1976. [Ps.]

26344 - **HELLEBUST, J.A.** : Effect of salinity on photosynthesis and mannitol synthesis in the green flagellate *Platymonas suecica*. - Can. J. Bot. *54* : 1735 - 1741, 1976.

26345 - **HELLEBUST, J.A.** : Osmoregulation. - Annu. Rev. Plant Physiol. *27* : 485 - 505, 1976. [Ps.]

26346 - **HELLER, S.** : Phytochromreaktionen bei *Mougeotia* mit unterschiedlicher spektraler Wirkungsverteilung. - Z. Pflanzenphysiol. *79* : 8 - 22, 1976. [Chloroplast.]

26347 - **HELLINGWERF, K.J., ARENTS, J.C., van DAM, K.** : Light-stimulated oxygen uptake by vesicles containing cytochrome *c* oxidase and bacteriorhodopsin. - FEBS Lett. *67* : 164 - 166, 1976.

26348 - **HELMS, J.A.** : Factors influencing net photosynthesis in trees : an ecological viewpoint. - In : **CANNELL, M.G.R., LAST, F.T.** (ed.) : Tree Physiology and Yield Improvement. Pp. 55 - 78. Academic Press, London - New York - San Francisco 1976.

26349 - **HELMS, K., WARDLAW, I.F.** : Movement of viruses in plants : long distance movement of tobacco mosaic virus in *Nicotiana glutinosa*. - In : **WARDLAW, I.F., PASSIOURA, J.B.** (ed.) : Transport and Transfer Processes in Plants. Pp. 283 - - 293. Academic Press, New York - San Francisco - London 1976. [Photosynthates.]

26350 - **HENDRICH, W., KUBIAK, Z., JURAJDA, K., PAWLASZYK-SZPILOWA, M.** : Effect of herbicides on photosynthetic electron transport and on growth of alga *Scenedesmus quadricauda*. - Acta Soc. Bot. Pol. *45* : 101 - 110, 1976.

26351 - **HENRIKSEN, L., GASSMAN, M.** : Protoheme synthesis in etiolated and greening red kidney bean leaves. - Plant Physiol. *57* (Suppl.) : 45, 1976.

26352 - **HENRIQUES, F., PARK, R.B.** : Compositional characteristics of a chloroform/methanol soluble protein fraction from spinach chloroplast membranes. - Biochim. biophys. Acta *430* : 312 - 320, 1976.

26353 - **HENRIQUES, F., PARK, R.B.** : Development of the photosynthetic unit in lettuce. - Proc. nat. Acad. Sci. USA *73* : 4560 - 4564, 1976.

26354 - **HENRIQUES, F., PARK, R.B.** : Identification of chloroplast membrane peptides with subunits of coupling factor and ribulose-1,5 diphosphate carboxylase. - Arch. Biochem. Biophys. *176* : 472 - 478, 1976.

26355 - **HENRY, E.W.** : The ultrastructural localization of polyphenol oxidase in chloroplasts of *Brassica napus* cv. Zephyr. - Z. Pflanzenphysiol. *78* : 446 - 452, 1976.

26356 - **HENRY, L.E.A., HALLIWELL, B., HALL, D.O.** : The superoxide dismutase activity of various photosynthetic organisms measured by a new and rapid assay technique. - FEBS Lett. *66* : 303 - 306, 1976.

26357 - **HERBERT, R.A.** : Isolation and identification of photosynthetic bacteria (*Rhodospirillaceae*) from antarctic marine and freshwater sediments. - J. appl. Bacteriol. *41* : 75 - 80, 1976.

26358 - **HERBLAND, A.** : *In situ* utilization of urea in the euphotic zone of the tropical Atlantic. - J. exp. mar. Biol. Ecol. *21* : 269 - 277, 1976. [Chl.]

26359 - **HERMAN, E.M., SWEENEY, B.M.** : *Cachonina illdefina* sp. nov. (*Dinophyceae*) : chloroplast tubules and degeneration of the pyrenoid. - J. Phycol. *12* : 198 - - 205, 1976.

26360 - **HEROLD, A., LEWIS, D.H., WALKER, D.A.** : Sequestration of cytoplasmic ortho-phosphate by mannose and its differential effect on photosynthetic starch syn-thesis in C_3 and C_4 species. - New Phytol. *76* : 397 - 407, 1976.

26361 - **HERRERA, A., BRADBEER, J.W., KEMBLE, R.J.** : The effects of 2-(4-methyl-2,6- -dinitroanilino)-N-methyl propionamide on bean chloroplast development. - In: Acides Nucléiques et Synthèse des Protéines chez les Végétaux. Colloq. Int. C.N.R.S. No.261. Pp. 457 - 461. C.N.R.S., Paris 1976.

26362 - **HERRMANN, F.H., SCHUMANN, B., BÖRNER, T., KNOTH, R.** : Struktur und Funktion der genetischen Information in den Plastiden. XII. Die plastidalen Lamellar-proteine der photosynthesedefekten Plastommutante en:gil-1 („Mrs. Pollock") und der Genmutante „Cloth of Gold" von *Pelargonium zonale* AIT. - Photosynthe-tica *10* : 164 - 171, 1976.

26363 - **HERTZBERG, S., BORCH, G., LIAAEN-JENSEN, S.** : Bacterial carotenoids. L. Abso-lute configuration of zeaxanthin dirhamnoside. - Arch. Microbiol. *110* : 95 - - 99, 1976.

26364 - **HESKETH, J.D., JONES, J.W.** : Some comments on computer simulators for plant growth - 1975. - Ecol. Model. *2* : 235 - 247, 1976.

26365 - **HESS, B.** : Technology of biomolecular design, with experiments on light con-trol of the photochemical cycle in *Halobacterium halobium*. - FEBS Lett. *64* : 26 - 28, 1976.

26366 - **HESSE, H., JANK-LADWIG, R., STROTMANN, H.** : On the reconstitution of photo-phosphorylation in CF_1-extracted chloroplasts. - Z. Naturforsch. *31C* : 445 - - 451, 1976.

26367 - **HESSE, M., BLEY, P., BÖGER, P.** : Photosynthetischer Elektronentransport der Alge *Bumilleriopsis filiformis* während der Zellentwicklung. - Planta *132* : 53 - 59, 1976.

26368 - **HEW, C.-S.** : Patterns of CO_2 fixation in tropical orchid species. - Proc. 8th World Orchid Conf. *1975* : 426 - 430, 1976.

26369 - **HEWITT, H.G., AYRES, P.G.** : Effect of infection by *Microsphaera alphitoides* (powdery mildew) on carbohydrate levels and translocation in seedlings of *Quercus robur*. - New Phytol. *77* : 379 - 390, 1976.

26370 - **HICKLENTON, P.R., OECHEL, W.C.** : Physiological aspects of the ecology of *Di-cranum fuscescens* in the subarctic. I. Acclimation and acclimation potential of CO_2 exchange in relation to habitat, light and temperature. - Can. J. Bot. *54* : 1104 - 1119, 1976.

26371 - **HICKMAN, M.** : Phytoplankton population efficiency studies. - Int. Rev. ges. Hydrobiol. *61* : 279 - 295, 1976. [Chl.]

26372 - **HIEBSCH, C.K., KANEMASU, E.T.** : Soil reflectance effects on net carbon dioxide exchange rates of sorghum. - Crop Sci. *16* : 113 - 116, 1976.

26373 - **HIEBSCH, C.K., KANEMASU, E.T., NICKELL, C.D.** : Effects of soybean leaflet ty-pe on net carbon dioxide exchange, water use, and water-use efficiency. - Can. J. Plant Sci. *56* : 455 - 458, 1976.

26374 - **HIGASHIDA, M., MUKOHATA, Y.** : Magnesium ion-induced changes in the binding mode of adenylates to chloroplast coupling factor 1. - J. Biochem. (Tokyo) *80* : 1177 - 1179, 1976.

26375 - **HIGGINBOTHAM, K.O., STRAIN, B.R.** : A climatized assimilation chamber for use with whole-plants or whole-branches. - Photosynthetica *10* : 54 - 58, 1976.

26376 - **HILL, R., CROFTS, A.R., PRINCE, R.C., EVANS, E.H., GOOD, N.E., WALKER, D.A.** : Uncoupling of electron transport by anionic quinonoid redox indicator dyes. - New Phytol. *77* : 1 - 9, 1976.

26377 - **HILLMAN, W.S.** : Calibrating duckweeds : light, clocks, metabolism, flowering. Special characteristics of *Lemnaceae* may offer unique insights into plant de-velopment. - Science *193* : 453 - 458, 1976. [Chl.]

26378 - HIRAYAMA, O., MATSUI, T. : Effects of lipolytic enzymes on the photochemical activities of spinach chloroplasts. - Biochim. biophys. Acta *423* : 540 - 547, 1976.

26379 - HIRSCH, M.D., MARCUS, M.A., LEWIS, A., MAHR, H., FRIGO, N. : A method for measuring picosecond phenomena in photolabile species. The emission lifetime of bacteriorhodopsin. - Biophys. J. *16* : 1399 - 1409, 1976.

26380 - HIYAMA, T., McSWAIN, B.D., ARNON, D.I. : Analysis of electron flow through P700. - Plant Physiol. *57* (Suppl.) : 54, 1976.

26381 - HIYAMA, T., McSWAIN, B.D., UFERT, W. : Simultaneous measurement of oxygen concentration and absorbance changes. - Anal. Biochem. *76* : 365 - 368, 1976.

26382 - HO, C.-H., IKAWA, T., NISIZAWA, K. : Nitrite-reducing activity of modified cytochrome *c*-553 from the red alga *Porphyra yezoensis* UEDA. - Plant Cell Physiol. *17* : 431 - 438, 1976.

26383 - HO, L.C. : Variation in the carbon/dry matter ratio in plant material. - Ann. Bot. *40* : 163 - 165, 1976.

26384 - HO, L.C. : The effect of current photosynthesis on the origin of translocates in old tomato leaves. - Ann. Bot. *40* : 1153 - 1162, 1976.

26385 - HO, L.C. : The relationship between the rates of carbon transport and of photosynthesis in tomato leaves. - J. exp. Bot. *27* : 87 - 97, 1976.

26386 - HOBSON, L.A. : Effects of interactions of daylength and temperature on net photosynthetic, dark respiratory, and excretory rates of two species of marine unicellular algae. - J. Phycol. *12* (Suppl.) : 11, 1976.

26387 - HOBSON, L.A., MORRIS, W.J., PIRQUET, K.T. : Theoretical and experimental analysis of the ^{14}C technique and its use in studies of primary production. - J. Fish. Res. Board Can. *33* : 1715 - 1721, 1976.

26388 - HOCHMAN, Y., LANIR, A., CARMELI, C. : Relations between divalent cation binding and ATPase activity in coupling factor from chloroplast. - FEBS Lett. *61* : 255 - 259, 1976.

26389 - HODDINOTT, J., GORHAM, P.R. : The effects of light quality and non-steady-state, localized ^{14}CO$_2$ pulse labelling on net assimilation and ^{14}C translocation profiles in *Heracleum lanatum*. - Can. J. Bot. *54* : 1206 - 1213, 1976.

26390 - HODGKINSON, K.C., QUINN, J.A. : Adaptive variability in the growth of *Danthonia caespitosa* GAUD. populations at different temperatures. - Aust. J. Bot. *24* : 381 - 396, 1976. [Growth analysis.]

26391 - HODSON, R.E., HOLM-HANSEN, O., AZAM, F. : Improved methodology for ATP determination in marine environments. - Mar. Biol. *34* : 143 - 149, 1976.

26392 - HOFÄCKER, W. : Untersuchungen über den Einfluß wechselnder Bodenwasserversorgung auf die Photosyntheseintensität und den Diffusionswiderstand bei Rebblättern. - Vitis *15* : 171 - 182, 1976.

26393 - HOFER-SIEGRIST, L. : Eine verbesserte Methode zur Bestimmung von ATP in Seewasser. - Schweiz. Z. Hydrol. *38* : 49 - 54, 1976.

26394 - HOFF, A.J. : De triplettoestand in de fotosynthese. [The triplet state in photosynthesis.] - Chem. Weekbl. *1976* (12) : 644 - 645, 1976. [In Dutch.]

26395 - HOFF, A.J. : Kinetics of populating and depopulating of the components of the photoinduced triplet state of the photosynthetic bacteria *Rhodospirillum rubrum*, *Rhodopseudomonas spheroides* (wild type) and its mutant R-26 as measured by ESR in zero-field. - Biochim. biophys. Acta *440* : 765 - 771, 1976.

26396 - HOFF, A.J., van der WAALS, J.H. : Zero field resonance and spin alignment of the triplet state of chloroplasts at 2°K. - Biochim. biophys. Acta *423* : 615 - 620, 1976.

26397 - HOFFMAN, G.J., HALL, A.E. : Performance of silver-foil psychrometer for measuring leaf water potential *in situ*. - Agron. J. *68* : 872 - 875, 1976. [Ps.]

26398 - HOFFMANN, P. : Das Licht als Energiequelle und Signal in der pflanzlichen Primärproduktion. - Wiss. Z. Humboldt-Univ. Berlin, math.-naturwiss. Reihe *25* : 685 - 696, 1976.

26399 - HOFFMANN, P., HIEKE, B., GIENAPP, C. : Die Entwicklung des Photosyntheseappa-
        rates bei Keimpflanzen von *Triticum aestivum* unter besonderer Berücksichti-
        gung energetischer Aspekte. - Wiss. Z. Humboldt-Univ. Berlin, math.-naturwiss.
        Reihe *25* : 713 - 722, 1976.

26400 - HOFFMANN, P., MICHAELIS, G. : Physiologische Gradienten in Primärblättern von
        Triticum aestivum L. - Wiss. Z. Humboldt-Univ. Berlin, math.-naturwiss. Reihe
        25 : 787 - 795, 1976.

26401 - HOFRICHTER, J., EATON, W.A. : Linear dichroism of biological chromophores. -
        Annu. Rev. Biophys. Bioeng. *5* : 511 - 560, 1976. [Chl.]

26402 - HOGETSU, D., MIYACHI, S. : [CO_2 and photosynthesis.] - Yuki Gosei Kagaku Kyo-
        kai Shi [J. synth. org. Chem. Jap.] *34* : 279 - 286, 1976. [In Jap., ab : E.]

26403 - HÖHLER, T., GROTHUS, R., SCHAUB, H., EGLE, K. : Über den Einfluss des Sauer-
        stoffs auf Stoffproduktion und tagesperiodische Schwankungen der CO_2-Aufnahme.
        2. Trockengewichtszunahme, Nettophotosynthese und Transpiration von *Amaranthus
        paniculatus* und *Zea mays* bei Anzucht unter 4% Sauerstoff im Vergleich zu nor-
        maler Luft. - Photosynthetica *10* : 59 - 70, 1976.

26404 - HOLDEN, M. : Chlorophylls. - In : GOODWIN, T.W. (ed.) : Chemistry and Bio-
        chemistry of Plant Pigments. 2nd Ed. Vol. 2. Pp. 1 - 37. Academic Press, Lon-
        don - New York - San Francisco 1976.

26405 - HOLDER, A.A. : Ribulose 1,5-diphosphate carboxylase from *Oenothera*. Purifica-
        tion and a peptide mapping procedure for the subunits. - Carlsberg Res. Com-
        mun. *41* : 321 - 334, 1976.

26406 - HOLMES, N.G., van GRONDELLE, R., HOFF, A.J., DUYSENS, L.N.M. : Changes of *in
        vivo* bacteriochlorophyll fluorescence yield in *Rhodopseudomonas sphaeroides*
        at low temperature and low redox potential. - FEBS Lett. *70* : 185 - 190, 1976.

26407 - HOLM-HANSEN, O., GOLDMAN, C.R., RICHARDS, R., WILLIAMS, P.M. : Chemical and
        biological characteristics of a water column in Lake Tahoe. - Limnol. Oceano-
        gr. *21* : 548 - 562, 1976. [Chl.]

26408 - HOLST, R.W., YOPP, J.H. : An algal polyphenol oxidase : characterization of
        the *o*-diphenol-oxidase from the charophyte, *Nitella mirabilis*. - Phycologia
        15 : 119 - 124, 1976. [Chl.]

26409 - HOLT, D.A., DOUGHERTY, C.T., BULA, R.J., SCHREIBER, M.M., PEART, R.M. : Water
        relations in SIMED, the Purdue model of alfalfa growth. - In : Proceedings of
        the Symposium on Modeling : Climate - Plants - Soils. Pp. 32 - 45. Univ.
        Guelph 1976. [Ps.]

26410 - HOLTEN, D., GOUTERMAN, M., PARSON, W.W., WINDSOR, M.W., ROCKLEY, M.G. : Elec-
        tron transfer from photoexcited singlet and triplet bacteriopheophytin. -
        Photochem. Photobiol. *23* : 415 - 423, 1976.

26411 - HOMANN, P.H. : Interaction of N-methylphenazinium methyl sulfate with the thy-
        lakoids of illuminated chloroplasts in the presence of 3-(3,4-dichlorophenyl)-
        -1,1-dimethylurea. - Plant Physiol. *57* : 387 - 392, 1976.

26412 - HOMANN, P.H. : The "energy dependent" lowering of chloroplast fluorescence. -
        Plant Physiol. *57* (Suppl.) : 23, 1976.

26413 - HONG, F.T. : Charge transfer across pigmented bilayer lipid membrane and its
        interfaces. - Photochem. Photobiol. *24* : 155 - 189, 1976. [Chl.]

26414 - HOOBER, J.K. : Synthesis of the major thylakoid polypeptides during greening
        of *Chlamydomonas reinhardtii* Y-1. - In : BUCHER, T., NEUPERT, W., SEBALD, W.,
        WERNER, S. (ed.) : Genetics and Biogenesis of Chloroplasts and Mitochondria.
        Pp. 87 - 94. North-Holland Publ. Co., Amsterdam - New York - Oxford 1976.

*26415 - HOOBER, J.K., STEGEMAN, W.J. : Regulation of chloroplast membrane synthesis.
        - In : BIRKY, C.W., Jr., PERLMAN, P.S., BYERS, T.J. (ed.) : Genetics and Bio-
        genesis of Mitochondria and Chloroplasts. Pp. 225 - 251. Ohio State Univ.
        Press, Columbus 1975.

26416 - HOOBER, J.K., STEGEMAN, W.J. : Kinetics and regulation of synthesis of the
        major polypeptides of thylakoid membranes in *Chlamydomonas reinhardtii* y-1 at
        elevated temperatures. - J. Cell Biol. *70* : 326 - 337, 1976.

26417 - HOOPER, N.M., ROBINSON, G.G.C. : Primary production of epiphytic algae in a
marsh pond. - Can. J. Bot. *54* : 2810 - 2815, 1976.

26418 - HORAK, A., ZALIK, S. : Studies on photophosphorylation utilizing methylene
diphosphate analogs of ADP and ATP. - Biochim. biophys. Acta *430* : 135 - 144,
1976.

26419 - HORNBACH, D., GEIGER, D.R. : A model-based method of steady-state labeling
for studying plant productivity. - Plant Physiol. *57* (Suppl.) : 28, 1976.

26420 - HORNBERGER, G.M., KELLY, M.G., ELLER, R.M. : The relationship between light
and photosynthetic rate in a river community and implications for water quality modelling. - Water Resour. Res. *12* : 723 - 730, 1976.

26421 - HORTON, B.D., EDWARDS, J.H. : Diffusive resistance rates and stomatal aperture
of peach seedlings as affected by aluminum concentration. - HortScience *11* :
591 - 593, 1976.

26422 - HORTON, G.L., HUYSER, E.S., AKAGI, J.M. : The behavior of alloxan in the Hill
reaction. - Experientia *32* : 861 - 862, 1976.

26423 - HORTON, P. : Organization and function of chloroplast photosystems. - Int. J.
Biochem. *7* : 597 - 605, 1976.

26424 - HORTON, P., CRAMER, W.A. : Stimulation of photosystem I-induced oxidation of
chloroplast cytochrome *b*-559 by pre-illumination and by low pH. - Biochim.
biophys. Acta *430* : 122 - 134, 1976.

26425 - HORTON, P., WHITMARSH, J., CRAMER, W.A. : On the specific site of action of
3-(3,4-dichlorophenyl)-1,1-dimethylurea in chloroplasts : inhibition of a
dark acid-induced decrease in midpoint potential of cytochrome *b*-559. - Arch.
Biochem. Biophys. *176* : 519 - 524, 1976.

26426 - HORVÁTH, I. : Effect of light rhythmicity on the utilization of photosynthetic energy. - Acta biochim. biophys. *11* : 157, 1976.

26427 - HOUGH, R.A. : Light and dark respiration and release of organic carbon in marine macrophytes of the Great Barrier Reef region. - Aust. J. Plant Physiol.
3 : 63 - 68, 1976.

26428 - HOUGH, R.A., WETZEL, R.G. : Evaluation of photosynthetic status (C_3 vs. C_4)
of some aquatic plants. - Plant Physiol. *57* (Suppl.) : 33, 1976.

26429 - HOWARD, R.J., WRIGHT, S.W., GRANT, B.R. : Structure and some properties of
soluble 1,3-β-glucan isolated from the green alga *Caulerpa simpliciuscula*. -
Plant Physiol. *58* : 459 - 463, 1976. [Ps.]

26430 - HOWELL, S., HEIZMANN, P., GELVIN, S. : Localization of the gene coding for
the large subunit of ribulose bisphosphate carboxylase on the chloroplast
genome of *Chlamydomonas reinhardi*. - In : BÜCHER, T., NEUPERT, W., SEBALD,
W., WERNER, S. (ed.) : Genetics and Biogenesis of Chloroplasts and Mitochondria. Pp. 625 - 628. North-Holland Publ. Co., Amsterdam - New York - Oxford
1976.

26431 - HOWELL, S.H., WALKER, L.L. : Informational complexity of the nuclear and chloroplast genomes of *Chlamydomonas reinhardi*. - Biochim. biophys. Acta *418* :
249 - 256, 1976.

26432 - HØYER-HANSEN, G., MACHOLD, O., KAHN, A. : Polypeptide composition of internal
membranes from barley etioplasts. - Carlsberg Res. Commun. *41* : 349 - 357,
1976.

26433 - HOZYO, Y., KATO, S. :[The interrelationship between source and sink of the
grafts of wild type and improved variety of *Ipomoea*.]- Proc. Crop Sci. Soc.
Jap. *45* : 117 - 123, 1976. [Photosynthates; in Jap., ab : E.]

26434 - HOZYO, Y., KATO, S. : [Thickening growth inhibition and re-thickening growth
of tuberous roots of sweet potato plants (*Ipomoea batatas* POIRET).] - Proc.
Crop Sci. Soc. Jap. *45* : 131 - 138, 1976. [Photosynthates; in Jap., ab : E.]

26435 - HŘIB, J. : Kinetics of chloroplast contraction and negative phototaxis in *Mougeotia* sp. induced by high radiant flux density. - Biol. Plant. *18* : 234 -
- 236, 1976.

26436 - HSIAO, T.C. : Stomatal ion transport. - In : LÜTTGE, U., PITMAN, M.G. (ed.): Transport in Plants II. Part B. Tissues and Organs. Pp. 195 - 221. Springer--Verlag, Berlin - Heidelberg - New York 1976. [Ps.]

26437 - HSIAO, T.C., ACEVEDO, E., FERERES, E., HENDERSON, D.W. : Water stress, growth and osmotic adjustment. - Phil. Trans. roy. Soc. London B *273* : 479 - 500, 1976. [Ps, Chl.]

26438 - HSIAO, T.C., FERERES, E., ACEVEDO, E., HENDERSON, D.W. : Water stress and dynamics of growth and yield of crop plants. - In : LANGE, O.L., KAPPEN, L., SCHULZE, E.-D. (ed.) : Water and Plant Life. Pp. 281 - 305. Springer-Verlag, Berlin - Heidelberg - New York 1976. [Ps.]

26439 - HUBBARD, J.S., RINEHART, C.A. : Bacteriorhodopsin formation in *Halobacterium halobium.* - Can. J. Microbiol. *22* : 1274 - 1281, 1976.

26440 - HUBER, D.J., NEWMAN, D.W. : Relationships between lipid changes and plastid ultrastructural changes in senescing and regreening soybean cotyledons. - J. exp. Bot. *27* : 490 - 511, 1976.

26441 - HUBER, S.C., EDWARDS, G.E. : A high-activity ATP translocator in mesophyll chloroplasts of *Digitaria sanguinalis,* a plant having the C-4 dicarboxylic acid pathway of photosynthesis. - Biochim. biophys. Acta *440* : 675 - 687, 1976.

26442 - HUBER, S.C., EDWARDS, G.E. : Studies on the pathway of cyclic electron flow in mesophyll chloroplasts of a C_4 plant. - Biochim. biophys. Acta *449* : 420 - - 433, 1976.

26443 - HUBER, S.C., EDWARDS, G.E. : The pathway of cyclic electron flow in C_4 mesophyll chloroplasts. - Plant Physiol. *57* (Suppl.) : 32, 1976.

26444 - HUBER, S.C., HALL, T.C., EDWARDS, G.E. : Differential localization of Fraction 1 protein between chloroplast types. - Plant Physiol. *57* : 730 - 733, 1976.

26445 - HUBER, S.C., HALL, T.C., EDWARDS, G.E. : Differential localization of fraction I protein between chloroplast types. - Plant Physiol. *57* (Suppl.) : 32, 1976.

26446 - HUBER, W., SANKHLA, N. : C_4 pathway and regulation of the balance between C_4 and C_3 metabolism. - In : LANGE, O.L., KAPPEN, L., SCHULZE, E.-D. (ed.) : Water and Plant Life. Pp. 335 - 363. Springer-Verlag, Berlin - Heidelberg - New York 1976.

26447 - HUDÁK, J. : Electronmicroscopical study of the influence of boron on the submicroscopical structure of the chloroplasts of *Zea mays* (L.) LSP-D. - Acta Fac. Rerum nat. Univ. Comenianae, Ser. Physiol. Plant. *12* : 47 - 53, 1976.

26448 - HUDÁK, J. : Influence of boron on chloroplasts differentiation in the primary leaves of *Vicia faba* (L.) cv. Považský. - Acta Fac. Rerum nat. Univ. Comenianae, Ser. Physiol. Plant. *12* : 55 - 61, 1976.

26449 - HUDÁK, J., HERICH, R. : Effect of boron on the ultrastructure of sunflower chloroplasts. - Photosynthetica *10* : 463 - 465, 1976.

26450 - HUDSON, J.P. : Food crops for the future. - J. roy. Soc. Arts *124* : 572 - 585, 1976. [Production.]

26451 - HULL, R.J. : A carbon-14 technique for measuring photosynthate distribution in field grown turf. - Agron. J. *68* : 99 - 102, 1976.

26452 - HULL, R.J., SULLIVAN, D.M., LYTLE, R.W., Jr. : Photosynthate distribution in natural stands of salt water cordgrass. - Agron. J. *68* : 969 - 972, 1976.

26453 - HUMPHREY, G.F. : The concentration of phytoplankton pigments in Australian waters. - Annu. Rep. CSIRO mar. Biochem. Unit *1975-76* : 16 - 21, 1976.

26454 - HUMPHREY, G.F. : Visible absorption spectra of whole algal cells. - Annu. Rep. CSIRO mar. Biochem. Unit *1975-76* : 24 - 32, 1976.

26455 - HUNER, N.P.A., MACDOWALL, F.D.H. : Chloroplastic proteins of wheat and rye grown at warm and cold-hardening temperatures. - Can. J. Biochem. *54* : 848 - - 853, 1976.

26456 - HUNTER, C.N., JONES, O.T.G. : Effect of added light-harvesting pigments on the reconstruction of photosynthetic reactions in membranes from bacteriochlorophyll-less mutants of *Rhodopseudomonas sphaeroides*. - Biochem. Soc. Trans. 4 : 669 - 670, 1976.

26457 - HUNTER, F., THORNBER, J.P. : Further characterization of the P700 a-protein from blue-green algae. - Plant Physiol. 57 (Suppl.) : 95, 1976.

26458 - HUNTSINGER, K.R. (G.), MASLIN, P.E. : Contribution of phytoplankton, periphyton, and macrophytes to primary production in Eagle Lake, California. - Calif. Fish Game 62 : 187 - 194, 1976.

26459 - HUPPERT, D., RENTZEPIS, P.M., TOLLIN, G. : Picosecond kinetics of chlorophyll and chlorophyll-quinone solutions in ethanol. - Biochim. biophys. Acta 440 : 356 - 364, 1976.

26460 - HURD, E.A. : Plant breeding for drought resistance. - In : KOZLOWSKI, T.T. (ed.) : Water Deficits and Plant Growth. Vol. IV. Pp. 317 - 353. Academic Press, New York - London 1976. [Ps.]

26461 - HURD, R.G., ENOCH, H.Z. : Effect of night temperature on photosynthesis, transpiration, and growth of spray carnations. - J. exp. Bot. 27 : 695 - 703, 1976.

26462 - HURKMAN, W.J., KENNEDY, G.S. : Fine structure and development of proteoplasts in primary leaves of mung bean. - Protoplasma 89 : 171 - 184, 1976. [Chloroplast.]

26463 - HURKMAN, W.J., MORRÉ, D.J., BRACKER, C.E. : Biochemical and ultrastructural criteria for identifying plastid fragments in cell-free fractions from etiolated soybean hypocotyls. - Plant Physiol. 57 (Suppl.) : 10, 1976.

26464 - HURT, P., WRIGHT, R. : CO_2 compensation point for photosynthesis : Effect of variable CO_2 and soil moisture levels. - Amer. Midland Naturalist 95 : 450 - - 455, 1976.

26465 - HUXLEY, P.A., SUMMERFIELD, R.J. : Leaf area manipulation with vegetative cowpea plants (*Vigna unguiculata* (L.) WALP). - J. exp. Bot. 27 : 1223 - 1232, 1976. [Sink-source relationships.]

26466 - HUXLEY, P.A., SUMMERFIELD, R.J. : Photomorphogenetic effects of lamp type on growth of some species of tropical grain legumes in controlled environment growth cabinets. - Plant Sci. Lett. 6 : 25 - 33, 1976. [Growth analysis.]

26467 - IGNAT'EV, A.N., KHRUSLOVA, S.G., POLEVAYA, V.S. : K voprosu o delenii khloroplastov vysshikh rastenii *in vitro*. [Division of chloroplasts of higher plants *in vitro*.] - Fiziol. Rast. 23 : 676 - 680, 1976. [In R, ab : E.]

26468 - IKEGAMI, I. : Fluorescence changes related in the primary photochemical reaction in the P-700-enriched particles isolated from spinach chloroplasts. - Biochim. biophys. Acta 449 : 245 - 258, 1976.

26469 - IKENAGA, T., KAMOTO, Y., OHASHI, H. : [Studies on the physiology and ecology of *Amaranthus viridis*. III. On the growth and chlorophyll content in leaves of *Amaranthus viridis* with sowing the seed once a month through the year.] - Zasso Kenkyu 21 : 6 - 11, 1976. [In Jap., ab : E.]

26470 - IKENAGA, T., OKUBO, S., OHASHI, H. : [Studies on the physiology and ecology of *Amaranthus viridis*. IV. On the growth and chlorophyll content of *Amaranthus viridis* with sowing the seed in April and August.] - Zasso Kenkyu 21 : 11 - 16, 1976. [In Jap., ab : E.]

26471 - IL'YASHUK, E.M., MANUIL'SKII, V.D., OKANENKO, A.S. : Intensivnost' gazoobmena list'ev sakharnoi svekly v svyazi s izmeneniem v nikh soderzhaniya kaliya. [Rate of gas-exchange in sugar-beet leaves connected with change in their potassium content.] - Fiziol. Biokhim. kul't. Rast. 8 : 415 - 423, 1976. [In R, ab : E.]

26472 - IMAFUKU, H. : Anaerobic ATP levels in the blue-green alga *Anabaena* : Dark--light transients and effects of light intensity. - Physiol. Plant. 38 : 191- - 195, 1976.

26473 - IMAFUKU, H., KATOH, T. : Intracellular ATP level and light-induced inhibition of respiration in a blue-green alga, *Anabaena variabilis*. - Plant Cell Physiol. *17* : 515 - 524, 1976.

26474 - IMAI, K., MURATA, Y. : [Effect of carbon dioxide concentration on growth and dry matter production of crop plant. 1. Effects on leaf area, dry matter, tillering, dry matter distribution ratio, and transpiration.] - Proc. Crop Sci. Soc. Jap. *45* : 598 - 606, 1976. [In Jap., ab : E.]

26475 - IMBAMBA, S.K. : The influence of light and temperature on photosynthesis and transpiration in some Kenyan plants. - Plant Physiol. *57* (Suppl.) : 106, 1976.

26476 - INADA, K. : Action spectra for photosynthesis in higher plants. - Plant Cell Physiol. *17* : 355 - 365, 1976.

26477 - INAMDAR, J.A., GANGADHARA, M., BHAT, R.B. : Leaf surface imprinting by gruel - a new technique. - Microscop. Acta *78* : 39 - 42, 1976.

26478 - INGLE, R.K., COLMAN, B. : The relationship between carbonic anhydrase activity and glycolate excretion in the blue-green alga *Coccochloris peniocystis*. - Planta *128* : 217 - 223, 1976.

26479 - INOUE, Y. : Manganese catalyst as a possible cation carrier in thermoluminescence from green plants. - FEBS Lett. *72* : 279 - 282, 1976.

26480 - INOUE, Y., FURUTA, S., OKU, T., SHIBATA, K. : Light-dependent development of thermoluminescence, delayed emission and fluorescence variation in dark- -grown spruce leaves. - Biochim. biophys. Acta *449* : 357 - 367, 1976.

26481 - INOUE, Y., ICHIKAWA, T., SHIBATA, K. : Development of thermoluminescence bands during greening of wheat leaves under continuous and intermittent illumination. - Photochem. Photobiol. *23* : 125 - 130, 1976.

26482 - INOUE, Y., OKU, T., FURUTA, S., SHIBATA, K. : Multiple-flash development of thermoluminescence bands in dark-grown spruce leaves. - Biochim. biophys. Acta *440* : 772 - 776, 1976.

26483 - INTYKBAEVA, B.B. : Opticheskie svoĭstva khlorofilonosnykh organov ozimoĭ pshenitsy. [Optical properties of chlorophyll-containing organs of winter wheat.] - In : Fotosintez i Produktivnost' Ozimoĭ Pshenitsy na Yugo-Vostoke Kazakhstana. Pp. 40 - 49, 130. Nauka kazakh. SSR, Alma-Ata 1976. [In R.]

26484 - INUYAMA, S., MUSICK, J.T., DUSEK, D.A. : Effect of plant water deficits at various growth stages on growth, grain yield and leaf water potential of irrigated grain sorghum. - Proc. Crop Sci. Soc. Jap. *45* : 298 - 307, 1976. [Ps.]

26485 - IORDANOV, I.T., VASIL'EVA, V.S. : Vliyanie povyshennykh temperatur na intensivnost' fotosinteza i aktivnost' ribulozodifosfat- i fosfoenolpiruvat-karboksilaz. [Effect of elevated temperatures on photosynthetic rate and activity of ribulose-diphosphate- and phosphoenolpyruvate carboxylases.] - Fiziol. Rast. *23* : 812 - 817, 1976. [In R, ab : E.]

26486 - IRSCHIK, H., OELZE, J. : The effect of transfer from low to high light intensity on electron transport in *Rhodospirillum rubrum* membranes. - Arch. Microbiol. *109* : 307 - 313, 1976.

26487 - ISAAKIDOU, J., PAPAGEORGIOU, G. : Effects of imidoester crosslinking on structural and functional characteristics of isolated spinach chloroplasts. - Biophys. J. *16* (2, Part 2) : 161 a, 1976.

26488 - ISAAKIDOU, J., PAPAGEORGIOU, G. : Interactions of metal cations with lipid-depleted chloroplasts. - Arch. Biochem. Biophys. *175* : 541 - 548, 1976.

26489 - ISEBRANDS, J.G., DICKSON, R.E., LARSON, P.R. : Translocation and incorporation of ^{14}C into the petiole from different regions within developing cottonwood leaves. - Planta *128* : 185 - 193, 1976.

26490 - ISÉPY, I. : A gyepszint-fitomassza mérése gyertyános-tölgyesekben. [Estimation of the herb layer phytomass in oak-hornbeam forests.] - Bot. Közlem. *63* : 205- - 212, 1976. [In Hung., ab : E.]

26491 - ISHITANI, T., UMEDA, K., KIMURA, S. : [Studies on the degradation of natural pigments. Part 1. Photodegradation of lycopene and β-carotene.] - Nippon Shokuhin Kogyo Gakkaishi [J. Food Sci. Technol. (Tokyo)] *23* (10) : 480 - 485, 1976. [In Jap.]

*26492 - ISHIZAKI, A., HASEGAWA, M. : [Physiological and morphological properties of
the Japanese black and red pine trees (*Pinus thunbergii* PARL. & *P. densiflora*
SIEB.*et* ZUCC.) on the resistance of the air pollution.] - Bull. Fac. Agr. Tamagawa Univ. *14* : 40 - 50, 1974. [Ps; in Jap., ab : E.]

26493 - ISMAIL, A.M.A., OBEID, M. : A study of assimilation and translocation in *Cus-*
cuta hyalina HEYNE ex ROTH., *Orobanche ramosa* L. and *Striga hermonthica*
BENTH. - Weed Res. *16* : 87 - 92, 1976. [Chl, Ps.]

26494 - ITAI, C., BENZIONI, A. : Water stress and hormonal response. - In : LANGE, O.
L., KAPPEN, L., SCHULZE, E.-D. (ed.) : Water and Plant Life. Pp. 225 - 242.
Springer-Verlag, Berlin - Heidelberg - New York 1976. [Ps, Chl.]

26495 - IVAKIN, A.P. : Vliyanie vysokikh temperatur na soderzhanie pigmentov v list'-
yakh tomatov v estestvennykh usloviyakh. [Effect of high temperature on pigment content in tomato leaves in natural conditions.] - Byull. vses. nauch.
issled. Inst. Rastenievod. Im. N.I. Vavilova *63* : 66 - 70, 1976. [In R.]

26496 - IVANCHANKA, V.M., URBANOVICH, T.A., KRUCHYNINA, S.S., MARSHAKOVA, M.I. :
Strukturna-funktsyyanal'naya rèaktsyya khlaraplastaŭ na ŭzdzeyanne roznymi
kantsèntratsyyami izaverynu. [Structural and functional reaction of chloroplasts on different concentrations of isoverin.] - Vestsi Akad. Navuk belarus. SSR, Ser. biyal. Navuk *1976* (5) : 40 - 42, 1976. [In Belorus.]

26497 - IVANCHENKO, V.M., MIKUL'SKAYA, S.A., GONCHARIK, M.N. : Vliyanie dimetilksan-
tina na intensivnost' fotosinteza v protsesse obezvozhivaniya assimilyatsion-
noĭ tkani. [Dimethylxanthine effect on photosynthetic rate during dehydration
of assimilatory tissue.] - Dokl. Akad. Nauk belorus. SSR *20* : 656 - 658, 1976.
[In R.]

26498 - IVANCHEV, V. : Biokhimichni pokazateli na rastezha i izpolzuvaneto im za opre-
delyane studoustoĭchivostta na lozata. [Biochemical growth indices and their
use in determining grapevine cold resistance.] - Fiziol. Rast. (Sofia) *2* (2):
94 - 100, 1976. [Photosynthates; in Bulg., ab : E, R.]

26499 - IVANOV, A.F., DERYUGINA, T.F., KRAVCHENKO, L.V., NOVIKOVA, A.A., RAKHTEENKO,
L.I. : Izuchenie fiziologo-biologicheskikh zakonomernosteĭ rosta i razvitiya
drevesnykh porod v zavisimosti ot usloviĭ mineral'nogo pitaniya, svetovogo i
vodnogo rezhima. [Physiological and biological laws of growth and development
of trees in connection with mineral nutrition, irradiance and water regime.]
- In : Regulyatsiya Rosta i Pitanie Rasteniĭ. Pp. 231 - 235. Zinatne, Riga
1976. [Ps; in R.]

26500 - IVANOV, B.N., AKULOVA, E.A. : Vliyanie ATF i usloviĭ fosforilirovaniya na ot-
noshenie H$^+$/e$^-$ i protonnyĭ obmen izolirovannykh khloroplastov. [Effect of ATP
and conditions of phosphorylation on the H$^+$/e$^-$ ratio and proton exchange in
isolated chloroplasts.] - Dokl. Akad. Nauk SSSR *227* : 999 - 1002, 1976. [In
R.]

26501 - IVE, J.R. : Growth and competition in annual legume-perennial grass pasture
in a dry monsoonal climate. - Aust. J. Ecol. *1* : 185 - 196, 1976. [Growth
analysis.]

26502 - IVE, J.R., ROSE, C.W., WALL, B.H., TORSSELL, B.W.R. : Estimation and simulation of sheet run-off. - Aust. J. Soil Res. *14* : 129 - 138, 1976. [Growth
analysis.]

26503 - IVERSON, R.L., BITTAKER, H.F., MYERS, V.B. : Loss of radiocarbon in direct
use of Aquasol for liquid scintillation counting of solutions containing ^{14}C-
-NaHCO$_3$. - Limnol. Oceanogr. *21* : 756 - 758, 1976.

26504 - IVLEV, A.A. : Voprosy teorii fraktsionirovaniya izotopov ugleroda v fotosinteziruyushchikh organizmakh. [Theory of fractionation of carbon isotopes in
photosynthesizing organisms.] - Uspekhi sovrem. Biol. *81* (1) : 84 - 105,
1976. [In R.]

26505 - IWAI, S., TANABE, Y., KAWASHIMA, N. :Origin of sequence heterogeneity of the
small subunit of Fraction 1 protein from *Nicotiana tabacum*. - Biochem. biophys. Res. Commun. *73* : 993 - 996, 1976.

26506 - **IWAKI, H., TAKEDA, G., UDAGAWA, T.** : Ecological studies on the photosynthesis of winter cereals II. Photosynthesis of wheat and rye plants under field conditions. - Proc. Crop Sci. Soc. Jap. *45* : 32 - 39, 1976.

26507 - **IZAWA, S., BERG, S.P.** : Phosphorylation associated with the DCMU-insensitive Hill reaction. - Biochem. biophys. Res. Commun. *72* : 1512 - 1518, 1976.

26508 - **JACKSON, A.H.** : Structure, properties and distribution of chlorophylls. - In: GOODWIN, T.W. (ed.) : Chemistry and Biochemistry of Plant Pigments. 2nd Ed. Vol. 1. Pp. 1 - 63. Academic Press, London - New York - San Francisco 1976.

26509 - **JACKSON, D.S., GIFFORD, H.H., CHITTENDEN, J.** : Environmental variables influencing the increment of *Pinus radiata* : (2) Effects of seasonal drought on height and diameter increment. - New Zeal. J. Forest Sci. *6* : 265 - 286, 1976.

26510 - **JACQUES, G., FIALA, M., NEVEUX, J., PANOUSE, M.** : Fertilisation de communautés phytoplanctoniques. II. Cas d'un milieu eutrophe : upwelling des côtes de Sahara espagnol. - J. exp. mar. Biol. Ecol. *24* : 165 - 176, 1976. [Chl.]

26511 - **JAHN, O.L., YOUNG, R.** : Changes in chlorophyll *a*, *b*, and the *a*/*b* ratio during color development in citrus fruit. - J. amer. Soc. hort. Sci. *101* : 416 - 418, 1976.

26512 - **JAMALE, B.B., JOSHI, G.V.** : Physiological studies in senescent leaves of mangroves. - Indian J. exp. Biol. *14* : 697 - 699, 1976. [Ps, Chl.]

26513 - **JANAUER, G.A., KINZEL, H.** : Die Wirkung von Atrazin auf den Stoffwechsel von *Phaseolus vulgaris* L. II. Kohlenhydrate und organische Säuren. - Z. Pflanzenphysiol. *77* : 383 - 394, 1976. [Ps.]

26514 - **JARVIS, P.G.** : The interpretation of the variations in leaf water potential and stomatal conductance found in canopies in the field. - Phil. Trans. roy. Soc. London B *273* : 593 - 610, 1976.

26515 - **JARVIS, P.G., JAMES, G.B., LANDSBERG, J.J.** : Coniferous forest. - In : MONTEITH, J.L. (ed.) : Vegetation and the Atmosphere. Vol. 2. Case Studies. Pp. 171 - 240. Academic Press, London - New York - San Francisco 1976. [Ps.]

26516 - **JASSBY, A.D., PLATT, T.** : Mathematical formulation of the relationship between photosynthesis and light for phytoplankton. - Limnol. Oceanogr. *21* : 540 - 547, 1976.

26517 - **JEANJEAN, R.** : The effect of metabolic poisons on ATP level and on active phosphate uptake in *Chlorella pyrenoidosa*. - Physiol. Plant. *37* : 107 - 110, 1976.

26518 - **JEFFREY, S.W.** : A report of green algal pigments in Central North Pacific Ocean. - Mar. Biol. *37* : 33 - 37, 1976.

26519 - **JEFFREY, S.W.** : Chlorophyll methodology in oceanography. - Annu. Rep. CSIRO mar. Biochem. Unit *1975-76* : 13 - 15, 1976.

26520 - **JEFFREY, S.W.** : Some effects of blue-green light on photosynthetic pigments and chloroplast structure in uni-cellular marine algae. - Annu. Rep. CSIRO mar. Biochem. Unit *1975-76* : 22 - 23, 1976.

26521 - **JEFFREY, S.W.** : The occurrence of chlorophyll c_1 and c_2 in algae. - J. Phycol. *12* : 349 - 354, 1976.

26522 - **JENNER, C.F.** : Wheat grains and spinach leaves as accumulators of starch. - In : WARDLAW, I.F., PASSIOURA, J.B. (ed.) : Transport and Transfer Processes in Plants. Pp. 73 - 83. Academic Press, New York - San Francisco - London 1976. [Ps.]

26523 - **JENNINGS, R.C., GARLASCHI, F.M., FORTI, G.** : Studies on the slow fluorescence decline in isolated chloroplasts. - Biochim. biophys. Acta *423* : 264 - 274, 1976.

26524 - **JENSEN, C.R.** : Effects of salinity in the root medium. III. Photosynthesis and leaf diffusive resistance in relation to leaf temperature and in relation to pre-salinity treatment. - Acta Agr. scand. *26* : 196 - 202, 1976.

26525 - JENSEN, R.G. : Effect of phosphate on starch formation during photosynthesis
with intact chloroplasts. - Plant Physiol. 57 (Suppl.) : 58, 1976.

26526 - JENSEN, R.G., BAHR, J.T. : Regulation of CO_2 incorporation via the pentose
phosphate pathway. - In : BURRIS, R.H., BLACK, C.C. (ed.) : CO_2 Metabolism
and Plant Productivity. Pp. 3 - 18. Univ. Park Press, Baltimore - London -
Tokyo 1976.

26527 - JENSEN, T.E., AYALA, R.P. : The fine structure of a tri-lamellar body in various species of Anabaena. - Protoplasma 89 : 91 - 103, 1976. [Chromatophore.]

*26528 - JENSEN, T.E., SICKO, L.M. : The fine structure of Chlorogloea fritschii cul-
tured in sodium acetate enriched medium. - Cytologia 38 : 381 - 391, 1973.
[Chromatophore.]

26529 - JEWSON, D.H. : The interaction of components controlling net phytoplankton
photosynthesis in a well-mixed lake (Lough Neagh, Northern Ireland). - Freshwater Biol. 6 : 551 - 576, 1976.

26530 - JITTS, H.R., MOREL, A., SAIJO, Y. : The relation of oceanic primary production to available photosynthetic irradiance. - Aust. J. mar. Freshwater Res.
27 : 441 - 454, 1976.

26531 - JOHNSON, C.E., BISCOE, P.V., CLARK, J.A., LITTLETON, E.J. : Turbulent transfer in a barley canopy. - Agr. Meteorol. 16 : 17 - 35, 1976. [Ps, CO_2 fluxes.]

26532 - JOHNSON, D.A., CALDWELL, M.M. : Water potential components, stomatal function,
and liquid phase water transport resistances of four arctic and alpine species in relation to moisture stress. - Physiol. Plant. 36 : 271 - 278, 1976.
[Stomatal resistance.]

26533 - JOHNSON, D.A., TIESZEN, L.L. : Aboveground biomass allocation, leaf growth,
and photosynthesis patterns in tundra plant forms in arctic Alaska. - Oecologia 24 : 159 - 173, 1976.

26534 - JOHNSON, F.L., BELL, D.T. : Plant biomass and net primary production along a
flood-frequency gradient in the streamside forest. - Castanea 41 : 156 - 165,
1976.

26535 - JOHNSON, H.B., TING, I.P. : Simultaneous measurements of photosynthesis and
transpiration in the field with a dual-label porometer Zac Hanscom, III. -
Plant Physiol. 57 (Suppl.) : 106, 1976.

26536 - JOHNSON, K.D., RAYLE, D.L. : Enhancement of CO_2 uptake in Avena coleoptiles
by fusicoccin. - Plant Physiol. 57 : 806 - 811, 1976.

26537 - JOHNSON, R.R., MOSS, D.N. : Effect of water stress on $^{14}CO_2$ fixation and
translocation in wheat during grain filling. - Crop Sci. 16 : 697 - 701,
1976.

26538 - JOLIOT, P., JOLIOT, A. : Localisation de la chlorophylle all dans la membrane
du thylakoïde. Effet de l'hydroxylamine. - Compt. rend. Acad. Sci. Paris, Sér.
D 283 : 393 - 396, 1976.

26539 - JOLIVET, E. : Les différentes modalités de carboxylation photosynthétique chez
les espèces de type C_3 et de type C_4. - Physiol. vég. 14 : 563 - 594, 1976.

26540 - JONES, A.K., BILLINGTON, C.A. : Observations on carbon dioxide fixation in
the Nant-y-Moch reservoir, Ceredigion, Wales. - Hydrobiologia 50 : 33 - 42,
1976.

26541 - JONES, H.G. : Crop characteristics and the ratio between assimilation and
transpiration. - J. appl. Ecol. 13 : 605 - 622, 1976.

26542 - JONES, J.H. : $^{13}C/^{12}C$ ratios in compressions of specific fossil plants. -
Plant Physiol. 57 (Suppl.) : 6, 1976.

26543 - JONES, J.R., BACHMANN, R.W. : Prediction of phosphorus and chlorophyll levels
in lakes. - J. Water Pollut. Contr. Fed. 48 : 2176 - 2182, 1976.

26544 - JONES, L.W., BISHOP, N.I. : Simultaneous measurement of oxygen and hydrogen
exchange from the blue-green alga Anabaena. - Plant Physiol. 57 : 659 - 665,
1976.

26545 - JONES, O.T.G. : Chlorophyll *a* biosynthesis. - Phil. Trans. roy. Soc. London
B *273* : 207 - 225, 1976.

26546 - JONES, W.T., MANGAN, J.L. : Large-scale isolation of fraction 1 leaf protein
(18S) from lucerne (*Medicago sativa* L.). - J. agr. Sci. *86* :495 - 501, 1976..

26547 - JØRGENSEN, S.E., MEJER, H. : Modelling the global cycle of carbon, nitrogen
and phosphorus and their influence on the global heat balance. - Ecol. Model.
2 : 19 - 31, 1976. [Production.]

26548 - JØRGENSEN, S.E., MEJER, H. : Modelling the global heat balance. - Ecol. Model.
2 : 273 - 277, 1976.

26549 - JOSHI, G.V. : Early products of photosynthesis in plants exposed to salt and
water stresses. - Plant biochem. J. *3* : 33 - 43, 1976.

26550 - JOSHI, G.V. : Photosynthesis under conditions of stress. - Proc. Indian nat.
Sci. Acad. *42 B* : 279 - 289, 1976.

B26551 - JOSHI, G.V. : Studies in Photosynthesis under Saline Conditions. - Shivaji
Univ., Kolhapur 1976.

26552 - JOŠT, M., GLATKI-JOŠT, M., HRUST, V., MILOHNIĆ, J. : Effects of *T. timopheevi*
cytoplasm on some traits of male-sterile common wheat. - Poljoprivred. znanstv.
Smotra *38* (48) : 39 - 57, 1976. [Chl.]

26553 - JOYARD, J., DOUCE, R. : Mise en evidence et role des diacylglycerols de l'enveloppe des chloroplastes d'epinard. - Biochim. biophys. Acta *424* : 125 - 131,
1976.

26554 - JOYARD, J., DOUCE, R. : Préparation et activités enzymatiques de l'enveloppe
des chloroplastes d'Épinard. - Physiol. vég. *14* : 31 - 48, 1976.

26555 - JULG, A., FRANÇOIS, P., RAJZMANN, M.A. : A theoretical and comparative CNDO
study of chlorophylls *a* and *b* and their beryllium homologs. - J. theor. Biol.
57 : 391 - 394, 1976.

26556 - JUNGE, W. : Flash kinetic spectrophotometry in the study of plant pigments.
- In : GOODWIN, T.W. (ed.) : Chemistry and Biochemistry of Plant Pigments.
2nd Ed. Vol. 2. Pp. 233 - 333. Academic Press, London - New York - San Francisco 1976.

26557 - JUPIN, H., PICAUD, A., GARNIER, J., HAUSWIRTH, N. : Relations entre l'ultrastructure plastidale et l'état du photosystème I chez *Chlamydomonas Reinhard-
ti*. - Compt. rend. Acad. Sci. Paris, Sér. D *283* : 627 - 630, 1976.

26558 - JURGENSON, J.E., BEALE, S.I., TROXLER, R.F. : Biosynthesis of δ-aminolevulinic acid in the unicellular rhodophyte, *Cyanidium caldarium*. - Biochem. biophys. Res. Commun. *69* : 149 - 157, 1976.

26559 - JURSINIC, P., WARDEN, J., GOVINDJEE : A major site of bicarbonate effect in
System II reaction. Evidence from EST signal II $_{vf}$; fast fluorescence yield
changes and delayed light emission. - Biochim. biophys. Acta *440* : 322 - 330,
1976.

26560 - JYUNG, W.H., CAMP, M.E. : The effect of zinc on the formation of ribulose diphosphate carboxylase of *Phaseolus vulgaris*. - Physiol. Plant. *36* : 350 - 355,
1976.

26561 - KAFALIEVA-BOEVA, D.N. : Spektralni kharakteristiki na vtora pigmentna sistema, polucheni chrez elektronen spinov rezonans. [Spectral characteristics of
the second pigment system obtained by the electron spin resonance method.]
- Fiziol. Rast. (Sofia) *2* (3) : 19 - 23, 1976. [In Bulg., ab : E, R.]

26562 - KAHN, A., AVIVI-BLEISER, N., von WETTSTEIN, D. : Genetic regulation of chlorophyll synthesis analyzed with double mutants in barley. - In : BÜCHER, T.,
NEUPERT, W., SEBALD, W., WERNER, S. (ed.) : Genetics and Biogenesis of Chloroplasts and Mitochondria. Pp. 119 - 131. North-Holland Publ. Co., Amsterdam
- New York - Oxford 1976.

26563 - **KAIRESALO, T.** : Measurements of production of epilithiphyton and littoral plankton in Lake Pääjärvi, southern Finland. - Ann. bot. fenn. *13* : 114 - 118, 1976.

26564 - **KAISER, J.A.C.** : The use of pyranometers for underwater total radiant energy flux measurements. - Deep-Sea Res. *23* : 881 - 887, 1976.

26565 - **KAISER, W.** : The effect of hydrogen peroxide on CO_2 fixation of isolated intact chloroplasts. - Biochim. biophys. Acta *440* : 476 - 482, 1976.

26566 - **KAISER, W., URBACH, W.** : Rates and properties of endogenous cyclic photophosphorylation of isolated intact chloroplasts measured by CO_2 fixation in the presence of dihydroxyacetone phosphate. - Biochim. biophys. Acta *423* : 91 - - 102, 1976.

26567 - **KALER, V.L., FRIDLYAND, L.E.** : Model' fotosinteza, imitiruyushchaya izmenenie khoda uglekislotnykh krivykh pri var'irovanii temperatury i kontsentratsii O_2. [Model of photosynthesis simulating change in CO_2 curves under varying temperature and oxygen concentration.] - In : Gazometricheskoe Issledovanie Fotosinteza i Dykhaniya Rasteniĭ. Pp. 51 - 53. Akad. Nauk SSSR, Tartu 1976. [In R.]

*26568 - **KALFF, J.** : A diel periodicity in the optimum light intensity for maximum photosynthesis in natural phytoplankton populations. - J. Fish. Res. Board Can. *26* : 463 - 468, 1969.

26569 - **KALIMULLINA, Kh.K., GOLOVATYĬ, A.G., GOLOVATYĬ, V.G.** : Izmeneniya v. soderzhanii sakharov i aminokislot v ezhe sbornoĭ v zavisimosti ot temperatury vozdukha, vlazhnosti pochvy i urovnya udobreniĭ. [Changes in the content of sugars and amino acids in orchard grass depending on air temperature, soil moisture and level of fertilizers.] - Fiziol. Biokhim. kul't. Rast. *8* : 209 - - 214, 1976. [Photosynthates; in R, ab : E.]

26570 - **KALINICHENKO, R.A.** : Otsenka produktsionnykh vozmozhnosteĭ fitobentosa v kanale sev. Donets - Donbass. [Estimation of the production potentials of the phytobenthos in the northern Donets - Donbass Basin Canal.] - Gidrobiol. Zh. *12* (1) : 98 - 102, 1976. [In R.]

26571 - **KALININA, S.G.** : Sezonnye izmeneniya intensivnosti fotosinteza v melkovodnoĭ i glubokovodnoĭ zonakh Tsimlyanskogo vodokhranilishcha. [Seasonal changes in photosynthetic rate in shallow- and deep-water zones of Tsimlyansk reservoir.] - Izv. gos. nauch.-issled. Inst. ozer. rech. ryb. Khoz. *94* (Sostoyanie i Perspektivy Rybokhozyaĭstvennogo Osvoeniya Ozer i Vodokhranilishch):62 - 66, 1976. [In R, ab : E.]

26572 - **KALLIS, A., OYA, V., LAĬSK, A.** : Vozmozhnaya izmenchivost' sostavlyayushchikh temnovogo dykhaniya v techenie vegetatsionnogo perioda u yachmenya. [Possible variability of dark respiration components during vegetation period in barley.] - In : Gazometricheskoe Issledovanie Fotosinteza i Dykhaniya Rasteniĭ. Pp. 54 - 57. Akad. Nauk SSSR, Tartu 1976. [Growth, model; in R.]

26573 - **KAN, K.-S., THORNBER, J.P.** : The light harvesting chlorophyll *a/b*-protein complex of *Chlamydomonas reinhardii*. - Plant Physiol. *57* : 47 - 52, 1976.

26574 - **KANIUGA, Z.** : Regulacja włączenia CO_2 w procesie fotosyntezy. [Regulatory mechanisms in photosynthetic CO_2 fixation.] - Postępy Biochem. *22* : 247 - 305, 1976. [In Pol.]

*26575 - **KANNANGARA, C.G., STUMPF, P.K.** : Fat metabolism in higher plants. LVI. Distribution and nature of biotin in chloroplasts of different plant species. - Arch. Biochem. Biophys. *155* : 391 - 399, 1973.

26576 - **KAO, O.H.W., EDWARDS, M.R., Mac COLL, R., BERNS, D.S.** : Thermophilic phycocyanin. - In : ZUBER, H. (ed.) : Enzymes and Proteins from Thermophilic Microorganisms. Pp. 291 - 305. Birkhäuser Verlag, Basel - Stuttgart 1976.

26577 - **KAPLAN, A., GALE, J., POLJAKOFF-MAYBER, A.** : Simultaneous measurement of oxygen, carbon dioxide, and water vapour exchange in intact plants. - J. exp. Bot. *27* : 214 - 219, 1976.

26578 - KAPLAN, A., GALE, J., POLJAKOFF-MAYBER, A. : Resolution of net dark fixation
of carbon dioxide into its respiration and gross fixation components in *Bryo-phyllum daigremontianum*. - J. exp. Bot. *27* : 220 - 230, 1976.

26579 - KAPLAN, D.T., ROHDE, R.A., TATTAR, T.A. : Leaf diffusive resistance of sun-flowers infected by *Pratylenchus penetrans*. - Phytopathology *66* : 967 - 969,
1976.

26580 - KAPPEN, L., LANGE, O.L., SCHULZE, E.-D., EVENARI, M., BUSCHBOM, U. : Distri-butional pattern of water relations and net photosynthesis of *Hammada scopa-ria* (POMEL) ILJIN in desert environment. - Oecologia *23* : 323 - 334, 1976.

26581 - KAR, M., MISHRA, D. : Catalase, peroxidase, and polyphenoloxidase activities
during rice leaf senescence. - Plant Physiol. *57* : 315 - 319, 1976. [Chl.]

26582 - KARABASHEV, G.S., SOLOV'EV, A.N. : Sutochnyĭ ritm fluorestsentsii khlorofilla
fitoplanktona v deyatel'nom sloe okeana. [Diurnal rhythm of phytoplankton
chlorophyll fluorescence in the active layer of ocean.] - Okeanologiya *16* :
316 - 323, 1976. [In R, ab : E.]

26583 - KARAKASHIAN, M.W., SCHWEIGER, H.G. : Evidence for a cycloheximide-sensitive
component in the biological clock of *Acetabularia*. - Exp. Cell Res. *98* : 303-
- 312, 1976. [Ps.]

26584 - KARAKASHIAN, M.W., SCHWEIGER, H.G. : Circadian properties of the rhythmic
system in individual nucleated and enucleated cells of *Acetabularia mediter-ranea*. - Exp. Cell Res. *97* : 366 - 377, 1976. [Ps.]

26585 - KARAKASHIAN, M.W., SCHWEIGER, H.G. : Temperature dependence of cycloheximide-
-sensitive phase of circadian cycle in *Acetabularia mediterranea*. - Proc. nat.
Acad. Sci. USA *73* : 3216 - 3219, 1976. [Ps.]

26586 - KARANOV, E., PAVLOVA, A. : Wirkung einiger Cytokinin-analoge Derivate des N-
-alkyl-N'-phenylharnstoffs auf die Verzögerung des Chlorophyllabbaus und auf
die Nitratreduktase-Aktivität in jungen Blättern von *Hordeum vulgare*. - Bio-chem. Physiol. Pflanzen *170* : 479 - 485, 1976.

26587 - KARANOV, E.N., POGONCHEVA, E.M. : Vliyanie na abstsisinovata kiselina v"rkhu
razrushavaneto na khlorofila v otk"snati lista i vzaimodeĭstvieto ĭ s drugi
rastezhni. [Effect of abscisic acid on chlorophyll destruction in detached
leaves and its interaction with other growth regulators.] - Fiziol. Rast.
(Sofia) *2* (1) : 3 - 12, 1976. [In R, ab : E.]

26588 - KARAVAEV, V.A., KUKUSHKIN, A.K. : Issledovanie sostoyaniya élektronno-trans-portnoĭ tsepi v list'yakh vysshikh rasteniĭ metodom bystroĭ induktsii fluores-tsentsii. [Application of the fast fluorescence induction to the study of the
electron transport chain states in leaves of higher plants.] - Biofizika *21* :
862 - 866, 1976. [In R, ab : E.]

26589 - KARIMOV, Kh.Kh., NIKOLAEVA, M.I. : O sinteze DNK v protsesse assimilyatsii
prorostkami gorokha $C^{14}O_2$. [DNA synthesis in the process of $^{14}CO_2$ assimila-tion by pea seedlings.] - Dokl. Akad. Nauk tadzh. SSR *19* (5) : 56 - 58, 1976.
[In R, ab : Tajik.]

26590 - KARL, D.M., HOLM-HANSEN, O. : Effects of luciferin concentration on the quan-titative assay of ATP using crude luciferase preparations. - Anal. Biochem.
75 : 100 - 112, 1976.

26591 - KARMANOV, V.G., ERMAKOV, E.I., SERGEEVA, E.A. : Gazoobmen rasteniĭ tomatov
v usloviyakh intensivnoĭ svetokul'tury. [Gas exchange in tomato in intensive
light culture.] - In : Gazometricheskoe Issledovanie Fotosinteza i Dykhaniya
Rasteniĭ. Pp. 60 - 62. Akad. Nauk SSSR, Tartu 1976. [In R.]

26592 - KAROLIN, A.Yu. : Apparatura i metodika opredeleniya transpiratsii i CO_2-obme-na nad- i podzemnykh chasteĭ otdel'nogo rasteniya v konditsionirovannykh uslo-viyakh. [Apparatus and procedure for determining transpiration and CO_2 exchan-ge in shoots and underground parts of an individual plant in controlled con-ditions.] - In : Gazometricheskoe Issledovanie Fotosinteza i Dykhaniya Raste-niĭ. Pp. 57 - 59. Akad. Nauk SSSR, Tartu 1976. [In R.]

26593 - KAROLIN, A.Yu., MOLDAU, Kh.A. : Faktorostatnaya kamera s registratsieĭ trans-piratsii i CO_2-obmena nadzemnykh i podzemnykh chasteĭ rasteniya. [An air conditioned chamber for recording transpiration and CO_2 exchange of shoots and underground plant parts.] - Fiziol. Rast. *23* : 630 - 634, 1976. [In R, ab : E.]

26594 - **KARPAVA, T.A., SAŬCHANKA, G.Ya., SHLYK, A.A.** : Spektral'nyya peratvarênni khlarafilidu i rêgeneratsyya protakhlarafilidu ŭ postêtyyaliravanykh listsyakh yachmenyu z blakiravanym byalkovym sintêzam. [Spectral transformations of chlorophyllide and regeneration of protochlorophyllide in postetiolated barley plant leaves with blocked protein synthesis.] - Vestsi Akad. Navuk belarus. SSR, Ser. biyal. Navuk *1976* (5) : 25 - 33, 138, 1976. [In Belorus., ab : R.]

26595 - **KARPILOV, Yu.S.** : Organizatsiya i regulyatsiya uglerodnogo metabolizma pri fotosinteze C_3- i C_4-rasteniĭ. [Organization and regulation of carbon metabolism in photosynthesis of C_3- and C_4-plants.] - In : Itogi Issledovaniya Mekhanizma Fotosinteza. Pp. 58 - 62. Pushchino 1976. [In R.]

26596 - **KARPILOV, Yu.S., BIL', K.Ya.** : Transport promezhutochnykh produktov fotosinteza pó tsitoplazme kletok assimilyatsionnykh tkaneĭ C_4-rasteniĭ. [Transport of photosynthesis intermediates in cytoplasm of assimilation tissue cells of C_4-plants.] - Dokl. Akad. Nauk SSSR *226* : 1469 - 1471, 1976. [In R.]

26597 - **KARPILOV, Yu.S., KUZNETSOVA, L.G., PERSANOV, V.M., BLINOVA, I.V., GOSTIMSKIĬ, S.A.** : Aktivnost' fermentov uglerodnogo metabolizma i sostav produktov fotosinteza v list'yakh mutantov gorokha. [Activity of enzymes of carbon metabolism and composition of photosynthetic products in leaves of pea mutants.] - Fiziol. Rast. *23* : 460 - 466, 1976. [In R, ab : E.]

26598 - **KARPILOV, Yu.S., NOVITSKAYA, I.L., LYUBIMOV, V.Yu., BELOBRODSKAYA, L.K., KARPILOVA, I.F., POPOVA, E.I.** : Okislenie promezhutochnykh produktov fotosinteza i tsikla Krebsa khloroplastami, kletkami i list'yami kukuruzy na svetu. [Oxidation of intermediate products of photosynthesis and Krebs cycle by chloroplasts, cells and leaves of maize in light.] - In : Gazometricheskoe Issledovanie Fotosinteza i Dykhaniya Rasteniĭ. Pp. 63 - 64. Akad. Nauk SSSR, Tartu 1976. [In R.]

26599 - **KARPUSHKIN, L.T.** : Fotosinteticheskaya deyatel'nost' ogurtsov v zakrytom grunte. [Photosynthetic activity of cucumbers in protected ground.] - In : Itogi Issledovaniya Mekhanizma Fotosinteza. Pp. 79 - 80. Pushchino 1976. [In R.]

26600 - **KARPUSHKIN, L.T., POLYAKOV, M.A.** : Opredelenie kineticheskikh parametrov gazovoĭ fazy CO_2- i H_2O-gazoobmena list'ev rasteniĭ. [Determination of the kinetic parameters of the gas phase of CO_2- and water vapour exchange in plant leaves.] - In : Gazometricheskoe Issledovanie Fotosinteza i Dykhaniya Rasteniĭ. Pp. 65 - 66. Akad. Nauk SSSR, Tartu 1976. [In R.]

26601 - **KARUNEN, P., VALANNE, N., WILKINSON, R.E.** : Influence of *S*-ethyl dipropylthiocarbamate on growth, chlorophyll and carotenoid production and chloroplast ultrastructure of germinating *Polytrichum commune*. - Bryologist *79* : 332 - - 338, 1976.

26602 - **KASAMO, K.** : The role of the epidermis in kinetin-induced retardation of chlorophyll degradation in tobacco leaf discs during senescence. - Plant Cell Physiol. *17* : 1297 - 1307, 1976.

26603 - **KASEMIR, H., HUBER, P., MOHR, H.** : Timing of the initial action of phytochrome with regard to protochlorophyll synthesis in the mustard seedling. - Planta *132* : 157 - 160, 1976.

26604 - **KASEMIR, H., PREHM, G.** : Control of chlorophyll synthesis by phytochrome. III. Does phytochrome regulate the chlorophyllide esterification in mustard seedlings ? - Planta *132* : 291 - 295, 1976.

26605 - **KASZUBIAK, H.** : Correlation between determinations of chlorophyll-type compounds, nitrogen available for plants and number of algae in the soil. - Pol. J. Soil Sci. *9* : 47 - 51, 1976.

26606 - **KATAYAMA, M.** : [Electrophoresis of plant carbonic anhydrase.] - Nippon Nôgei-kagaku Kaishi [J. agr. Chem. Soc. Jap.] *50* : 621 - 623, 1976. [In Jap., ab : E.]

26607 - **KATO, S., HOZYO, Y.** : [The interrelationship between translocation of ^{14}C-photosynthate and $^{14}CO_2$ exposed leaf position on the grafts of *Ipomoea*.] - Proc. Crop Sci. Soc. Jap. *45* : 351 - 356, 1976. [In Jap., ab : E.]

26608 - **KATOH, T., OHKI, K.** : Regeneration of photosystem II and phycobilin pigments in photoorganotrophically grown *Anabaena variabilis*. - Plant Cell Physiol. *17* : 525 - 536, 1976.

26609 - **KATZ, J.J., OETTMEIER, W., NORRIS, J.R.** : Organization of antenna and photo--reaction centre chlorophylls on the molecular level. - Phil. Trans. roy. Soc. London B *273* : 227 - 253, 1976.

26610 - **KAUFMANN, K.J., PETTY, K.M., DUTTON, P.L., RENTZEPIS, P.M.** : Picosecond kinetics in reaction centers of *Rps. sphaeroides* and the effects of ubiquinone extraction and reconstitution. - Biochem. biophys. Res. Commun. *70* : 839 -
- 845, 1976.

26611 - **KAUFMANN, K.J., RENTZEPIS, P.M., STOECKENIUS, W., LEWIS, A.** : Primary photochemical processes in bacteriorhodopsin. - Biochem. biophys. Res. Commun. *68* : 1109 - 1115, 1976.

26612 - **KAUFMANN, M.R.** : Water transport through plants : current perspectives. - In: WARDLAW, I.F., PASSIOURA, J.B. (ed.) : Transport and Transfer Processes in Plants. Pp. 313 - 327. Academic Press, New York - San Francisco - London 1976. [Stomatal resistance.]

26613 - **KAWAMURA, Y., MIYAKE, H., MAEDA, E.** : Effects of oxadiazon on the ultrastructure of barnyardgrass chloroplasts. - Proc. Crop Sci. Soc. Jap. *45* : 538 -
- 544, 1976.

26614 - **KAZAKOVA, A.A., KISELEV, B.A., EVSTIGNEEV, V.B.** : Okislitel'no-vosstanovitel'-nye potentsialy khlorofillovykh pigmentov fotosinteziruyushchikh organizmov, stoyashchikh na raznykh urovnyakh ěvolyutsionnogo razvitiya. [Redox potentials of chlorophyll pigments in photosynthesizing organisms of different evolutionary stages.] - Biofizika *21* : 434 - 438, 1976. [In R, ab : E.]

26615 - **KE, B., HAWKRIDGE, F.M., SAHU, S.** : Redox titration of fluorescence yield of photosystem II. - Proc. nat. Acad. Sci. USA *73* : 2211 - 2215, 1976.

26616 - **KE, B., SUGAHARA, K., SAHU, S.** : Light-induced absorption changes in photosystem I at low temperatures. - Biochim. biophys. Acta *449* : 84 - 94, 1976.

26617 - **KEAST, J.F., GRANT, B.R.** : Chlorophyll *a:b* ratios in some siphonous green algae in relation to species and environment. - J. Phycol. *12* : 328 - 331, 1976.

26618 - **KECK, R.W., OGREN, W.L.** : Differential oxygen response of photosynthesis in soybean and *Panicum milioides*. - Plant Physiol. *57* (Suppl.) : 59, 1976.

26619 - **KECK, R.W., OGREN, W.L.** : Differential oxygen response of photosynthesis in soybean and *Panicum milioides*. - Plant Physiol. *58* : 552 - 555, 1976.

26620 - **KĚERBERG, O., VYARK, Ě., KĚERBERG, Kh., PYARNIK, T.** : Spektral'naya zavisimost' glikolatnogo puti i gazoobmena u list'ev fasoli. [Spectral dependence of glycollate pathway and gas exchange in bean leaves.] - In : Gazometricheskoe Issledovanie Fotosinteza i Dykhaniya Rasteniǐ. Pp. 83 - 85. Akad. Nauk SSSR, Tartu 1976. [In R.]

26621 - **KELLER, T.** : Der Einfluss von Schwefeldioxid als Luftverunreinigung auf die Assimilation der Fichte. - Beiheft Z. schweiz. Forstvereins *57* : 48 - 53, 1976.

26622 - **KELLY, G.J., LATZKO, E.** : Inhibition of spinach-leaf phosphofructokinase by 2-phosphoglycollate. - FEBS Lett. *68* : 55 - 58, 1976.

26623 - **KELLY, G.J., LATZKO, E., GIBBS, M.** : Regulatory aspects of photosynthetic carbon metabolism. - Annu. Rev. Plant Physiol. *27* : 181 - 205, 1976.

26624 - **KELLY, G.J., ZIMMERMANN, G., LATZKO, E.** : Light induced activation of fructose-1,6-bisphosphatase in isolated intact chloroplasts. - Biochem. biophys. Res. Commun. *70* : 193 - 199, 1976.

26625 - KEMPH, G.S. : Measuring fibrous roots with a leaf area meter. - J. Range Manage. *29* : 85 - 86, 1976.

26626 - KENNEDY, R.A. : Relationship between leaf development, carboxylase enzyme activities and photorespiration in the C_4-plant *Portulaca oleracea* L. - Planta *128* : 149 - 154, 1976.

26627 - KENNEDY, R.A. : The effect of different killing techniques on primary photosynthetic product identification in C_4 plants. - Plant Physiol. *57* (Suppl.) : 33, 1976.

26628 - KENNEDY, R.A. : Photosynthesis and photorespiration in C_3 and C_4 plant tissue cultures : significance of Kranz anatomy to operation of the C_4 pathway. - Plant Physiol. *57* (Suppl.) : 53, 1976.

26629 - KENNEDY, R.A. : Photorespiration in C_3 and C_4 plant tissue cultures. Significance of Kranz anatomy to low photorespiration in C_4 plants. - Plant Physiol. *58* : 573 - 575, 1976.

26630 - KENNER, G.W., RIMMER, J., SMITH, K.M., UNSWORTH, J.F. : Studies on the biosynthesis of the *Chlorobium* chlorophylls. - Phil. Trans. roy. Soc. London B *273* : 255 - 276, 1976.

26631 - KERBER, N.L., PUCHEU, N.L., GARCIA, A.F. : Isolation of a membrane-bound protein having coupling factor capacity as well as ADP-P_i exchange and ADPase activities from *Rhodopseudomonas viridis*. - FEBS Lett. *72* : 63 - 66, 1976.

26632 - KERESZTES, Á., DAVEY, M.R., LÅNG, F. : A membrán-szerkezet, valamint a két fotokémiai rendszer aktivitásának összefüggése normális és mutáns *Tradescantia* kloroplasztiszokban. [Freeze etched membrane faces and photosynthetic activity in normal and mutant *Tradescantia* chloroplasts.] - Biologia (Budapest) *24* (2) : 133 - 147, 1976. [In Hung., ab : E.]

26633 - KERESZTES, Á., DAVEY, M.R., LÅNG, F. : Freeze-etched membrane faces and photosynthetic activity in normal and mutant *Tradescantia* chloroplasts. - Protoplasma *90* : 1 - 14, 1976.

26634 - KERIN, V. : Vliyanie na retardanta khlorkholinkhlorid v"rkhu biosintezata na plastidni pigmenti i zdravinata na khlorofilbelt"chniya kompleks. [Influence of the retardant chlorocholine chloride on plastid pigment biosynthesis and on the strength of the chlorophyll-protein complex.] - Fiziol. Rast. (Sofia) *2* (2) : 88 - 93, 1976. [In Bulg., ab : E, R.]

26635 - KERNER, H., KOCH, W. : Struktur und Funktion des Assimilationsapparates einer mitherrschenden Fichte (*Picea abies* (L.) KARST.) in einem Altbestand des Ebersberger Forstes bei München. 1. Methodik der Gaswechselmessung. - Photosynthetica *10* : 324 - 334, 1976.

26636 - KERSCHER, L., OESTERHELT, D. : A ferredoxin from halobacteria. - FEBS Lett. *67* : 320 - 322, 1976.

26637 - KERSCHER, L., OESTERHELT, D., CAMMACK, R., HALL, D.O. : A new plant-type ferredoxin from Halobacteria. - Europe. J. Biochem. *71* : 101 - 107, 1976.

26638 - KESSICK, M.A. : The calibration of closed-end manometric biochemical oxygen demand respirometers. - Biotechnol. Bioeng. *18* : 595 - 598, 1976.

26639 - KESSLER, E. : Comparative physiology, biochemistry, and the taxonomy of *Chlorella (Chlorophyceae)*. - Plant Syst. Evol. *125* : 129 - 138, 1976. [Car.]

26640 - KESSLER, E. : Hydrogen metabolism of eukaryotic organisms. - In : Microbial Production and Utilization of Gases. Pp. 247 - 254. Akad. Wiss., Göttingen 1976.

26641 - KHADKE, V., MERRETT, M.J. : The physiology of *Neospongiococcum ovatum* DEASON. - New Phytol. *77* : 635 - 639, 1976. [Ps.]

26642 - KHAILOV, K.M. : Fiziologicheskaya raznokachestvennost' élementov sloevishcha *Ascophyllum nodosum, Fucus vesiculosus* i *Fucus inflatus* iz Barentseva morya. [Physiological differences of thallus elements in *Ascophyllum nodosum, Fucus vesiculosus* and *Fucus inflatus* from the Barents Sea.] - Fiziol. Rast. *23* : 835 - 839, 1976. [Ps; in R, ab : E.]

26643 - KHAILOV, K.M. : The relationships between weight, length, age and intensity
         of photosynthesis and organotrophy of macrophytes in the Barents Sea. - Bot.
         Mar. *19* : 335 - 339, 1976.

26644 - KHAILOV, K.M., FIRSOV, Yu.K. : The relationships between weight, length, age
         and intensity of photosynthesis and organotrophy in thallus of *Cystoseira
         barbata* from the Black Sea. - Bot. Mar. *19* : 329 - 334, 1976.

26645 - KHAĬLOV, K.M., FIRSOV, Yu.K. : Fotosintez i organotrofiya morskikh makrofitov
         kak funktsiya individual'nogo vesa ikh tallomov. [Photosynthesis and organo-
         trophy in seaweeds as a function of individual weight of their thalli.] Biol.
         Morya *1976* (6) : 47 - 51, 1976. [In R, ab : E.]

26646 - KHAIRI, M.M.A., HALL, A.E. : Temperature and humidity effects on net photo-
         synthesis and transpiration of citrus. - Physiol. Plant. *36* : 29 - 34, 1976.

26647 - KHAIRI, M.M.A., HALL, A.E. : Comparative studies of net photosynthesis and
         transpiration of some citrus species and relatives. - Physiol. Plant. *36* :
         35 - 39, 1976.

26648 - KHAVARI-NEJAD, R.A., HANNAY, J.W. : Growth of tomato plants in different oxy-
         gen concentrations. - Plant Physiol. *57* (Suppl.) : 7, 1976.

26649 - KHERA, P.K., TILNEY-BASSETT, R.A.E. : Fine structural observations of embryo
         development in *Pelargonium* x *Hortorum* BAILEY : with normal and mutant plas-
         tids. - Protoplasma *88* : 7 - 23, 1976.

26650 - KHETCH, M.D. : Varianty fotosinteticheskogo metabolizma ugleroda i ikh fizio-
         logicheskie preimushchestva. [Variants of photosynthetic carbon metabolism
         and their physiological advantages.] - Fiziol. Biokhim. kul't. Rast. *8* : 473-
         - 482, 1976. [In R, ab : E.]

*26651 - KHODZHAEV, A.S., RODIMTSEVA, N.E. : Gazoobmen i chistaya produktivnost' foto-
         sinteza u khlopchatnika pri povyshennoĭ norme udobreniya. [Effect of increas-
         ed fertilizer rate on gas exchange and net productivity of photosynthesis in
         cotton.] - Uzb. biol. Zh. *19* (3) : 26 - 28, 76, 1975. [In R, ab : Uz .]

26652 - KHOKHAR, M.F.K., PANDEY, H.N. : Biomass, productivity and growth analysis of
         two varieties of paddy. - Trop. Ecol. *17* : 125 - 131, 1976.

*26653 - KHRISTIN, M.S., AKULOVA, E.A. : Obnaruzhenie ferredoksinov i plastotsianina
         v rannem ontogeneze i pri zelenenii ėtiolirovannykh prorostkov gorokha. [De-
         tection of ferredoxins and plastocyanin in the early ontogenesis and in green-
         ing etiolated pea sprouts.] - Dokl. Akad. Nauk SSSR *223* : 758 - 761, 1975.
         [In R.]

26654 - KHRISTIN, M.S., AKULOVA, E.A. : Dve formy ferredoksina list'ev gorokha. [Two
         forms of ferredoxin in pea leaves.] - Biokhimiya *41* : 500 - 505, 1976. [In R,
         ab : E.]

26655 - KIEFER, D.A., ENNS, T. : A steady-state model of light-, temperature-, and
         carbon-limited growth of phytoplankton. - In : CANALE, R.P. (ed.) : Modeling
         Biochemical Processes in Aquatic Ecosystems. Pp. 319 - 336. Ann Arbor Sci.
         Publ., Ann Arbor 1976.

26656 - KIEFER, D.A., HOLM-HANSEN, O., BERMAN, T. : Phytoplankton and primary produc-
         tion in the Gulf of California. - J. Phycol. *12* (Suppl.) : 33, 1976. [Chl.]

26657 - KIM, V.A., ELFIMOV, E.I., VOZNYAK, V.M., EVSTIGNEEV, V.B. : Issledovanie mek-
         hanizma fotokhimicheskogo okisleniya bakterioviridina. [Mechanism of photo-
         chemical oxidation of bacterioviridin.] - Biofizika *21* : 50 - 54, 1976. [In
         R, ab : E.]

26658 - KIM, V.A., VOZNYAK, V.M., ELFIMOV, E.I., EVSTIGNEEV, V.B. : Issledovanie mek-
         hanizma fotokhimicheskogo okisleniya i vosstanovleniya bakteriokhlorofilla
         "c" (bakterioviridina). [Mechanism of photochemical oxidation and reduction
         of bacteriochlorophyll *a* (bacterioviridin).] - Zh. prikl. Spektroskop. *25* :
         836 - 840, 1976. [In R.]

26659 - KING, D., GIBBS, M. : Hydrogen metabolism in photosynthetic organisms : the
         role of carbon metabolism. - Plant Physiol. *57* (Suppl.) : 60, 1976.

26660 - **KING, M.-T., DREWS, G.** : Isolation and partial characterization of the cytochrome oxidase from *Rhodopseudomonas palustris*. - Europe. J. Biochem. *68* :
5 - 12, 1976.

26661 - **KING, R.J., SCHRAMM, W.** : Determination of photosynthetic rates for the marine algae *Fucus vesiculosus* and *Laminaria digitata*. - Mar. Biol. *37* : 209 -
- 213, 1976.

26662 - **KING, R.J., SCHRAMM, W.** : Photosynthetic rates of benthic marine algae in relation to light intensity and seasonal variations. - Mar. Biol. *37* : 215 -
- 222, 1976.

26663 - **KIRBY, C.J., GOSSELINK, J.G.** : Primary production in a Louisiana gulf coast
Spartina alterniflora marsh. - Ecology *57* : 1052 - 1059, 1976.

26664 - **KIRCHANSKI, S.J.** : Copper ferricyanide localization of Photosystem II in glutaraldehyde fixed and unfixed chloroplasts. - J. Ultrastruct. Res. *57* : 113 -
- 119, 1976.

26665 - **KIRCHANSKI, S.J., PARK, R.B.** : Comparative studies of the thylakoid proteins
of mesophyll and bundle sheath plastids of *Zea mays*. - Plant Physiol. *58* :
345 - 349, 1976.

26666 - **KIRICHENKO, A.B., KIRICHENKO, E.B., CHEBOTAR', A.A., SMOLYGINA, L.D., SERDYUK,
O.P.** : Sostav pigmentov reproduktivnykh organov *Zea mays*. [Pigment composition of reproductive organs of *Zea mays* L.] - Fiziol. Rast. *23* : 697 - 701,
1976. [In R, ab : E.]

26667 - **KIRICHENKO, E.B., SMOLYGINA, L.D., SERDYUK, O.P., VASIL'EVA, V.T.** : Sintez i
nakoplenie pigmentov pri razvitii lamellyarnoĭ sistemy dimorfnykh plastid
Zea mays L. [Synthesis and accumulation of pigments during development of the
lamellar system of dimorphous plastids in *Zea mays* L.] - Fiziol. Rast. *23* :
25 - 30, 1976. [In R, ab : E.]

26668 - **KIRK, M.R., HEBER, U.** : Rates of synthesis and source of glycolate in intact
chloroplasts. - Planta *132* : 131 - 141, 1976.

26669 - **KIRST, G.O., KELLER, H.-J.** : Der Einfluß unterschiedlicher NaCl-Konzentratio-
nen auf die Atmung der einzelligen Alge *Platymonas subcordiformis* HAZEN. -
Bot. Mar. *19* : 241 - 244, 1976. [Ps.]

26670 - **KISELEV, B.A., EVSTIGNEEV, V.B., TSYGANKOVA, I.G.** : Pigment-pigmentnyĭ perenos élektrona i razdelenie zaryadov v agregatakh khlorofilla pri fotovozbuzh-
denii. [Pigment-pigment electron transfer and charge separation in chlorophyll
aggregates during photostimulation.] - Dokl. Akad. Nauk SSSR *230* : 726 - 728,
1976. [In R.]

26671 - **KISELEV, B.A., KAZAKOVA, A.A., EVSTIGNEEV, V.B., GINS, V.K., MUKHIN, E.N.** :
O vosstanovlenii ferredoksina polyarograficheskimi metodami. [Ferredoxin reduction by polarographic methods.] - Biofizika *21* : 35 - 39, 1976. [In R,
ab : E.]

26672 - **KISELEV, B.A., KOZLOV, Yu.N., EVSTIGNEEV, V.B.** : Fotokhimicheskie reaktsii
khlorofilla v aspekte okislitel'no-vosstanovitel'nykh potentsialov. [Photo-
chemical reactions of chlorophyll from the point of view of redox potentials.]
- In : Itogi Issledovaniya Mekhanizma Fotosinteza. Pp. 16 - 19. Pushchino
1976. [In R.]

26673 - **KISELEV, B.A., KOZLOV, Yu.N., EVSTIGNEEV, V.B.** : Ob okislitel'no-vosstanovitel'nykh potentsialakh vozbuzhdennogo sostoyaniya khlorofilla. [Oxidation-re-
duction potentials of the excited state of chlorophyll.] - Dokl. Akad. Nauk
SSSR *228* : 210 - 213, 1976. [In R.]

26674 - **KISLYAKOVA, T.E., BOGACHEVA, I.I., GOLUBKOVA, B.M., SHLYKOVA, I.M.** : Struktura i funktsiya fotosinteticheskogo apparata u nekotorykh vidov paporotniko-
vidnykh i golosemennykh. [Structure and function of photosynthetic apparatus
of some species of *Pteropsida* and *Gymnospermae*.] - Zh. obshch. Biol. *37* :
870 - 886, 1976. [In R, ab : E.]

26675 - **KITAJIMA, M.** : Light-induced redistribution of excitation energy in leaves as
observed in terms of fluorescence induction. - Plant Cell Physiol. *17* : 921 -
- 930, 1976.

*26676 - KITAJIMA, M., BUTLER, W.L. : Excitation spectra of PSI and PSII of chloro-
         plasts. - Plant Physiol. *56* (Suppl.) : 47, 1975.

26677 - KITAJIMA, M., BUTLER, W.L. : Microencapsulation of chloroplast particles. -
        Plant Physiol. *57* : 746 - 750, 1976.

26678 - KJØSEN, H., NORGÅRD, S., LIAAEN-JENSEN, S., SVEC, W.A., STRAIN, H.H., WEG-
        FAHRT, P., RAPOPORT, H., HAXO, F.T. : Algal carotenoids. XV. Structural stu-
        dies on peridinin. Part 2. Supporting evidence. - Acta chem. scand. B *30* :
        157 - 164, 1976.

26679 - KLEIBEUKER, J.F., PLATENKAMP, R.J., SCHAAFSMA, T.J. : Optically induced elec-
        tron spin polarization in the triplet state of chlorophyll and its model com-
        pounds. - Chem. phys. Lett. *41* : 557 - 561, 1976.

26680 - KLEIBEUKER, J.F., SCHAAFSMA, T.J. : Spin polarization in the lowest triplet
        state of some photosynthetic pigments. - In : BIRKS, J.B. (ed.) : Excited
        States of Biological Molecules. Pp. 314 - 326. Wiley, Chichester - New York -
        Brisbane - Tokyo 1976.

26681 - KLEIN, N.C., MINDICH, L. : Isolation and characterization of a glycerol auxo-
        troph of *Rhodopseudomonas capsulata* : Effect of lipid synthesis on the syn-
        thesis of photosynthetic pigments. - J. Bacteriol. *128* : 337 - 346, 1976.

26682 - KLEINEN HAMMANS, J.W., RABOU, L.P.L.M., PIETERSEN, H.Q. : Participation of
        pigment complexes in uptake and incorporation of mercury ions by *Anacystis*.
        - Photosynthetica *10* : 440 - 446, 1976.

26683 - KLEINEN HAMMANS, J.W., THOMAS, J.B. : On the correlation between the amounts
        of chlorophyll *b* and chlorophyll *a* forms in various plants. - Acta bot. neer.
        25 : 63 - 69, 1976.

26684 - KLEINKOPF, G.E., WALLACE, A., HARTSOCK, T. : Lime chlorosis on photosynthesis
        and transpiration of iron-inefficient soybeans. - Commun. Soil Sci. Plant
        Anal. *7* : 97 - 99, 1976.

26685 - KLEINKOPF, G.E., WALLACE, A., HARTSOCK, T.L. : *Galenia pubescens* - salt-tole-
        rant, drought-tolerant potential source of leaf protein. - Plant Sci. Lett.
        7 : 313 - 320, 1976. [Ps.]

26686 - KLEMME, J.-H. : Unidirectional inhibition of phosphoenolpyruvate carboxykinase
        from *Rhodospirillum rubrum* by ATP. - Arch. Microbiol. *107* : 189 - 192, 1976.

26687 - KLIMOV, V.V., SHUVALOV, V.A., KRAKHMALEVA, I.N.,, KARAPETYAN, N.V., KRASNOV-
        SKIĬ, A.A. : Izmeneniya vykhoda fluorestsentsii bakteriokhlorofilla pri foto-
        vosstanovlenii bakteriofeofitina v khromatoforakh purpurnykh sernykh bakteriĬ.
        [Bacteriochlorophyll fluorescence changes related to the bacteriopheophytin
        photoreduction in the chromatophores of purple sulfur bacteria.] - Biokhimiya
        41 : 1435 - 1441, 1976. [In R, ab : E.]

26688 - KLINCK, H.R., SIM, S.L. : The influence of source of photosynthate and sink
        size on grain yield in oats (*Avena sativa* L.). - Ann. Bot. *40* : 785 - 793,
        1976.

26689 - KLUGE, M. : Carbon and nitrogen metabolism under water stress. - In : LANGE,
        O.L., KAPPEN, L., SCHULZE, E.-D. (ed.) :Water and Plant Life. Pp. 243 - 252.
        Springer-Verlag, Berlin - Heidelberg - New York 1976. [Ps.]

26690 - KLUGE, M. : Crassulacean acid metabolism (CAM) : CO_2 and water economy. - In :
        LANGE, O.L., KAPPEN, L., SCHULZE, E.-D. (ed.) : Water and Plant Life. Pp. 313-
        - 322. Springer-Verlag, Berlin - Heidelberg - New York 1976.

26691 - KLUGE, M. : Metabolism of organic acids. - Progress Bot. *38* : 100 - 107, 1976.

26692 - KLUGE, M. : Models of CAM regulation. - In : BURRIS, R.H., BLACK, C.C. (ed.):
        CO_2 Metabolism and Plant Productivity. Pp. 205 - 216. Univ. Park Press, Balti-
        more - London - Tokyo .1976.

26693 - KLYACHKO-GURVICH, G.L., SEMENOVA, A.N. : Soderzhanie i zhirnokislotnyĬ sostav
        monogalaktozildiglitseridov v zavisimosti ot osveshchennosti i fazy rosta
        khlorelly v nakopitel'noĬ kul'ture. [Effect of illumination and growth phase
        of *Chlorella* in the accumulative culture on the contents and fatty acid compo-
        sition of monogalactosyl diglycerides.] - Fiziol. Rast. *23* : 726 - 733, 1976.
        [Ps; in R, ab : E.]

26694 - KNAFF, D.B., MALKIN, R. : Photosystem II reactions in oxidant-treated chloroplast fragments. - Arch. Biochem. Biophys. *174* : 414 - 419, 1976.

26695 - KNAFF, D.B., MALKIN, R. : Iron-sulfur proteins of the green photosynthetic bacterium *Chlorobium*. - Biochim. biophys. Acta *430* : 244 - 252, 1976.

26696 - KNOBLOCH, K. : *Rhodopseudomonas palustris* and *Rhodopseudomonas spheroides* as
comparative objects for the study of energy-dependent reactions. - Veröffent-
lichungen Univ. Innsbruck *108* : 77 - 79, 1976.

26697 - KNOECHEL, R., KALFF, J. : The applicability of grain density autoradiography
to the quantitative determination of algal species production : A critique. -
Limnol. Oceanogr. *21* : 583 - 590, 1976.

26698 - KNOECHEL, R., KALFF, J. : Track autoradiography : A method for the determination of phytoplankton species productivity. - Limnol. Oceanogr. *21* : 590 -
- 596, 1976.

26699 - KNOTH, R. .: Über den Zusammenhang zwischen Lamellarproteinen und Thylakoidmorphologie bei der Biosynthese grüner, etiolierter und mutierter Plastiden von
Hordeum, Pelargonium und *Lycopersicon*. II. Elektronenmikroskopische Untersuchungen der Plastidensubstruktur. - Acta histochem. *17* (Suppl.) : 157 - 162,
1976.

26700 - KNOTH, R., HERRMANN, F.H., BÖRNER, T. : Struktur und Funktion der genetischen
Information in Plastiden XV. Beziehungen zwischen Chlorophyllgehalt, Photosyn-
theseverhalten und Plastidenfeinstruktur in Kerngen-bedingten Aureamutanten
von *Antirrhinum majus* (Mutante "Aurea") und *Pelargonium zonale* (Sorte "Cloth
of Gold"). - Biochem. Physiol. Pflanzen *170* : 433 - 442, 1976.

26701 - KNYPL, J.S., WITEK, S., OŚWIECIMSKA, M. : Growth retarding effect of N,N-di-
methylmorpholinium chloride and CCC in *Spirodela oligorrhiza*. - Z. Pflanzenphysiol. *79* : 53 - 61, 1976. [Ch1.]

26702 - KOBAYASHI, M. : Utilization and disposal of wastes by photosynthetic bacteria.
- In : Microbial Energy Conversion. Pp. 443 - 453. Inst. Mikrobiol., Göttingen
1976. [Mass cultures.]

26703 - KOBAYASHI, Y., INOUE, Y., SHIBATA, K. : Inhibitory effect of *p*-nitrothiophenol
in the light on the photosystem II activity of spinach chloroplasts. - Biochim. biophys. Acta *423* : 80 - 90, 1976.

26704 - KOBAYASHI, Y., INOUE, Y., SHIBATA, K. : Light-dependent inhibitory action of
p-nitrothiophenol on Photosystem II in relation to the redox state of electron carriers. - Biochim. biophys. Acta *440* : 600 - 608, 1976.

26705 - KOBLENTZ-MISHKE, O.J., VEDERNIKOV, V.I. : A tentative comparison of primary
production and phytoplankton quantities at the ocean surface. - Mar. Sci.
Commun. *2* : 357 - 374, 1976.

26706 - KOCH, H. : Zebrastreifung, eine Chlorose keimender Monokotylen, verursacht
durch Belichtung und Temperatur. - Angew. Bot. *50* : 233 - 250, 1976.

26707 - KOCH, W. : Blattfarbstoffe von Fichte (*Picea abies* (L.) KARST.) in Abhängigkeit vom Jahresgang, Blattalter und -typ. - Photosynthetica *10* : 280 - 290,
1976.

26708 - KOCH, W., ROTH, H. : Eine große Präzisions-Gaswechselmeßanlage mit getrenn-
ter Grün- und Bodenzone und ihre Leistungsfähigkeit dargestellt am Beispiel
der Fichte. - Photosynthetica *10* : 71 - 82, 1976.

26709 - KOCHUBEĬ, S.M., ABARSUA, A.Z. : Ob izmenenii sostoyaniya pigmentlipoproteidno-
go kompleksa fotosistemy I pri okislenii tsentrov. [Conformational change of
photosystem-1 pigment-lipoprotein complex associated with centres oxidation.]
- Studia biophys. *58* : 1 - 5, 1976. [In R, ab : E.]

26710 - KOCHUBEĬ, S.M., SAMOKHVAL, E.G., MYULLER, I. : Temperaturnye zavisimosti spektrov fluorestsentsii khloroplastov i legkikh fragmentov. [Temperature dependence of fluorescence spectra of chloroplasts and their light fragments.] -
Studia biophys. *54* : 217 - 224, 1976. [In R, ab : E.]

26711 - KOENIG, F., SCHMID, G.H., RADUNZ, A., PINEAU, B., MENKE, W. : The isolation
of further polypeptides from the thylakoid membrane, their localization and
function. - FEBS Lett. *62* : 342 - 346, 1976.

26712 - **KOESTER, V.J., FONG, F.K.** : Exciton interactions in the symmetrical dimeric aggregate of chlorophyll a monohydrate. - J. phys. Chem. *80* : 2310 - 2312, 1976.

26713 - **KOESTER, V.J., POLLES, J.S., KOREN, J.G., GALLOWAY, L., ANDREWS, R.A., FONG, F.K.** : Optical and redox properties of dimeric chlorophyll-water aggregates : an *in vitro* approach toward characterizing the primary molecular adduct in photosynthesis. - J. Luminescence *12-13* : 781 - 786, 1976.

26714 - **KÖHLER, G.H., RINDT, K.-P., OHMANN, E.** : Die Bildung aktiver Pyruvat-Carboxylase auf Apoenzym und Biotin in *Rhodopseudomonas spheroides*. - Biochem. Physiol. Pflanzen *169* : 99 - 104, 1976.

26715 - **KOHN, S., KLEIN, S.** : Light-induced structural changes during incubation of isolated maize etioplasts. - Planta *132* : 169 - 175, 1976.

26716 - **KOIWAI, A., KISAKI, T.** : Effect of ozone on photosystem II of tobacco chloroplasts in the presence of piperonyl butoxide. - Plant Cell Physiol. *17* : 1199- - 1207, 1976.

26717 - **KOJIMA, K., TSUCHITANI, Y.** : [A simple limit test of free ionizable copper in copper chlorophyll and in sodium copper chlorophyllin.] - Bunseki Kagaku *25* : 476 - 478, 1976. [In Jap., ab : E.]

26718 - **KOK, B.** : Photosynthesis : the path of energy. - In : **BONNER, J., VARNER, J. E.** (ed.) : Plant Biochemistry. Third Edition. Pp. 845 - 885. Academic Press, New York - San Francisco - London 1976.

26719 - **KOK, B., RADMER, R.** : Energy requirements of a biosphere. - In : **PONNAMPERU-MA, C.** (ed.) : Chemical Evolution of the Giant Planets. Pp. 183 - 197. Academic Press, New York - San Francisco - London 1976.

26720 - **KOK, B., RADMER, R.** : Mechanisms in photosynthesis. - In : **SANADI, D.R.** (ed.): Chemical Mechanisms in Bioenergetics. ACS Monogr. Ser. Vol. 172. Pp. 172 - 220. Amer.' chem. Soc., Washington 1976.

26721 - **KOLESNIKOV, P.A., PETROCHENKO, E.L., ZORÉ, S.V., MUTUSKIN, A.A., PSHENOVA, K. V.** : Pogloshchenie kisloroda khloroplastami list'ev gorokha v prisutstvii ribozo-5-fosfata na svetu. [Absorption of oxygen by pea leaf chloroplasts in the presence of ribose-5-phosphate in the light.] - Dokl. Akad. Nauk SSSR *227* : 236 - 239, 1976. [In R.]

26722 - **KOLESNIKOV, P.A., ZORÉ, S.V., PETROCHENKO, E.I.** : Okislitel'nye puti obrazovaniya 3-fosfoglitserinovoĭ kisloty v ékstraktakh iz khloroplastov. [Oxidative ways of 3-phosphoglyceric acid formation in extracts from chloroplasts.] - Dokl. Akad. Nauk SSSR *228* : 981 - 984, 1976. [In R.]

26723 - **KOLESNIKOV, V.A., AZAMATOV, M.A.** : Vliyanie orosheniya i udobreniya na rost, razvitie i plodonoshenie yabloni v usloviyakh stepnoĭ zony Kabardino-Balkarii. [Effect of irrigation and fertilization on the growth, development and fruit bearing of the apple tree in the Kabardino-Balkarian steppe zone.] - Izv. timiryazev. sel'.-khoz. Akad. *1976* (2) : 117 - 127, 1976. [Growth analysis; in R, ab : E.]

26724 - **KOLLMAN, A.L., WALI, M.K.** : Intraseasonal variations in environmental and productivity relations of *Potamogeton pectinatus* communities. - Arch. Hydrobiol. *50* (Suppl.4) :439 - 472, 1976.

26725 - **KOMAROVA, Yu.M., TEREKHOVA, I.V., DOMAN, N.G., AL'BITSKAYA, O.N.** : Karboangidraza sine-zelenoĭ vodorosli *Spirulina platensis*. [Carboanhydrase of a blue-green alga *Spirulina platensis*.] - Biokhimiya *41* : 183 - 187, 1976. [In R, ab : E.]

26726 - **KOMATSU, M., MURAKAMI, S.** : Inhibition of photophosphorylation by ATP and the role of magnesium in photophosphorylation. - Biochim. biophys. Acta *423* : 103 - 110, 1976.

26727 - **KONDRATIEVA, E.N., ZHUKOV, V.G., IVANOVSKY, R.N., PETUSHKOVA, Yu.P., MONOSOV, E.Z.** : The capacity of phototrophic sulfur bacterium *Thiocapsa roseopersicina* for chemosynthesis. - Arch. Microbiol. *108* : 287 - 292, 1976.

26728 - **KONISHI, T., PACKER, L.** : Light-dark conformational states in bacteriorhodopsin. - Biochem. biophys. Res. Commun. *72* : 1437 - 1442, 1976.

26729 - **KONNO, S.** : [Physiological study on the mechanism of seed production of soybeans.] - Bull. nat. Inst. agr. Sci., Ser. D *27* : 139 - 295, 1976. [Ps; in
Jap., ab : E.]

26730 - **KONO, Y., TAKAHASHI, M.-A., ASADA, K.** : Oxidation of manganous pyrophosphate
by superoxide radicals and illuminated spinach chloroplasts. - Arch. Biochem.
Biophys. *174* : 454 - 462, 1976.

26731 - **KONONENKO, A.A., KNOX, P.P., ADAMOVA, N.P., PASCHENKO, V.Z., TIMOFEEV, K.N.,
RUBIN, A.B.** : Spectral and photochemical properties of photosynthetic reac-
tion centre preparations from *Rhodopseudomonas spheroides*, strain 1760-1. -
Studia biophys. *55* : 183 - 198, 1976.

*26732 - **KONOVALOV, B.V., BEKASOVA, O.D.** : K metodike opredeleniya soderzhaniya pigmentov morskogo fitoplanktona bez ékstragirovaniya. [Methods for determining the
amount of pigments of the sea phytoplankton without extraction.] - Okeanologiya *9* : 883 - 892, 1969. [In R, ab : E.]

26733 - **KONZE, J.R., ELSTNER, E.F.** : Ethylene- and ethane-formation in leaf disks,
plastids and mitochondria. - Ber. deut. bot. Ges. *89* : 547 - 553, 1976.

26734 - **KOPELMAN, R.** : Excitons in ternary mixed molecular crystals : a prototype for
the primary step of photosynthesis ? - J. Luminescence *12/13* : 775 - 780,
1976.

26735 - **KORBUT, V.L., VIL'YAMS, M.V., RUMYANTSEVA, V.B.** : K voprosu vzaimosvyazi vidimogo fotosinteza i produktivnosti rasteniĭ. [Interrelation between net photosynthesis and productivity.] - In : Gazometricheskoe Issledovanie Fotosinteza i Dykhaniya Rasteniĭ. Pp. 67 - 68. Akad. Nauk SSSR, Tartu 1976. [In R.]

26736 - **KORDAN, H.A.** : Anaerobiosis-induced etiolation in light-germinated rice seedlings. - Ann. Bot. *40* : 347 - 350, 1976. [Chl.]

26737 - **KORDAN, H.A.** : Normal pigment development in germinating rice seedlings is
oxygen dependent. - Ann. Bot. *40* : 1329 - 1332, 1976.

26738 - **KORDAN, H.A.** : Oxygen as an environmental factor in influencing normal mor-
phogenetic development in germinating rice seedlings. - J. exp. Bot. *27* :
947 - 952, 1976. [Chl.]

26739 - **KÖRNER, C.** : A semi-automatic, recording diffusion porometer and its performance under alpine field conditions. - Photosynthetica *10* : 172 - 181, 1976.

26740 - **KORNYUSHENKO, G.A., EVDOKIMOVA, I.V., VASIL'EVA, V.E.** : Indukovani svitlom
i askorbinovoyu kyslotoyu vzaemoperetvorennya ksantofiliv u khloroplastakh
gorokhu. [Xanthophylls interconversions induced by light and ascorbic acid
in pea chloroplasts.] - Ukr. bot. Zh. *33* : 257 - 260, 332, 1976. [In Ukr.,
ab : R.]

26741 - **KORSAK, M.N., NAKANI, D.V.** : Dinamika produktsii fitoplanktona i bakteriĭ na
melkovod'e Rybinskogo vodokhranilishcha v 1974 g. [Dynamics of phytoplankton
and bacteria production under low water level in the Rybinsk reservoir in
1974.] - Biol. vnutr. Vod *31* : 16 - 18, 1976. [In R.]

26742 - **KORSHUNOV, A.V., POPOV, B.A.** : Vysokie urozhai kartofelya : optimal'nye normy
udobreniĭ i oroshenie. [High yield of potato : optimal fertilization and irrigation.] - Vestn. sel'skokhoz. Nauki *1976* (2) : 92 - 96, 1976. [Ps; in R.]

26743 - **KORZH, B.V.** : K voprosu ob oshibkakh raschetov skorosteĭ CO_2-gazoobmena, iz-
merennykh na IK gazoanalizatorakh. [Errors of calculation of CO_2 exchange rates measured with IRGAs.] - In : Gazometricheskoe Issledovanie Fotosinteza i
Dykhaniya Rasteniĭ. Pp. 69 - 72. Akad. Nauk SSSR, Tartu 1976. [In R.]

26744 - **KORZH, B.V.** : Primenenie impul'snogo osveshcheniya dlya differentsirovaniya
protsessov fotosinteza i dykhaniya zelenykh rasteniĭ. [Use of impulse illumination for distinguishing processes of photosynthesis and respiration in green
plants.] - In : Gazometricheskoe Issledovanie Fotosinteza i Dykhaniya Rasteniĭ. Pp. 73 - 75. Akad. Nauk SSSR, Tartu 1976. [In R.]

26745 - **KORZH, B.V.** : Nekotorye osobennosti izmereniĭ CO_2-gazoobmena rasteniĭ na in-frakrasnom gazoanalizatore. [Measurement of plant CO_2 exchange with an infra-red gas analyzer.] - Fiziol. Rast. *23* : 413 - 420, 1976. [In R, ab : E.]

26746 - **KORZH, B.V.** : Novye dannye o fotosinteze i dykhanii zelenykh rasteniĭ v uslo-viyakh impul'snogo osveshcheniya. [New data on the photosynthesis and respi-ration of green plants under conditions of flash illumination.] - Dokl. Akad. Nauk SSSR *227* : 1014 - 1017, 1976. [In R.]

26747 - **KORZUKHINA, A.F.** : Fotosinteticheskaya deyatel'nost' podsolnechnika pri raz-lichnoĭ gustote poseva i vnesenii udobreniĭ. [Photosynthetic activity of sun-flower in stands of different densities and fertilizer application.] - Sibir. Vestn. sel'skokhoz. Nauki *1976* (6) : 82 - 87, 1976. [In R.]

26748 - **KOSITSIN, A.V., KHALIDOVA, G.B.** : O znachenii tsinka dlya aktivnosti karboan-gidrazy rastitel'nogo proiskhozhdeniya. [Importance of zinc for activity of carboanhydrase of plant origin.] - Fiziol. Biokhim. kul't. Rast. *8* : 182 - - 187, 1976. [In R, ab : E.]

26749 - **KOSMAKOVA, V.E., ZVEREVA, E.G.** : Aktivnost' fotosinteticheskogo apparata raste-niĭ soi v usloviyakh pereuvlazhneniya pochvy. [Activity of photosynthetic ap-paratus of soybean plants under conditions of waterlogging.] - In : Ustoĭchi-vost' Rasteniĭ k Pereuvlazhneniyu Pochvy v Usloviyakh Dal'nego Vostoka. Pp. 3 - 25. Dal'nevost. gosud. Univ., Vladivostok 1976. [In R.]

26750 - **KOSSATZ, V.C., van HUYSTEE, R.B.** : The specific activities of peroxidase and aminolevulinic acid dehydratase during the growth cycle of peanut suspension culture. - Can. J. Bot. *54* : 2089 - 2094, 1976.

26751 - **KOSTIKOV, A.P., SADOVNIKOVA, N.A., EVSTIGNEEV, V.B.** : Issledovanie vykhoda kation-radikalov pigmenta v reaktsii fotookisleniya khlorofilla khinonami. [Yield of pigment cation-radicals during photooxidation of chlorophyll with quinones.] - Biofizika *21* : 803 - 807, 1976. [In R, ab : E.]

26752 - **KOSTIKOVA, L.E.** : Pervichnaya produktsiya zelenykh nitchatykh vodorosleĭ Kiev-skogo vodokhranilishcha. [Primary production of green filamentous algae from the Kiev Reservoir.] - Gidrobiol. Zh. *12* (1) : 62 - 70, 132, 1976. [In R, ab : E.]

26753 - **KOVÁČ, J., HENSELOVÁ, M.** : A rapid method for detection of Hill reaction in-hibitors. - Photosynthetica *10* : 343 - 344, 1976.

26754 - **KOWAL, R.R., LECHOWICZ, M.J., ADAMS, M.S.** : The use of canonical analysis to compare response curves in physiological ecology. - Flora *165* : 29 - 46, 1976. [Ps.]

26755 - **KOWALLIK, W., RUYTERS, G.** : Über Aktivitätssteigerungen der Pyruvatkinase durch Blaulicht oder Glucose bei einer chlorophyllfreien *Chlorella*-Mutante. - Planta *128* : 11 - 14, 1976.

26756 - **KOWALLIK, W., SCHEIL, I.** : Lichtbedingte Veränderungen des ATP-Spiegels einer chlorophyllfreien *Chlorella*-Mutante. - Planta *131* : 105 - 108, 1976.

26757 - **KOZHUSHKO, N.N., CHERNYSHEVA, S.V.** : Sostoyanie pigmentnogo kompleksa plastid-nogo apparata list'ev pshenitsy v reproduktivnyĭ period razvitiya. [State of the pigment complex of plastid apparatus of wheat leaves in the reproductive period of development.] - Byull. vses. nauch. issled. Inst. Rastenievod. im. N.I. Vavilova *63* : 15 - 18, 1976. [In R.]

26758 - **KOZHUSHKO, N.N., RUTMAN, G.I.** : Izmenenie sostoyaniya pigment-belkovogo kom-pleksa pri deĭstvii zasukhi u raznykh po stepeni zasukhoustoĭchivosti sortov zernovykh kul'tur. [Drought-induced changes in the state of pigment-protein complex in cereal crops distinguished by their drought resistance.] - Tr. pri-kl. Bot., Genet. Selektsii *57* (2) : 83 - 90, 1976. [In R, ab : E.]

26759 - **KOZLOWSKI, T.T.** : Drought and transplantability of trees. - USDA Forest Serv. gen. tech. Rep. *NE-22* : 77 - 90, 1976. [Ps.]

26760 - **KRAPF, G.** : Der Einbau von $^{14}CO_2$ im Licht bei suspendierten Blattstücken von langfristig verdunkelten Spinatpflanzen. - Ber. deut. bot. Ges. *89* : 631 - - 641, 1976.

26761 - **KRASNOVSKIĬ, A.A.** : Issledovanie fotobiokhimii pigmentov v fotosinteziruyush-chikh organizmakh. [Studies of photobiochemistry of pigments in photosynthes-izing organisms.] - In : Itogi Issledovaniya Mekhanizma Fotosinteza. Pp. 6 - - 8. Pushchino 1976. [In R.]

26762 - **KRASNOVSKIĬ, A.A., BRIN, G.P., NIKANDROV, V.V.** : Fotovosstanovlenie kisloroda i fotoobrazovanie vodoroda na neorganicheskikh fotokatalizatorakh. [Photore-duction of oxygen and photoformation of hydrogen on inorganic photocatalysts.] - Dokl. Akad. Nauk SSSR *229* : 990 - 993, 1976. [In R.]

26763 - **KRASNOVSKIĬ, A.A., LUGANSKAYA, A.N.** : Spektral'nye svoĭstva khlorofilla i ego analogov v vodnykh rastvorakh poverkhnostno-aktivnykh veshchestv. [Spectral properties of chlorophyll and its analogues in aqueous solutions of deter-gents.] - Izv. Akad. Nauk SSSR, Ser. biol. *1976* : 182 - 192, 1976. [In R, ab: E.]

26764 - **KRASNOVSKIĬ, A.A. ml.** : Fotosensibilizirovannaya lyuminestsentsiya singletnogo kisloroda v rastvore. [Photosensitized luminescence of singlet oxygen in solu-tion.] - Biofizika *21* : 748 - 749, 1976. [In R, ab : E.]

26765 - **KRASNOVSKY, A.A.** : Chemical evolution of photosynthesis. - Origins Life *7* : 133 - 143, 1976.

*26766 - **KRASZNER-BERNDORFER, E., TELEGDY KOVÁTS, L.** : Some observations on the change of bioquinones in cereals due to exogenous agents. - In : MORTON, I., RHODES, D.N. (ed.) : The Contribution of Chemistry to Food Supplies. Pp. 399 - 409. Butterworths, London 1974. [Plastoquinones.]

26767 - **KRAUSE, G.H., HEBER, U.** : Energetics of intact chloroplasts. - In : BARBER, J. (ed.) : The Intact Chloroplasts. Pp. 171 - 214. Elsevier, Amsterdam - New York - Oxford 1976.

26768 - **KRAUSPE, R., PARTHIER, B.** : Zur Regulation der Synthese plastiden-spezifis-cher Aminoacyl-tRNA-Synthetasen in *Euglena gracilis*. - Acta histochem. *17* (Suppl.) : 129 - 139, 1976. [Chl.]

26769 - **KREĬTSBERG, O.É., PAVULINYA, D.A.** : Vliyanie svetovogo rezhima na zhirnokis-lotnyĭ sostav polyarnykh lipidov i aktivnost' perekisnogo okisleniya zhirnykh kislot v khloroplastakh rasteniĭ ogurtsov. [Effect of light regime on fatty acid composition of lipids and activity of the peroxidation of fatty acids in cucumber chloroplasts.] - In : Regulyatsiya Rosta i Pitanie Rasteniĭ. Pp. 177 - 184. Zinatne, Riga 1976. [In R.]

26770 - **KREMER, B.P.** : ^{14}C-assimilate pattern and kinetics of photosynthetic $^{14}CO_2$- -assimilation of the marine red alga *Bostrychia scorpioides*. - Planta *129* : 63 - 67, 1976.

26771 - **KREMER, B.P.** : Photosynthetic carbon metabolism of chloroplasts symbiotic with a marine opisthobranch. - Z. Pflanzenphysiol. *77* : 139 - 145, 1976.

26772 - **KREMER, B.P., SCHMITZ, K.** : Aspects of $^{14}CO_2$-fixation by endosymbiotic rhodo-plasts in the marine opisthobranchiate *Hermaea bifida*. - Mar. Biol. *34* : 313- - 316, 1976.

26773 - **KRENDELEVA, T.E., KAUROV, B.S., TULBU, G.V., RUBIN, A.B.** : Vzaimosvyaz' in-dutsirovannogo svetom tusheniya fluorestsentsii atebrina i fotosintetichesko-go transporta élektronov v izolirovannykh khloroplastakh gorokha. [Relation-ship between light-induced quenching of atebrin fluorescence and photosynthe-tic electron transport in isolated chloroplasts of pea.] - Biofizika *21* : 524 - 528, 1976. [In R, ab : E.]

26774 - **KRETZER, F., OHAD, I., BENNOUN, P.** : Ontogeny, insertion, and activation of two thylakoid peptides required for photosystem II activity in the nuclear, temperature sensitive T4 mutant of *Chlamydomonas reinhardi*. - In : BÜCHER, T., NEUPERT, W., SEBALD, W., WERNER, S. (ed.) : Genetics and Biogenesis of Chlo-roplasts and Mitochondria. Pp. 25 - 32. North-Holland Publ. Co., Amsterdam - New York - Oxford 1976.

26775 - **KRIEDEMANN, P.E., LOVEYS, B.R., POSSINGHAM, J.V., SATOH, M.** : Sink effects on stomatal physiology and photosynthesis. - In : WARDLAW, I.F., PASSIOURA, J.B. (ed.) : Transport and Transfer Processes in Plants. Pp. 401 - 414. Academic Press, New York - San Francisco - London 1976.

26776 - **KRIEDEMANN, P.E., SWARD, R.J., DOWNTON, W.J.S.** : Vine response to carbon dioxide enrichment during heat therapy. - Aust. J. Plant Physiol. *3* : 605 - 618,
1976. [Ps.]

26777 - **KRISHNAMANI, M.R.S., LAKSHMANAN, M.** : Photosynthetic changes in *Fusarium*-in-
fected cotton. - Can. J. Bot. *54* : 1257 - 1263, 1976.

26778 - **KRISHNAMURTHY, K., RAJASHEKARA, B.G., RAGHUNATHA, G., JAGANNATH, M.K.** : Pattern of dry-matter accumulation and distribution in sorghums (*Sorghum vulgare* PERS.). - Mysore J. agr. Sci. *10* : 161 - 168, 1976.

26779 - **KRSTIĆ, B.** : Uticaj svetlosti različitih talasnih dužina na zastupljenost
pigmenata hloroplasta u nekim biljnim vrstama. [Effect of light of various
wavelengths on the content of chloroplast pigments in some plant species.]
- Arh. poljopriv. Nauke *29* (107) : 75 - 96, 1976. [In Croat., ab : E.]

26780 - **KRUPENKO, A.N., VOLKOVA, T.V.** : Fotosinteticheskiĭ transport èlektrona i ne-
sootvetstvie mezhdu ènergo- i gazoobmenom pri fotosinteze khlorelly. [Photo-
synthetic electron transport and discrepancy of energy and gas exchange during photosynthesis in *Chlorella*.] - In : Gazometricheskoe Issledovanie Fotosinteza i Dykhaniya Rasteniĭ. P. 76. Akad. Nauk SSSR, Tartu 1976. [In R.]

26781 - **KRUPENKO, A.N., VOLKOVA, T.V., SHUVALOVA, N.P., BELL, L.N.** : Vliyanie diurona
na sootnoshenie mezhdu ènergo- i gazoobmenom pri fotosinteze khlorelly na si-
nem svetu. [The effect of DCMU on the relation between energy storage and gas
exchange in photosynthesis of *Chlorella* in blue light.] - Dokl. Akad. Nauk
SSSR *228* : 1001 - 1004, 1976. [In R.]

26782 - **KRUŠKOVÁ-NOVOTNÁ, L., NEUBERG, J.** : Účinnost chlorcholinchloridu (CCC) na jarní
pšenici pri regulovaném obsahu vody v půdě. [The efficiency of chlorocholine-
chloride (CCC) on spring wheat at a controlled moisture content in soil.] -
Rostlinná Výroba (Praha) *22* : 343 - 351, 1976. [Growth analysis; in Czech,
ab : E, R.]

26783 - **KRZYSCH, G.** : CO_2-Haushalt und Stoffbildung eines *Beta*-Rübenbestandes. Teil
II : Die CO_2-Quellen Atmosphäre und Boden und ihre Dynamik. - Z. Acker-
Pflanzenbau *143* : 109 - 133, 1976.

26784 - **KU, H.S.** : Growth and development of plants in different oxygen concentrations.
- Plant Physiol. *57* (Suppl.) : 58, 1976. [Ps, photorespiration, photosynthate
translocation.]

26785 - **KU, S.B., EDWARDS, G.E.** : Effects of light, CO_2 and temperature on photosynthetic characteristics in potato, a high yielding C_3 crop. - Plant Physiol.
57 (Suppl.) : 105, 1976.

26786 - **KU, S.B., EDWARDS, G.E., KANAI, R.** : Distribution of enzymes related to C_3
and C_4 pathway of photosynthesis between mesophyll and sheath cells of *Panicum hians* and *Panicum milioides*. - Plant Cell Physiol. *17* : 615 - 620, 1976.

26787 - **KUBÍČEK, F.** : Growth and development of *Asperula odorata* L. in an oak-hornbeam
ecosystem. - Biológia (Bratislava) *31* : 41 - 53, 1976. [Growth analysis.]

26788 - **KUBÍČEK, F.** : The retraction and reduction index for estimation the leaf area
index and leaf production from litterfall in an oak-hornbeam woodland. - Biológia (Bratislava) *31* : 263 - 268, 1976.

26789 - **KÜHN, W., SCHÄTZLER, H.P.** : Non-destructive determination of the biomass of a
plot of cereals by *gamma*-radiation. - Z. Pflanzenzücht. *76* : 240 - 249, 1976.

26790 - **KULANDAIVELU, G., HALL, D.O.** : Stabilization of the photosynthetic activities
of isolated spinach chloroplasts during prolonged ageing. - Z. Naturforsch.
31 C : 452 - 455, 1976.

26791 - **KULANDAIVELU, G., HALL, D.O.** : Ultrastructural changes in *in vitro* ageing
spinach chloroplasts. - Z. Naturforsch. *31 C* : 82 - 84, 1976.

26792 - **KULANDAIVELU, G., HALL, D.O.** : Site specific inhibition by α-benzyl-α-bromo-
malodinitrile (BBMD) of electron transport in spinach chloroplasts. - Biochim.
biophys. Acta *430* : 46 - 52, 1976.

26793 - **KULANDAIVELU, G., SENGER, H.** : The 520 nm absorbance changes in *Scenedesmus obliquus* and its relation to photosystem I. - Biochim. biophys. Acta *430* : 94 - 104, 1976.

26794 - **KULANDAIVELU, G., SENGER, H.** : Changes in the reactivity of the photosynthetic apparatus in heterotrophic ageing cultures of *Scenedesmus obliquus*. I. Changes in the photochemical activities. - Physiol. Plant. *36* : 157 - 164, 1976.

26795 - **KULANDAIVELU, G., SENGER, H.** : Changes in the reactivity of the photosynthetic apparatus in heterotrophic ageing cultures of *Scenedesmus obliquus*. II. Changes in ultrastructure and pigment composition. - Physiol. Plant. *36* : 165 - 168, 1976.

26796 - **KULANDAIVELU, G., SENGER, H.** : Changes in the reactivity of the photosynthetic apparatus in heterotrophic ageing cultures of *Scenedesmus obliquus*. III. Recovery of the photosynthetic capacity in aged cells. - Physiol. Plant. *36* : 169 - 173, 1976.

26797 - **KULANDAIVELU, G., SPILLER, H., BÖGER, P.** : Action of nitrite on fluorescence induction in algae. - Plant Sci. Lett. *7* : 225 - 231, 1976.

26798 - **KULASEGARAM, S., KATHIRAVETPILLAI, A.** : Effect of shade and water supply on growth and apical dominance in tea (*Camellia sinensis* (L.) O. KUNTZE). - Trop. Agr. *53* : 161 - 172, 1976. [Growth analysis.]

26799 - **KULIKOV, G.V., CHEMARIN, N.G., YAROSLAVTSEVA, Z.P.** : Intensivnost' fotosinteza raznovozrastnykh list'ev u nekotorykh vechnozelenykh drevesnykh rasteniĭ. [Photosynthetic rate of leaves of different ages in some evergreen woody plants.] - Byull. gos. nikitsk. bot. Sada *1976* (1[29]): 61 - 64, 1976. [In R, ab : E.]

26800 - **KULIKOV, G.V., IVANTSOVA, Z.V.** : Pigmenty plastid v raznovozrastnykh list'-yakh nekotorykh vechnozelenykh drevesnykh rasteniĭ. [Plastid pigments in leaves of different age of some sempervirent woody plants.] - Nauch. Dokl. vyssh. Shkoly, biol. Nauki *19* (3) : 86 - 90, 1976. [In R.]

26801 - **KULIKOV, G.V., YAROSLAVTSEVA, Z.P., CHEMARIN, N.G.** : Sutochnaya intensivnost' fotosinteza u drevesnykh introdutsentov Kryma. [Diurnal changes in photosynthetic rate in introduced woody plants in Crimea.] - Byul. glav. bot. Sada Akad. Nauk SSSR *99* : 48 - 52, 1976. [In R.]

26802 - **KULL', K., LAĬSK, A.** : Opredelenie fotosinteticheskikh parametrov lista s u-chetom neodnorodnoĭ otkrytosti ust'its. [Determination of leaf photosynthetic parameters with respect to unequal stomata opening.] - In : Gazometricheskoe Issledovanie Fotosinteza i Dykhaniya Rasteniĭ. Pp. 77 - 79. Akad. Nauk SSSR, Tartu 1976. [In R.]

26803 - **KUMAR, A.** : Dry matter production and growth rates of three arid zone grasses in culture. - Comp. Physiol. Ecol. *1* : 23 - 26, 1976.

26804 - **KUMAR, A.** : The influence of varying density on the dry matter production of *Dactyloctenium aegyptium* LINN. (P.) BEAUV. - Comp. Physiol. Ecol. *1* : 101 - - 104, 1976.

26805 - **KUMAZAWA, S., FRANK, J., SKJOLDAL, H.R., MITSUF, A.** : Hydrogen production by tropical marine blue-green algae and photosynthetic bacteria. - Plant Physiol. *57* (Suppl.) : 60, 1976.

26806 - **KUNDU, A., PALIT, P., MANDAL, R.K., SIRCAR, S.M.** : C_3-type photosynthetic carbon dioxide fixation in the rice plant. - Plant biochem. J. *3* : 111 - 118, 1976.

26807 - **KUNERT, K.-J., BÖHME, H., BÖGER, P.** : Reactions of plastocyanin and cytochrome 553 with Photosystem I of *Scenedesmus*. - Biochim. biophys. Acta *449* : 541- - 553, 1976.

26808 - **KUNG, M.C., DeVAULT, D.** : Carotenoid triplet state in *R. spheroides* Ga chromatophores. - Photochem. Photobiol. *24* : 87 - 91, 1976.

26809 - **KUNG, S.** : Tobacco fraction I protein : A unique genetic marker. - Science *191* : 429 - 434, 1976.

26810 - **KUNG, S.D., MARSHO, T.V.** : Regulation of RuDP carboxylase/oxygenase activity
and its relationship to plant photorespiration. - Nature *259* : 325 - 326,
1976.

26811 - **KÜNSTLE, E., ULLRICH, C.H.** : Vergleichende Gaswechselmessungen an Gemeiner
Kiefer (*Pinus silvestris* L.) und Schwarzkiefer (*Pinus nigra* ARN.) unter den
Extrembedingungen des Sommers 1975 im oberrheinischen Trockengebiet. - All-
gem. Forst- Jagdzeit. *147* (4) : 65 - 68, 1976.

26812 - **KUPKA, J., TAN, T.Q.** : Minerální výživa a fotosyntéza v ontogenezi kukuřice.
[Mineral nutrition and photosynthesis in maize ontogenesis.] - Rostlinná Výroba (Praha) *22* : 1047 - 1051, 1976. [In Czech, ab : E, R.]

26813 - **KUPRIN, S.P., KUKUSHKIN, A.K.** : Khromaticheskie perekhody fluorestsentsii
v khloroplastakh bobov. [Chromatic fluorescence transients of broad bean chloroplasts.] - Biofizika *21* : 113 - 117, 1976. [In R, ab : E.]

26814 - **KURAISHI, S.** : Ineffectiveness of cytokinin-induced chlorophyll retention in
hypostomatous leaf discs. - Plant Cell Physiol. *17* : 875 - 885, 1976.

B26815 - **KURSANOV, A.L.** : Transport Assimilyatov v Rastenii. [Assimilate Transport in
Plants.] - Nauka, Moskva 1976. [In R.]

26816 - **KURSANOV, A.L., PARAMONOVA, N.V.** : On the state of membranes in mesophyll
cells of *Beta vulgaris* in terms of assimilate transport. - In : **JACQUES, R.**
(ed.) : Etudes de Biologie Végétale. Hommage au Professeur Pierre Chouard.
Pp. 509 - 519. Paris 1976.

26817 - **KURSANOV, A.L., PARAMONOVA, N.V.** : Ul'trastrukturnye izmeneniya v mezofille
list'ev *Beta vulgaris, L* v svyazi s transportom assimilyatov. [Ultrastructu-
ral changes in the mesophyll of leaves of *Beta vulgaris* related to photosynthate transport.] - Fiziol. Rast. *23* : 286 - 291, 1976. [In R, ab : E.]

26818 - **KUSHNIRENKO, M.D., KRYUKOVA, E.V., PECHERSKAYA, S.N., KANASH, E.V.** : Vodnyĭ
i belkovyĭ obmen khloroplastov u rasteniĭ razlichnoĭ zasukhoustoĭchivosti.
[Water content and protein metabolism of chloroplasts in plants with different drought resistance.] - Fiziol. Rast. *23* : 473 - 482, 1976. [In R, ab : E.]

26819 - **KUSHWAHA, S.C., KATES, M.** : Effect of nicotine in biosynthesis of C_{50} carotenoids in *Halobacterium cutirubrum*. - Can. J. Biochem. *54* : 824 - 829, 1976.

26820 - **KUSHWAHA, S.C., KATES, M., PORTER, J.W.** : Enzymatic synthesis of C_{40} carotenes by cell-free preparation from *Halobacterium cutirubrum*. - Can. J. Biochem.
54 : 816 - 823, 1976.

26821 - **KUSHWAHA, S.C., KATES, M., STOECKENIUS, W.** : Comparison of purple membrane
from *Halobacterium cutirubrum* and *Halobacterium halobium*. - Biochim. biophys.
Acta *426* : 703 - 710, 1976.

26822 - **KUTYURIN, V.M., ULUBEKOVA, M.V., ZAKHAROVA, N.I., AKSENOV, V.P., ANISIMOVA,
I.N.** : Deĭstvie defitsita margantsa na vydelenie kisloroda i sostav pigment-
-belkovykh kompleksov iz khloroplastov *Scenedesmus obliquus*. [Effect of manganese deficiency on oxygen evolution and composition of pigment-protein complexes from chloroplasts of *Scenedesmus obliquus*.] - Fiziol. Rast. *23* : 932 -
- 937, 1976. [In R, ab : E.]

26823 - **KWIATKOWSKI, R.E., ROFF, J.C.** : Effects of acidity on the phytoplankton and
primary productivity of selected northern Ontario lakes. - Can. J. Bot. *54* :
2546 - 2561, 1976.

26824 - **LABUS, B., NOBEL, W., SMETANA, R., KOHLER, A.** : Der Einfluss der Abwassersub-
stanzen Marlon A (anionenaktives Tensid) und Bor auf die Photosyntheserate
einiger submerser Makrophyten. - Verh. Ges. Ökol. (Göttingen) *1976* : 325 -
- 333, 1976.

26825 - **LACAZE, J.C., VILLEDON de NAÏCE, O.** : Influence of illumination on phytotoxi-
city of crude oil. - Mar. Pollut. Bull. *7* (4) : 73 - 76, 1976. [Primary production.]

26826 - **LACH, H.-J., BÖGER, P.** : Variable composition of cytochrome b_6-f particles. - Z. Naturforsch. *31 c* : 606 - 611, 1976.

26827 - **LAD, S.L., KALBHOR, P.N.** : To study the effects of irrigation scheduled according to water use factors on the yield and yield contributing characters of wheat (*Triticum aestivum* L.) var. NI. 747-19 under varying fertility levels of nitrogen and phosphate. - J. Maharashtra agr. Univ. *1* : 234 - 237, 1976. [Yield formation.]

26828 - **LADIGES, P.Y.** : Variation in drought resistance in adjacent edaphic populations of *Eucalyptus viminalis* LABILL. - Aust. J. Ecol. *1* : 67 - 76, 1976. [Growth analysis.]

26829 - **LADYGIN, V.G.** : Poluchenie i otbor mutantov odnokletochnykh vodorosleǐ s narusheniyami tsepi fotosinteticheskogo perenosa élektrona. [Production and selection of mutants of unicellular algae with damaged chain of photosynthetic electron transport.] - Fiziol. Rast. *23* : 877 - 884, 1976. [In R, ab : E.]

26830 - **LADYGIN, V.G., KLIMOV, V.V., SHUVALOV, V.A., TAGEEVA, S.V.** : Kharakteristika fotokhimicheskikh reaktsionnykh tsentrov trekh tipov mutantov *Chlamydomonas reinhardi* s neaktivnymi fotosistemami. [Characteristics of photochemical reaction sites in three types of mutants of *Chlamydomonas reinhardi* with damaged photosystems.] - Fiziol. Rast. *23* : 681 - 689, 1976. [In R, ab : E.]

26831 - **LADYGIN, V.G., SEMENOVA, G.A., TAGEEVA, S.V., POPOV, V.I.** : Éksperimental'nyǐ mutagenez v poluchenii model'nykh ob"ektov dlya izucheniya membrannoǐ organizatsii khloroplastov. [Experimental mutagenesis in the production of models for studying the membrane organization of chloroplasts.] - In : DUBININ, N.P. (ed.) : Molekulyarnye Mekhanizmy Geneticheskikh Protsessov. Pp. 52 - 57. Nauka, Moskva 1976. [In R.]

26832 - **LAFFERTY, J., LAND, E.J., TRUSCOTT, T.G.** : Electron-transfer reactions involving chlorophyll *a* and carotenoids. - J. chem. Soc. chem. Commun. *2* : 70 - 71, 1976.

26833 - **LAGOUTTE, B., DURANTON, J.** : Structure and morphogenesis of photosynthetic membranes. I. Is there a single basic peptide within *Zea mays* thylakoids ? - Biochim. biophys. Acta *427* : 141 - 152, 1976.

26834 - **LAING, W.A., CHRISTELLER, J.T.** : A model for the kinetics of activation and catalysis of ribulose 1,5-bisphosphate carboxylase. - Biochem. J. *159* : 563 - - 570, 1976.

26835 - **LAǏSK, A., OYA, V.** : Potentsial'naya intensivnost' fotosinteza lista opredelyaetsya reaktsiyami resinteza ribulozodifosfata. [Potential photosynthetic rate is limited by the reactions of ribulose diphosphate resynthesis.] - Izv. Akad. Nauk éston. SSR, Biol. *25* (2) : 146 - 150, 1976. [In R, ab : E, Eston.]

26836 - **LAǏSK, A., OYA, V.** : Gazometricheskoe issledovanie fotodykhaniya. [Gasometric study of photorespiration.] - In : Gazometricheskoe Issledovanie Fotosinteza i Dykhaniya Rasteniǐ. Pp. 89 - 91. Akad. Nauk SSSR, Tartu 1976. [In R.]

26837 - **LAǏSK, A.Kh.** : Primenenie gazometrii pri issledovaniyakh fotosinteza. [Use of gasometry in photosynthesis studies.] - In : Gazometricheskoe Issledovanie Fotosinteza i Dykhaniya Rasteniǐ. Pp. 86 - 88. Akad. Nauk SSSR, Tartu 1976. [In R.]

26838 - **LAKATOS, G.** : On the photosynthesis and chlorophyll efficiency in benthonic communities dominated by *Vaucheria dichotoma* AGH. in Lake Velence (Hungary). - Acta bot. Acad. Sci. hung. *22* : 381 - 391, 1976.

26839 - **LAKIN, M.B.** : Chemical catalyst interference in the Winkler titration determination of dissolved oxygen - a method for correction. - Water Res. *10* : 961 - 966, 1976.

26840 - **LAKSO, A.N., MUSSELMAN, R.C.** : Effects of cloudiness on interior diffuse light in apple trees. - J. amer. Soc. hort. Sci. *101* : 642 - 644, 1976. [Predicted Ps.]

26841 - **LAMONT, B.** : The effects of seasonality and waterlogging on the root systems of a number of *Hakea* species. - Aust. J. Bot. *24* : 691 - 702, 1976. [Dry-matter production.]

26842 - **LANCER, H.A., COHEN, C.E., SCHIFF, J.A.** : Changing ratios of phototransform-
able protochlorophyll and protochlorophyllide of bean seedlings developing in
the dark. - Plant Physiol. *57* : 369 - 374, 1976.

26843 - **LANDSBERG, J.J., BLANCHARD, T.W., WARRIT, B.** : Studies on the movement of
water through apple trees. - J. exp. Bot. *27* : 579 - 596, 1976. [Resistances.]

26844 - **LANG, A.R.G., SHELL, G.S.G.** : Sunlit areas and angular distributions of sunflower leaves for plants in single and multiple rows. - Agr. Meteorol. *16* :
5 - 15, 1976.

26845 - **LANG, V., KÜHBAUCH, W., VOIGTLÄNDER, G., IMHOFF, H.** : Wanderung von ^{14}C-Assi-
milaten in Wiesenkerbel (*Anthriscus silvestris* (L.) HOFFM.). - Z. Acker-
Pflanzenbau *143* : 287 - 293, 1976.

B26846 - **LANGE, O.L., KAPPEN. L., SCHULZE, E.-D.** (ed.) : Water and Plant Life. Problems and Modern Approaches. (Ecological Studies Vol. 19.). - Springer-Verlag,
Berlin - Heidelberg - New York 1976. [Ps.]

26847 - **LANGER, R.H.M., DOUGHERTY, C.T.** : Physiology of grain yield in wheat. - In :
SUNDERLAND, N. (ed.) : Perspectives in Experimental Biology. Vol. 2. Botany.
Pp. 59 - 67. Pergamon Press, Oxford - New York 1976. [Yield formation.]

26848 - **LANGFORD, K.J.** : Change in yield of water following a bushfire in a forest of
Eucalyptus regnans. - J. Hydrol. *29* : 87 - 114, 1976. [Resistances.]

26849 - **LÄNNERGREN, C.** : Primary production in Lindåspollene, a Norwegian land-locked
fjord. - Bot. mar. *19* : 259 - 272, 1976.

26850 - **LANYI, J.K., MacDONALD, R.E.** : Existence of electrogenic hydrogen ion/sodium
ion antiport in *Halobacterium halobium* cell envelope vesicles. - Biochemistry
15 : 4608 - 4614, 1976.

26851 - **LANYI, J.K., YEARWOOD-DRAYTON, V., MacDONALD, R.E.** : Light-induced glutamate
transport in *Halobacterium halobium* envelope vesicles. I. Kinetics of the
light-dependent and the sodium-gradient-dependent uptake. - Biochemistry *15* :
1595 - 1603, 1976.

26852 - **LAPINA, L.P., BIKMUKHAMETOVA, S.A., MURASHOV, I.N.** : Vliyanie Na_2SO_4 i NaCl
na aktivnost' fotosinteticheskogo fosforilirovaniya u razlichnykh po sole-
ustoĭchivosti rasteniĭ. [Effect of sodium sulphate and chloride on the activity of photosynthetic phosphorylation in plants with different salt resistance.] - Fiziol. Rast. *23* : 279 - 285, 1976. [In R, ab : E.]

26853 - **LAPTEV, Yu.P., MAKAROV, P.P., GLAZOVA, M.V., SHUGAEVA, E.V., MIKHAĬLOVA, S.P.,
ARKHANGEL'SKAYA, M.A., VLADIMIROVA, I.A.** : Ust'ichnyĭ apparat i pyl'tsa kak
pokazateli ploidnosti rasteniĭ. [Stomatal apparatus and pollen as indicators
of plant ploidy.] - Genetika *12* (1) : 47 - 55, 1976. [Chloroplast number; in
R, ab : E.]

26854 - **LARCHER, W., WAGNER, J.** : Temperaturgrenzen der CO_2-Aufnahme und Temperatur-
resistenz der Blätter von Gebirgspflanzen im vegetationsaktiven Zustand. -
Oecol. Plant. *11* : 361 - 374, 1976.

26855 - **LARPENT, J.-P., JACQUES, R., QUENNEMET, J., MONÉGER, R.** : Qualité spectrale
de la lumière et morphogénèse du thalle chez le *Draparnaldia mutabilis* (ROTH)
CEDERG. - Bull. Soc. phycol. Fr. *21* : 49 - 55, 1976. [Chl, Car.]

26856 - **LARSON, P.R.** : The leaf-cambium relation and some prospects for genetic improvement. - In : CANNELL, M.G.R., LAST, F.T. (ed.) : Tree Physiology and Yield
Improvement. Pp. 261 - 282. Academic Press, London - New York - San Francisco
1976. [Ps.]

26857 - **LASSOIE, J.P., DOUGHERTY, P.M.** : Fall and winter gas exchange rates in eastern redcedar. - Plant Physiol. *57* (Suppl.) : 45, 1976. [Chl.]

26858 - **LATZKO, E., KELLY, G.J.** : Photosynthesis. Biochemical and physiological aspects of carbon metabolism. - Progr. Bot. *38* : 81 - 99, 1976.

26859 - **LAVOREL, J.** : An alternative to Kok's model for the oxygen-evolving system in
photosynthesis. - FEBS Lett. *66* : 164 - 167, 1976.

26860 - **LAVOREL, J.** : Matrix analysis of the oxygen evolving system of photosynthesis. - J. theor. Biol. *57* : 171 - 185, 1976.

26861 - **LAVOREL, J., LEMASSON, C.** : Anomalies in the kinetics of photosynthetic oxygen emission in sequences of flashes revealed by matrix analysis. Effects of carbonyl cyanide *m*-chlorophenylhydrazone and variation in time parameters. - Biochim. biophys. Acta *430* : 501 - 516, 1976.

26862 - **LAWANSON, A.O.** : Effect of prior heat stress on photophosphorylation in seedlings of *Zea mays*. - Fyton *34* : 51 - 54, 1976.

26863 - **LAWLOR, D.W.** : Water stress induced changes in photosynthesis, photorespiration, respiration and CO_2 compensation concentration of wheat. - Photosynthetica *10* : 378 - 387, 1976.

26864 - **LAWLOR, D.W.** : Assimilation of carbon into photosynthetic intermediates of water-stressed wheat. - Photosynthetica *10* : 431 - 439, 1976.

26865 - **LAWLOR, D.W., LAKE, J.V.** : Evaporation rate, leaf water potential and stomatal conductance in *Lolium, Trifolium* and *Lysimachia* in drying soil. - J. appl. Ecol. *13* : 639 - 646, 1976.

26866 - **LAWRIE, A.C., CODD, G.A., STEWART, W.D.P.** : The incorporation of nitrogen into products of recent photosynthesis in *Anabaena cylindrica* LEMN. - Arch. Microbiol. *107* : 15 - 24, 1976.

26867 - **LAWS, E., CAPERON, J.** : Carbon and nitrogen metabolism by *Monochrysis lutheri*: Measurement of growth-rate-dependent respiration rates. - Mar. Biol. *36* : 85 - 97, 1976.

26868 - **LEBEDEV, N.N., NAUSH, Ya., KRASNOVSKIĬ, A.A. ml.** : Fosforestsentsiya khlorofilla *a* v polimernoĭ matritse triatsetattsellyulozy. [Phosphorescence of chlorophyll *a* in polymer matrix of triacetate cellulose.] - Biofizika *21* : 382 - - 384, 1976. [In R, ab : E.]

26869 - **LEBEDEV, V.M., SOKOLOV, O.A.** : Vliyanie urovnya azotnogo pitaniya na poglotitel'nuyu deyatel'nost' kornevoĭ sistemy i fotosinteticheskuyu aktivnost' listovogo apparata karlikovykh i polukarlikovykh rasteniĭ yabloni. [Effect of the nitrogen nutrition level on the absorptive activity of the root system and on the photosynthetic activity of the foliar apparatus of dwarf and semi-dwarf apple trees.] - Agrokhimiya *1976* (11) : 10 - 18, 1976. [In R.]

26870 - **LEBEDEV, V.V.** : Opredelenie potokov CO_2 v usloviyakh lesa. [Determination of CO_2 fluxes in a forest.] - In : Gazometricheskoe Issledovanie Fotosinteza i Dykhaniya Rasteniĭ. Pp. 92 - 97. Akad. Nauk SSSR, Tartu 1976. [In R.]

26871 - **LEBLANC, R.M., CHAPADOS, C.** : Aggregation of chlorophylls in monolayers. II. Chlorophyll-dioxane interaction. - Biophys. Chem. *6* : 77 - 85, 1976.

26872 - **LEBLOVÁ, S.** : Fotosyntetická fixace CO_2 u vyšších rostlin. [Photosynthetic fixation of CO_2 in higher plants.] - Chem. Listy *70* : 603 - 638, 1976. [In Czech, ab : E.]

26873 - **LECLERC, J.-C., COUTÉ, A.** : Révision de la position systématique de l'Algue parasite *Phyllosiphon arisari* KÜHN d'après l'analyse spectrophotométrique de ses chlorophylles. - Compt. rend. Acad. Sci. Paris, Sér. D *282* : 2067 - 2070, 1976.

26874 - **Le CLERC, M.H.** : Effects of nitrogen and rest periods on the seasonal production of a pure grass sward. 1. Dry matter production. - Ir. J. agr. Res. *15* : 247 - 255, 1976.

26875 - **Le CREN, E.D.** : The productivity of freshwater communities. - Phil. Trans. roy. Soc. London B *274* : 359 - 374, 1976.

26876 - **LEDENT, J.-F.** : Assimilation et translocation du $^{14}CO_2$ chez le froment (*Triticum aestivum* L.) en fonction de la densité et de la structure du couvert aérien. - Bull. Acad. roy. Belg. Cl. Sci. *60* : 790 - 798, 1974/1976.

26877 - **LEDENT, J.-F.** : Beam light interception by twisted leaf surfaces. - Agr. Meteorol. *17* : 271 - 280, 1976.

26878 - **LEDENT, J.F.** : Comment interpreter les fréquences de contact dans l'analyse des surfaces photosynthétiques par la méthode des points quadrats, spéciale- ment dans le cas d'une distribution binomiale négative ? - Biometrie Praxi- metrie *16* : 37 - 49, 1976.

26879 - **LEDENT, J.F.** : Un modèle simple pour étudier l'interception du rayonnement solaire en relation avec la géométrie du couvert végétal, dans le cas d'une culture en rangées. - Biometrie Praximetrie *16* : 104 - 112, 1976. [Ps.]

26880 - **LEDENT, J.-F.** : Photosynthèse et photorespiration. Quelques développements récents. - Rev. Quest. sci. *147* : 367 - 381, 1976.

26881 - **LEDENT, J.-F., ISERENTANT, R.** : L'exploitation de l'énergie solaire en agri- culture. - Rev. Agr. *29* : 815 - 841, 1976.

26882 - **LEDERMAN, T.C., HORNBERGER, G.M., KELLY, M.G.** : The calibration of a phyto- plankton growth model using batch culture data. - Water Air Soil Pollut. *5* : 431 - 442, 1976.

26883 - **LEDIG, F.T.** : Physiological genetics, photosynthesis and growth models. - In: CANNELL, M.G.R., LAST, F.T. (ed.) : Tree Physiology and Yield Improvement. Pp. 21 - 54. Academic Press, London - New York - San Francisco 1976.

26884 - **LEDIG, F.T., DREW, A.P., CLARK, J.G.** : Maintenance and constructive respira- tion, photosynthesis, and net assimilation rate in seedlings of Pitch pine (*Pinus rigida* MILL.). - Ann. Bot. *40* : 289 - 300, 1976.

26885 - **LEDOIGT, G., LEFORT-TRAN, M.** : Relationship between chloroplastic metabolism and cytoplasmic translation. - In : BÜCHER, T., NEUPERT, W., SEBALD, W., WER- NER, S. (ed.) : Genetics and Biogenesis of Chloroplasts and Mitochondria. Pp. 95 - 98. North-Holland Publ. Co., Amsterdam - New York - Oxford 1976.

26886 - **LEE, C.I., KUNG, S.D.** : Chemical nature of the polypeptides of tobacco frac- tion 1 protein. - Plant Physiol. *57* (Suppl.) : 4, 1976.

26887 - **LEE, E., MILES, D.** : Comparative quantitation of plastoquinones by extraction and *in vivo* absorption from photosynthesis mutants of maize. - Plant Physiol. *57* (Suppl.) : 53, 1976.

26888 - **LEE, S.S.** : Effect of light intensity on photosynthesis of plant cells. - Plan Physiol. *57* (Suppl.) : 25, 1976.

26889 - **LEE, S.S., FANG, S.C., FREED, V.H.** : Effect of DDT on photosynthesis of *Sele- nastrum capricornutum*. - Pesticide Biochem. Physiol. *6* : 46 - 51, 1976.

26890 - **LEECH, R.M.** : Plastid development in isolated etiochloroplasts and isolated etioplasts. - In : SUNDERLAND, N. (ed.) : Perspectives in Experimental Biolo- gy. Vol. 2. Botany. Pp. 145 - 162. Pergamon Press, Oxford - New York 1976.

26891 - **LEECH, R.M.** : The replication of plastids in higher plants. - In : YEOMAN, M. M. (ed.) : Cell Division in Higher Plants. Pp. 135 - 159. Academic Press, London - New York - San Francisco 1976.

26892 - **LEECH, R.M.** : The photosynthetic apparatus of higher plants. - In : HALL, M. A. (ed.) : Plant Structure, Function and Adaptation. Pp. 125 - 156. Macmillan, London 1976.

26893 - **LEECH, R.M., MURPHY, D.J.** : The cooperative function of chloroplasts in the biosynthesis of small molecules. - In : BARBER, J. (ed.) : The Intact Chloro- plast. Pp. 365 - 401. Elsevier, Amsterdam - New York - Oxford 1976.

26894 - **LEESE, B.M., LEECH, R.M.** : Sequential changes in the lipids of developing pro- plastids isolated from green maize leaves. - Plant Physiol. *57* : 789 - 794, 1976.

26895 - **LEGGE, A.H., AMUNDSON, R.G., JAQUES, D.R., WALKER, R.B.** : Field studies of pine, spruce and aspen periodically subjected to sulfur gas emissions. - In : Proceedings of the First International Symposium on Acid Precipitation and the Forest Ecosystem. USDA Forest Serv. Gen. Tech. Rep. *NE-23* : 1033 - 1061, 1976. [Ps.]

26896 - **LEHMAN, J.T.** : Photosynthetic capacity and luxury uptake of carbon during phosphate limitation in *Pediastrum duplex (Chlorophyceae)*. - J. Phycol. *12* : 190 - 193, 1976.

26897 - **LEHOCZKI, E., CSATORDAY, K., DEMETER, S.** : Light absorption in a chlorophyll--detergent photosynthetic model system. - Acta biochim. biophys. Acad. Sci. hung. *11* : 158, 1976.

26898 - **LEMEUR, R., ROSENBERG, N.J.** : Reflectant induced modification of soybean canopy radiation balance. III. A comparison of the effectiveness of celite and kaolinite reflectants. - Agron. J. *68* : 30 - 35, 1976.

26899 - **LEMOINE, Y.** : Étude de la composition protéique des thylakoïdes en fonction de l'ultrastructure des chloroplastes et de leur contenu pigmentaire chez un mutant chlorophyllien de Tabac. - Physiol. vég. *14* : 437 - 452, 1976..

26900 - **LENDZIAN, K., BASSHAM, J.A.** : NADPH/NADP⁺ ratios in photosynthesizing reconstituted chloroplasts. - Biochim. biophys. Acta *430* : 478 - 489, 1976.

26901 - **LENDZIAN, K., BASSHAM, J.A.** : Ribulose diphosphate carboxylase regulation in reconstituted spinach chloroplasts. - Plant Physiol. *57* (Suppl.) : 4, 1976.

26902 - **LENDZIAN, K.J., ZIEGLER, H.** : Effect of phosfon-D on photoreactions of spinach chloroplasts. - Plant Physiol. *57* (Suppl.) : 24, 1976.

*26903 - **LEONARDI, S., LUCIANI, F., POLI, E.** : Superficie fotosinteticamente attiva del *Quercus ilex* L. nella lecceta di M. Minardo (Etna). [Active photosynthetic surface of *Quercus ilex* L. in an oak forest on M. Minardo (Etna).] - Arch. bot. biogeograf. ital. *50*, Ser. 4 *20* (3/4) : 3 - 15, 1974. [In Ital., ab : G.]

26904 - **LERCH, G.** : Einfluss einiger Klima- und Anbaufaktoren auf die Ertragsbildung der Reissorte 'IR-8' in Cuba. I. Standweite und Stickstoffdüngung. - Kulturpflanze *24* : 13 - 51, 1976.

26905 - **LERCH, G.** : Einfluss einiger Klima- und Anbaufaktoren auf die Ertragsbildung der Reissorte 'IR-8' in Cuba. II. Standweite und CO₂-Assimilation. - Kulturpflanze *24* : 53 - 63, 1976.

26906 - **LERCH, G.** : Einfluss einiger Klima- und Anbaufaktoren auf die Ertragsbildung der Reissorte 'IR 8' in Cuba. III. Klimaverhältnisse. - Kulturpflanze *24* : 65 - 117, 1976.

26907 - **LEVI, C., GIBBS, M.** : Starch degradation in isolated spinach chloroplasts. - Plant Physiol. *57* : 933 - 935, 1976.

26908 - **LEVITT, J.** : Physiological basis of stomatal response. - In : **LANGE, O.L., KAPPEN, L., SCHULZE, E.-D.** (ed.) : Water and Plant Life. Pp. 160 - 168. Springer-Verlag, Berlin - Heidelberg - New York 1976. [Ps.]

26909 - **LEVY, Y., KAUFMANN, M.R.** : Cycling of leaf conductance in citrus exposed to natural and controlled environments. - Can. J. Bot. *54* : 2215 - 2218, 1976. [Ps.]

26910 - **LEWANDOWSKA, M., HART, J.W., JARVIS, P.G.** : Photosynthetic electron transport in plants of Sitka spruce subjected to differing light environments during growth. - Physiol. Plant. *37* : 269 - 274, 1976.

26911 - **LEWANTY, Z., MALESZEWSKI, S.** : Conversion of photosynthetic products in the light in CO₂-free O₂ and N₂ in leaves of *Zea mays* L. and *Phaseolus vulgaris* L. - Planta *131* : 121 - 123, 1976.

26912 - **LEWIS, M.C., CALLAGHAN, T.V.** : Tundra. - In : **MONTEITH, J.L.** (ed.) : Vegetation and the Atmosphere. Vol. 2. Case Studies. Pp. 399 - 433. Academic Press, London - New York - San Francisco 1976. [Ps, Chl.]

26913 - **LEWIS, T.J., PETHIG, R.** : The determination of localized energy states in β--carotene by A.C. conduction studies. - In : **BIRKS, J.B.** (ed.) : Excited States of Biological Molecules. Pp. 342 - 352. Wiley-Interscience, London - New York 1976.

26914 - **LEY, A.C., BUTLER, W.L.** : Efficiency of energy transfer from photosystem II to photosystem I in *Porphyridium cruentum*. - Proc. nat. Acad. Sci. USA *73* : 3957 - 3960, 1976.

26915 - **LIAAEN-JENSEN, S.** : New structures. - Pure appl. Chem. *47* : 129 - 145, 1976. [Car.]

26916 - **LICHTLÉ, C., THOMAS, J.C.** : Étude ultrastructurale des thylacoïdes des algues
à phycobiliprotéines, comparaison des résultats obtenus par fixation classi-
que et cryodécapage. - Phycologia *15* : 393 - 404, 1976.

26917 - **LIEBEREI, R., BIEHL, B.** : Freisetzung und Aktivierung von Polyphenoloxydasen
aus Thylakoidmembranen der Spinat-Chloroplasten. - Ber. deut. bot. Ges. *89* :
663 - 676, 1976.

26918 - **LIEPINYA, I.Ė., YAKOBSONE, G.Ė.** : Vliyanie khlorkholinkhlorida i alara na de-
korativnye i fiziologicheskie svoĭstva gvozdiki gruppy Sim. [Effect of chlorocholine chloride and Alar on the decorative and physiological properties of
carnations of the Sim group.] - In : Regulyatsiya Rosta i Pitanie Rasteniĭ.
Pp. 61 - 67. Zinatne, Riga 1976. [Chl; in R.]

*26919 - **LIETH, H.** : Primary productivity in ecosystems : comparative analysis of glo-
bal patterns. - In : Van DOBBEN, W.H., LOWE-McCONNELL, R.H. (ed.) : Unifying
Concepts in Ecology. Pp. 67 - 88. Dr. W. Junk b.v. Publ., The Hague; Pudoc,
Wageningen 1975.

26920 - **LIETH, H.** : The use of correlation models to predict primary productivity
from precipitation or evapotranspiration. - In : LANGE, O.L., KAPPEN, L.,
SCHULZE, E.-D. (ed.) : Water and Plant Life. Pp. 392 - 407. Springer-Verlag,
Berlin - Heidelberg - New York 1976.

26921 - **LIN, C.H., STOCKING, C.R.** : Glyceraldehyde 3-phosphate dehydrogenase in greening maize leaves. - Plant Physiol. *57* (Suppl.) : 33, 1976.

26922 - **LIN, K.L., NG, T.B., TAN, P.Y., TAN, S.F.** : A laboratory study of the recovery of carotenoids from the spent adsorbents used in decolorizing palm oil. -
J. Singapore nat. Acad. Sci. *5* : 54 - 59, 1976. [TLC.]

26923 - **LINACRE, E.** : Swamps. - In : MONTEITH, J.L. (ed.) : Vegetation and the Atmosphere. Vol. 2. Case Studies. Pp. 329 - 347. Academic Press, London - New
York - San Francisco 1976. [Ps production.]

26924 - **LINEK, V., VACEK, V.** : Oxygen electrode response lag induced by liquid film
resistance against oxygen transfer. - Biotechnol. Bioeng. *18* : 1537 - 1555,
1976.

26925 - **LIPSKAYA, G.A.** : Skorost' reaktsii Khilla khloroplastov, izolirovannykh iz
postėtiolirovannykh prorostkov yachmenya, pri neodinakovom snabzhenii ikh kobal'tom. [Rate of Hill reaction of chloroplasts, isolated from postetiolated
barley seedlings under various supply of cobalt.] - Vest. belorus. Univ., Ser.
2, *1976* (1) : 52 - 55, 1976. [In R.]

26926 - **LIPSKAYA, G.A.** : Sovremennoe sostoyanie voprosa o roli kobal'ta v formirovanii i funktsionirovanii fotosinteticheskogo apparata i zhiznedeyatel'nosti
rasteniĭ. [Current state of the problem of the role of cobalt in the formation
and functioning of the photosynthetic apparatus and the vital activity of
plants.] - In : KAKHNOVICH, L.V. (ed.) : Optimizatsiya Fotosinteticheskogo
Apparata Vozdeĭstviem Razlichnykh Faktorov. Pp. 63 - 81. Izd. BGU, Minsk 1976.
[In R.]

26927 - **LIPSKAYA, G.A., CHERNETSKIĬ, S.S.** : Nekotorye pokazateli fotosinteticheskogo
apparata i produktivnost' yachmenya vo vtorom semennom pokolenii v svyazi s
udobreniem materinskikh rasteniĭ kobal'tom. [Some characteristics of photosynthetic apparatus and productivity of barley in the second seed generation
in connection with cobalt nutrition of maternal plants.] - Botanika (Issledovaniya) *18* : 107 - 115, 1976. [In R.]

26928 - **LIPSKAYA, G.A., IVANOV, N.P., DOLGORUKOVA, T.V.** : Formirovanie fotosinteticheskogo apparata i produktivnost' yachmenya pri vnesenii kobal'ta na melio-
rirovannoĭ mineral'noĭ pochve. [Formation of the photosynthetic apparatus and
the productivity of barley after addition of cobalt to reclaimed mineral
soil.] - In : KAKHNOVICH, L.V. (ed.) : Optimizatsiya Fotosinteticheskogo Apparata Vozdeĭstviem Razlichnykh Faktorov. Pp. 93 - 101. Izd. BGU, Minsk 1976.
[In R.]

26929 - **LIPSKAYA, G.A., LAZAREVICH, Z.I.** : Geterogennost' khlorofilla *a* i *b* v ontogeneze rasteniĭ yachmenya pri raznoĭ prodolzhitel'nosti vyderzhivaniya list'ev
v temnote i neodinakovom snabzhenii rasteniĭ kobal'tom. [Heterogeneity of chlo-

rophyll a and b in the ontogenesis of barley plants after various durations of exposure of the leaves to the dark and after supplying cobalt to the plants at different rates.] - In : KAKHNOVICH, L.V. (ed.) : Optimizatsiya Fotosinteticheskogo Apparata Vozdeľstviem Razlichnykh Faktorov. Pp. 81 - 93. Izd. BGU, Minsk 1976. [In R.]

26930 - LIPSKAYA, G.A., ZELENAYA, L.A. : Geterogennost' khlorofillov a i b v usloviyakh temnoty pri neodinakovom snabzhenii prorostkov yachmenya kobal'tom. [Heterogeneity of chlorophylls a and b in darkness under different supply of barley seedlings with cobalt.] - Fiziol. Rast. *23* : 792 - 798, 1976. [In R, ab : E.]

26931 - LIPTAY, A., DAVIDSON, D., ORMROD, D.P. : Differential effects of carbon dioxide on the growth rates of coleoptiles and the enclosed leaves of *Hordeum vulgare* L. seedlings. - Ann. Bot. *40* : 1029 - 1032, 1976.

26932 - LITTLE, E.C.S. : The potential of maize and sorghum. - New Zeal. J. Agr. *132* (5) : 17 - 18, 1976. [Ps.]

26933 - LITTLER, M.M., MURRAY, S.N. : Seasonal variations in primary productivity of seaweeds with different morphological forms. - J. Phycol. *12* (Suppl.) : 16, 1976.

26934 - LITTON, J.R., Jr., GILBERT, J.J. : Assimilation and retention of tocopherol and chlorophylls in rotifers *Brachionus calyciflorus* and. *Asplancha sieboldi*. - Experientia *32* : 1530 - 1532, 1976.

26935 - LITVIN, F.F., EFIMTSEV, E.I., IGNATOV, N.V. : Kinetika fotokhimicheskikh reaktsiľ biosinteza khlorofilla i migratsiya ėnergii mezhdu pigmentami v ėtiolirovannykh list'yakh. [Kinetics of photochemical reactions of chlorophyll biosynthesis and energy migration between pigments in etiolated leaves.] - Fiziol. Rast. *23* : 653 - 659, 1976. [In R, ab : E.]

26936 - LITVIN, F.F., EFIMTSEV, E.I., IGNATOV, N.V. : Migratsiya ėnergii v fotoaktivnykh kompleksakh predshestvennika khlorofilla v ėtiolirovannykh list'yakh i spektroskopicheskie kharakteristiki pigmentnykh form. [Energy migration in photoactive complexes of chlorophyll precursor in etiolated leaves and spectroscopic characteristics of pigment forms.] - Biofizika *21* : 307 - 312, 1976. [In R, ab : E.]

26937 - LITVIN, F.F., EFIMTSEV, E.I., IGNATOV, N.V., BELYAEVA, O.B. : Dokazatel'stvo sushchestvovaniya dvukh fotokhimicheskikh reaktsiľ v protsesse biosinteza khlorofilla i issledovanie perenosa ėnergii mezhdu nimi. [Confirmation of the existence of two photochemical reactions during biosynthesis of chlorophyll and investigation of energy transfer between them.] - Fiziol. Rast. *23* : 17 - - 24, 1976. [In R, ab : E.]

26938 - LITVIN, F.F., SINESHCHEKOV, V.A., SHUBIN, V.V. : Issledovanie migratsii ėnergii mezhdu nativnymi formami khlorofilla pri -196 °C metodom sensibilizirovannoľ fluorestsentsii. [Energy migration between native chlorophyll forms studied at -196 °C by the method of sensitized fluorescence.] - Biofizika *21* : 669 - 675, 1976. [In R, ab : E.]

26939 - LIU, M.S., HELLEBUST, J.A. : Effects of salinity changes on growth and metabolism of the marine centric diatom *Cyclotella cryptica*. - Can. J. Bot. *54* : 930 - 937, 1976. [Ps, Chl.]

26940 - LLOYD, E.J. : The influence of shading on enzyme activity in seedling leaves of barley. - Z. Pflanzenphysiol. *78* : 1 - 12, 1976. [Ps, Chl.]

26941 - LLOYD, N.D.H., WOOLHOUSE, H.W. : The effect of temperature on photosynthesis and transpiration in populations of *Sesleria caerulea* (L.) ARD. - New Phytol. *77* : 553 - 559, 1976.

26942 - LOACH, P.A. : Chemical properties of the phototrap in bacterial photosynthesis. - In : KAISER, E.T., KEZDY, F.J. (ed.) : Progress in Bioorganic Chemistry. Vol. 4. Pp. 89 - 192. J. Wiley & Sons, New York - London - Sydney - Toronto 1976.

26943 - LOACH, P.A., HALES, B.J. : Free radicals in photosynthesis. - In : PRYOR, W. A. (ed.) : Free Radicals in Biology. Vol. 1. Pp. 199 - 237. Academic Press, New York - London - San Francisco 1976.

26944 - LOCH, D.S., HOPKINSON, J.M., ENGLISH, B.H. : Seed production of *Stylosanthes guyanensis*. 1. Growth and development. - Aust. J. exp. Agr. anim. Husb. *16* : 218 - 225, 1976. [Growth analysis.]

26945 - LOCH, D.S., HOPKINSON, J.M., ENGLISH, B.H. : Seed production of *Stylosanthes guyanensis*. 2. The consequences of defoliation. - Aust. J. exp. Agr. anim. Husb. *16* : 226 - 230, 1976. [Growth analysis.]

26946 - LOCKAU, W., SELMAN, B.R. : Correlation of the photosynthetic reduction of p- -(diazonium-) benzenesulfonic acid with the increased binding of the probe to the thylakoid membrane. - Z. Naturforsch. *31 c* : 48 - 54, 1976.

26947 - LOF, H. : Water use efficiency and competition between arid zone annuals, especially the grasses *Phalaris minor* and *Hordeum murinum*. - Agr. Res. Rep. (Wageningen) *853* : 1 - 107, 1976. [Ps.]

26948 - LÖFFELHARDT, W., KINDL, H. : Formation of benzoic acid and p-hydroxybenzoic acid in the blue green alga *Anacystis nidulans* : a thylakoid-bound enzyme complex analogous to the chloroplast system. - Z. Naturforsch. *31 c* : 693 - - 699, 1976.

26949 - LOHR, J.B., FRIEDMANN, H.C. : New pathwayfor δ-aminolevulinic acid biosynthe- sis : formation from α-ketoglutaric acid by two partially purified plant en- zymes. - Biochem. biophys. Res. Commun. *69* : 908 - 913, 1976.

26950 - LOISEAUX, S. : Ultrastructural modifications of plastids after action of chlo- ramphenicol, lincomycin and rifampicin. - Physiol. vég. *14* : 1 - 10, 1976.

26951 - LOMAGIN, A.G. : Otsenka teploustoĭchivosti rastitel'nykh kletok po dvizheniyu tsitoplazmy i fototaksisu khloroplastov. [Assessment of heat resistance of plant cells from cytoplasm movement and chloroplast phototaxis.] - In : Meto- dy Otsenki Ustoĭchivosti Rasteniĭ k Neblagopriyatnym Usloviyam Sredy. Pp. 60- - 70. Kolos, Leningrad 1976. [In R.]

*26952 - LONG, J.N., TURNER, J. : Aboveground biomass of understorey and overstorey in an age sequence of four Douglas-fir stands. - J. appl. Ecol. *12* : 179 - 188, 1975. [Production.]

26953 - LONGHURST, A.R. : Interactions between zooplankton and phytoplankton profiles in the eastern tropical Pacific Ocean. - Deep-Sea Res. *23* : 729 - 754, 1976. [Chl.]

26954 - LOOMIS, R.S., NG, E., HUNT, W.F. : Dynamics of development in crop production systems. - In : BURRIS, R.H., BLACK, C.C. (ed.) : CO_2 Metabolism and Plant Productivity. Pp. 269 - 286. Univ. Park Press, Baltimore - London - Tokyo 1976. [Growth analysis.]

26955 - LOOS, E. : The effect of magnesium ions on action spectra for reactions me- diated by Photosystems I and II in spinach chloroplasts. - Biochim. biophys. Acta *440* : 314 - 321, 1976.

26956 - LORD, C.E.C., TIRIMANNA, A.S.L. : A qualitative study of the carotenoid pig- ments of Sri Lanka chillies (*Capsicum annuum*). - Microchim. Acta *1976 I* : 469 - 476, 1976.

26957 - LORENZEN, H., KAUSHIK, B.D. : Experiments withsynchronous *Anacystis nidulans*. - Ber. deut. bot. Ges. *89* : 491 - 498, 1976. [Ps, Chl.]

26958 - LORIMER, G.H., BADGER, M.R., ANDREWS, T.J. : The activation of ribulose-1,5- -bisphosphate carboxylase by carbon dioxide and magnesium ions. Equilibria, kinetics, a suggested mechanism, and physiological implications. - Biochemist- ry *15* : 529 - 536, 1976.

26959 - LORIMER, G.H., OSMOND, C.B., AKAZAWA, T. : [^{18}O]-oxygen incorporation during glycolate synthesis by *Chromatium*. - Plant Physiol. *57* (Suppl.) : 6, 1976.

26960 - LOSADA, M. : Reducing power and the regulation of photosynthesis. - In : KORN- BERG, A., HORECKER, B.L., CORNUDELLA, L. (ed.) : Reflections on Biochemistry. Pp. 73 - 84. Pergamon, Oxford - New York 1976.

*26961 - LOTT, J.N.A. : Dynamics of the nuclear envelope in the cotyledons of *Cucurbi- ta maxima*. - Amer. J. Bot. *61* (5, Suppl.) : 59, 1974. [Ps.]

26962 - **LOWRY, J., ELLIS, R.** : Effects of amitrole (3-amino-1,2,4-triazole) on the
growth and pigment content of the green alga *Golenkinia*. - J. Phycol. *12*
(Suppl.) : 30, 1976.

26963 - **LOZIER, R.H., NIEDERBERGER, W., BOGOMOLNI, R.A., HWANG, S.-B., STOECKENIUS,
W.** : Kinetics and stoichiometry of light-induced proton release and uptake
from purple membrane fragments, *Halobacterium halobium* cell envelopes, and
phospholipid vesicles containing oriented purple membrane. - Biochim. biophys.
Acta *440* : 545 - 556, 1976.

26964 - **LUCAS, E.O., MILBOURN, G.M.** : The effect of density of planting on the growth
of two *Phaseolus vulgaris* varieties in England. - J. agr. Sci. *87* : 89 - 99,
1976. [Growth analysis.]

26965 - **LUCAS, E.O., MILBOURN, G.M., WHITFORD, P.N.** : The translocation of ^{14}C photosynthate from leaves and pods in *Phaseolus vulgaris*. - Ann. appl. Biol. *83* :
285 - 290, 1976.

26966 - **LUCAS, W.J.** : Plasmalemma transport of HCO_3^- and OH^- in *Chara corallina* : non-
-antiporter systems. - J. exp. Bot. *27* : 19 - 31, 1976.

26967 - **LUCERO, H.A., ANDREO, C.S., VALLEJOS, R.H.** : Sulphydryl groups in photosynthetic energy conservation. III. Inhibition of photophosphorylation in spinach
chloroplasts by $CdCl_2$. - Plant Sci. Lett. *6* : 309 - 313, 1976.

26968 - **LUCERO, H.A., RAVIZZINI, R.A., VALLEJOS, R.H.** : Inhibition of spinach chloroplasts photophosphorylation by the antibiotics leucinostatin and efrapeptin.
- FEBS Lett. *68* : 141 - 144, 1976.

26969 - **LÜCKE, F.-K., KLEMME, J.-H.** : Coupling factor adenosine-5'-triphosphatase from
Rhodospirillum rubrum : A simple and rapid procedure for its purification. -
Z. Naturforsch. *31 c* : 272 - 279, 1976.

26970 - **LUDLOW, M.M.** : Ecophysiology of C_4 grasses. - In : **LANGE, O.L., KAPPEN, L.,
SCHULZE, E.-D.** (ed.) : Water and Plant Life. Pp. 364 - 386. Springer-Verlag,
Berlin - Heidelberg - New York 1976.

26971 - **LUDLOW, M.M., NG, T.T.** : Effect of water deficit on carbon dioxide exchange
and leaf elongation rate of *Panicum maximum* var. *trichoglume*. - Aust. J. Plant
Physiol. *3* : 401 - 413, 1976.

26972 - **LUDLOW, M.M., NG, T.T.** : Photosynthetic light response curves of leaves from
controlled environment facilities, glasshouses or outdoors. - Photosynthetica
10 : 457 - 462, 1976.

26973 - **LUDLOW, M.M., TROUGHTON, J.H., JONES, R.J.** : A technique for determining the
proportion of C_3 and C_4 species in plant samples using stable natural isotopes of carbon. - J. agr. Sci. *87* : 625 - 632, 1976.

26974 - **LUKASHEV, E.P., TIMOFEEV, K.N., KONONENKO, A.A., VENEDIKTOV, P.S., RUBIN, A.
B.** : Temperature dependence of the reduction kinetics of photooxidized reaction centre bacteriochlorophyll in dark and light adapted chromatophores
of purple bacteria. - Photosynthetica *10* : 423 - 430, 1976.

*26975 - **LUK'YANOVA, L.M.** : Osobennosti vzaimoprevrashcheniya ksantofillov v list'yakh
svetolyubivykh i tenevynoslivykh rastenii na Krainem Severe. [Peculiarities
of xanthophyll transformations in leaves of sun and shade plants in Extreme
North.] - In : Issledovaniya po Fiziologii Rastenii v Zapolyar'e. Pp. 38 - 42.
Izd. kol'skogo Filiala Akad. Nauk SSSR, Apatity 1975. [In R.]

*26976 - **LUK'YANOVA, L.M.** : O vliyanii uslovii, predshestvuyushchikh opytam, na soderzhanie pigmentov plastid v list'yakh rastenii. [Effect of pretreatment on the
content of plastid pigments in plant leaves.] - In : Issledovanie po Fiziologii Rastenii v Zapolyar'e. Pp. 84 - 96. Izd. kol'skogo Filiala Akad. Nauk
SSSR, Apatity 1975. [In R.]

*26977 - **LUK'YANOVA, L.M., BULYCHEVA, T.M., RYZHOVA, E.F.** : O sezonnykh izmeneniyakh
sostoyaniya pigmentnogo kompleksa list'ev nekotorykh drevesnykh i kustarniko-
vykh rastenii v usloviyakh Zapolyar'ya. [Seasonal changes in the state of the
pigment complex in the leaves of some trees and shrubs under transpolar region
conditions.] - In : Estestvennaya Sreda i Biologicheskie Resursy Krainego Se-
vera. Pp. 113 - 120. Geogr. Obshch. SSSR, Sev. Filial, Leningrad 1975. [In R.]

26978 - LUMSDEN, J., CAMMACK, R., HALL, D.O. :Purification and physicochemical properties of superoxide dismutase from two photosynthetic microorganisms. - Biochim. biophys. Acta *438* : 380 - 392, 1976.

26979 - LUSH, W.M. : Leaf structure and translocation of dry matter in a C_3 and a C_4
grass. - Planta *130* : 235 - 244, 1976.

26980 - LÜTTGE, U., BALL, E. : ATP levels and energy requirements of ion transport in
cells of slices of greening barley leaves. - Z. Pflanzenphysiol. *80* : 50 - 59,
1976.

B26981 - LÜTTGE, U., PITMAN, M.G. (ed.) : Transport in Plants II. Part A. Cells. -
Springer-Verlag, Berlin - Heidelberg - New York 1976. [Ps, photosynthates.]

B26982 - LÜTTGE, U., PITMAN, M.G. (ed.) : Transport in Plants II. Part B. Tissues and
Organs. - Springer-Verlag, Berlin - Heidelberg - New York 1976. [Photosyntha-
tes.]

26983 - LÜTTGE, U., PITMAN, M.G. : Transport and energy. - In : LÜTTGE, U., PITMAN,
M.G. (ed.) : Transport in Plants II. Part A. Cells. (Encycl. Plant Physiol.
N.S. Vol. 2.). Pp. 251 - 259. Springer-Verlag, Berlin - Heidelberg - New York
1976. [Photosynthates.]

26984 - LÜTTGE, U., SCHNEPF, E. : Organic substances. - In : LÜTTGE, U., PITMAN, M.G.
(ed.) : Transport in Plants II. Part B. Tissues and Organs. Pp. 244 - 277.
Springer-Verlag, Berlin - Heidelberg - New York 1976. [Ps.]

26985 - LUTZ, M. : Resonance Raman scattering of photosynthetic pigments *in vivo*.
Light harvesting structures and reaction centers. - In : SCHMID, E., BRANDMÜL-
LER, J., KIEFER, W. (ed.) : Proceedings of the 5th International Conference
on Raman Spectroscopy. Pp. 200 - 201. H.F. Schulz Verlag, Freiburg/Br. 1976.

26986 - LUTZ, M., KLEO, J., GILET, R., HENRY, M., PLUS, R., LEICKNAM, J.P. : Vibra-
tional spectra of chlorophylls *a* and *b* labeled with ^{26}Mg and ^{15}N. - In :
KLEIN, E.R., KLEIN, P.D. (ed.) : Proceedings of the Second International Conference on Stable Isotopes. Pp. 462 - 469. NTIS, Springfield, Va. 1976.

26987 - LUUKKANEN, O. : Relationship between the CO_2 compensation point and carbon
fixation efficiency in trees. - In : CANNELL, M.G.R., LAST, F.T. (ed.) : Tree
Physiology and Yield Improvement. Pp. 111 - 118. Academic Press, London - New
York - San Francisco 1976.

26988 - LUUKKANEN, O., BHUMIBHAMON, S., PELKONEN, P. : Photosynthesis in three prove-
nances of *Pinus merkusii*. - Silvae Genet. *25* : 7 - 10, 1976.

26989 - LYALIN, O.O., KTITOROVA, I.N. : Èksperimental'nye sposoby smeshcheniya vnutri-
kletochnoĭ kislotnosti i vliyanie vnutrikletochnogo pH na èlektrogennyĭ vodo-
rodnyĭ nasos rastitel'noĭ kletki. [Experimental techniques of shifting intracellular acidity and effect of intracellular pH on electrogenic hydrogen pump
of plant cells.] - Fiziol. Rast. *23* : 305 - 314, 1976. [In R, ab : E.]

26990 - LYMAN, H.,ALBERTE, R.S.,THORNBER, J.P. : Photosynthetic mutants of *Euglena* :
chloroplast lamellar characteristics. - Plant Physiol. *57* (Suppl.) : 73, 1976.

26991 - ŁYSZCZ, S., RUSZKOWSKA, M., WOJCIESKA, U., ZIENKIEWICZ, E. : The activity of
ascorbic acid and catechol oxidase, the rate of photosynthesis and respiration as related to plant organs, stage of development and copper supply. -
Acta agrobot. *29* : 99 - 105, 1976.

26992 - MacCOLL, R., BERNS, D.S., GIBBONS, O. : Characterization of cryptomonad phy-
coerythrin and phycocyanin. - Arch. Biochem. Biophys. *177* : 265 - 275, 1976.

26993 - MACHARDY, W.E., BUSCH, L.V., HALL, R. : *Verticillium* wilt of chrysanthemum :
quantitative relationship between increased stomatal resistance and local vascular dysfunction preceding wilt. - Can. J. Bot. *54* : 1023 - 1034, 1976.

26994 - MACHE, R. : Les acides ribonucléiques des ribosomes chloroplastiques. - Physiol. vég. *14* : 49 - 66, 1976.

26995 - MACHOLD, O., HØYER-HANSEN, G. : Polypeptide composition of thylakoids from
viridis and xantha mutants in barley. - Carlsberg Res. Commun. *41* : 359 - 366,
1976.

26996 - **MACIEJEWSKA, U., MALESZEWSKI, S.** : Embryonal dormancy and photosynthetic carbon metabolism in apple seedlings. - Z. Pflanzenphysiol. *79* : 300 - 309, 1976.

26997 - **MACNICOL, P.K.** : Rapid metabolic changes in the wounding response of leaf discs following excision. - Plant Physiol. *57* : 80 - 84, 1976. [Chl.]

26998 - **MACNICOL, P.K., DUDZIŃSKI, M.L., CONDON, B.N.** : Estimation of chlorophyll in tobacco leaves by direct photometry. - Ann. Bot. *40* : 143 - 152, 1976.

26999 - **MACRÌ, F., VIANELLO, A.** : Sucrose synthetase and neutral invertase depression in sugar beet roots affected by *Rizomania*. - Phytopathol. Z. *88* : 327 - 334, 1976. [Ps, Chl, Car.]

27000 - **MADIGAN, M.T., BROCK, T.D.** : Quantitative estimation of bacteriochlorophyll *c* in the presence of chlorophyll *a* in aquatic environments. - Limnol. Oceanogr. *21* : 462 - 467, 1976.

B27001 - **MADSEN, E.** : Effect of CO_2-Concentration on Morphological, Histological, Cytological and Physiological Processes in Tomato Plants. - State Seed Testing Station, Lyngby 1976. [Ps, Chl.]

27002 - **MAEDA, E., SUGIURA, T.** : [Ultrastructure of barnyardgrass callus cells cultured under an aseptic condition.] - Proc. Crop Sci. Soc. Jap. *45* : 591 - 597, 1976. [Chloroplast ; in Jap., ab : E.]

27003 - **MAGALHÃES, A.C., PETERS, D.B., HAGEMAN, R.H.** : Influence of temperature on nitrate metabolism and leaf expansion in soybean (*Glycine max* L. MERR.) seedlings. - Plant Physiol. *58* : 12 - 16, 1976. [Chloroplast.]

27004 - **MAGNUSSON, R.P., McCARTY, R.E.** : Illumination of chloroplast thylakoids in the presence of ATP causes the binding of ADP to one of the large subunits of coupling factor 1. - Biochem. biophys. Res. Commun. *70* : 1283 - 1289, 1976.

27005 - **MAGNUSSON, R.P., McCARTY, R.E.** : Acid-induced phosphorylation of adenosine 5'-diphosphate bound to coupling factor 1 in spinach thylakoids. - J. biol. Chem. *251* : 6874 - 6877, 1976.

27006 - **MAGNUSSON, R.P., McCARTY, R.E.** : Light-induced exchange of nucleotides into coupling factor 1 in spinach chloroplast thylakoids. - J. biol. Chem. *251* : 7417 - 7422, 1976.

27007 - **MAGOMEDOV, I.M., KOVALEVA, L.B., CHERNYAD'EV, I.I., DOMAN, N.G.** : Fiksatsiya uglekisloty i aktivnost' nekotorykh fermentov uglerodnogo metabolizma v list'-yakh kukuruzy. [Carbon dioxide fixation and activity of some enzymes of carbon metabolism in maize leaves.] - Fiziol. Rast. *23* : 36 - 41, 1976. [In R, ab : E.]

27008 - **MAGOMEDOV, I.M., TISHCHENKO, N.N.** : Izozimnyĭ sostav osnovnykh fermentov C_4-puti ugleroda fotosinteza. [Isozyme composition of the main enzymes involved in the C_4-pathway of photosynthesis.] - Fiziol. Rast. *23* : 1241 - 1247, 1976. [In R, ab : E.]

*27009 - **MAGRISO, Yu., SLAVCHEVA, T.** : Vliyanie na pochvenata vlaga i toreneto v"rkhu intenziteta na fotosintezata na lozata. [Effect of soil moisture and nutrient level on photosynthesis of *Vitis vinifera*.] - Gradin. lozar. Nauka *12* (3) : 64 - 73, 1975. [In Bulg., ab : F, R.]

27010 - **MAGYAROSY, A.C., SCHÜRMANN, P., BUCHANAN, B.B.** : Effect of powdery mildew infection on photosynthesis by leaves and chloroplasts of sugar beets. - Plant Physiol. *57* : 486 - 489, 1976.

27011 - **MAHON, J.D., LOWE, S.B., HUNT, L.A.** : Photosynthesis and assimilate distribution in relation to yield of cassava grown in controlled environments. - Can. J. Bot. *54* : 1322 - 1331, 1976.

27012 - **MAĬCHEKINA, R.M.** : Anatomiya fotosinteziruyushchikh organov ozimykh pshenits v svyazi s sortovymi osobennostyami rasteniĭ. [Anatomy of photosynthesizing organs of winter wheat in relation to cultivar characteristics of plants.] - In : Fotosintez i Produktivnost' Ozimoĭ Pshenitsy na Yugo-Vostoke Kazakhstana. Pp. 114 - 121, 133 - 134. Nauka kazakh. SSR, Alma-Ata 1976. [In R.]

27013 - **MAIER, R.** : Primärproduktionsuntersuchungen im ufernahen Schilfgürtel des Neusiedlersees. - Ber. biol. Forschungsinst. Burgenland *13* : 43 - 59, 1976.

27014 - **MAIER, R.** : Untersuchungen zur Primärproduktion im Grüngürtel des Neusiedler Sees Teil I : *Carex riparia* CURT. - Pol. Arch. Hydrobiol. *23* : 377 - 390, 1976.

27015 - **MAĬRANOVSKIĬ, V.G., MARINOVA, R.I., IOFFE, N.T., ENGOVATOV, A.A.** : Osushchestvlenie kataliticheskogo perenosa ėlektrona v sistemakh porfirin⁻-karotin, khlorofill⁻-karotin protiv gradienta standartnogo potentsiala. [Catalytic electron transfer in porphyrin⁻-carotene, chlorophyll⁻-carotene system against the standard potential gradient.] - Bioorg. Khim. *2* : 1266 - 1267, 1976. [In R, ab : E.]

27016 - **MAJOR, D.J., CHARNETSKI, W.A.** : Distribution of ^{14}C-labeled assimilates in rape plants. - Crop Sci. *16* : 530 - 532, 1976.

27017 - **MAKAROV, A.D.** : Osobennosti detal'nogo mekhanizma protsessa fosforilirovaniya v sopryazhennykh sistemakh razlichnoĭ stepeni slozhnosti. [Peculiarities of the detailed mechanism of phosphorylation process in coupled systems of various complexity.] - In : Itogi Issledovaniya Mekhanizma Fotosinteza. Pp. 39 - - 43. Pushchino 1976. [In R.]

27018 - **MAKAROV, A.D., KARTASHOV, I.M., KUZNETSOV, V.P., RUDENKO, T.I.** : O svyazi ob"emnykh izmeneniĭ khloroplastov so svetoindutsirovannym pogloshcheniem protonov. Rol' gidratatsii protonov. [Correlation between volume changes of chloroplasts and light-induced absorption of protons. The role of proton hydration.] - Fiziol. Rast. *23* : 893 - 898, 1976. [In R, ab : E.]

27019 - **MAKAROV, A.D., KARTASHOV, I.M., ZAKHAROV, S.D., MAL'YAN, A.N., TATARINTSEV, N.P.** : Vzaimodeĭstvie ferredoksina s adenilnukleotidami i ego vliyanie na fermentativnyĭ gidroliz i sintez ATF pri kislotno-osnovnykh perekhodakh v khloroplastakh. [Reaction of ferredoxin with adenine nucleotides and its effect on enzymic hydrolysis and synthesis of ATP during acid-base transitions in chloroplasts.] - Dokl. Akad. Nauk SSSR *231* : 746 - 749, 1976. [In R.]

27020 - **MAKAROV, A.D., STAKHOV, L.F.** : Zhelezosoderzhashchiĭ pterin-belkovyĭ kompleks; uchastie v vosstanovlenii NADP i transporte ėlektronov. [Participation of the iron-containing pterin-protein complex in NADP reduction and electron transport.] - Biochimiya *41* : 1380 - 1386, 1976. [In R, ab : E.]

27021 - **MAKEDONSKA, Ts.** : Zam"rsyavaneto na v"zdukha ot transportni sredstva i fotosintezata i s"d"rzhanieto na pigmenti v listata na *Platanus orientalis* L., *Aceŕ negundo* L. i *Betula alba* L. [Air pollution due to motor vehicles and photosynthesis and leaf pigment content of *Platanus orientalis* L., *Acer negundo* L., and *Betula alba* L.]-Fiziol. Rast. (Sofia) 2 (4) : 47 - 54, 1976. [In Bulg., ab : E, R.]

*27022 - **MAKHAMADZHANOV, I.** : Vliyanie vneseniya mineral'nogo udobreniya na soderzhanie vitamina C i karotina. [Effect of mineral fertilization on the contents of vitamin C and carotene.] - Uzb. biol. Zh. *19* (2) : 35 - 37, 91, 1975. [In R, ab : Uz.]

27023 - **MAKHMADBEKOVA, L.M., SHAGADAEVA, L.M., NASYROV, Yu.S.** : Deĭstvie ingibitorov translyatsii i transkriptsii geneticheskoĭ informatsii na fotosintez atsetabulyarii. [Effect of inhibitors of translation and transcription of the genetic information on *Acetabularia* photosynthesis.]- Dokl. Akad. Nauk tadzh. SSR *19* (6) : 61 - 63, 1976. [In R, ab : Tajik.]

*27024 - **MAKINO, H., NAKASONE, K., UEMURA, S., OKIMOTO, Y.** :[Effect of N⁶-benzylaminopurine on the growth of horticultural crops.] - Tamagawa Daigaku Nogakubu Kenkyu Hokoku [Bull. Fac. Agr. Tamagawa Univ.] *9* : 79 - 89, 1969. [Chi; in Jap., ab : E.]

27025 - **MAKOVETZKI, S., GOLDSCHMIDT, E.E.** : A requirement for cytoplasmic protein synthesis during chloroplast senescence in the aquatic plant *Anacharis canadensis*. - Plant Cell Physiol. *17* : 859 - 862, 1976.

27026 - **MAKSIMOVA, I.V., GORSKAYA, N.V., DAL', E.S.** : Izmenenie soderzhaniya vnekletochnykh produktov i ikh sostava v sinkhronnoĭ kul'ture *Chlorella pyrenoidosa* na svetu pri povyshenii kontsentratsii O_2. [Changes in the content and composition of extracellular products synthesized by synchronous culture of *Chlorella pyrenoidosa* in the light at elevated oxygen concentration.] - Mikrobiologiya *45* : 54 - 59, 1976. [In R, ab : E.]

27027 - **MALASHEVICH, A.V., RADYUK, M.S.** : Labilizatsiya khlorofilla v khloroplastakh pri deistvii gidroliticheskikh fermentov. [Labilization of chlorophyll in chloroplasts under the effect of hydrolytic enzymes.] - In : **KAKHNOVICH, L.V.** (ed.) : Optimizatsiya Fotosinteticheskogo Apparata Vozdeistviem Razlichnykh Faktorov. Pp. 141 - 147. Izd. BGU, Minsk 1976. [In R.]

27028 - **MALHOTRA, S.S.** : Effects of sulphur dioxide on biochemical activity and ultrastructural organization of pine needle chloroplasts. - New Phytol. *76* : 239 - 245, 1976.

27029 - **MALHOTRA, S.S., HOCKING, D.** : Biochemical and cytological effects of sulphur dioxide on plant metabolism. - New Phytol. *76*:227 - 237, 1976. [Ps, Chl.]

27030 - **MALINOVSKII, A.V., KLESHNIN, A.F., KORBUT, V.L.** : K probleme avtomaticheskoi optimizatsii fotosinteza tsenoza vysshikh rastenii. [Automatic optimization of photosynthesis in a stand of higher plants.] - In : Gazometricheskoe Issledovanie Fotosinteza i Dykhaniya Rastenii. Pp. 98 - 99. Akad. Nauk SSSR, Tartu 1976. [In R.]

27031 - **MALKIN, R., BEARDEN, A.J.** : The effect of alkaline pH on chloroplast photosystem I reactions at cryogenic temperature. - FEBS Lett. *69* : 216 - 220, 1976.

27032 - **MALKIN, R., BEARDEN, A.J.** : Thr primary electron acceptor of chloroplast photosystem I. - Biophys. J. *16* (2, Part 2) : 159, 1976.

27033 - **MALKIN, R., BEARDEN, A.J., HUNTER, F.A., ALBERTE, R.S., THORNBER, J.P.** : Properties of the low-temperature Photosystem I primary reaction in the *P*-700-chlorophyll *a*-protein. - Biochim. biophys. Acta *430* : 389 - 394, 1976.

27034 - **MALKIN, R., KNAFF, D.B.** : The nature of the primary electron acceptor of plant photosystem II. - Biophys. J. *16* (2, Part 2) : 159, 1976.

27035 - **MALKINA, I.S.** : Izmenenie svetovykh krivykh fotosinteza s vozrastom lista klena ostrolistnogo. [Changes in light curves of photosynthesis with aging of the leaf of Norway maple.] - Fiziol. Rast. *23* : 247 - 253, 1976. [In R, ab : E.]

27036 - **MALOFEEV, V.M.** : Vremennoi khod fotosinteza kak diagnosticheskoe sredstvo stepeni zasukhoustoichivosti rastenii. [Time course of photosynthesis as a diagnostic symptom of drought resistance of plants.] - Sel'skokhoz. Biol. *11* : 373 - 377, 1976. [In R, ab : E.]

27037 - **MAL'YAN, A.N.** : Svoistva sopryagayushego faktora fotofosforilirovaniya CF_1 i mekhanizm ego uchastiya v reaktsiyakh gidroliza i sinteza ATF. [Properties of the coupling factor CF_1 and mechanism of its participation in reactions of ATP hydrolysis and synthesis.] - In : Itogi Issledovaniya Mekhanizma Fotosinteza. Pp. 43 - 47. Pushchino 1976. [In R.]

27038 - **MAL'YAN, A.N., MAKAROV, A.D.** : Nekotorye osobennosti kinetiki reaktsii gidroliza ATP Ca-zavisimoi ATPazoi khloroplastov. [Reaction kinetics of ATP hydrolysis by Ca^{2+}-dependent ATPase of chloroplasts.] - Biokhimiya *41* : 1837 - 1839, 1976. [In R, ab : E.]

27039 - **MAL'YAN, A.N., MAKAROV, A.D.** : Issledovanie kinetiki i mekhanizma gidroliza ATP rastvorimoi ATPazoi khloroplastov (CF_1) v prisutstvii ionov Mg^{2+}. [Kinetics and mechanism of ATP hydrolysis by soluble ATPase of chloroplasts (CF_1) in the presence of Mg^{2+}.] - Biokhimiya *41* : 1087 - 1093, 1976. [In R, ab : E.]

27040 - **MAMLEEVA, N.A., NEKRASOV, L.I.** : Izuchenie vliyaniya parov vody na strukturu poverkhnostnykh sloev khlorofilla *a* metodom adsorbtsii kisloroda. [Effect of water vapour on the structure of surface layers of chlorophyll *a* by the method of oxygen adsorption.]-Zh. fiz. Khim. *50*: 1794 - 1797, 1976. [In R.]

27041 - **MANCINELLI, A.L., YANG, C.-P.H., RABINO, I., KUZMANOFF, K.M.** : Photocontrol of anthocyanin synthesis. V. Further evidence against the involvement of photosynthesis in high irradiance reaction anthocyanin synthesis of young seedlings. - Plant Physiol. *58* : 214 - 217, 1976.

27042 - **MANDAHAR, C.L., GARG, I.D.** : Cytokin activity of powdery mildew infected leaves of *Abelmoschus esculentus*. - Phytopathol. Z. *85* : 86 - 89, 1976. [Chl.]

27043 - **MANETAS, Y., AKOYUNOGLOU, G.A.** : Enzymic nature of the protein moiety of protochlorophyllide holochrome. - Plant Physiol. *58* : 43 - 46, 1976.

27044 - **MANGAN, J.L., VETTER, R.L., JORDAN, D.J., WRIGHT, P.C.** : The effect of the condensed tannins of sainfoil (*Onobrychis viciaefolia*) on the release of soluble leaf protein into the food bolus of cattle. - Proc. nutr. Soc. *35* : 95A - 96A, 1976. [Chl.]

27045 - **MANGEL, M.** : Properties of liposomes that contain chloroplast pigments : photosensitivity and efficiency of energy conversion. - Biochim. biophys. Acta *430* : 459 - 466, 1976.

B27046 - **MANSFIELD, T.A.** (ed.) : Effects of Air Pollutants on Plants. (Society for Experimental Biology. Seminar Series. Vol. 1.) - Cambridge University Press, Cambridge - London - New York - Melbourne 1976. [Resistances.]

27047 - **MANSFIELD, T.A.** : Mechanisms involved in turgor changes of guard cells. - In : SUNDERLAND, N. (ed.) : Perspectives in Environmental Biology. Vol. 2. Botany. Pp. 453 - 462. Pergamon Press, Oxford - New York - Toronto - Sydney - Paris - Braunschweig 1976. [Ps, chloroplast.]

27048 - **MANZANO, C., CANDAU, P., GOMEZ-MORENO, C., RELIMPIO, A.M., LOSADA, M.** : Ferredoxin-dependent photosynthetic reduction of nitrate and nitrite by particles of *Anacystis nidulans*. - Mol. cell. Biochem. *10* : 161 - 169, 1976.

27049 - **MAOTANI, T., MACHIDA, Y.** : [Studies on leaf water stress in fruit trees. V. Seasonal changes in leaf water potential and leaf diffusion resistance of satsuma mandarin trees.] - J. jap. Soc. hort. Sci. *45* : 261 - 266, 1976. [In Jap., ab : E.]

27050 - **MAR, T., GINGRAS, G.** : Photodichroic studies of the photoreaction center from *Rhodospirillum rubrum*. I. Attribution of P_{870} to two non parallel dipoles. - Biochim. biophys. Acta *440* : 609 - 621, 1976.

27051 - **MARCHANT, H.J.** : Actin in the green algae *Coleochaete* and *Mougeotia*. - Planta *131* : 119 - 120, 1976. [Chloroplast movement.]

27052 - **MARCO, G. di, GIOVANNOZZI-SERMANNI, G., GREGO, S., VERI, G., PIETROSANTI, T., TRICOLI, D., CATARCIONE, M., DI PIETRO, A.** : Daily fluctuations of some metabolic characters of durum wheat under field conditions. - Crop Sci. *16* : 343 - 347, 1976.

27053 - **MARCO, G. di, GREGO, S., PIETROSANTI, T., TRICOLI, D.** : Seasonal trends of nitrate reductase, carboxylating enzymes, and water-soluble proteins in two field-grown cultivars of *Triticum*. - J. exp. Bot. *27* : 725 - 734, 1976.

27054 - **MAREK, J., HRAŠKA, Š., PETROVIC, J.** : Príspevok k otázke degradácie plastidov. [Plastid degradation.] - Biológia (Bratislava) *31* : 493 - 500, 1976. [In Slovak, ab : E.]

27055 - **MARETZKI, A., THOM, M., MOORE, P.H.** : Growth patterns and carbohydrate distribution in sugarcane plants treated with an amine salt of glyphosate. - Hawaiian Planters' Record *59* (3) : 21 - 32, 1976.

27056 - **MARGULIES, M.M., WEISTROP, J.** : A chloroplast membrane fraction enriched in chloroplast ribosomes. - In : BÜCHER, T., NEUPERT, W., SEBALD, W., WERNER, S. (ed.) : Genetics and Biogenesis of Chloroplasts and Mitochondria. Pp. 657 - 660. North-Holland Publ. Co., Amsterdam - New York - Oxford 1976.

27057 - **MARIANI COLOMBO, P., ORSENIGO, M., SOLAZZI, A., TOLOMIO, C.** : Effetti della profondità sull'apparato fotosintetico delle alghe. IV° - Osservazioni sull'apparato fotosintetico di *Halimeda tuna* (*Siphonales*) a profondità comprese tra 7 e 16 m. [Sea depth effects on the algal photosynthetic apparatus IV. Observations on the photosynthetic apparatus of *Halimeda tuna* (*Siphonales*) at sea depths between 7 and 16 m.] - Mem. Biol. mar. Oceanogr. N.S. *6* : 197 - 208, 1976. [In Ital., ab : E.]

27058 - **MARIANI COLOMBO, P., RASCIO, N., SOLAZZI, A., TOLOMIO, C.** : Effetti della profonditá sull'apparato fotosintetico delle alghe. III° - Un nuovo metodo di fissazione in profonditá di alghe per microscopia elettronica. [Sea depth effects on the algal photosynthetic apparatus. III. A new method for under--water fixation of algae for electron microscopy.]-Mem. Biol. mar. Oceanogr. N.S. *6* : 183 - 195, 1976. [In Ital., ab : E.]

27059 - **MARKAROVA, E.N., VESELOVSKIĬ, V.A., VESELOVA, T.V., RAZAKOV, A.** : Teploustoĭchivost' fotosinteticheskogo apparata khlopchatnika pri zabolevanii viltom. [Heat resistance of the photosynthetic apparatus of *Verticillium*-wilted cotton.] - Sel'skokhoz. Biol. *11* : 695 - 699, 1976. [In R, ab : E.]

27060 - **MARÓTI, I.** : Photosynthetical pigments in the spongy and palisade parenchymas and the alternative ways of photosynthesis. - Acta biol. (Szeged) *22* (1 - 4) : 7 - 14, 1976.

27061 - **MARÓTI, J., GÁBOR, G.** : Thylakoid aggregation and pigment ratios in the spongy and palisade parenchymas. - Acta biol. (Szeged) *22* (1 - 4) : 15 - 27, 1976.

27062 - **MARSHALL, P.E., KOZLOWSKI, T.T.** : Importance of photosynthetic cotyledons for early growth of woody angiosperms. - Physiol. Plant. *37* : 336 - 340, 1976.

27063 - **MARSHO, T.V., KUNG, S.D.** : Oxygenase properties of crystallized Fraction 1 protein from tobacco. - Arch. Biochem. Biophys. *173* : 341 - 346, 1976.

27064 - **MARSHO, T.V., SOKOLOVE, P.M.** : Comparison of slow changes in fluorescence quenching and oxygen evolution in intact and osmotically shocked chloroplasts.- Plant Physiol. *57* (Suppl.) : 23, 1976.

27065 - **MARTIN, D.F., REID, G.A. Jr.** : Uptake of manganese in *Hydrilla verticillata* ROYLE. - Agr. Food Chem. *24* : 1161 - 1165, 1976. [Ps.]

27066 - **MASAROVIČOVÁ, E., DUDA, M.** : Some ecophysiological aspects of quantitative changes of carotenoids in the leaves of *Pulmonaria officinalis* L. ssp. *maculosa* (HAYNE) GAMS. - Biológia (Bratislava) *31* : 15 - 23, 1976.

27067 - **MASSOL, R.H., BALLESTER, A.** : Nueva metodología para la determinación en continuo de la actividad fotosintética de las algas fitoplanctónicas. [New methodology for continuous determination of photosynthetic activity of phytoplanktonic algae.] - Invest. Pesquera *40* : 111 - 123, 1976. [In Spanish, ab : E.]

27068 - **MATHERON, R., BAULAIGUE, R.** : Bactéries fermentatives, sulfato-réductrices et phototrophes sulfureuses en cultures mixtes. - Arch. Microbiol. *109* : 319 - 320, 1976.

*27069 - **MATHEW, C., RAMADASAN, A.** : Photosynthetic efficiency in relation to annual yield and chlorophyll content in the coconut palm. - J. Plant Crops *3* : 26 - 28, 1975.

27070 - **MATHIEU, Y.** : Étude des activités photochimiques des étiochloroplastes isolés de plantules verdies en lumière intermittente. I. Absorption d'oxygène. - Physiol. Plant. *37* : 49 - 54, 1976.

27071 - **MATHIEU, Y.** : Étude des activités photochimiques des étiochloroplastes isolés de plantules verdies en lumière intermittente. II. Photodécolorations pigmentaires. - Physiol. Plant. *37* : 55 - 61, 1976.

27072 - **MATHIS, P., BRETON, J., VERMEGLIO, A., YATES, M.:** Orientation of the primary donor chlorophyll of photosystem I.I in chloroplast membranes. - FEBS Lett. *63* : 171 - 173, 1976.

27073 - **MATHUR, D.D., VINES, H.M.** : Photoperiod and thermoperiod effects on Crassulacean Acid Metabolism. - Indian J. Plant Physiol. *19* : 47 - 52, 1976.

27074 - **MATHUR, P.N., DUNGARWAL, H.S., SINGH, H.G.** : Metabolic changes associated with the prevention of chlorosis by the application of elemental sulphur and foliar sprays of sequestrene 138-Fe in *Pisum sativum* L. - Ann. Bot. *40* : 833 - 836, 1976.

27075 - **MATORIN, D.N., VENEDIKTOV, P.S., GASHIMOV, R.M., RUBIN, A.B.** : Millisecond delayed fluorescence activated by reduced DPIP in DCMU-treated chloroplasts and in subchloroplast particles. - Photosynthetica *10* : 266 - 273, 1976.

27076 − **MATSON, R.S., KIMURA, T.** : Ferredoxin biosynthesis in *Euglena gracilis*. −
Biochim. biophys. Acta *442* : 76 − 87, 1976.

27077 − **MATSUDA, Y.** : Studies on chloroplast development in *Chlamydomonas reinhardtii*
IV. Control of rapid chlorophyll formation in greening y-1 cells. − Plant
Cell Physiol. *17* : 887 − 897, 1976.

*27078 − **MATSUI, T., AIGA, I.** : Biological studies on light quality in environment
control. III. Effect of wavelength width of monochromatic light on protochlorophyllide phototransformation. − Rep. Environ. Control Biol. *10* : 18 − 20,
1972.

27079 − **MATSUI, T., EGUCHI, H., SOEJIMA, Y.** : Control of artificial light for plants.
III. Simulation and monitoring of spectral composition. − Environ. Control
Biol. *14* : 41 − 50, 1976.

*27080 − **MATSUI, T., SOEJIMA, Y., EGUCHI, H.** : Control of artificial light for plants.
I. Measurement and control of light. − Environ. Control Biol. *12* : 53 − 68,
1974.

27081 − **MATTHEIS, J.R., REBEIZ, C.C., REBEIZ, C.A.** : Factors limiting chloroplast
differentiation *in vitro*. − Plant Physiol. *57* (Suppl.) : 72, 1976.

27082 − **MAU, A.W.-H.** : Fluorescence of chlorophyll *a* in EPA. − Chem. Phys. Lett.
38 : 279 − 282, 1976.

27083 − **MAUZERALL, D.** : Chlorophyll and photosynthesis. − Phil. Trans. roy. Soc.
London *B 273* : 287 − 294, 1976.

27084 − **MAUZERALL, D.** : Multiple excitations in photosynthetic systems. − Biophys.
J. *16* : 87 − 91, 1976.

27085 − **MAXWELL, N.P., WUTSCHER, H.K.** : Yield fruit size and chlorosis of grapefruit
on 10 rootstocks. − HortScience *11* : 496 − 498, 1976.

27086 − **MAXWELL, P.C., BIGGINS, J.** : Role of cyclic electron transport in photosynthesis as measured by the photoinduced turnover of P_{700} *in vivo*. − Biochemistry *15* : 3975 − 3981, 1976.

27087 − **MAYER, H.** : Die Windverhältnisse in und über einem Fichtenwald. − Forstwiss.
Centralbl. *95* : 333 − 345, 1976.

27088 − **MAYHEW, S.G., LUDWIG, M.L.** : Flavodoxins and electron-transferring flavoproteins. − In : **BOYER, P.D.** (ed.) : The Enzymes. Vol. 12. Pp. 57 − 118. Academic Press, New York − San Francisco − London 1976.

27089 − **MAYNE, B.C., RAY, T.B., BLACK, C.C.** : A comparison of the photosynthetic characteristics of the halophytic grasses of the Atlantic coast of North America.
− Plant Physiol. *57* (Suppl.) : 105, 1976.

27090 − **MAZELIS, M., MIFLIN, B.J., PRATT, H.M.** : A chloroplast-localized diaminopime-
late decarboxylase in higher plants. − FEBS Lett. *64* : 197 − 200, 1976.

27091 − **M'BAKU, S.B., FRITZ, G.J., BOWES, G.** : Photosynthetic carbon metabolism in
isolated leaf cells of *Digitaria pentzii* var Slenderstem Digitgrass. − Plant
Physiol. *57* (Suppl.) : 31, 1976.

27092 − **McCARTY, R.E.** : Ion transport and energy conservation in chloroplasts. −
In : **STOCKING, C.R., HEBER, U.** (ed.) : Transport in Plants III. Intracellu-
lar Interactions and Transport Processes. Pp. 347 − 376. Springer-Verlag,
Berlin − Heidelberg − New York 1976.

27093 − **McCARTY, R.E., PORTIS, A.R. Jr.** : A simple, quantitative approach to the coupling of photophosphorylation to electron flow in terms of proton fluxes. −
Biochemistry *15* : 5110 − 5114, 1976.

27094 − **McCLAIN, K.M., ARMSON, K.A.** : Effect of water supply, nitrogen, and seedbed
density on white spruce seedling growth. − J. Soil Sci. Soc. Amer. *40* : 443 −
446, 1976. [Dry-matter distribution.]

27095 − **McCREE, K.J.** : A comparison of experimental and theoretical spectra for photosynthetically active radiation at various atmospheric turbidities. − Agr.
Meteorol. *16* : 405 − 412, 1976.

27096 - McCREE, K.J. : A rational approach to light measurements in plant ecology. -
In : SMITH, H. (ed.) : Commentaries in Plant Science. Pp. 45 - 50. Pergamon
Press, Oxford - New York - Toronto - Sydney - Paris - Frankfurt 1976.

27097 - McCREE, K.J. : The role of dark respiration in the carbon economy of a plant.
- In : BURRIS, R.H., BLACK, C.C. (ed.) : CO_2 Metabolism and Plant Productivity. Pp. 177 - 184. University Park Press, Baltimore - London - Tokyo 1976.

27098 - McINTIRE, G.L., DUNSTAN, W.M. : Non-structural carbohydrates in *Spartina alterniflora* LOISEL. - Bot. mar. *19* : 93 - 96, 1976. [Photosynthates.]

27099 - McINTOSH, A.R., BOLTON, J.R. : Electron spin resonance spectrum of species
"X" which may function as the primary electron acceptor in Photosystem I of
green plant photosynthesis. - Biochim. biophys. Acta *430* : 555 - 559, 1976.

27100 - McINTOSH, A.R., BOLTON, J.R. : Triplet state involvement in primary photochemistry of photosynthetic photosystem II. - Nature *263* : 443 - 445, 1976.

27101 - McLAREN, J.S., SMITH, H. : The effect of abscisic acid on growth, photosynthetic rate and carbohydrate metabolism in *Lemna minor* L. - New Phytol. *76* :
11- 20, 1976.

27102 - McMICHAEL, B.L., ELMORE, C.D., CATHEY, G.W. : Nondestructive method for the
determination of boll surface area in cotton. - Agron. J. *68* : 426 - 427,
1976.

27103 - McROY, C.P., GOERING, J.J. : Annual budget of primary production in the Bering sea. - Mar. Sci. Commun. *2* : 255 - 267, 1976.

27104 - McSWAIN, B.D., TSUJIMOTO, H.Y., ARNON, D.I. : Effects of magnesium and chloride ions on light-induced electron transport in membrane fragments from a
blue-green alga. - Biochim. biophys. Acta *423* : 313 - 322, 1976.

27105 - MEDEGHINI-BONATTI, P., BONETTA CONTE, M.D. : Ultrastrutture plastidiali in
foglioline di gemme di "*Picea excelsa*" e di "*Larix decidua*" nel corso della
germogliazione. [Ultrastructural modifications of plastids in leaflets of
Picea excelsa and *Larix decidua* during sprouting of buds.] - G. bot. ital.
110 : 9 - 20, 1976.

27106 - MEDINA, E., BIFANO, T.DE, DELGADO, M. : *Paspalum repens* BERG., a truly aquatic C_4 plant. - Acta cient. venez. *27*· : 258 - 260, 1976.[Ps.]

27107 - MEDINA, E., DELGADO, M. : Photosynthesis and night CO_2 fixation in *Echeveria
columbiana* v. POELLNITZ. - Photosynthetica *10* : 155 - 163, 1976.

27108 - MEGO, V. : Vplyv vlhkosti substrátu na rast a metabolizmus základných živín
lucerny siatej (*Medicago sativa* L.). 1.Dynamika rastu nadzemných orgánov.
[Effect of substrate humidity on growth and metabolism of basic nutrients
of *Medicago sativa* L. 1. Rate of growth of above-ground organs.] - Ved. Práce výsk. Úst. rast. Výr. Piešť. (Piešťany) *13* : 49 - 60, 1976. [Growth analysis; in Slovak, ab : E, R.]

27109 - MEIDNER, H. : Aspects of leaf water relations and stomatal functioning. -
In : SUNDERLAND, N. (ed.) : Perspectives in Experimental Biology. Vol.2. Bo-
tany. Pp. 445 - 452. Pergamon Press, Oxford - New York 1976.
[Resistances.]

27110 - MELIKSETYAN, F.Z., ATABEKYAN, E.A., AGADZHANYAN, Zh.G. : [Carotenoids in
young *Asparagus* shoots.] - Izv. sel'skokhoz. Nauk (Ereyan) *19* (8) : 68 - 71,
1976. [In Arm., ab : R.]

27111 - MELIS, A., HOMANN, P.H. : Heterogeneity of the photochemical centers in system II of chloroplasts. - Photochem. Photobiol. *23* : 343 - 350, 1976.

27112 - MEL'NIKOV, I.A. : Sravnenie velichin biomassy mikroplanktona opredelennykh
po ATF i metodu pryamogo mikroskopirovaniya. [Comparison of microplankton
biomass values determined by the ATP method and direct microscopy.] - Okeanologiya *16* : 324 - 328, 1976. [In R, ab : E.]

27113 - MENDIOLA-MORGENTHALER, L.R., MORGENTHALER, J.-J., PRICE, C.A. : Synthesis of
coupling factor CF_1 protein by isolated spinach chloroplasts. - FEBS Lett.
62 : 96 - 100, 1976.

27114 - MENKE, W., SCHMID, G.H. : Cyclic photophosphorylation in the mykotrophic orchid *Neottia nidus-avis*. - Plant Physiol. *57* : 716 - 719, 1976.

27115 - MENZEL, E.R. : On the energy upconversion mechanism of light utilization by chlorophyll in photosynthesis. - J. theor. Biol. *56* : 401 - 406, 1976.

27116 - MERRETT, M.J. : Enzyme levels in relation to obligate phototrophy in *Chlamydobotrys*. - Plant Physiol. *58* : 179 - 181, 1976.

27117 - MERTA, K., DUNN, I.J. : Oxygen electrode characteristics. - Biotechnol. Bioeng. *18* : 591 - 593, 1976.

27118 - MESQUITA, J.F. : La différenciation des plastes dans les fleurs de *Narcissus* L. I - Modifications ultrastructurales et pigmentaires pendant la morphogénèse des chromoplastes chez *N. bulbocodium* L. - Rev. Biol. *10* : 127 - 150, 1974-1976. [Chl, Car.]

27119 - MESQUITA, J.F. : La différenciation des plastes dans les fleurs de *Narcissus* L. II - Modifications ultrastructurales et pigmentaires pendant la morphogénèse des leucoplastes chez *N. triandrus* L. var. *cernuus* (SALISB.) BAK. - Rev. Biol. *10* : 151 - 164, 1974-1976. [Chl, Car.]

27120 - MESQUITA, J.F., SANTOS, M.F. : Études cytologiques sur les algues jaunes (*Chrysophyceae*). I. Ultrastructure de *Chrysocapsa epiphytica* LUND. - Bol. Soc. broter., Sér. 2.a. *50* : 63 - 99, 1976. [Chloroplast.]

27121 - MESQUITA, J.F., SANTOS, M.F. : Cytological studies in golden algae (*Chrysophyceae*). II. First cytochemical demonstration of peroxisomes in *Chrysophyceae* (*Chrysocapsa epiphytica* LUND). - Cytobiologie *14* : 38 - 48, 1976.

27122 - METZNER, H. : The electron source in photosynthesis. - Bioelectrochem. Bioenerg. *3* : 573 - 581, 1976.

27123 - METZNER, H., GERSTER, R. : Energy conservation in photosynthesis models III. Role of bicarbonate anions in oxygen evolution. - Photosynthetica *10* : 302 - 306, 1976.

27124 - MEZENTSEVA, V.T., DERYUZHKIN, R.I., SKOROBOGATOVA, T.I., PROKHOROV, P.I., SAGALAEVA, A.P. : Sezonnaya dinamika khlorofilla v khvoe otdel'nykh vidov i ēkotipov listvennitsy. [Seasonal dynamics of chlorophyll in the needles of different larch species and ecotypes.] - Izv. vyssh. ucheb. Zaved., Lesn. Zh. *19* (6) : 132 - 135, 1976. [In R.]

27125 - MGALOBLISHVILI, M.P., KANDELAKI, R.A., SANADZE, G.A. : Fotofosforilirovanie v izolirovannykh khloroplastakh topolya. [Photophosphorylation in isolated chloroplasts of poplar leaves.] - Izv. Akad. Nauk gruz. SSR, Ser. biol. *2* : 422 - 428, 1976. [In R, ab : E, Georg.]

27126 - MGALOBLISHVILI, M.P., TARKHNISHVILI, G.M., KANDELAKI, R.A., SANADZE, G.A. : Optimal'nye usloviya dlya reaktsii Khilla. [Optimum conditions for Hill reaction.] - Tr. tbilis. gos. Univ. *167* : 103 - 109, 1976. [In R, ab : E, Georg.]

27127 - MICHEL, H.-J., BARTH, A. : Aktuelle Probleme der Wirkstofforschung. - Pharmazie *31* : 753 - 766, 1976. [Ps inhibitors.]

27128 - MICHEL-VILLAZ, M. : Fluorescence polarization : Pigment orientation and energy transfer in photosynthetic membranes. - J. theor. Biol. *58* : 113 - 129, 1976.

27129 - MIKHAĬLOVA, S., SLAVCHEVA, T. : Vliyanie na kontsentratsiyata na khranitel'nite veshchestva v″rkhu intenziteta na fotosintezata i nyakoi biokhimichni protsesi pri lozata. [Effect of the concentration of nutrient substances on the intensity of photosynthesis and some biochemical processes in grape vines.] - Gradinar. lozar. Nauka *13* (4) : 90 - 95, 1976. [In Bulg., ab : F,R.]

27130 - MIKULSKA, E., ŻOŁNIEROWICZ, H. : Ultrastructure of plastids in differentiating epithelial cells of *Abies homolepis* SIB. and ZUCC. leaves. - Biochem. Physiol. Pflanz. *170* : 355 - 362, 1976.

27131 - MILES, C.D. : Manganese stimulation of oxygen consumption in chloroplasts with dibromothymoquinone. - FEBS Lett. *61* : 251 - 254, 1976.

27132 - MILES, C.D. : Selection of diquat resistance photosynthesis mutants from maize. - Plant Physiol. *57* : 284 - 285, 1976.

*27133 - MILFORD, G.F.J., LAWLOR, D.W. : Effects of varying air and soil moisture on the water relations and growth of sugar beet. - Ann. appl. Biol. *80* : 93 - 102, 1975. [Ps.]

*27134 - MILKUS, B.N. : The influence of the grapevine infectious chlorosis virus on chloroplast ultrastructure. - Acta phytopathol. Acad. Sci. hung.*10* : 179 - 184, 1975.

27135 - MILLER, K.R. : A particle spanning the photosynthetic membrane. - J. Ultrastruct. Res. *54* : 159 - 167, 1976.

27136 - MILLER, K.R., BLOODGOOD, R.A., STAEHELIN, L.A. : Crystals within thylakoids : A structural analysis. - J. Ultrastruct. Res. *54* : 29 - 36, 1976.

27137 - MILLER, K.R., MILLER, G.J., McINTYRE, K.R. : The light-harvesting chlorophyll-protein complex of photosystem II. Its location in the photosynthetic membrane. - J. Cell Biol. *71* : 624 - 638, 1976.

27138 - MILLER, K.R., STAEHELIN, L.A. : Analysis of the thylakoid outer surface. Coupling factor is limited to unstacked membrane regions. - J. Cell Biol. *68* : 30 - 47, 1976.

27139 - MILLER, P.C., STONER, W.A., TIESZEN, L.L. : A model of stand photosynthesis for the wet meadow tundra at Barrow, Alaska. - Ecology *57* : 411 - 430, 1976.

27140 - MILLS, J.D., TELFER, A., BARBER, J. : Cation control of chlorophyll a fluorescence yield in chloroplasts. Location of cation sensitive sites. - Biochim. biophys. Acta *440* : 495 - 505, 1976.

*27141 - MILOSAVLJEVIĆ, M., BOJOVIĆ-CVETIĆ, D., VUJIČIĆ, R. : Uticaj benomyla na intenzitet fotosinteze, udeo aminokiselina i ultrastrukturu ćelija mezofila lista vinove loze. [Effect of benomyl on the photosynthetic rate, amino acid level, and ultrastructure of mesophyll cells of grape leaves.] - Arh. poljopr. Nauke *28* : 3 - 14, 1975. [In Croat., ab : E.]

27142 - MILTHORPE, F.L. : Quantitative aspects of leaf growth. - In : SUNDERLAND, N. (ed.) : Perspectives in Experimental Biology. Vol. 2.Botany. Pp. 33 - 40. Pergamon Press, Oxford - New York - Toronto - Sydney - Paris - Braunschweig 1976. [Ps.]

27143 - MINAS, M. : Production organique primaire dans un milieu saumâtre eutrophe (Etang de Berre). Effets d'une forte dilution (dérivation des eaux de la Durance). - Mar. Biol. *35* : 13 - 29, 1976.

27144 - MINAS, M. : Relations entre la production photosynthétique et la pénétration de la lumière dans les eaux de l'Étang de Berre. - Tethys *7* : 131 - 135, 1975 (1976).

27145 - MIŠOVIĆ, M. : Uticaj herbicida linurona i prometrina na suzbijanje korova i produktivnost fotosinteze krompira. [Effect of the herbicides linuron and prometryne on weed control and photosynthetic productivity of potatoes.] - Agrohemija *1976* (5-6) : 191 - 197, 1976. [In Croat., ab : E.]

27146 - MISRA, H.P., FRIDOVICH, I. : A convenient calibration of the Clark oxygen electrode. - Anal. Biochem. *70* : 632 - 634, 1976.

27147 - MISRA, J.B., NARULA, R., MADAN, V.K. : Diurnal variations in activities of phosphoenolpyruvate carboxylase, malate dehydrogenase (decarboxylating) and acidity in *Cuscuta reflexa* filaments. - Z. Pflanzenphysiol. *79* : 272 - 275, 1976.

27148 - MITCHELL, C.A., DOSTAL, H.C., SEIPEL, T.M. : Analysis of dry weight gain reduction in mechanically-dwarfed tomato plants. - HortScience *11* : 299, 1976. [Ps.]

27149 - MITROFANOV, B.A., GOĬSA, N.I. : Issledovanie gazoobmena ozimoĭ pshenitsy v polevykh usloviyakh. [Gas exchange of winter wheat in the field conditions.]- In : Gazometricheskoe Issledovanie Fotosinteza i Dykhaniya Rastenĭĭ. Pp. 104 - 105. Akad. Nauk SSSR, Tartu 1976. [In R.]

27150 - MITROFANOV, B.A., GULYAEV, B.I., MITROFANOVA, S.V., BELOUS, I.I., ALIEV, É.A., FERENTS, A.F. : Uglekislotnyĭ rezhim gidroponnykh teplits. [Carbon dioxide regime in hydroponic warmhouses.] - Fiziol. Biokhim. kul't. Rast *8* : 293 - 298, 1976. [Ps; in R, ab : E.]

27151 - **MITSUI, A.** : A survey of hydrogen producing photosynthetic organisms in tropical and subtropical marine environments. - Annu. Rep. Univ. Miami *AER75-
11171* : 1 - 68, 1976.

27152 - **MITSUI, A.** : Bioconversion of solar energy in salt water photosynthetic hy-
drogen production systems. - In : VEZIROGLU, T.N. (ed.) : Proceedings of the
First World Hydrogen Energy Conference. Vol. II. Pp. 4B 77 - 99. Univ. Miami
Press, Miami 1976.

27.153 - **MITSUI, A.** : Long range concepts : applications of photosynthetic hydrogen
production and nitrogen fixation research. - In : Proceedings. A Conference
on Capturing the Sun through Bioconversion. Pp. 653 - 672. Washington 1976.

27154 - **MITSULOV, N.** : Mnogotochkov registrirashch termomet"r za eksperimentalni iz-
sledovaniya v"v fitokameri. [Multipoint recording thermometer for experimental studies in growth chambers.] - Fiziol. Rast. (Sofia) *2* (2) : 101 - 108,
1976. [In Bulg., ab : E,R.]

27155 - **MIYACHI, S.** : [Chloroplasts.] - Tampakushitsu Kakusan Koso, Bessatsu : 64 -
69, 1976. [In Jap.]

27156 - **MIYACHI, S., OKABE, K.** : Oxygen enhancement of photosynthesis in *Anacystis
nidulans* cells. - Plant Cell Physiol. *17* : 973 - 986, 1976.

27157 - **MIYAKE, H., MAEDA, E.** : Development of bundle sheath chloroplasts in rice
seedlings. - Can. J. Bot. *54* : 556 - 565, 1976.

27158 - **MIYAKE, H., MAEDA, E.** : The fine structure of plastids in various tissues in
the leaf blade of rice. - Ann. Bot. *40* : 1131 - 1138, 1976.

27159 - **MIZRAHI, Y., DOSTAL, H.C., CHERRY, J.H.** : Descriptive physiology and biochemistry of the abnormally ripening tomato fruit (*Lycopersicon esculentum*)
cv. Snowball. - Physiol. Plant. *38* : 309 - 312, 1976. [Chl, Car.]

*27160 - **MŁODZIANOWSKI, F., SZWEYKOWSKA, A.** : Fine structure of kinetin-treated pro-
tonema and kinetin-induced gametophore buds in *Funaria hygrometrica*. - Acta
Soc. Bot. Pol. *40* : 549 - 555, 1971. [Chloroplast.]

27161 - **MOGILEVA, G.A., SHITOVA, I.P., ZELENSKIĬ, M.I., FATTAKHOVA, F.Z.** : Fotokhi-
micheskaya aktivnost' khloroplastov pshenits raznogo urovnya ploidnosti i
proiskhozhdeniya. [Photochemical activity of chloroplasts of wheat of different ploidy and origin.] - Nauch. Dokl. vyssh. Shkoly, biol. Nauki *19*(12) :
100 - 108, 1976. [In R.]

27162 - **MOHAMMAD, A.M.S., AL-MASHHADANI, Y.** : The effect of root formation on the
levels of protein, chlorophyll, RNA, DNA and carbohydrates in excised cotyledons of *Cucurbita pepo*. - Physiol. Plant. *37* : 195 - 199, 1976.

27163 - **MOHANTY, P., BOYER, J.S.** : Chloroplast response to low water potentials. IV.
Quantum yield is reduced. - Plant Physiol. *57* : 704 - 709, 1976.

27164 - **MOHAPATRA, S.C., JOHNSON, W.H.** : Microscopic studies in bright leaf tobacco.
V. Scanning electron microscopic observations on chloroplast structure and
distribution. - Trans. amer. microsc. Soc. *95* : 228 - 232, 1976.

27165 - **MOK, M.C., GABELMAN, W.H., SKOOG, F.** : Carotenoid synthesis in tissue cultures of *Daucus carota* L. - J. amer. Soc. hort. Sci. *101* : 442 - 449, 1976.

27166 - **MOLCHANOV, A.G.** : Opredelenie balansa CO_2 nasazhdeniya po intensivnosti fotosinteza i soderzhaniya CO_2 v kronovom prostranstve na primere sosnovogo
nasazhdeniya. [Determination of CO_2 balance of a stand by means of photosynthetic rate and CO_2 concentration in crown stratum as demonstrated on pine
stand.] - In : Gazometricheskoe Issledovanie Fotosinteza i Dykhaniya Rasteniĭ. Pp. 110 - 112. Akad. Nauk SSSR, Tartu 1976. [In R.]

27167 - **MOLCHANOV, M.I., TRUSOVA, V.M.** : K izucheniyu svoĭstv belkovykh komponentov
membrannoĭ sistemy plastid pri biogeneze khloroplastov. [Properties of protein components of the membrane system of plastids in the biogenesis of chloroplasts.] - Dokl. Akad. Nauk SSSR *226* : 968 - 971, 1976. [In R.]

27.168 - **MOLCHANOV, M.I., TRUSOVA, V.M., KOTOVSKAYA, A.P.** : Glikolipoproteidy membrannoĭ sistemy plastid pri biogeneze khloroplastov. [Glycolipoproteins of plastid membrane system in chloroplast biogenesis.] - Biokhimiya *41* : 1928 - 1933, 1976. [In R, ab : E.]

27169 - **MOLCHANOV, M.I., TRUSOVA, V.M., SHAPOSHNIKOV, G.L.** : Izuchenie aminokislotnogo sostava i biosinteza belkovykh komponentov membrannoĭ sistemy plastid pri biogeneze khloroplastov. [Amino acid composition and biosynthesis of membrane proteins in chloroplast biogenesis.] - Biokhimiya *41* : 926 - 932, 1976. [In R, ab : E.]

27170 - **MOLCHANOV, M.I., TRUSOVA, V.M., SHAPOSHNIKOV, G.L., KOTOVSKAYA, A.P.** : O biosinteze i aminokislotnom sostave proteolipidov lamellarnoĭ sistemy khloroplastov. [Biosynthesis and amino acid composition of proteolipids of the lamellar system of chloroplasts.] - Dokl. Akad. Nauk SSSR *226* : 711 - 714, 1976. [In R.]

27171 - **MOLDAU, Kh.A., KAROLIN, A.Yu.** : Podavlenie fotosinteza fondom assimilyatov osushchestvlyaetsya ust'itsami. [Inhibition of photosynthesis by photosynthates is realized by means of stomata.] - In : Gazometricheskoe issledovanie fotosinteza i Dykhaniya Rasteniĭ. Pp. 106 - 109. Akad. Nauk SSSR, Tartu 1976. [In R.]

27172 - **MONAGHAN, J.L., HULL, P.** : Differences in vegetative characteristics among four populations of *Senecio vulgaris* L. possibly due to. interspecific hybridization. - Ann. Bot. *40* : 125 - 128, 1976. [Leaf area measurement.]

27173 - **MONGER, T.G., COGDELL, R.J., PARSON, W.W.** : Triplet states of bacteriochlorophyll and carotenoids in chromatophores of photosynthetic bacteria. - Biochim. biophys. Acta *449* : 136 - 153, 1976.

27174 - **MONICA, R.F. LA, MARRS, B.L.** : The branched respiratory system of photosynthetically grown *Rhodopseudomonas capsulata*. - Biochim. biophys. Acta *423* : 431 - 439, 1976.

27175 - **MONTENY, B., GOSSE, G.** : Analyse et estimation du rayonnement net d'une culture de *Panicum maximum* en zone tropicale humide. - Oecol. Plant.*11* : 173 - 191, 1976. [Growth analysis.]

27176 - **MONTES, G., BRADBEER, J.W.** : An association of chloroplasts and mitochondria in *Zea mays* and *Hyptis suaveolens*. - Plant Sci. Lett. *6* : 35 - 41, 1976.

27177 - **MOON, R.E., DAWES, C.J.** : Pigment changes and photosynthetic rates under selected wavelengths in the growing tips of *Eucheuma isiforme* (C. AGARDH) J. AGARDH var. *denudatum* CHENEY during vegetative growth. - J. Phycol. *12* (Suppl.) : 12, 1976.

27178 - **MOON, R.E., DAWES, C.J.** : Pigment changes and photosynthetic rates under selected wavelengths in the growing tips of *Eucheuma isiforme* (C. AGARDH) J. AGARDH var. *denudatum* CHENEY during vegetative growth. - Brit. phycol. J. *11* : 165 - 174, 1976.

27179 - **MOONEY, H.A., BJÖRKMAN, O., EHLERINGER, J., BERRY, J.** : Photosynthetic capacity of *in situ* Death Valley plants. - Carnegie Inst. Year Book *75* : 410 - 413, 1976.

27180 - **MOONEY, H.A., EHLERINGER, J., BERRY, J.A.** : High photosynthetic capacity of a winter annual in Death Valley. - Science *194* : 322 - 324, 1976.

27181 - **MOORE, T.C., COOLBAUGH, R.C.** :.Conversion of geranylgeranyl pyrophosphate to *ent*-kaurene in enzyme extracts of sonicated chloroplasts. - Phytochemistry *15* : 1241 - 1247, 1976.

27182 - **MORADSHAHI, A., VINES, H.M., BLACK, C.C.** : Light metabolism in detached leaves of *Ananas comosus*. - Plant Physiol. *57* (Suppl.) : 32, 1976.

27183 - **MOREL, C.** : Rythmes circadiens : interactions entre voies métaboliques. - In : JACQUES, R. (ed.) : Etudes de Biologie Végétale. Hommage au Professeur P. CHOUARD. Pp. 457 - 466. Paris 1976. [Ps.]

27184 - **MORELAND, D.E., HILTON, J.L.** : Action on photosynthetic systems. - In : AUDUS, L.J. (ed.) : Herbicides. Physiology, Biochemistry, Ecology. Vol. 1. Pp. 493 - 523. Academic Press, London - New York - San Francisco 1976.

*27185 - **MORGAN, K., KALFF, J.** : The winter dark survival of an algal flagellate - *Cryptomonas erosa* (SKUJA). - Verh. internat. Verein. Limnol. *19* : 2734 - 2740, 1975. [Ps.]

27186 - **MORGENTHALER, J.-J., MENDIOLA-MORGENTHALER, L.** : Synthesis of soluble, thylakoid, and envelope membrane proteins by spinach chloroplasts purified from gradients. - Arch. Biochem. Biophys. *172* : 51 - 58, 1976.

27187 - **MOROT-GAUDRY, J.-F., BETHENOD, O., CHARTIER, M., CHARTIER, P.** : Photosynthèse comparée d'un Maïs normal (W 64 A) et d'un Maïs mutant opaque 2 (W 64 A o_2). - Physiol. vég. *14* : 595 - 606, 1976.

27188 - **MOROZOV, V.L.** : Rezhim solnechnoĭ radiatsii v travyanistykh soobshchestvakh Kamchatki. [Solar radiation regime in herbaceous phytocoenoses of the Kamchatka Peninsula.]-Ekologiya *4* (6) : 44 - 48, 1976.[Growth analysis; in R.]

27189 - **MORRISON, I.N.** : The structure of the chlorophyll-containing cross cells and tube cells of the inner pericarp of wheat during grain development. - Bot. Gaz. *137* : 85 - 93, 1976.

27190 - **MOSS, D.N.** : Studies on increasing photosynthesis in crop plants. - In : BURRIS, R.H., BLACK, C.C. (ed.) : CO_2 Metabolism and Plant Productivity. Pp. 31 - 41. University Park Press, Baltimore - London - Tokyo 1976.

27191 - **MOSS, G.I.** : Thinning "Washington" navel and "Late Valencia" sweet orange fruits with photosynthetic inhibitors. - HortScience *11* : 48 - 50, 1976. [Ps.]

27192 - **MOSS, G.P.** : Carbon-13 NMR spectra of carotenoids. - Pure appl. Chem. *47* : 97 - 102, 1976.

27193 - **MOSS, G.P., WEEDON, B.C.L.** : Chemistry of the carotenoids. - In : GOODWIN, T.W. (ed.) : Chemistry and Biochemistry of Plant Pigments. 2nd Ed. Vol.1. Pp. 149 - 224. Academic Press, London - New York - San Francisco 1976.

27194 - **MOUDRIANAKIS, E.N., TIEFERT, M.A.** : Synthesis of bound adenosine triphosphate from bound adenosine diphosphate by the purified coupling factor 1 of chloroplasts. Evidence for direct involvement of the coupling factor in this "adenylate kinase-like" reaction. - J. biol. Chem. *251* : 7796 - 7801, 1976.

27195 - **MOURSI, M.A., EL-HABBASHA, K.M., SHAHEEN, A.M.** : Air temperature affecting photosynthetic efficiency, water and nitrogen contents of *Vicia faba* L. seedlings. - Z. Acker- Pflanzenbau *142* : 249 - 255, 1976.

27196 - **MOUSSEAU, M., LOUASON, G.** : Action de la photopériode sur la respiration nocturne. Cas d'une plante de jours courts, le *Chenopodium polyspermum* L. - Compt. rend. Acad. Sci. Paris, Sér. D *282* : 719 - 722, 1976. [Ps.]

27197 - **MOUTOUNET, M.** : Les caroténoïdes de la Prune d'Ente et du Pruneau d'Agen. - Ann. Technol. agr. *25* : 73 - 84, 1976.

27198 - **MOYSE, A.** : Les types métaboliques des plantes : C_4 et CAM. Comparaison avec les plantes C_3. - Physiol. vég. *14* : 533 - 550, 1976.

27199 - **MUCHOW, R.C., WILSON, G.L.** : Photosynthetic and storage limitations to yield in *Sorghum bicolor* (L. MOENCH). - Aust. J. agr. Res. *27* : 489 - 500, 1976.

27200 - **MÜHLE, H.** : Beziehungen zwischen Wasserangebot, Stoffproduktion und Ertragsbildung bei Winterweizen. - Wiss. Z. Humboldt-Univ. Berlin, math.-naturwiss. Reihe *25* : 818 - 823, 1976. [Growth analysis.]

27201 - **MUJEEB, K.A., GREIG, J.K.** : Growth stimulation in *Phaseolus vulgaris* L. induced by gamma irradiation of seeds. - Biol. Plant. *18* : 301 - 303, 1976. [Chl.]

27202 - **MUKHERJEE, D.** : Keto acid changes in starved leaves of *Portulaca grandiflora* and *Centaurea moschata* on exposure to light. - Plant biochem. J. *3* : 24 - 32, 1976. [Ps.]

27203 - **MUKHIN, E.N.** : Issledovanie protsessa obrazovaniya vosstanovitelya za schet énergii sveta v khloroplastakh rasteniĭ. [Process of reductant formation on account of light energy in plant chloroplasts.] - In : Itogi Issledovaniya Mekhanizma Fotosinteza. Pp. 34 - 39. Pushchino 1976. [In R.]

27204 - **MUKOHATA, Y., SUGIYAMA, Y.** : Effects of 1,N[6]-ethenoadenylates on the regulation of photosynthetic activities by adenylates; a study of the nucleotide
binding sites on chloroplast coupling factor 1. - Plant Cell Physiol. *17* :
733 - 742, 1976.

27205 - **MÜLLER, H., SCHUPHAN, W.** : Biochemische Untersuchungen zum Wirkungsmechanis-
mus des Wachstumregulators Chlorcholinchlorid (CCC) bei der Tomatenpflanze
(*Lycopersicon esculentum*). 3.Mitt. : Der Einfluss des CCC auf den Cholin-Stoff-
wechsel und den Gehalt an Chlorophyll und Protein. - Qualitas Plant.- Plant
Foods hum. Nutr. *25* : 297 - 309, 1976.

27206 - **MUNAWAR, M., BURNS, N.M.** : Relationships of phytoplankton biomass with soluble nutrients, primary production, and chlorophyll *a* in Lake Erie, 1970. -
J. Fish. Res. Board Can. *33* : 601 - 611, 1976.

27207 - **MURAKAMI, T.** : Studies on the time course of respiratory rate of mulberry
leaves after putting out the illumination. - J. sericult. Sci. Jap. *45* : 161 -
- 165, 1976.

27208 - **MURDOCH, S., SINCLAIR, J.** : The electrical potential difference and resistance
of *Chara vulgaris*. - Can. J. Bot. *54* : 2187 - 2192, 1976.

27209 - **MUREĬ, I.A.** : Zatraty na dykhanie v period vegetativnoĭ fazy rosta tomatov.
[Expenditures for respiration during vegetative growth phase of tomato.] -
Fiziol. Rast. *23* : 964 - 971, 1976. [Photosynthates; in R, ab : E.]

27210 - **MURTY, K.S., NAYAK, S.K., SAHU, G., RAMAKRISHNAYYA, G., JANARDHAN, K.V.,
RAI, R.S.V.** : Efficiency of ^{14}C photosynthesis and translocation in local
and high yielding rice varieties. - Plant biochem. J. *3* : 63 - 71, 1976.

27211 - **MUSTÁRDY, L.A., MACHOWICZ, E., FALUDI-DÁNIEL, Á.** : Light-induced structural
changes of thylakoids in normal and carotenoid deficient chloroplasts of maize.
- Protoplasma *88* : 65 - 73, 1976.

27212 - **MUTSAERS, H.J.W.** : Growth and assimilate conversion of cotton bolls (*Gossypium hirsutum* L.) 1. Growth of fruits and substrate demand. - Ann. Bot. *40* :
301 - 315, 1976.

27213 - **MUTSAERS, H.J.W.** : Growth and assimilate conversion of cotton bolls (*Gossypium hirsutum* L.) 2. Influence of temperature on boll maturation period and
assimilate conversion. - Ann. Bot. *40* : 317 - 324, 1976.

*27214 - **MUZAFAROVA, S.M.** : Sintez lamellyarnykh belkov khloroplastov. [Synthesis of
lamellar proteins of chloroplasts.] - Dokl. Akad. Nauk tadzh. SSR *18* (1) :
63 - 66, 1975. [In R , ab : Tajik.]

27215 - **NAFZIGER, E.D., KOLLER, H.R.** : Influence of leaf starch concentration on CO_2
assimilation in soybean. - Plant Physiol. *57* : 560 - 563, 1976.

27216 - **NAGARAJAH, S.** : The effects of increased illumination and shading on the
low-light-induced decline in photosynthesis in cotton leaves. - Physiol.
Plant. *36* : 338 - 342, 1976. [Chl.]

27217 - **NAGL, W., KÜHNER, S.** : Early embryogenesis in *Tropaeolum majus* L. : Diversi-
fication of plastids. Planta *133* : 15 - 19, 1976.

27218 - **NAIK, M.S.** : Role of photosynthesis in nitrate assimilation in crop plants. -
Plant biochem. J. *3* : 91 - 95, 1976.

27219 - **NAIMAN, R.J.** : Primary production, standing stock, and export of organic matter in a Mohave Desert thermal stream. - Limnol. Oceanogr. *21* : 60 - 73, 1976.

27220 - **NAIR, P.K.R., BALAKRISHNAN, T.K.** : Pattern of light interception by canopies
in a coconut-cacao crop combination. - Indian J. agr. Sci. *46* : 453 - 462,
1976.

27221 - **NAKAMURA, H., OTA, Y., HASHIMOTO, S., OKINO, H.** : [Photochemical oxidants injury in rice plants. 2. The effect of filtered ambient air on the growth and
yield of rice plants.] - Proc. Crop Sci. Soc. Jap. *45* : 630 - 636, 1976.
[Ps; in Jap., ab : E.]

27222 - **NAKAMURA, K., OGAWA, T., SHIBATA, K.** : Chlorophyll and peptide compositions in the two photosystems of marine green algae. - Biochim. biophys. Acta *423* : 227 - 237, 1976.

27223 - **NAKANI, D.V., KORSAK, M.N.** : Deĭstvie tsinka, khroma i kadmiya na intensivnost' fotosinteza v kratkosrochnykh éksperimentakh. [Effect of zinc, chromium and cadmium on photosynthetic rate in short-time experiments.] - Nauch. Dokl. vyssh. Shkoly, biol. Nauki *19* (9) : 84 - 86, 1976. [In R.]

27224 - **NAKANI, D.V., KORSAK, M.N.** : K izucheniyu kombinirovannogo vozdeĭstviya ionov tsinka i khroma na pervichnuyu produktsiyu Rybinskogo vodokhranilishcha. [Combined action of zinc and chromium on primary production of the Rybinsk reservoir.] - Biol. vnutr. Vod. *31* : 7 - 11, 1976. [In R.]

27225 - **NAKANI, D.V., KORSAK, M.N.** : Pervichnaya produktsiya fitoplanktona Uchinskogo vodokhranilishcha. [Primary production of phytoplankton of the Uchinskoe reservoir.] - Gidrobiol. Zh. *12* (2): 87 - 90, 1976. [In R.]

27226 - **NAKANI, D.V., KORSAK, M.N.** : Pervichnaya produktsiya fitoplanktona v éksperimentakh po biologicheskomu monitoringu na Uchinskom vodokhranilishche. [The primary production of phytoplankton in experiments on biomonitoring at the Uchinskoe reservoir.] - Vestn. mosk. Univ., Ser. VI Biol. Pochvoved. *31* (3): 74 - 77, 1976. [In R, ab : E.]

27227 - **NAKANISHI, M., MONSI, M.** : Factors that control the species composition of freshwater phytoplankton, with special attention to nutrient concentrations. - Int. Rev. ges. Hydrobiol. *61*:439 - 470, 1976. [Ps, Chl.]

27228 - **NAKASEKO, K., GOTOH, K.** : Physio-ecological studies on prolificacy in maize. I. Differences in dry matter accumulation among one-, two- and three-eared plants. - Proc. Crop Sci. Soc. Japan *45* : 263 - 269, 1976.

27229 - **NAKAYAMA, A.** : Utilization of light quality in shading culture of tea plants.- JARQ-Jap. agr. Res. Quart. *10* : 193 - 196, 1976.

27230 - **NALBANDYAN, R.M.** : Kristallicheskiĭ ferredoksin tipa *b* iz *Chenopodium album*. [Crystalline ferredoxin *b* from *Chenopodium album*.] - Biokhimiya *41* : 188 - 191, 1976. [In R, ab : E.]

27231 - **NALEWAJKO, C., SCHINDLER, D.W.** : Primary production, extracellular release, and heterotrophy in two lakes in the ELA, northwestern Ontario. - J. Fish. Res. Board Can. *33*: 219 - 226, 1976.

27232 - **NASAR, S.A.K., NASAR, S.K.T.** : An unusual record of oxygen production in a freshwater pond. - Curr. Sci. *45* : 676 - 677, 1976.

27233 - **NASYROV, J.S., BÖRNER, T.** : Zur Autonomie und Evolution der Chloroplasten und Mitochondrien. - Wiss. Fortschr. *26* : 496 - 500, 1976.

27234 - **NÄTR, L.** : Vliv sponu rostlin na asimilační aparát a výnos jarního ječmene. [The effect of plant spacing on the assimilating apparatus and yield of spring barley.] - Rostl. Výroba (Praha) *22* : 577 - 591, 1976. [Growth analysis; in Czech, ab : E, R.]

27235 - **NAUMANN, W.-D., PLANCHER, B.** : Untersuchungen zur klimatisierenden Beregnung von Obstgehölzen. I. Wachstum und Mineralstoffgehalte im Blatt bei Apfelsämlingen als Modellpflanzen. - Gartenbauwissenschaft *41* : 205 - 212, 1976. [Growth analysis.]

27236 - **NAZAROV, S.K., P'YANKOV, V.I.** : Vliyanie temperatury na pervichnye reaktsii biosinteza uglevodov u rasteniĭ Arktiki i umerennoĭ zony. [Effect of temperature on the primary reactions of biosynthesis of saccharides in plants of Arctic and temperate zone.] - In : Gazometricheskoe Issledovanie Fotosinteza i Dykhaniya Rasteniĭ. Pp. 113 - 115. Akad. Nauk SSSR, Tartu 1976. [In R.]

27237 - **NEAMŢU, G., LÁSZLO, T., BILAUS, C., ILLYES, G.** : Cercetări carotenotaxonomice la plante superioare. VIII. Pigmenţi carotenoidici şi clorofilieni din unele plante leguminoase. [Carotenotaxonomic investigations on higher plants. VIII. Carotenoid and chlorophyll pigments of some leguminous plants.] - Stud. Cercet. biochim. *19* : 81 - 85, 1976. [In Roum., ab : E.]

*27238 - **NECHAEVA, E.P.** : Deĭstvie zelenogo sveta na soderzhanie khlorofilla v zeleneyushchikh i zelenykh prorostkakh yachmenya. [Effect of green radiation on
the content of chlorophyll in greening and green barley seedlings.] - Uch.
Zapiski perm. gosud. pedag. Inst., Kafedra Bot. *124* (Nekotorye Voprosy Botaniki) : 19 - 25, 1974. [In R.]

*27239 - **NECHAEVA, E.P.** : Vilyanie dekapitatsii koleoptile na morfogenez i soderzhanie pigmentov u vskhodov ovsa i ozimoĭ rzhi. [Effect of coleoptile decapitation on morphogenesis and pigment content in seedlings of oat and winter rye.]
- Uch. Zapiski perm. gosud. pedag. Inst., Kafedra Bot. *141* (Voprosy Botaniki,
Ėkologii i Fiziologii Rasteniĭ) : 96 - 102, 1975. [In R.]

27240 - **NECHIPORENKO, G.A., GRINEVA, G.M.** : Vilyanie raznykh srokov zatopleniya na
raspredelenie C^{14}-sakharozy-U u rasteniĭ kukuruzy. [Effect of duration of
inundation on distribution of ^{14}C-saccharose-U in maize plants.] - Fiziol.
Rast. *23* : 984 - 989, 1976. [In R, ab : E.]

27241 - **NEDUKHA, E.M.** : Pory khloroplastov funarii vlagomernoĭ. [Pores of the *Funaria
hygrometrica* chloroplasts.] - Tsitol. Genet. *10* (1) : 43 - 44, 1976. [In R,
ab : E.]

27242 - **NELSON, C.J.** : Agronomic potential for biomass production. - In : Third Annual UMR-MEC Conference on Energy. Pp. 1 - 11. University of Missouri, Rolla
1976. [Ps.]

27243 - **NELSON, N.** : Structure and function of chloroplast ATPase. - Biochim. biophys.
Acta *456* : 314 - 338, 1976.

27244 - **NELSON, N., BROZA, R.** : Salt inactivation as a mechanistic probe of membrane-
-bound chloroplast coupling factor 1. - Europe. J. Biochem. *69* : 203 - 208,
1976.

27245 - **NELSON, N., KARNY, O.** : The role of δ subunit in the coupling activity of
chloroplasts coupling factor 1. - FEBS Lett. *70* : 249 - 253, 1976.

27246 - **NELSON, P.E., SURZYCKI, S.J.** : A mutant strain of *Chlamydomonas reinhardi*
exhibiting altered ribulosebisphosphate carboxylase. - Europe. J. Biochem.
61 : 465 - 474, 1976.

27247 - **NELSON, P.E., SURZYCKI, S.J.** : Characterization of the oxygenase activity in
a mutant of *Chlamydomonas reinhardi* exhibiting altered ribulosebisphosphate
carboxylase. - Europe. J. Biochem. *61* : 475 - 480, 1976.

*27248 - **NESTSYAROVICH, M.D., DZYARUGINA, T.F.** : Nakaplenie sukhoga rêchyva sadzhantsami
vyaza gladkaga i yasenya zvychaĭnaga ŭ zalezhnastsi ad pratsyaglastsi zataple-
nnya. [Accumulation of dry matter by elm and ash seedlings depending on duration of flooding.] - Vestsi Akad. Navuk belaruss. SSR, Ser. biyal. Navuk
1975 (5) : 12 - 14, 1975. [Ps; in Beloruss.]

27249 - **NEUMANN, D., PARTHIER, B.** : Wirkungen einiger Hemmstoffe der Genexpression
auf die Ausbildung der Organellenstrukturen in *Euglena gracilis*. - Acta histochem. *17* (Suppl.) : 95 - 106, 1976.

27250 - **NEVÉ, R.A., CLASBY, R.C., GOERING, J.J., HOOD, D.W.** : Enhancement of primary
productivity by artificial upwelling. - Marine Sci. Commun. *2* : 109 - 124,
1976.

27251 - **NEVEUX, J.** : Dosage de la chlorophylle *a* et de la phéophytine *a* par la fluori-
métrie. - Ann. Inst. océanogr. (Paris) *52* : 165 - 174, 1976.

27252 - **NEWTON, J.W.** : Photoproduction of molecular hydrogen by a plant-algal symbiotic system. - Science *191* : 559 - 561, 1976.

*27253 - **NICHIPOROVICH, A.A.** : Energy and mass transfer in plant communities. - In :
DE VRIES, D.A., AFGAN, N.H. (ed.) : Advances in Thermal Engineering. Vol.1.
Heat and Mass Transfer in the Biosphere. Part 1. Transfer Processes in Plant
Environment. Pp. 427 - 441. Halsted Press, New York; Scripta Book Company,
Washington 1975.

27254 - **NIEDERMAN, R.A., MALLON, D.E., LANGAN, J.J.** : Membranes of *Rhodopseudomonas
sphaeroides*. IV. Assembly of chromatophores in low-aeration cell suspensions.
- Biochim. biophys. Acta *440* : 429 - 447, 1976.

27255 - **NIEMAN, R.H., CLARK, R.A.** : Interactive effects of salinity and phosphorus nutrition of the concentrations of phosphate and phosphate esters in mature photosynthesizing corn leaves. - Plant Physiol. *57* : 157 - 161, 1976.

27256 - **NIGEL, B.** : Light as a chemical reagent. - Crucible (Toronto) *7* (3) : 42 - - 45, 1976. [Ps.]

27257 - **NIKITINA, K.A., GUSEV, M.V.** : Nekotorye osobennosti razlichnykh stadiĭ razvitiya sine-zelenoĭ vodorosli *Anacystis nidulans (Synechococcus)*. [Characteristics of different growth stages of the blue-green alga *Anacystis nidulans (Synechococcus)*.] - Mikrobiologiya *45* : 490 - 496, 1976. [In R, ab : E.]

27258 - **NIKITINA, K.A., GUSEV, M.V.** : Vliyanie nekotorykh organicheskikh veshchestv na rost na svetu i otmiranie v temnote sine-zelenoĭ vodorosli *Anabaena variabilis*. [Effect of some organic compounds on growth of the blue-green alga *Anabaena variabilis* in the light and its dying in the darkness.] - Fiziol. Rast. *23* : 1219 - 1224, 1976. [Chl, biliproteins; in R, ab : E.]

27259 - **NIKITINA, K.A., KHODZHAEV, M.N., GUSEV, M.V.** : Vozrastnye izmeneniya populyatsii kletok sine-zelenoĭ vodorosli *Anacystis nidulans* pri raznykh temperaturakh. [Growth changes in population of blue-green alga *Anacystis nidulans* cells under various temperatures.] - Nauch. Dokl. vyssh. Shkoly, biol. Nauki *19* (10) : 99 - 104, 1976. [Chl, biliproteins; in R.]

27260 - **NIKOLAEVA, L.F., RUBIN, A.B., KAZAKOVA, A.S., KONONENKO, A.A.** : Izuchenie kinetiki fotoindutsirovannogo perekhoda protokhlorofillida v khlorofillid u nekotorykh golosemennykh rasteniĭ i kukuruzy. [Kinetics of photoinduced transition of protochlorophyllide into chlorophyllide in some gymnosperms and maize.] - Nauch. Dokl. vyssh. Shkoly, biol. Nauki *19* (8) : 91 - 95, 1976. [In R.]

27261 - **NILSHAMMAR-HOLMVALL, M.** : Cyclic variation in thylakoid aggregation in normal and calcium deficient cells of *Scenedesmus*. - Protoplasma *87* : 263 - 271, 1976.

*27262 - **NISHIMURA, M.** : [Energy transfer in solid-state and membrane systems in photosynthesis.] - Seikagaku *40* : 347 - 356, 1968. [In Jap.]

27263 - **NISHIMURA, M., GRAHAM, D., AKAZAWA, T.** : Isolation of intact chloroplasts and other cell organelles from spinach leaf protoplasts. - Plant Physiol. *58* : 309 - 314, 1976.

27264 - **NISHIZAKI, Y.** : Relation between slow delayed light emission and acid-base triggered luminescence in chloroplasts. - Biochim. biophys. Acta *449* : 368 - 375, 1976.

27265 - **NITSCH, A.** : Genetische und physiologische Untersuchungen an Polyenfettsäure- -Mutanten von Raps. II. Entwicklung und Polyenfettsäuregehalt von reifenden Samen. - Angew. Bot. *50* : 31 - 42, 1976.

27266 - **NIVAL, P., GOSTAN, J., MALARA, G., CHARRA, R.** : Évolution du plancton dans la baie de Villefranche-sur-Mer à la fin du printemps (Mai et Juin 1971). II - Biomasse de phytoplancton, production primaire. - Vie Milieu, Sér. B Océanogr. *26* : 47 - 76, 1976. [Chl.]

27267 - **NIXON, S.W., OVIATT, C.A., GARBER, J., LEE, V.** : Diel metabolism and nutrient dynamics in a salt marsh embayment. - Ecology *57* : 740 - 750, 1976. [Ps.]

*27268 - **NIYAZMUKHAMEDOVA, M.B.** : Osobennosti ul'trastruktury khloroplastov i soderzhaniya uglevodov u pigmentnogo mutanta *Arabidopsis thaliana* (L.). [Peculiarities of chloroplast ultrastructure and saccharide contents in a pigment mutant of *Arabidopsis thaliana* (L.).] - Dokl. Akad. Nauk tadzh. SSR *18* (10) : 59 - 62, 1975. [In R, ab : Tajik.]

27269 - **NKEMDIRIM, L.C.** : Crop development and water loss - a case study over a potato crop. - Agr. Meteorol. *16* : 371 - 388, 1976.

27270 - **NOBEL, P.S.** : Photosynthetic rates of sun *versus* shade leaves of *Hyptis emoryi* TORR. - Plant Physiol. *58* : 218 - 223, 1976.

27271 - **NOBEL, P.S.** : Water relations and photosynthesis of a desert CAM plant, *Agave deserti*. - Plant Physiol. *58* : 576 - 582, 1976.

27272 - **NOBLE, R.D.** : Photosynthetic and developmental responses of soybean chlorophyll mutants to two light intensities. - Plant Physiol. *57* (Suppl.) : 73,
1976.

27273 - **NOKS, P.P., ADAMOVA, N.P., PASHCHENKO, V.Z., TIMOFEEV, K.N., KONONENKO, A.A.,
RUBIN, A.B.** : Vydelenie i issledovanie fotokhimicheskikh svoĭstv fotosinteticheskikh reaktsionnykh tsentrov iz kletok *Rhodopseudomonas spheroides* ,
shtamm 1760-1. [Isolation and characterization of photochemical properties
of photosynthetic reaction centres from *Rhodopseudomonas spheroides*, strain
1760-1.] - Mol. Biol. (Moskva) *10* : 641 - 651, 1976. [In R, ab : E.]

27274 - **NOKS, P.P., KONONENKO, A.A., RUBIN, A.B.** : Funktsional'naya aktivnost' reaktsionnykh tsentrov fotosinteticheskikh edinits v khromatoforakh *Rhodospirillum
rubrum* pri obezvozhivanii. [Effects of desiccation on the functional activity
of reaction centres of photosynthetic units in *Rhodospirillum rubrum* chromatophores.] - Vestn. mosk. Univ., Biol., Pochvoved. *1976* (2) : 62 - 66, 1976.
[In R.]

27275 - **NOLAN, W.G., BISHOP, D.G.** : The effect of amphotericin B on cytochrome *f*,
C550 and the 515 nm shift in maize chloroplasts. - Plant Physiol. *57* (Suppl.)
: 14, 1976.

27276 - **NOLAN, W.G., SMILLIE, R.M.** : Changes in the activation energy for photosynthetic electron transport by isolated barley chloroplasts. - Plant Physiol.
57 (Suppl.) : 15, 1976.

27277 - **NOLAN, W.G., SMILLIE, R.M.** : Multi-temperature effects on Hill reaction activity of barley chloroplasts. - Biochim. biophys. Acta *440* : 461 - 475, 1976.

27278 - **NORCIO, N.V., SULLIVAN, C.Y.** : Stomatal and non-stomatal inhibition of photosynthesis at high temperatures in sorghum. - Plant Physiol. *57* (Suppl.) : 44,
1976.

27279 - **NORDHORN, G., WEIDNER, M., WILLENBRINK, J.** : Isolation and photosynthetic
activities of chloroplasts of the brown alga *Fucus serratus* L. - Z. Pflanzenphysiol. *80* : 153 - 165, 1976.

27280 - **NORDIN, Å.** : Water flow in wheat seedlings after small water deficits. -
Physiol. Plant. *37* : 157 - 162, 1976. [Stomatal resistance.]

27281 - **NORDLIE, F.G.** : Plankton communities of three Central Florida lakes. - Hydrobiologia *48* : 65 - 78, 1976.

27282 - **NORRIS, J.R.** : Triplet states and photosynthesis. - Photochem. Photobiol.
23 : 449 - 450, 1976.

27283 - **NORTON, J.E., AUNG, L.H.** : Photoperiods on NAR, RGR and shoot growth of tomato seedlings. - HortScience *11* (3, Sect.2) : 299, 1976.

27284 - **NOSTICZIUS, A.** : Indirect examination of the role of formaldehyde and gly-
colaldehyde in carbon metabolism. - Acta agron. Acad. Sci. hung. *25* : 183 -
208, 1976.

27285 - **NOVAK, V.A., IVANKINA, N.G.** : Vzaimosvyaz' mezhdu svetoindutsirovannoĭ vnutri-
kletochnoĭ i vnekletochnoĭ bioėlektricheskoĭ reaktsieĭ rasteniĭ. [Relationship
between intracellular and extracellular light-induced bioelectrical reaction
of plants.] - Biofizika *21* : 519 - 523, 1976. [In R, ab : E.]

27286 - **NOVER, L.** : Density labeling of chloroplast-specific leucyl-t-RNA synthetase
in greening cells of *Euglena gracilis*. - Plant Sci. Lett. *7* : 403 - 407,
1976.

27287 - **NOVIKOVA, A.A.** : Vliyanie razlichnoĭ prodolzhitel'nosti osveshcheniya na soderzhanie khlorofilla v list'yakh rasteniĭ korotkogo, srednego i dlinnogo
dnya. [Effect of different duration of illumination on chlorophyll content
in leaves of short-, medium-, and long-day plants.] - Dokl. Akad. Nauk belorus. SSR *20* : 272 - 273, 1976. [In R.]

27288 - **NOVITSKAYA, G.V., RUTSKAYA, L.A.** : Kolichestvennoe opredelenie lipidov membran
khloroplastov. [Quantitative determination of lipids in chloroplast membranes.]
- Fiziol. Rast. *23* : 899 - 905, 1976. [In R, ab : E.]

27289 - **NUNES, M.A.** : Water relations in coffee. Significance of plant water deficits
to growth and yield : A review. - J. Coffee Res. *6* : 4 - 21, 1976. [Ps.]

27290 - **NURITDINOV, N., VARTAPETYAN, B.B.** : Transport ^{14}C-sakharozy u khlopchatnika
v usloviyakh kislorodnogo golodaniya korneĭ. [Transport of ^{14}C-saccharose
in cotton plant under root oxygen deficiency.] - Dokl. Akad. Nauk SSSR *228* :
509 - 511, 1976. [In R.]

27291 - **NUS, J.L.** : The influence of high light and chilling temperatures on chlorophyll degradation. - HortScience *11* (3, Sect. 2) : 301, 1976.

27292 - **NUTBEAM, A.R., DUFFUS, C.M.** : Evidence for C_4 photosynthesis in barley pericarp tissue. - Biochem. biophys. Res. Commun. *70* : 1198 - 1203, 1976.

27293 - **NYFFELER, A.** : Der Einfluß des Bestandesalters und des Genotyps auf die Er-
tragsleistung von schweizerischem Mattenklee. - Mitt. schweiz. Landwirtsch.
24 (4) : 98 - 103, 1976. [Production.]

27294 - **NYFFELER, A., KOBLET, R., NÖSBERGER, J.** : Der Einfluß des Bestandesalters
auf die Ertragsbildung beim schweizerischen Mattenklee (*Trifolium pratense*
L.). - Z. Acker- Pflanzenbau *142* : 214 - 225, 1976. [Dry-matter production.]

27295 - **NYFFELER, A., KOBLET, R., NÖSBERGER, J.** : Der Einfluß der Ploidiestufe auf die
Ertragsbildung beim schweizerischen Mattenklee (*Trifolium pratense* L.). -
Z. Acker- Pflanzenbau *142* : 226 - 236, 1976. [Growth analysis.]

27296 - **OBRAZTSOV, A.S., KOVALEV, V.M.** : Ob"emnyĭ sposob opredeleniya ploshchadi listovoĭ poverkhnosti rasteniĭ v posevakh.[Volumetric technique for determining
the area of leaf surface of plants in stands.] - Fiziol. Rast. *23* : 1084 -
1087, 1976. [In R, ab : E.]

27297 - **O'BRIEN, M.J., EASTERBY, J.S., POWLS, R.** : Algal glyceraldehyde-3-phosphate
dehydrogenases. Conversion of the NADH-linked enzyme of *Scenedesmus obliquus*
into a form which preferentially uses NADPH as coenzyme. - Biochim. biophys.
Acta *449* : 209 - 223, 1976.

27298 - **O'BRIEN, W.J., DENOYELLES, F. Jr.** : Response of three phytoplankton bioassay
techniques in experimental ponds of known limiting nutrient. - Hydrobiologia
49 : 65 - 76, 1976.

27299 - **Ó CARRA, P., Ó hEOCHA, C.** : Algal biliproteins and phycobilins. - In :
GOODWIN, T.W. (ed.) : Chemistry and Biochemistry of Plant Pigments. 2nd Ed.
Vol. 1. Pp. 328 - 376. Academic Press, London - New York - San Francisco
1976.

27300 - **ODINTSOVA, M.S., YURINA, N.P.** : Ribosomy khloroplastov. [Chloroplast ribosomes.] - Biokhimiya *41* : 1915 - 1927, 1976. [In R, ab : E.]

27301 - **OECHEL, W.C.** : Seasonal patterns of temperature response of CO_2 flux and acclimation in arctic mosses growing *in situ*. - Photosynthetica *10* : 447 - 456,
1976.

27302 - **OECHEL, W.C., COLLINS, N.J.** : Comparative CO_2 exchange patterns in mosses
from two tundra habitats at Barrow, Alaska. - Can. J. Bot. *54* : 1355 - 1369,
1976.

27303 - **OELZE, J.** : Early formation of intracytoplasmic membranes in *Rhodospirillum
rubrum*. - Biochim. biophys. Acta *436* : 95 - 100, 1976. [Chl.]

27304 - **OELZE, J., PAHLKE, W.** : The early formation of the photosynthetic apparatus
in *Rhodospirillum rubrum*. - Arch. Microbiol. *108* : 281 - 285, 1976.

27305 - **OESTERHELT, D.** : Bacteriorhodopsin als Beispiel einer lichtgetriebenen Proto-
nenpumpe. - Angew. Chem. *88* : 16 - 24, 1976.

27306 - **OESTERHELT, D.** : Bacteriorhodopsin as an example of a light-driven proton
pump. - Angew. Chem., int. Ed. *15* : 17 - 24, 1976.

27307 - **OESTERHELT, D.** : Bacteriorhodopsin as a light-driven ion exchanger ? - FEBS
Lett. *64* : 20 - 22, 1976.

27308 - **OETTMEIER, W., NORRIS, J.R., KATZ, J.J.** : Evidence for the localization of chlorophyll in lipid vesicles : A spin label study. - Biochem. biophys. Res. Commun. *71* : 445 - 451, 1976.

23709 - **OETTMEIER, W., NORRIS, J.R., KATZ, J.J.** : Photo-induced electron transfer in chlorophyll containing liposomes. - Z. Naturforsch. *31 C* : 163 - 168, 1976.

27310 - **OGANEZOVA, É.P., NALBANDYAN, R.M.** : Ochistka i svoĭstva plastotsianina i ferredoksina iz *Ceratophyllum demersum* L. [Purification and properties of plastocyanin and ferredoxin from *Ceratophyllum demersum* L.] - Biokhimiya *41* : 794 - 800, 1976. [In R , ab : E.]

27311 - **OGAWA, M., OTA, Y.** : [3-hydroxy-5-methylisoxazole as a plant growth stimulant.] - Bull. nat. Inst. agr. Sci., Ser. D (Physiol. Genet.) *27* : 103 - 137, 1976. [Ps, Chl; In Jap., ab : E.]

27312 - **OGAWA, T.** : [Separation and purification of two photosystems.] - Tampakushitsu Kakusan Koso, Bessatsu [Protein, Nucleic Acid, Enzyme, Supplement] *1976* : 70 - 75, 1976. [In Jap.]

27313 - **OGREN, W.L.** : Search for higher plants with modifications of the reductive pentose phosphate pathway of CO_2 assimilation. - In : **BURRIS, R.H., BLACK, C.C.** (ed.) : CO_2 Metabolism and Plant Productivity. Pp. 19 - 29. University Park Press, Baltimore - London - Tokyo 1976.

27314 - **OH-HAMA, T., HASE, E.** : Enhancing effects of CO_2 on chloroplast regeneration in glucose-bleached cells of *Chlorella protothecoides* I. Role of non-photosynthetic CO_2-fixation in chloroplast regeneration. - Plant Cell Physiol. *17* : 45 - 53, 1976.

27315 - **OH-HAMA, T., HASE, E.** : Enhancing effects of CO_2 on chloroplast regeneration in glucose-bleached cells of *Chlorella protothecoides* II. Role of carbamylphosphate in chloroplast regeneration. - Plant Cell Physiol. *17* : 55 - 62, 1976.

27316 - **OHIWA, T.** : Observations on chloroplast growth and pyrenoid formation in *Spirogyra*. A study by means of uncoiled picture of chloroplast. - Bot. Mag. (Tokyo) *89* : 259 - 266, 1976.

27317 - **OHKI, K.** : Effect of zinc nutrition on photosynthesis and carbonic anhydrase activity in cotton. - Physiol. Plant. *38* : 300 - 304, 1976.

27318 - **OHKI, K.** : Photosynthesis and carbonic anhydrase activity related to zinc nutrition of cotton. - Plant Physiol. *57* (Suppl.) : 85, 1976.

27319 - **OHKI, K., KATOH, T.** : Incompetence of dark-synthesized phycocyanin in excitation transfer to photosystem II chlorophyll. - Planta *129* : 249 - 251, 1976.

27320 - **OHNO, M.** : [A preliminary report on the photosynthetic activity of *Phyllophora antarctica* A. et E.S.GEPP antarctic red alga.] - Antarctic Record (Tokyo) *57* : 141 - 145, 1976. [In Jap., ab : E.]

27321 - **OHNO, M.** : Some observations on the influence of salinity on photosynthetic activity and chloride ion loss in several seaweeds. - Int. Rev. ges. Hydrobiol. *61* : 665 - 672, 1976.

27322 - **OHTAKI, E., SEO, T.** : Infrared device for measurement of carbon dioxide fluctuations under field conditions. I. Single beam system. - Ber. Ôhara Inst. landwirtsch. Biol. Okayama Univ. *16* : 175 - 182, 1976.

27323 - **OHTAKI, E., SEO, T.** : Infrared device for measurement of carbon dioxide fluctuations under field conditions. II. Double beam system. - Ber. Ôhara Inst. landwirtsch. Biol. Okayama Univ. *16* : 183 - 190, 1976.

27324 - **OKADA, M., KITAJIMA, M., BUTLER, W.L.** : Inhibition of photosystem I and photosystem II in chloroplasts by UV radiation. - Plant Cell Physiol. *17* : 35 - 43, 1976.

27325 - **OKAFOR, L.I., De DATTA, S.K.** : Competition between upland rice and purple nutsedge for nitrogen, moisture, and light. - Weed Sci. *24* : 43 - 46, 1976. [Radiation and LAI in canopy.]

27326 - OKAMURA, M.Y., ACKERSON, L.C., ISAACSON, R.A., PARSON, W.W., FEHER, G. :
The primary electron acceptor in *Chromatium vinosum* (strain D). - Biophys.
J. *16* (2, Part 2) : 223, 1976.

27327 - OKANENKO, A.S., TONKAL', E.A., BISOVETSKIǏ, T.Ya., BOGACHUK, G.K., BERSHTEǏN,
B.I., IL'YASHCHUK, E.M., IVANISHCHEVA, S.Yu., PSHENICHNAYA, A.K. :
Produktivnost' fotosinteza u svekly na pochvakh s razlichnym soderzhaniem
podvizhnogo kaliya. [Productivity of photosynthesis in sugar beet in soils
with different content of mobile potassium.] - Fiziol. Biokhim. kul't. Rast.
8 : 12 - 19, 1976. [In R, ab : E.]

27328 - OKAYAMA, S. : Redox potential of plastoquinone A in spinach chloroplasts. -
Biochim. biophys. Acta *440* : 331 - 336, 1976.

27329 - OKAZAKI, M., YOSHIDA, T., FERUYA, K. : The effects of light on carbonic an-
hydrases in the etiolated leaves of *Phaseolus vulgaris* and in intact chloroplasts from spinach. - Bull. Tokyo Gakugei Univ., Ser. IV - Math. nat. Sci.
28 : 199 - 206, 1976.

27330 - OKITA, T., SULLIVAN, C.E., VOLCANI, B.E. : Gel electrophoresis of ion-stimu-
lated ATPases from protoplast membranes of the diatom *Nitzschia alba*. -
Plant Sci. Lett. *6* : 129 - 134, 1976.

27331 - OKORO, O.O., GRACE, J. : The physiology of rooting *Populus* cuttings. I. Carbohydrates and photosynthesis. - Physiol. Plant. *36* : 133 - 138, 1976.

27332 - OKU, T., TOMITA, G. : Photoactivation of oxygen-evolving system in dark-grown
spruce seedlings. - Physiol. Plant. *38* : 181 - 185, 1976.

*27333 - OKUBO, T., OIZUMI, H., HOSHINO, M. : [Efficiencies of energy conversion in
pasture ecosystem. I. Dry matter production and efficiencies of light utilization in primary canopies of several pasture species.] - J. jap. Soc.
Grassl. Sci. *15* : 138 - 149, 1969. [In Jap., ab : E.]

27334 - OLIVA, M., STEUBING, L. : Untersuchungen über die Beeinflussung von Photosynthese, Respiration und Wasserhaushalt durch H₂S bei *Spinacia oleracea*. -
Angew. Bot. *50* : 1 - 17, 1976.

27335 - OLIVER, D., JAGENDORF, A. : Exposure of free amino groups in the coupling
factor of energized spinach chloroplasts. - J. biol. Chem. *251* : 7168 -
7175, 1976.

27336 - OLIVER, I., FRANS, R.E., TALBERT, R.E. : Field competition between tall mor-
ningglory and soybean. I. Growth analysis. - Weed Sci. *24* : 482 - 488, 1976.

27337 - OLLERENSHAW, J.H., STEWART, W.S., GALLIMORE, J., BAKER, R.H. : Low-temperature growth in grasses from northern latitudes. - J. agr. Sci. *87* : 237 - 239,
1976. [Ps.]

27338 - OLOFINBOBA, M.O., FAWOLE, M.O. : The effect of MCPP on aspects of the metabolism of 17-day old seedlings of *Theobroma cacao*, variety F₃ Amazon. - Turrialba *26* : 167 - 173, 1976. [Photosynthates.]

27339 - OLOFINBOBA, M.O., FAWOLE, M.O. : Response of six-month-old seedlings of *Theobroma cacao* to foliar treatment with Spruce Seal. - Turrialba *26* : 365 -
370, 1976. [Ps, Chl.]

27340 - OLSON, J.M., GIDDINGS, T.H. Jr., SHAW, E.K. : An enriched reaction center
preparation from green photosynthetic bacteria. - Biochim. biophys. Acta
449 : 197 - 208, 1976.

27341 - OLSON, J.M., KE, B., THOMPSON, K.H. : Exciton interaction among chlorophyll
molecules in becteriochlorophyll *a* proteins and bacteriochlorophyll *a* reaction center complexes from green bacteria. - Biochim. biophys. Acta *430* :
524 - 537, 1976.

27342 - OLSON, J.M., SHAW, E.K., ENGLBERGER, F.M. : Comparison of bacteriochlorophyll
a-proteins from two green bacteria. - Biochem. J. *159* : 769 - 774, 1976.

27343 - OLUGBEMI, L.B., AUSTIN, R.B., BINGHAM, J. : Effects of awns on the photosynthesis and yield of wheat, *Triticum aestivum*. - Ann. appl. Biol. *84* : 241 -
250, 1976.

27344 - OLUGBEMI, L.B., BINGHAM, J., AUSTIN, R.B. : Ear and flag leaf photosynthesis of awned and awnless *Triticum* species. - Ann. appl. Biol. *84* : 231 - 240, 1976.

27345 - ÖQUIST, G., HELLGREN, N.O. : The photosynthetic electron transport capacity of chloroplasts prepared from needles of unhardened and hardened seedligs of *Pinus silvestris*. - Plant Sci. Lett. *7* : 359 - 369, 1976.

27346 - ORAV, I., ZOZ, N., SEREBRYANYĬ, A., RANDALU, K. : Chastota khlorofil'nykh mutatsiĭ u yachmenya posle obrabotki khimicheskimi mutagenami pri raznykh pH. [Frequency of chlorophyll mutations in barley following treatment with chemical mutagens at various pH.] - Izv. Akad. Nauk êst. SSR, Biologiya *25* : 249 - 251, 1976. [In R.]

27347 - ORLOVIUS, K., HÖFNER, W. : Einfluß einer variierten Stickstoffdüngung auf Assimilationsleistung und Ertragsbildung von Sommerweizen. - Z. Pflanzenern. Bodenkunde *5* : 631 - 640, 1976.

27348 - ORMSBEE, P., BAZZAZ, F.A., BOGGESS, W.R. : Physiological ecology of *Juniperus virginiana* in oldfields. - Oecologia *23* : 75 - 82, 1976. [Ps, Chl.]

27349 - ORR, A.R., KESSLER, J.E., TEPASKE, E.R. : DCMU induced inhibition of growth, photosynthesis and motility in *Eudorina elegans* cultures. - J. Phycol. *12* (Suppl.) : 31, 1976.

27350 - ORR, A.R., KESSLER, J.E., TEPASKE, E.R. : DCMU induced inhibition of growth, photosynthesis and motility in *Eudorina elegans* cultures. - Amer. J. Bot. *63* : 973 - 978, 1976.

27351 - ORSENIGO, M., RASCIO, N. : Chloroplast fine structure in the japonica-2 maize mutant exposed to continuous illumination. 1 The green tissues. - Cytobios *16* : 171 - 182, 1976.

27352 - ORSENIGO, M., RASCIO, N., BONATTI, P.M. : Fine structure of the etioplast in two mutants of maize. - J. Ultrastructure Res. *55* : 42 - 49, 1976.

27353 - ORT, D. : On the mechanism of control of photosynthetic electron transport by phosphorylation. - FEBS Lett. *69* : 81 - 85, 1976.

27354 - ORT, D.R., DILLEY, R.A. : Photophosphorylation as a function of illumination time. I. Effects of permeant cations and permeant anions. - Biochim. biophys. Acta *449* : 95 - 107, 1976.

27355 - ORT, D.R., DILLEY, R.A., GOOD, N.E. : Photophosphorylation as a function of illumination time. II. Effects of permeant buffers. - Biochim. biophys. Acta *449* : 108 - 124, 1976.

27356 - ORTON, P.J., MANSFIELD, T.A. : Studies of mechanism by which daminozide (B9) inhibits stomatal opening. - J. exp. Bot. *27* : 125 - 133, 1976. [Ps.]

27357 - OSHCHEPKOV, V.P., KRASNOVSKIĬ, A.A. : Fotoobrazovanie molekulyarnogo vodoroda zelenymi vodoroslyami. [Photoproduction of molecular hydrogen by green algae.] - Izv. Akad. Nauk SSSR, Ser. biol. *1976* : 87 - 100, 1976. [In R, ab : E.]

27358 - OSMOND, C.B. : Ion absorption and carbon metabolism in cells of higher plants. - In : LÜTTGE, U., PITMAN, M.G. (ed.) : Transport in Plants II. Part A. Cells. Pp. 347 - 372. Springer-Verlag, Berlin - Heidelberg - New York 1976.

27359 - OSMOND, C.B. : CO_2 assimilation and dissimilation in the light and dark in CAM plants. - In : BURRIS, R.H., BLACK, C.C. (ed.) : CO_2 Metabolism and Plant Productivity. Pp. 217 - 233. University Park Press, Baltimore - London - Tokyo 1976.

27360 - OSMOND, C.B., BENDER, M.M., BURRIS, R.H. : Pathways of CO_2 fixation in the CAM plant *Kalanchoë daigremontiana*. III. Correlation with $\delta^{13}C$ value during growth and water stress. - Aust. J. Plant Physiol. *3* : 787 - 799, 1976.

27361 - OSMOND, C.B., SMITH, F.A. : Symplastic transport of metabolites during C_4-photosynthesis. - In : GUNNING, B.E.S., ROBARDS, A.W. (ed.) : Intercellular Communication in Plants : Studies on Plasmodesmata. Pp. 229 - 241. Springer-Verlag, Berlin - Heidelberg - New York 1976.

27362 - OSZLÁNYI, O. : The distribution of leaves in the canopy of the oak-hornbeam stand at Báb (I. B. P.). - Oecol. Plant. *11* : 277 - 289, 1976.

27363 - O'TOOLE, J.C., CROOKSTON, E.K., TREHARNE, K.J., OZBUN, J.L. : Mesophyll resistance and carboxylase activity. A comparison under water stress conditions. - Plant Physiol. *57* : 465 - 468, 1976.

27364 - OUTLAW, W.H.,Jr., SCHMUCK, C.L., TOLBERT, N.E. : Photosynthetic carbon metabolism in the palisade parenchyma and spongy parenchyma of *Vicia faba* L. - Plant Physiol. *58* : 186 - 189, 1976.

27365 - OVCHINNIKOVA, M.F. : Komplementatsiya khloroplastov kukuruzy. [Complementation of maize chloroplasts.] - Sel'skokhoz. Biol. *11* : 675 - 679, 1976. [In R, ab : E.]

27366 - OVERDIECK, D., RAGHI-ATRI, F. : CO_2-Netto-Assimilation von *Phragmites australis* (CAV.) TRIN. ex STEUD. Blättern bei unterschiedlichen Mengen an Stickstoff und Phosphor im Nährsubstrat. - Angew. Bot. *50* : 267 - 283, 1976.

27367 - OVERNELL, J. : Inhibition of marine algal photosynthesis by heavy metals. - Mar. Biol. *38* : 335 - 342, 1976.

27368 - OWENS, O.v.H., ESAIAS, W.E. : Physiological responses of phytoplankton to major environmental factors. - Annu. Rev. Plant Physiol. *27* : 461 - 483, 1976. [Ps, Chl.]

27369 - OYA, V. : Metrologiya izmerenii optiko-akusticheskim analizatorom CO_2 . [Metrology of the measurements with an optical-acoustic analyser of CO_2.] - In : Gazometricheskoe Issledovanie Fotosinteza i Dykhaniya Rastenii. Pp. 116 - 119. Akad. Nauk SSSR, Tartu 1976. [In R.]

27370 - OYA, V.M., LAĬSK, A.Kh. : Adaptatsiya fotosinteticheskogo apparata k profilyu sveta v liste. [Adaptation of photosynthetic apparatus to the light profile in the leaf.] - Fiziol. Rast. *23* : 445 - 451, 1976. [In R, ab : E.]

27371 - OZIMEK, T., PREJS, A., PREJS, K. : Biomass and distribution of underground of *Potamogeton perfoliatus* L. and *P. lucens* L. in Mikołajskie Lake, Poland. - Aquat. Bot. *2* : 309 - 316, 1976.

27372 - PABIC, C. LE : Modification de la croissance et de la pigmentation de *Spirodela polyrrhiza* (SCHLEID.) induites par une cytokinine, la 6-benzyl-aminopurine. - Physiol. vég. *14* : 77 - 88, 1976.

27373 - PABIC, C. LE : Modifications de la pigmentation et de l'ultrastructure des chloroplastes de *Spirodela polyrrhiza*. - J. Microsc. Biol. cell. *25* : 181 - 186, 1976.

27374 - PACE, G.W., YANG, H.S., TANNENBAUM, S.R., ARCHER, M.C. : Photosynthetic regeneration of ATP using bacterial chromatophores. - Biotechnol. Bioeng. *18* : 1413 - 1423, 1976.

27375 - PACKER, L. : Problems in the stabilization of the *in vitro* photochemical activity of chloroplasts used for H_2 production. - FEBS Lett. *64* : 17 - 22, 1976.

27376 - PAERL, H.W., TILZER, M.M., GOLDMAN, C.R. : Chlorophyll *a* versus adenosine triphosphate as algal biomass indicators in lakes. - J. Phycol. *12* : 242 - 246, 1976.

27377 - PAGE, D.L., COLMAN, B. : The effect of methoxychlor on carbon assimilation by freshwater phytoplankton. - Can. J. Bot. *54* : 2848 - 2851, 1976.

27278 - PAILLOTIN, G. : Capture frequency of excitations and energy transfer between photosynthetic units in the photosystem II. - J. theor. Biol. *58* : 219 - 235, 1976.

27379 - PAILLOTIN, G. : Movement of excitations in the photosynthetic domains of photosystem II. - J. theor. Biol. *58* : 237 - 252, 1976.

27380 - PAIS, M.S.S., NOVAIS, M.C., ABREU, I. : Sur le reverdissement de la spathe de *Zantedeschia aethiopica* au cours de la frutification. I - Hormones du type kinine. II - Effets sur le taux en proteines; taxes photosynthetique et respiratoire. - Portugal. Acta biol., Sér. *A 15* : 1 - 22, 1976.

*27381 - PAKSHINA, E.V., KRASNOVSKIĬ, A.A. : Feofitinizatsiya tsinkovykh i kadmievykh proizvodnykh khlorofilla i ego analogov, deĭstvie sveta. [Pheophytinization of zinc and cadmium derivatives of chlorophyll and its analogues; the action of light.] - Dokl. Akad. Nauk SSSR *224* : 1216 - 1219, 1975. [In R.]

27382 - PALIS, R.K., BUSTRILLOS, A.R. : The effect of limited light on the carbohydrate and protein content of grain sorghum. - Philippine J. Crop Sci. *1* (3) : 161 - 166, 1976. [Ps.]

27383 - PALISANO, J.R., WALNE, P.L. : Light and electron microscopy of two permanently bleached cell lines of *Euglena gracilis (Euglenophyceae)*. - Nova Hedwigia *27* : 455 - 466 1976. [Chloroplast.]

27384 - PALIT, P., KUNDU, A., MANDAL, R.K., SIRCAR, S.M. : Photosynthetic efficiency and productivity of tropical rice. - Plant biochem. J. *3* : 54 - 62, 1976.

27385 - PALIT, P., KUNDU, A., MANDAL, R.K., SIRCAR, S.M. : Growth and yield parameters of two dwarf and two tall varieties of rice under different fertilizer combinations. - Indian J. agr. Sci. *46* : 292 - 299, 1976. [Growth analysis.]

27386 - PALIT, P., KUNDU, A., MANDAL, R.K., SIRCAR, S.M. : Varietal and seasonal differences in growth and yield parameters of dwarf and tall varieties of rice grown with constant doses of fertilizers. - Indian J. agr. Sci. *46* : 327 - 337, 1976. [Chl.]

27387 - PALLETT, K.E., DODGE, A.D. : Experiments into the mechanism of action of the photosynthetic inhibitor herbicide, monuron. - Proc. brit. Crop Protection Conf. - Weeds *13* (1) : 235 - 240, 1976.

27388 - PANDELIEV, S., TSANKOV, B., BRAĬKOV, D. : Vliyanie na svetlinniya rezhim v"rkhu strukturata na listata, fotosintetichnata im aktivnost i stepenta na diferentsiatsiya na zimnite ochi pri lozata. II. S"d"rzhanie na plastidni pigmenti i intenzivnost na fotosintezata na listata pri sort Bolgar v zavisimost ot stepenta na zasenchvane. [Effect of light regime on leaves structure their photosynthetic activity, and degree of differentiation of winter buds in grapevine. II. Content of plastid pigments and photosynthetic rate in leaves of cultivar Bolgar in relation to the amount of shade.]- Gradinar. lozar. Nauka *13* (8) : 97 - 105, 1976. [In Bulg., ab : F, R.]

27389 - PANDEY, R.K., SAXENA, M.C., KALUBARME, M.H., SINGH, V.B., PRASAD, V.V.S.S. : Genotypic variations in photosynthetic rate and respiratory losses in some grain legumes. - Plant biochem. J. *3* : 72 - 80, 1976.

27390 - PANET, R., SANADI, D.R. : Soluble and membrane ATPases of mitochondria, chloroplasts, and bacteria : Molecular structure, enzymatic properties, and functions. - In : BRONNER, F., KLEINZELLER, A. (ed.) : Current Topics in Membranes and Transport. Vol. 8. Pp. 99 - 160. Academic Press, New York 1976.

27391 - PANOVA, M.G., SINESHCHEKOV, V.A., KARAPETYAN, N.V. : Zavisimost' spektra fluorestsentsii khloroplastov ot aktivnosti fotosistemy 2. [The dependence of fluorescence spectra of chloroplasts on the activity of photosystem 2.] - Mol. Biol. (Moskva) *10* : 1175 - 1182, 1976. [In R, ab : E.]

27392 - PANT, A., BHARGAVA, R.M.S., GOSWAMI, S.C. : Nannoplankton, total phytoplankton & zooplankton standing stock measurements in Goa waters. - Indian J. mar. Sci. *5* : 103 - 106, 1976. [Chl.]

27393 - PANT, A., DEVASSY, V.P. : Release of extracellular matter during photosynthesis by a *Trichodesmium* bloom. - Curr. Sci. *45* : 487 - 489, 1976.

27394 - PARAMONAVA, T.K., SHLYK, A.A. : Zmyanenni naboru strukturnykh kampanentaŭ fotasintètychnaĭ membrany pad uplyvam khloramfenikolu. [Changes in the assembly of structural components of a photosynthetic membrane under the effect of chloramphenicol.] - Vestsi Akad. Nauk belorus. SSR, Ser. biyaĭ. Navuk *1976* (6) : 57 - 63, 140, 1976. [In Belorus., ab : R.]

27395 - PARDY, R.L., DIECKMANN, C. : Oxygen consumption in the symbiotic hydra *Hydra viridis*. - J. exp. Zool. *194* : 373 - 378, 1976. [Ps.]

27396 - **PÂRJOL, L., MOGA, I.** : Acțiunea îngrășămintelor asupra principalilor indici
fiziologici și biochimici la raigrasul aristat (*Lolium multiflorum*) și golomăț
(*Dactylis glomerata*) în condiții de irigare. [The influence of fertilizers
on main physiological and biochemical indices of Italian ryegrass (*Lolium
multiflorum*) and orchard grass (*Dactylis glomerata*) under irrigation conditions.] - An. Inst. Cercet. Cereale Plante teh. - Fundulea *41* : 639 - 649,
1976. [Ps, Chl, Car; in Roum., ab : E, R.]

27397 - **PÂRJOL, L., ȘARPE, N.** : Influența erbicidelor asupra soiei în condiții de
irigare. [The effect of herbicides on soybeans under irrigation conditions.]
- An. Inst. Cercet. Cereale Plante teh. - Fundulea *41* : 581 - 591, 1976.
[Ps; in Roum., ab : E, R.]

27398 - **PÂRJOL, L., VARGA, P., MILICĂ, C.I.** : Cercetări asupra rezistenței la secetă
la lucerna. [Studies upon the drought resistance at alfalfa.] - An.Inst. Cercet. Cereale Plante teh. - Fundulea *41* : 617 - 626, 1976. [Ps, Chl, Car;
in Roum., ab : R, E.]

27399 - **PARK, R.B.** : The chloroplast. - In : **BONNER, J., VARNER, J.E.** (ed.) : Plant
Biochemistry. Third Ed. Pp. 115 - 145. Academic Press, New York - San Francisco - London 1976.

27400 - **PARKER, R.R., SIBERT, J.** : Responses of phytoplankton to renewed solar radiation in a stratified inlet. - Water Res. *10* : 123 - 128, 1976.

27401 - **PARKHURST, D.F.** : Effects of *Verbascum thapsus* leaf hairs on heat and mass
transfer : A reassessment. - New Phytol. *76* : 453 - 457, 1976. [Ps.]

27402 - **PÄRNIK, T., KEERBERG, O., VILL, J.** : Estimation of photorespiration in bean
and maize leaves. - Newslett. Appl. nuclear Methods Biol. Agr. *1976* (6) :
5 - 7, 1976.

27403 - **PARRISH, D.J., DAVIES, P.J.** : Light-dependent elongation of anaerobically
maintained green pea stem segments and its implications. - Plant Physiol.
58 : 757 - 760, 1976. [Ps.]

27404 - **PARRISH, J.A.D., BAZZAZ, F.A.** : Underground niche separation in successional
plants. - Ecology *57* ; 1281 - 1288, 1976. [Dry-matter production.]

27405 - **PARSHINA, Z.S.** : Dinamika zelenykh i zheltykh pigmentov v ontogeneze ozimoĭ
pshenitsy. [Dynamics of green and yellow pigments in the ontogenesis of winter wheat.] - In : Fotosintez i Produktivnost' Ozimoĭ Pshenitsy na Yugo-Vostoke Kazakhstana. Pp. 83 - 94, 132. Nauka kaz. SSR, Alma-Ata 1976. [In R.]

27406 - **PARSONS, T.R., LI, W.K.W., WATERS, R.** : Some preliminary observations on the
enhancement of phytoplankton growth by low levels of mineral hydrocarbons. -
Hydrobiologia *51* : 85 - 89, 1976.

27407 - **PARTHIER, B.** : Lichtabhängige Transformation von Proplastiden zu Chloroplasten. - Acta histochem. *17* (Suppl.) : 77 - 93, 1976.

27408 - **PARYS, E., OSTROWSKA, E.** : Działanie regulatorów wzrostu na fotosyntezę, foto-
oddychanie i oddychanie. Cz. I. Fotosynteza. [Action of growth regulators on
photosynthesis, photorespiration and respiration. Part I. Photosynthesis.] -
Wiadom. bot. *20* : 17 - 37, 1976. [In Pol.]

27409 - **PASSIOURA, J.B.** : Physiology of grain yield in wheat growing on stored water. - Aust. J. Plant Physiol. *3* : 559 - 565, 1976. [Ps.]

27410 - **PASSIOURA, J.B.** : The control of water movement through plants. - In :
WARDLAW, I.F., PASSIOURA, J.B. (ed.) : Transport and Transfer Processes in
Plants. Pp. 373 - 380. Academic Press, New York - San Francisco - London
1976. [Stomatal resistance, Ps.]

27411 - **PASTERNAK, D., WILSON, G.L.** : Photosynthesis and transpiration in the heads
of droughted grain sorghum. - Austr. J. exp. Agr. anim. Husb. *16* : 272 - 275,
1976.

27412 - **PATRICK, J.W., WAREING, P.F.** : Auxin-promoted transport of metabolites in
stems of *Phaseolus vulgaris* L. - J. exp. Bot. *27* : 969 - 982, 1976.

27413 - **PATRICK, W.H. Jr., FONTENOT, W.J.** : Growth and mineral composition of rice
at various soil moisture tensions and oxygen levels. - Agron. J. *68* : 325 -
329, 1976. [Production.]

27414 - **PATTERSON, D.T.** : C_4 photosynthesis in smooth pigweed. - Weed Sci. *24* : 127 -
130, 1976.

27415 - **PATTERSON, D.T., BUNCE, J.A., PEET, M.M., KRAMER, P.J.** : Analysis of the
growth and variability of cotton and soybean in controlled and field environments. - Plant Physiol. *57* (Suppl.) : 104, 1976.

27416 - **PAUL, J.S., VOLCANI, B.E.** : Photorespiration in diatoms. IV. Two pathways of
glycolate metabolism in synchronized cultures of *Cylindrotheca fusiformis*. -
Arch. Microbiol. *110* : 247 - 252, 1976.

27417 - **PAUL, R., IMPENS, R., THUNIS, M.** : Mesure du métabolisme des plantes à l'aide
d'un analyseur de CO_2 de type U.R.A.S. 1. - Bull. Rech. agron. Gembloux *11* :
259 - 264, 1976. [Ps.]

27418 - **PAVLETIĆ, Z., MATONIČKIN, I., STILINOVIĆ, B., HABDIJA, I.** : A method for determining the P/R ratio in running waters. - Hydrobiologia *48* : 51 - 57, 1976.

27419 - **PAYNE, W.J., WILLIAMS, M.L.** : Carbon assimilation from simple and complex
media by prototrophic heterotrophic bacteria. - Biotechnol. Bioeng. *18* :
1653 - 1655, 1976.

27420 - **PEACH, C., RYAN, F.J., McCURRY, S.D., TOLBERT, N.E.** : Reaction of ribulose
bisphosphate carboxylase with pyridoxal-5'- phosphate. - Plant Physiol. *57*
(Suppl.) : 54, 1976.

27421 - **PEARCE, R.B., CROSSBIE, T.M., MOCK, J.J.** : A rapid method for measuring net
photosynthesis of excised leaves by using air-sealed chambers. - Iowa State
J. Res. *51* : 25 - 33, 1976.

27422 - **PEET, M.M., OZBUN, J.L.** : Physiological and anatomical effects of growth
temperature on *Phaseolus vulgaris* L. - Plant Physiol. *57* (Suppl.) : 105, 1976

27423 - **PEISER, G., YANG, S.F.** : Chlorophyll destruction in the presence of bisulfite
and linoleic acid hydroperoxide. - Plant Physiol. *57* (Suppl.) : 47, 1976.

27424 - **PEISKER, M.** : Ein Modell der Sauerstoffabhängigkeit des photosynthetischen
CO_2-Gaswechsels von C_3-Pflanzen.   - Kulturpflanze *24* : 221 - 235, 1976.

27425 - **PEISKER, M., APEL, P.** : Influence of oxygen on photosynthesis and photorespiration in leaves of *Triticum aestivum* L. 2. Response of CO_2 gas exchange to
oxygen at various leaf ages and its variability. - Photosynthetica *10* : 140 -
146, 1976.

27426 - **PELROY, R.A., KIRK, M.R., BASSHAM, J.A.** : Photosystem II regulation of macro-
molecule synthesis in the blue-green alga *Aphanocapsa* 6714. - J. Bacteriol.
128 : 623 - 632, 1976.

27427 - **PELROY, R.A., LEVINE, G.A., BASSHAM, J.A.** : Kinetics of light-dark CO_2 fixation and glucose assimilation by *Aphanocapsa* 6714. - J. Bacteriol. *128* :
633 - 643, 1976.

27428 - **PEMADASA, M.A., LOVELL, P.H.** : Effects of the timing of the life-cycle on the
vegetative growth of some dune annuals. - J. Ecol. *64* : 213 - 222, 1976.
[Growth analysis.]

27429 - **PENEDA-SARAIVA, M.C.** : L' utilisation d'une algue nanoplanctonique comme or-
ganisme-test en molysmologie marine. Quelques réponses de *Dunaliella biocula-
ta* a l'irradiation gamma et a la contamination par le chrome et le cadmium. -
Rev. int. Océanogr. méd. *43* : 111 - 115, 1976. [Ps, Chl.]

27430 - **PENNY, M.G., MOORE, K.G., LOVELL, P.H.** : The effects of inhibition of cotyledon photosynthesis on seedling development in *Cucumis sativus* L. - Ann. Bot.
40 : 815 - 824, 1976.

27431 - **PEREIRA, J.F., SPLITTSTOESSER, W.E.:**Physiological studies of cassava (*Manihot esculenta* CRANTZ). - Plant Physiol. *57* (Suppl.) : 6, 1976.

27432 - **PEREIRA, J.S., KOZLOWSKI, T.T.** : Diurnal and seasonal changes in water balance of *Abies balsamea* and *Pinus resinosa*. - Oecol. Plant. *11* : 397 - 412, 1976.
[Resistances.]

27433 - **PEREIRA, J.S., KOZLOWSKI, T.T.** : Leaf anatomy and water relations of *Eucalyptus camaldulensis* and *E. globosus* seedlings. - Can. J. Bot. *54* : 2868 - 2880, 1976. [Resistances.]

27434 - **PERRIER, A., ITIER, B., BERTOLINI, J.M., KATERJI, N.** : A new device for continuous recording of the energy balance of natural surfaces. - Agr. Meteorol. *16* : 71 - 84, 1976.

27435 - **PERSANOV, V.M., VORONOVA, E.A., KARPILOV, Yu.S.** : Issledovanie svoĭstv NADP-malatdegidrogenazy iz list'ev kukuruzy. [Properties of NADP-malic enzyme from maize leaves.] - Biokhimiya *41* : 1014 - 1022, 1976. [In R, ab : E.]

27436 - **PERSON, P.** : A leaf disc mounting method for measurement of net photosynthesis under field conditions. - Photosynthetica *10* : 264 - 265, 1976.

27437 - **PEŠEK, J.** : Růstové funkce, alometrie a modely růstu ve fyziologii rostlin. [Growth functions, allometry, and growth models in plant physiology.] - Genet. Šlechtění (Praha) *12* (2, Suppl.) : I - XV, 1976. [Growth analysis; in Czech, ab : E,G,R.]

27438 - **PESHEKHOD'KO, V.M., TITLYANOV, É.A., BELIKOV, I.F.** : Fotosintez i prizhiznennoe vydelenie organicheskikh veshchestv tallomami nekotorykh morskikh prikreplennykh vodorosleĭ. [Photosynthesis and metabolic liberation of organic substances by thallomes of attached marine algae.] - Fiziol. Rast. *23* : 1047 - 1051, 1976. [In R, ab : E.]

27439 - **PESTEMER, W.** : Quantitativer Biotest zur Bestimmung von Photosynthesehemmern im Boden. - Weed Res. *16* : 357 - 363, 1976.

27440 - **PETERS, G.A., EVANS, W.R., TOIA, R.E. Jr.** : *Azolla-Anabaena azollae* relationship. IV. Photosynthetically driven, nitrogenase-catalyzed H_2 production. - Plant Physiol. *58* : 119 - 126, 1976.

27441 - **PETERS, J., PETERS, R., STOECKENIUS, W.** : A photosensitive product of sodium borohydride reduction of bacteriorhodopsin. - FEBS Lett. *61* : 128 - 134, 1976.

27442 - **PETERSON, É.K., IRBE, I.K.** : Vliyanie doposevnoĭ obrabotki semyan khlorkholinkhloridom na rost i zimostoĭkost' ozimoĭ pshenitsy. [Influence of pre-sowing treatment of seeds by chlorcholine chloride on the growth and cold resistance in wheat.] - In : Regulyatsiya Rosta i Pitanie Rasteniĭ. Pp. 36 - 43. Zinatne, Riga 1976. [In R.]

27443 - **PETRENKO, A.V.** : Formirovanie plastidnogo kompleksa list'ev pri predposevnoĭ induktsii prorastaniya semyan geteroauksinom. [Formation of leaf plastid complex following the presowing induction of seed germination with heteroauxin.] - In : KAKHNOVICH, L.V. (ed.) : Optimizatsiya Fotosinteticheskogo Apparata Vozdeĭstviem Razlichnykh Faktorov. Pp. 49 - 56. Izd. BGU, Minsk 1976. [In R.]

27444 - **PETRENKO, A.V., BIRYUKOVA, E.P.** : Uroven' soderzhaniya fotosinteziruyushchikh pigmentov v list'yakh kukuruzy pod vliyaniem ékzogennykh gibberellina i kinetina. [Level of photosynthesizing pigments in the leaves of corn plants under the effect of exogenous gibberellin and kinetin.] - In : KAKHNOVICH, L.V. (ed.) : Optimizatsiya Fotosinteticheskogo Apparata Vozdeĭstviem Razlichnykh Faktorov. Pp. 40 - 49. Izd. BGU, Minsk 1976. [In R.]

27445 - **PETROV, A., MANOLOV, P.** : Pridvizhvane na belyazaniya v"glerod v razklonenite endogodishni prirasti na praskovata sled asimilatsiyata na $^{14}CO_2$ ot otdelnite p"rvichni lista i vtorichni letorasti. [Translocation of labelled carbon in ramified peach shoots followig $^{14}CO_2$ assimilation by single primary leaves and secondary shoots.] - Fiziol. Rast. (Sofia) *2* (2) : 39 - 48, 1976. [In Bulg., ab : R, E.]

27446 - **PETROVA, V.N.** : Abstsizovaya kislota - gormon rasteniĭ. [Abscisic acid - the hormone of plants.] Bot. Zh. *61* : 1004 - 1016, 1976. [Chl; in R.]

27447 - **PETTY, K.M., DUTTON, P.L.** : Properties of the flash-induced proton binding encountered in membranes of *Rhodopseudomonas sphaeroides* : A functional pK on the ubisemiquinone ? - Arch. Biochem. Biophys. *172* : 335 - 345, 1976.

27448 - **PETTY, K.M., DUTTON, P.L.** : Ubiquinone-cytochrome *b* electron and proton transfer : a functional pK on cytochrome b_{50} *Rhodopseudomonas sphaeroides* membranes. - Arch. Biochem. Biophys. *172* : 346 - 353, 1976.

27449 - PETTY, K.M., DUTTON, P.L., PRINCE, R.C. : Multiple physical-chemical roles of ubiquinones in photosynthetic electron and proton transfer. - Plant Physiol. *57* (Suppl.) : 15, 1976.

27450 - PEYRIÈRE, M. : Étude infrastructurale du plastidome de deux Rhodophycées parasites *Harveyella mirabilis* (REINSCH) SCHMITZ et REINKE et *Holmsella pachyderma* (REINSCH) STURCH. - Compt. rend. Acad. Sci. Paris, Sér. D *283* : 1169 - 1171, 1976.

27451 - PHILIPPOVICH, I.I., NOZDRINA, V.N., BEZSMERTNAYA, I.N., OPARIN, A.I. : Polyribosomes bound to chloroplast membranes (localization, relation to the formation of thylakoids of the grana). - Acta histochem. *17* (Suppl.) : 141 - 152, 1976.

27452 - PHUNG NHU HUNG, S., HOULIER, B., MOYSE, A. : Mn content in wheat etioplast membranes during greening under intermittent or continuous light. - Plant Sci. Lett. *6* : 243 - 251, 1976.

27453 - PICCIONI, R.G., MAUZERALL, D.C. : Increase effected by calcium ion in the rate of oxygen evolution from preparation of *Phormidium luridum*. - Biochim. biophys. Acta *423* : 605 - 609, 1976.

27454 - PICK, U., AVRON, M. : A method for measuring the internal pH in illuminated chloroplasts based on the stimulation of proton uptake by amines. -Europe. J. Biochem. *70* : 569 - 576, 1976.

27455 - PICK, U., AVRON, M. : Measurement of transmembrane potentials in *Rhodospirillum rubrum* chromatophores with an oxacarbocyanine dye. - Biochim. biophys. Acta *440* : 189 - 204, 1976.

27456 - PICK, U., AVRON, M. : Neutral red response as a measure of the pH gradient across chloroplast membranes in the light. - FEBS Lett. *65* : 348 - 353, 1976.

27457 - PICKETT, S.T.A., BAZZAZ, F.A. : Divergence of two co-occurring successional annuals on a soil moisture gradient. - Ecology *57* : 169 - 176, 1976. [Growth analysis.]

*27458 - PILARSKI, J. : The application of infra-red gas analyser for measurements of gas exchange of aquatic plants. - Acta hydrobiol. *15* : 275 - 293, 1973.

27459 - PINGREE, R.D., HOLLIGAN, P.M., MARDELL, G.T., HEAD, R.N. : The influence of physical stability on spring, summer and autumn phytoplankton blooms in the Celtic Sea. - J. mar. biol. Assoc. U. K. *56* : 845 - 873, 1976. [Chl.]

27460 - PLANCHON, C. : Essai de détermination de critères physiologiques en vue de l'amélioration du blé tendre : les facteurs de la photosynthèse de la dernière feuille. - Ann. Amélior. Plant. *26* : 717 - 744, 1976.

27461 - PLATT, S.G., PLAUT, Z., BASSHAM, J.A. : Analysis of steady state photosynthesis in alfalfa leaves. - Plant Physiol. *57* : 69 - 73, 1976.

27462 - PLATT, S.G., PLAUT, Z., BASSHAM, J.A. : Perturbed steady state alfalfa photosynthesis : origins of glycine and serine. - Plant Physiol. *57* (Suppl.) : 60, 1976.

27463 - PLATT, T., JASSBY, A.D. : The relationship between photosynthesis and light for natural assemblages of coastal marine phytoplankton. - J. Phycol. *12* : 421 - 430, 1976.

27464 - PLAUT, Z., PLATT, S., BASSHAM, J.A. : Nitrate and ammonium regulation of carbon metabolism in photosynthesizing alfalfa leaf discs. - Plant Physiol. *57* (Suppl.) : 58, 1976.

27465 - PLESNIČAR, M., BOGDANOVIĆ, M. : Fotohemijske aktivnosti hloroplasta crnog bora. [Photochemical activities of the black pine chloroplasts.] - Acta bot. croat. *35* : 71 - 75, 1976. [In Croat., ab : E.]

27466 - POCHINOK, Kh.N., OKANENKO, A.S., GOLIK, K.N., POGOL'SKAYA, V.I. : Vliyanie khlorkholinkhlorida na intensivnost' fotosinteza, urozhaĭ i sakharistost' sakharnoĭ svekly. [Effect of chlorcholine chloride on photosynthesis rate, yield and sugar content of sugar beet.] - Fiziol. Biokhim. kul't. Rast. *8* : 273 - 279, 1976. [In R, ab : E.]

27467 - **POHLHEIM, F., POHLHEIM, E.** : Herstellung von Plastommutanten bei *Saintpaulia ionaatha* H. WENDL. - Biochem. Physiol.Pflanzen *169* : 377 - 383, 1976.

27468 - **POINCELOT, R.P.** : Lipid and fatty acid composition of chloroplast envelope membranes from species with differing net photosynthesis. - Plant Physiol. *58* : 595 - 598, 1976.

27469 - **POINCELOT, R.P., DAY, P.R.** : Isolation and bicarbonate transport of chloroplast envelope membranes from species of differing net photosynthetic efficiency. - Plant Physiol. *57* : 334 - 338, 1976.

27470 - **POKROVSKAYA, T.N.** : Ėkologicheskie usloviya fotosinteza litoral'nykh gidrofitov. [Ecological factors of photosynthesis of litoral hydrophytes.] - In : Antropogennoe Evtrofirovanie Ozer. Pp. 17 - 44, 119. Nauka, Moskva 1976. [In R.]

27471 - **POLEVAYA, V.S., SMOLOV, A.P., IGNAT'EV, A.R.** : Fotosinteticheskaya aktivnost' kul'tur tkaneĭ rasteniĭ. [Photosynthetic activity of plant tissue cultures.] - In : Itogi Issledovaniya Mekhanizma Fotosinteza. Pp. 71 - 74. Pushchino 1976. [In R.]

27472 - **POLING, S.M., HSU, W.-J., YOKOYAMA, H.** : Synthetic regulators of carotenoid biosynthesis in *Citrus paradisi.* - Phytochemistry *15* : 1685 - 1687, 1976.

27473 - **POLLOCK, C.J.** : Changes in the activity of sucrose-synthesizing enzymes in developing leaves of *Lolium temulentum.* - Plant Sci. Lett. *7* : 27 - 31, 1976. [Chl.]

27474 - **POLONSKIĬ, V.I.** : Vegetatsionnyĭ shkaf dlya opytov s rasteniyami pri vysokikh intensivnostyakh FAR. [Vegetation chamber for experiments with plants at high PhAR flux densities.] - In : Osnashchenie Selektsionnykh Tsentrov Svetotekhnicheskim Oborudovaniem. (Materialy I Vsesoyuznogo Nauchno-Tekhnicheskogo Soveshachaniya. Shortandy, 1975.) Pp. 53 - 54. Ėlektronika, Moskva 1976. [In R.]

27475 - **POLYAKOV, M.A.** : Gazometricheskoe issledovanie deĭstviya kachestva sveta na CO_2-gazoobmen vysshikh nadzemnykh rasteniĭ. [Gasometric study of the effect of light quality on CO_2 exchange in higher terrestrial plants.] - In : Gazometricheskoe Issledovanie Fotosinteza i Dykhaniya Rasteniĭ. Pp. 123 - 125. Akad. Nauk SSSR, Tartu 1976. [In R.]

27476 - **POMEROY, W.M., STOCKNER, J.G.** : Effects of environmental disturbance on the distribution and primary production of benthic algae on a British Columbia estuary. - J. Fish. Res. Board Can. *33* : 1175 - 1187, 1976.

27477 - **PONOMAREVA, R.P., ZUEV, N.D., KAL'CHENKO, V.A.** : Radiochuvstvitel'nost' fotosinteticheskikh sistem u sel'skokhozyaĭstvennykh rasteniĭ. [Radiosensitivity of the photosynthetic system of agricultural plants.] - Radiobiologiya *16* : 678 - 682, 1976. [In R, ab : E.]

27478 - **POOLE, R.J.** : Transport in cells of storage tissues. - In : LÜTTGE, U., PITMAN, M.G. (ed.) : Transport in Plants II. Part A. Cells. Pp. 229 - 248. Springer-Verlag, Berlin - Heidelberg - New York 1976. [Photosynthates.]

27479 - **POOVAIAH, B.W., LEOPOLD, A.C.** : Effects of inorganic salts on tissue permeability. - Plant Physiol. *58* : 182 - 185, 1976. [Chl.]

27480 - **POPESCU, V., TĂMAŞ, V., ALBU, M.** : La productivité et la valeur fourragère des certaines espèces de graminées vivaces. Note II. Sur le contenu en provitamine A. - Bul. Inst. agron. Cluj-Napoca *30* : 23 - 26, 1976. [Car.]

27481 - **POPOV, V.I., ALLAKHVERDOV, B.L., TAGEEVA, S.V.** : Issledovanie nadmolekulyarnoĭ organizatsii fotosinteticheskikh membran metodom zamorazhivaniya-skalyvaniya. [Investigation of supermolecular organization of photosynthetic membranes by the method of freeze-fracturing.] - Dokl. Akad. Nauk SSSR *231* : 1476 - 1478, 1976. [In R.]

27482 - **POPOV, V.I., SARANCHA, Yu.P., TAGEEVA, S.V., MAKAROV, A.D.** : Rol' sopryagayushchego faktora v formirovanii gran khloroplastov. [Role of coupling factor in the formation of chloroplast grana.] - Dokl. Akad. Nauk SSSR *229* : 1244 - 1247, 1976. [In R.]

27483 - **POPOV, V.I., TAGEEVA, S.V.** : Novye perspektivy primeneniya morfometrii dlya èlektronno-mikroskopicheskogo analiza nadmolekulyarnoĭ organizatsii membran, vyyavlyaemoĭ metodom zamorazhivaniya-skalyvaniya. [New perspectives of using morphometry for electron-microscopic analysis of supermolecular organization of membranes, studied by means of freeze-fracturing.] - In :.Simpozium. Metody Podgotovki Slozhnykh Ob"ektov i Analiz Èlektronno-Mikroskopicheskikh Izobrazheniĭ. Tezisy Dokladov. Pp. 162 - 163. Petrozavodsk 1976. [In R.]

*27484 - **POPOVA, N.M., POPOV, V.K.** : O nakoplenii khlorofilla v khvoe sosny v smeshchannykh sosnovo-berezovykh kul'turakh. [Chlorophyll accumulation in pine needles of mixed pine and birch cultures.] - In : Lesnaya Geobotanika i Biologiya Drevesnykh Rasteniĭ. Vol. 3. Pp. 118 - 123. Priokskoe knizh. izd., bryansk. Otd., Bryansk 1975. [In R, ab : E.]

27485 - **POPOVIĆ, Ž.** : Rad atomskih centara u Saclay-u i Cadarache-u u oblasti agronomije i fotosinteze. [Research of the atomic centres at Saclay and Cadarache in the field of agronomy and photosynthesis.] - Agrohemija *5-6* : 163 - 180, 1976. [In Croat.]

27486 - **PORAT, N., BEN-SHAUL, Y., FRIEDBERG, I.** : Membrane-bound ATPase in chloroplasts of *Euglena gracilis*. - Biochim. biophys. Acta *440* : 365 - 376, 1976.

27487 - **POROKHNEVICH, N.V.** : Sovremennye dannye o roli tsinka v strukturnoĭ organizatsii fotosinteticheskogo apparata, metabolizme plastidnykh pigmentov i reaktsiyakh fotosinteza. [Present data on the role of zinc in the structural organization of the photosynthetic apparatus, metabolism of plastid pigments and reactions of photosynthesis.] - In : **KAKHNOVICH, L.V.** (ed.) : Optimizatsiya Fotosinteticheskogo Apparata Vozdeĭstviem Razlichnykh Faktorov. Pp. 101 - 124. Izd. BGU, Minsk 1976. [In R.]

27488 - **POROKHNEVICH, N.V., IVANOV, N.P.** : Vliyanie tsinka na strukturnye i biokhimicheskie pokazateli plastidnogo apparata list'ev i urozhaĭ yachmenya na fone vysokoĭ dozy kaliĭnoĭ soli. [Effect of zinc on the structural and biochemical characteristics of leaf plastid apparatus and yield of barley under a high dose of potassium salt.] - Vestn. beloruss. gos. Univ. V.I.Lenina, Ser.II, *1976* (1) : 34 - 37, 1976. [In R.]

27489 - **POROKHNEVICH, N.V., VAKUL'CHIK, M.N.** : Mikrostruktura plastidnogo apparata lista, nakoplenie fotosinteticheskikh pigmentov i produktivnost' yachmenya v usloviyakh posledeĭstviya tsinka v pochve. [Microstructure of the plastid apparatus of leaves, accumulation of photosynthetic pigments and productivity of barley as an after-effect of zinc in the soil.] - Agrokhimiya *1976* (7) : 106 - 109, 1976. [In R.]

27490 - **PORTER, J., JOST, M.** : Physiological effects of the presence and absence of gas vacuoles in the blue-green alga, *Microcystis aeruginosa* KUETZ. emend. ELENKIN. - Arch. Microbiol. *110* : 225 - 231, 1976.

27491 - **PORTIS, A.R., McCARTY, R.E.** : Quantitative relationships between phosphorylation, electron flow, and internal hydrogen ion concentrations in spinach chloroplasts. - J. biol. Chem. *251* : 1610 - 1617, 1976.

27492 - **PORTIS, A.R. Jr., HELDT, H.W.** : Light-dependent changes of the Mg^{2+} concentration in the stroma in relation to the Mg^{2+} dependency of CO_2 fixation in intact chloroplasts. - Biochim. biophys. Acta *449* : 434 - 446, 1976.

27493 - **POSKUTA, J.** : Lichtatmung als Faktor der photosynthetischen Produktivität bei Getreide. - Wiss. Z. Humboldt-Univ. Berlin, math.-nat. Reihe *25* : 742 - 746, 1976.

27494 - **POSKUTA, J., WRÓBLEWSKA, B., MIKULSKA, M.** : $^{14}CO_2$ uptake and photosynthetic products of bean leaves after 0,5 or 10 hours pretreatment of plants with 100 % O_2 in light. - Z. Pflanzenphysiol. *78* : 396 - 402, 1976.

27495 - **POSORSKE, L., JAGENDORF, A.T.** : Nucleotide·and metal interactions affecting inactivation of spinach chloroplast coupling factor by NaCl in the cold. - Arch. Biochem. Biophys. *177* : 276 - 283, 1976.

27496 - **POSPÍŠILOVÁ, J., ZIMA, J., ŠESTÁK, Z.** : Effect of hydration level in primary bean leaves on the activity of photosystems 1 and 2 in isolated chloroplasts. - Biol. Plant. *18* : 473 - 479, 1976.

27497 - **POSSINGHAM, J.V.** : Controls to chloroplast division in higher plants. -
J. Microsc. Biol. cell. *25* : 283 - 288, 1976.

27498 - **POSSINGHAM, J.V., ROSE, R.J.** : Studies of the growth and replication of spinach chloroplasts and on the location and segregation of their DNA. - In :
BÜCHER, T., NEUPERT, W., SEBALD, W., WERNER, S. (ed.) : Genetics and Biogenesis of Chloroplasts and Mitochondria. Pp. 387 - 390. North-Holland Publ. Co.,
Amsterdam - New York - Oxford 1976.

27499 - **POSSINGHAM, J.V., ROSE, R.J.** : Chloroplast replication and chloroplast DNA
synthesis in spinach leaves. - Proc. roy. Soc. London, Ser. B *193* : 295 -305,
1976.

27500 - **POSTIUS, C., KLEMME, B., JACOBI, G.** : Dark starvation and plant metabolism.
V. Comparative studies on the alteration of enzyme activities during dark
starvation and senescence. - Z. Pflanzenphysiol. *78* : 122 - 132, 1976. [Chl.]

27501 - **POSTON, F.L., PEDIGO, L.P., PEARCE, R.B., HAMMOND, R.B.** :Effects of artificial and insect defoliation on soybean net photosynthesis. - J. econ. Entomol.
69 : 109 - 112, 1976.

27502 - **POTTER, J.R.** : Oxygen effect in partitioning of photosynthate into new leaf
area. - Plant Physiol. *57* (Suppl.) : 59, 1976.

27503 - **POULSEN, C., STRØBAEK, S., HASLETT, B.G.** : Studies on the primary structure
of the small subunit of ribulose-1,5-diphosphate carboxylase. - In : **BÜCHER,
T., NEUPERT, W., SEBALD, W., WERNER, S.** (ed.) : Genetics and Biogenesis of
Chloroplasts and Mitochondria. Pp. 17 - 24. North-Holland Publ. Co., Amsterdam - New York - Oxford 1976.

27504 - **PRADEL, J., CLÉMENT-METRAL, J.** : Contrôle du métabolisme par l'ATP chez *Rhodopseudomonas spheroides*. - Compt. rend. Acad. Sci. Paris, Sér. D *282* : 1449
- 1451, 1976.

27505 - **PRADEL, J., CLÉMENT-METRAL, J.D.** : A 4-vinylprotochlorophyllide complex
accumulated by "Phofil" mutant of *Rhodopseudomonas spheroides*. An authentic
intermediate in the development of the photosynthetic apparatus. - Biochim.
biophys. Acta *430* : 253 - 264, 1976.

27506 - **PRAKASH, N., SHEN, T.C., YAP, K.C., YIM, K.M.** : A survey of the leaf structure and its relationship to the photosynthetic pathways in certain Malaysian
plants. - Malays. J. Sci. *4* (A) : 67 - 73, 1976.

27507 - **PRÉCSÉNYI, I., CZIMBER, Gy., CSALA, G., SZÖCS, Z., MOLNÁR, E., MELKÖ, E.** :
Studies on the growth analysis of maize hybrids (OSSK-218 and DKXL-342). -
Acta bot. Acad. Sci. hung. *22* : 185 - 200, 1976.

27508 - **PREIL, W.** : Zur Beziehung zwischen dem Chlorophyllgehalt der Blätter und dem
Fruchtertrag bei Tomaten während der Gewächshaus-Winterkultur. - Gartenbauwissenschaft *41* : 248 - 252, 1976.

27509 - **PREISS, J., KOSUGE, T.** : Regulation of enzyme activity in metabolic pathways.-
In : **BONNER, J., VARNER, J.E.** (ed.) : Plant Biochemistry. Third Edition.
Pp. 277 - 336. Academic Press, New York - San Francisco - London 1976.[Ps.]

27510 - **PREMACHANDRA, B.R., VASANTHARAJAN, V.N., CAMA, H.R.** : Improvement of the nu-
tritive value of tomatoes. - Curr. Sci. *45* : 56 - 57, 1976. [Car.]

27511 - **PRÉZELIN, B.B.** : The role of peridinin-chlorophyll *a*-proteins in the photosynthetic light adaption   of the marine dinoflagellate, *Glenodinium* sp. -
Planta *130* : 225 - 233, 1976.

27512 - **PRÉZELIN, B.B., HAXO, F.T.** : Purification and characterization of peridinin-
-chlorophyll *a*-proteins from the marine dinoflagellates *Glenodinium* sp. and
Gonyaulax polyedra. - Planta *128* : 133 - 141, 1976.

27513 - **PRÉZELIN, B.B., LEY, A.C., HAXO, F.T.** : Effects of growth irradiance on the
photosynthetic action spectra of the marine dinoflagellate, *Glenodinium* sp. -
Planta *130* : 251 - 256, 1976.

27514 - **PRIESTLEY, C.A., CATLIN, P.B., OLSSON, E.A.** : The distribution of ^{14}C-labelled
assimilates in young apple trees as influenced by doses of supplementary nitrogen. I. Total ^{14}C radioactivity in extracts. - Ann. Bot. *40* : 1163 - 1170,
1976.

27515 - PRIESTLEY, C.A., CATLIN, P.B., OLSSON, E.A. : The distribution of ^{14}C-label-
led assimilates in young apple trees as influenced by doses of supplementary
nitrogen. II. Soluble carbohydrates and amino acids. - Ann. Bot. *40* : 1171 -
1176, 1976.

27516 - PRINCE, R.C., DUTTON, P.L. : Further studies on the Rieske iron-sulfur center in mitochondrial and photosynthetic systems : a pK on the oxidized form.
- FEBS Lett. *65* : 117 - 119, 1976.

27517 - PRINCE, R.C., DUTTON, P.L. : The primary acceptor of bacterial photosynthesis :
its operating midpoint potential ? - Arch. Biochem. Biophys. *172* : 329 -
334, 1976.

27518 - PRINCE, R.C., LEIGH, J.S. Jr., DUTTON, P.L. : Thermodynamic properties of
the reaction center of *Rhodopseudomonas viridis* : *in vivo* measurement of
the reaction center bacteriochlorophyll-primary acceptor intermediary electron carrier. - Biochim. biophys. Acta *440* : 622 - 636, 1976.

27519 - PRINCE, R.C., OLSON, J.M. : Some thermodynamic and kinetic properties of the
primary photochemical reactants in a complex from a green photosynthetic bacterium. - Biochim. biophys. Acta *423* : 357 - 362, 1976.

27520 - PROCTOR, J.T.A., LOUGHEED, E.C. : The effect of covering apples during development. - HortScience *11* : 108 - 109, 1976. [Chl.]

27521 - PROCTOR, J.T.A., WATSON, R.L., LANDSBERG, J.J. : The carbon budget of a young
apple tree. - J. amer. Soc. hort. Sci. *101* : 579 - 582, 1976.

27522 - PROKHORENKO, I.R., LOBACHEV, V.M., KUTYURIN, V.M. : K voprosu ob opticheskoĭ
aktivnosti khlorofilla. [Optical activity of chlorophyll.] - Zh. obshch. Khi-
mii *46* : 2147 - 2151, 1976. [In R, ab : E.]

27523 - PRONINA, N.B., MAKAROV, A.D., STAKHOV, L.F., KHOLODENKO, N.Ya., KALASHNIKOV,
Yu.E., BELONOG, N.P. : Ob uchastii nekotorykh nizkomolekulyarnykh kofaktorov
v fotofosforilirovanii. [Participation of some low-molecular cofactors in
photophosphorylation.] - Izv. Akad. Nauk SSSR, Ser. biol. *1976* : 520 - 529,
1976. [In R, ab : E.]

27524 - PRZYBYLLA, K.R., FOCK, H. : Der Weg des Kohlenstoffs während der CO_2-Frei-
setzung im Licht und Dunkeln bei Sonnenblumen. - Ber. deut. bot. Ges. *89* :
651 - 661, 1976.

27525 - PUCHEU, N.L., KERBER, N.L., GARCIA, A.F. : Isolation and purification of reaction center from *Rhodopseudomonas viridis* NHTC 133 by means of LDAO. -
Arch. Microbiol. *109* : 301 - 305, 1976.

27526 - PUECH, A.A., REBEIZ, C.A., CRANE, J.C. : Characterization of major plastid
pigments in skin of "Mission" fig fruits. - J. amer. Soc. hort. Sci. *101* :
392 - 394, 1976.

27527 - PUECH, A.A., REBEIZ, C.A., CRANE, J.C. : Pigment changes associated with application of ethephon ((2-chloroethyl phosphonic acid) to fig (*Ficus carica*
L.) fruits. - Plant Physiol. *57* : 504 - 509, 1976.

27528 - PULLES, M.P.J., VAN GORKOM, H.J., VERSCHOOR, G.A.M. : Primary reactions of
photosystem II at low pH. 2. Light-induced changes of absorbance and electron spin resonance in spinach chloroplasts. - Biochim. biophys. Acta *440* :
98 - 106, 1976.

27529 - PULLES, M.P.J., VAN GORKOM, H.J., WILLEMSEN, J.G. : Absorbance changes due to
the charge-accumulating species in System 2 of photosynthesis. - Biochim.
biophys. Acta *449* : 536 - 540, 1976.

27530 - PUNNETT, T., KELLY, J.H. : Environmental control over C_3 and C_4 photosynthesis in vascular plants. - Plant Physiol. *57* (Suppl.) : 59, 1976.

27531 - PUROHIT, K., McFADDEN, B.A. : Heterogeneity of large subunits of ribulose-
-1,5-bisphosphate carboxylase from *Hydrogenomonas eutropha*. - Biochem. biophys. Res. Commun. *71* : 1220 - 1227, 1976.

27532 - PUSHKINA, G.P., STARIKOVA, V.T. : Vliyanie regulyatorov i ingibitorov rosta
na protsess fotosinteticheskogo fosforilirovaniya v khloroplastakh, vydelennykh iz list'ev kukuruzy. [Effect of regulators and inhibitors of growth on

photosynthetic phosphorylation in chloroplasts isolated from maize leaves.] -
Fiziol. Rast. 23 : 202 - 203, 1976. [In R.]

27533 - PYARNIK, T., KĖĖRBERG, O., VIĬL, Yu. : Izmerenie vnutrilistovykh potokov CO_2
pri fotosinteze radioizotopnym metodom. [Determination of CO_2 fluxes inside
the leaf during photosynthesis by means of radioisotope method.] - In : Gazometricheskoe Issledovanie Fotosinteza i Dykhaniya Rasteniĭ. Pp. 126 - 128.
Akad. Nauk SSSR, Tartu 1976. [In R.]

27534 - PYRINA, I.L., RUTKOVSKAYA, V.A. : Zavisimost' intensivnosti fotosinteza vol-
zhskogo fitoplanktona ot pronikayushcheĭ v vodu summarnoĭ solnechnoĭ radiatsii. [Dependence of photosynthetic rate of the Volga phytoplankton on the
sum of solar radiation penetrating through water.] - In : Biologicheskie
Produktsionnye Protsessy v Basseĭne Volgi. Pp. 48 - 60. Nauka, Leningrad
1976. [In R.]

27535 - PYT'EVA, N.F., ANDREENKO, T.I., RUBIN, A.B. : Existence of cyclic and non-
-cyclic electron transport systems in bacterial photosynthetic apparatus.
2. Experimental evidence for the scheme of electron-transport reactions in
the chromatophores of photosynthesizing bacteria *Ectothiorhodospira shaposhnikovii* . - Photosynthetica 10 : 14 - 19, 1976.

27536 - PYT'EVA, N.F., KONONENKO, A.A., RATYNI, A.I., LUKASHEV, E.P., RUBIN, A.B. :
Teoreticheskoe issledovanie kinetiki temnovogo vosstanovleniya fotokhimiches-
ki okislennogo bakteriokhlorofilla reaktsionnogo tsentra v khromatoforakh
Rhodospirillum rubrum. [Theoretical study of the kinetics of dark reduction
of photochemically oxidized bacteriochlorophyll of the reaction centre in
chromatophores of *Rhodospirillum rubrum*.] - Biofizika 21 : 118 - 123, 1976.
[In R, ab : E.]

27537 - PYT'EVA, N.F., RATYNI, A.I., RUBIN, A.B. : Existence of cyclic and non-cyclic
electron transport systems in bacterial photosynthetic apparatus. 1. Theoretical differences in kinetic characteristics of cyclic and non-cyclic systems
of electron transport. - Photosynthetica 10 : 7 - 13, 1976.

27538 - QUAST, P. : Gibberellinbestimmung in Verbindung mit Kohlenhydratgehalten und
Trockensubstanzverteilung bei *Solanaceae*. - Angew. Bot. 50 : 311 - 313, 1976.

27539 - QUEBEDEAUX, B., CHOLLET, R. : Comparative growth analyses of *Panicum milioides*, *P. bisulcatum* and *P. miliaceum* determined at altered pO_2 and pCO_2. -
Plant Physiol. 57 (Suppl.) : 59, 1976.

27540 - QUEBEDEAUX, B., HARDY, R.W.F. : Oxygen concentration : regulation of crop
growth and productivity. - In : BURRIS, R.H., BLACK, C.C. (ed.) : CO_2 Metabolism and Plant Productivity. Pp. 185 - 204. University Park Press, Baltimore - London - Tokyo 1976. [Ps.]

27541 - QUEIROZ, O. : Un modèle pour les relations entre photopériodisme, horloge
biologique et régulation enzymatique. - In : JACQUES, R. (ed.) : Études de
Biologie Végétale. Hommage au Professeur P. CHOUARD. Pp. 467 - 482. Paris
1976. [Ps.]

27542 - QUEIROZ, O. : Chronobiologie du système CAM : les rythmes de capacité enzymatique sont-ils des mécanismes régulateurs? - Physiol. vég. 14 : 629 - 639,
1976.

27543 - QUIRÓS, C.F. : Effects of extra heterochromatin on the expression of genes
producing chlorophyll variegation in the tomato. - J. Hered. 67 : 141 - 145,
1976.

27544 - RABOY, B., PADAN, E., SHILO, M. : Heterotrophic capacities of *Plectonema boryanum*. - Arch. Microbiol. 110 : 77 - 85, 1976. [Ps.]

27545 - RACKER, E. : Reconstitution and mechanisms of ion translocation systems. -
J. Biochem. (Tokyo) 79 : 46p, 1976. [Bacteriorhodopsin.]

27546 - RACUSEN, D., POINCELOT, R.P. : Distribution of protein-bound hexosamine in
chloroplasts. - Plant Physiol. 57 : 53 - 54, 1976.

27547 - **RADEMACHER, E., FEIERABEND, J.** : Formation of chloroplast pigments and ste-
rols in rye leaves deficient in plastid ribosomes. - Planta *129* : 147 - 153,
1976.

27548 - **RADMER, R.J., KOK, B.** : Photoreduction of O_2 primes and replaces CO_2 assimilation. - Plant Physiol. *58* : 336 - 340, 1976.

27549 - **RADOSEVICH, S.R., DEVILLIERS, O.T.** : Studies on the mechanism of *s*-triazine
resistance in common groundsel. - Weed Sci. *24* : 229 - 232, 1976.

27550 - **RAGHAVENDRA, A.S., DAS, V.S.R.** : Inhibition of chloroplast photochemical reactions with 2-chloromercuri 4,6-dinitrophenol. - Arch. Biochem. Biophys.
175 : 355 - 356, 1976.

27551 - **RAGHAVENDRA, A.S., DAS, V.S.R.** : Phosphoenolpyruvate carboxylase from *Setaria italica* : Inhibition by oxalacetate and malate. - Z. Pflanzenphysiol.
78 : 434 - 437, 1976.

27552 - **RAGHAVENDRA, A.S., DAS, V.S.R.** : Photochemical activities of mesophyll and
bundle sheath chloroplasts from *Euphorbia hirta,* a dicotyledonous C_4 plant. -
Photosynthetica *10* : 345 - 347, 1976.

27553 - **RAGHAVENDRA, A.S., RAO, I.M., DAS, V.S.R.** : Characterisation of abscisic
acid inhibition of stomatal opening in isolated epidermal strips. - Plant
Sci. Lett. *6* : 111 - 115, 1976. [Ps.]

27554 - **RAGHAVENDRA, A.S., RAO, I.M., DAS, V.S.R.** : Adenosine triphosphatase in epidermal tissue of *Commelina benghalensis* : possible involvement of isozymes
in stomatal movement. - Plant Sci. Lett. *7* : 391 - 396, 1976.

27555 - **RAGHAVENDRA, A.S., RAO, I.M., DAS, V.S.R.** : Replacibility of potassium by sodium for stomatal opening in epidermal strips of *Commelina benghalensis.* -
Z. Pflanzenphysiol. *80* : 36 - 42 , 1976.

27556 - **RAGHAVENDRA, A.S., RAO, I.M., DAS, V.S.R.** : Shrinkage of guard cell chloroplasts in relation to stomatal opening in *Commelina benghalensis* L. - Ann.
Bot. *40* : 899 - 901, 1976.

27557 - **RAHAT, M.** : Direct development and symbiotic chloroplasts in *Elysia timida
(Mollusca : Opisthobranchia).* - Israel J. Zool. *25* : 186 - 193, 1976.

27558 - **RAKHIMOV, G.T., TADZHIEVA, F.N.** : Posledeĭstvie vysokikh temperatur na soderzhanie pigmentov i fotokhimicheskuyu aktivnost' khloroplastov (reaktsiya Khilla) nekotorykh pustynnykh vidov èfemerov. [After-effect of high
temperature on pigment content and photochemical activity of chloroplasts
(Hill reaction) of some ephemer desert species.] - Dokl. Akad. Nauk uz. SSR
1976 (3) : 51 - 52, 1976. [In R.]

27559 - **RAKHMANKULOV, S.A.** : Osobennosti fotosinteticheskogo apparata u gibridov
khlopchatnika razlichnogo geneticheskogo proiskhozhdeniya. [Characteristics
of the photosynthetic apparatus in cotton hybrids of different genetic origin.] - Sel'skokhoz. Biol. *11* : 556 - 558, 1976. [In R, ab : E.]

27560 - **RAKITIN, Yu.V., MIKHAĬLOVA, T.P.** : O vliyanii kinetina na starenie list'ev
tabaka. [The effect of kinetin on tobacco leaves aging.] - Fiziol. Biokhim.
kul't. Rast. *8* : 125 - 131, 1976. [Ps, Chl; in R, ab : E.]

*27561 - **RAKOW, G., McGREGOR, D.I.** : Oil, fatty acid and chlorophyll accumulation in
developing seeds of two "linolenic acid lines" of low erucic acid rapeseed.
- Can. J. Plant Sci. *55* : 197 - 203, 1975.

27562 - **RALPH, R.K., BULLIVANT, S., WOJCIK, S.J.** : Effects of kinetin on phosphorylation of leaf membrane proteins. - Biochim. biophys. Acta *421* : 319 - 327,
1976. [Chloroplast.]

27563 - **RAMAM, S.S.** : Biological productivity of *Shorea* plantations. - Indian Fores-
ter *102* : 174 - 184, 1976. [Growth analysis.]

27564 - **RAMASWAMY, N.K., BEHERE, A.G., NAIR, P.M.** : A novel pathway for the synthesis
of solanidine in the isolated chloroplast from greening potatoes. - Europe.
J. Biochem. *67* : 275 - 282, 1976. [Chl.]

27565 - **RAMASWAMY, N.K., NAIR, P.M.** : Pathway for the biosynthesis of delta-aminolevulinic acid in greening potatoes. – Indian J. Biochem. Biophys. *13* : 394 – 397, 1976.

27566 – **RAMSHAW, J.A.M., SCAWEN, M.D., JONES, E.A., BROWN, R.H., BOULTER, D.** : The amino acid sequence of plastocyanin from *Lactuca sativa* (lettuce). – Phytochemistry *15* : 1199 – 1202, 1976.

27567 – **RAMUS, J., BEALE, S.I., MAUZERALL, D.** : Correlation of changes in pigment content with photosynthetic capacity of seaweeds as a function of water depth. – Mar. Biol. *37* : 231 – 238, 1976.

27568 – **RAMUS, J., BEALE, S.I., MAUZERALL, D., HOWARD, K.L.** : Changes in photosynthetic pigment concentration in seaweeds as a function of water depth. – Mar. Biol. *37* : 223 – 229, 1976.

27569 – **RANDALL, D.D.** : Phosphoglycollate phosphatase in marine algae : isolation and characterization from *Halimeda cylindracea*. – Aust. J. Plant Physiol. *3* : 105 – 111, 1976.

27570 – **RANGANATHAN, D., RANGANATHAN, S.** : Chlorophyll. – In : RANGANATHAN, D., RANGANATHAN, S. : Art in Biosynthesis. Vol. I. The Synthetic Chemist's Challenge. Pp. 36   . Academic Press, New York – London 1976.

27571 – **RAO, K.K., ROSA, L., HALL, D.O.** : Prolonged production of hydrogen gas by a chloroplast biocatalytic system. – Biochem. biophys. Res. Commun. *68* : 21 – 28, 1976.

27572 – **RAO, P.S.N., TALPASAYI, E.R.S.** : Chlorophyll stability of brine and fresh-water algae. – Curr. Sci. *45* : 462 – 463, 1976.

27573 – **RASCHKE, K.** : Transfer of ions and products of photosynthesis to guard cells. – In : **WARDLAW, I.F., PASSIOURA, J.B.**(ed.) : Transport and Transfer Processes in Plants. Pp. 203 – 215. Academic Press, New York – San Francisco – London 1976.

27574 – **RASCHKE, K., PIERCE, M., POPIELA, C.C.** : Abscisic acid content and stomatal sensitivity to CO_2 in leaves of *Xanthium strumarium* L. after pretreatments in warm and cold growth chambers. – Plant Physiol. *57* : 115 – 121, 1976. [Ps.]

27575 – **RASCHKE, K., ZEEVAART, J.A.D.** : Abscisic acid content, transpiration, and stomatal conductance as related to leaf age in plants of *Xanthium strumarium* L. – Plant Physiol. *58* : 169 – 174, 1976.

27576 – **RASCIO, N., ORSENIGO, M.** : Chloroplast fine structure in the japonica-2 maize mutant exposed to continuous illumination. 2 The white tissues. – Cytobios *16* : 183 – 191, 1976

27577 – **RASCIO, N., ORSENIGO, M., ARBOIT, D.** : Prolamellar body transformation with increasing cell age in the maize leaf. – Protoplasma *90* : 253 – 263, 1976.

27578 – **RASHKE, K.** : Sistema obratnoĭ svyazi v ust'itsakh. Reaktsiya na CO_2 i abstsizovuyu kislotu. [The stomatal feedback system. Responses to CO_2 and abscisic acid.] – Fiziol. Biokhim. kul't. Rast. *8* : 242 – 246. [Stomatal resistance; in R, ab : E.]

27579 – **RASMUSSEN, O.S.** : Water stress in plants. I. Abscisic acid level in tomato leaves after a long period of wilting. – Physiol. Plant. *36* : 208 – 212, 1976. [Chl.]

27580 – **RATHNAM, C.K.M., DAS, V.S.R.** : Biophysical characterization of mesophyll and bundle sheath chloroplasts isolated from the leaves of *Eleusine coracana,* an aspartate-type C_4 plant. II. Photosynthetic electron transfer and energy conservation reactions. – Aust. J. Plant Physiol. *3* : 185 – 199, 1976.

27581 – **RATHNAM, C.K.M., DAS, V.S.R.** : Biophysical characterization of mesophyll and bundle sheath chloroplasts isolated from the leaves of *Eleusine coracana,* an aspartate-type C-4 plant III. Photochemical activities of subchloroplast fragments including grana and stroma lamellae. – Biochem. Physiol. Pflanzen *170* : 321 – 331, 1976.

27582 – **RATHNAM, C.K.M., EDWARDS, G.E.** : Distribution of nitrate-assimilating enzymes between mesophyll protoplasts and bundle sheath cells in leaves of three groups of C_4 plants. - Plant Physiol. *57* : 881 - 885, 1976.

27583 - **RATHNAM, C.K.M., EDWARDS, G.E.** : Isolation of chloroplasts from mesophyll protoplasts for photosynthetic studies. - Plant Physiol. *57* (Suppl.) : 32, 1976.

27584 - **RATHNAM, C.K.M., EDWARDS, G.E.** : Protoplasts as a tool for isolating functional chloroplasts from leaves. - Plant Cell Physiol. *17* : 177 - 186, 1976.

27585 - **RATHNAM, C.K.M., RAGHAVENDRA, A.S., DAS, V.S.R.** : Diversity in the arrangements of mesophyll cells among leaves of certain C_4 dicotyledons in relation to C_4 physiology. - Z. Pflanzenphysiol. *77* : 283 - 291,1976. [Ps, Chl.]

27586 - **RAU, W.** : Photoregulation of carotenoid biosynthesis in plants. - Pure appl. Chem. *47* : 237 - 243, 1976.

27587 - **RAUNER, J.L.** : Deciduous forest. - In : **MONTEITH, J.L.** (ed.) : Vegetation and the Atmosphere. Vol. 2. Case Studies. Pp. 241 - 264. Academic Press, London - New York - San Francisco 1976. [Growth analysis.]

27588 - **RAVEN, J.A.** : Active influx of hexose in *Hydrodictyon africanum*. - New Phytol. *76* : 189 - 194, 1976. [Ps.]

27589 - **RAVEN, J.A.** : Glucose metabolism in *Hydrodictyon africanum* in relation to cell energetics. - New Phytol. *76* : 195 - 204, 1976.

27590 - **RAVEN, J.A.** : The rate of cyclic and non-cyclic photophosphorylation and oxidative phosphorylation, and regulation of the rate of ATP consumption in *Hydrodictyon africanum*. - New Phytol. *76* : 205 - 212, 1976.

27591 - **RAVEN, J.A.** : Division of labour between chloroplast and cytoplasm. - In : **BARBER, J.** (ed.) : The Intact Chloroplast. Pp. 403 - 443. Elsevier, Amsterdam - New York - Oxford 1976.

27592 - **RAVEN, J.A.** : The quantitative role of 'dark' respiratory processes in heterotrophic and photolithotrophic plant growth. - Ann. Bot. *40* : 587 - 602, 1976.

27593 - **RAVEN, J.A.** : Transport in algal cells. - In : **LÜTTGE, U., PITMAN, M.G.** (ed.): Transport in Plants II. Part A. Cells. Pp. 129 - 188. Springer-Verlag, Berlin - Heidelberg - New York 1976.

27594 - **RAWSON, H.M., GIFFORD, R.M., BREMNER, P.M.** : Carbon dioxide exchange in relation to sink demand in wheat. - Planta *132* : 19 - 23, 1976.

27595 - **RAWSON, H.M., WOODWARD, R.G.** : Photosynthesis and transpiration in dicotyledonous plants. I. Expanding leaves of tobacco and sunflower. - Aust. J. Plant Physiol. *3* : 247 - 256, 1976.

27596 - **RAY, T.B., BLACK, C.C.** : Characterization of phosphoenolpyruvate carboxykinase from *Panicum maximum*. - Plant Physiol. *57* (Suppl.) : 31, 1976.

27597 - **RAY, T.B., BLACK, C.C.** : Inhibition of oxalacetate decarboxylation during C_4 photosynthesis by 3-mercaptopicolinic acid. - J. biol. Chem. *251* : 5824 - 5826, 1976.

27598 - **RAY, T.B., BLACK, C.C.Jr.** : Characterization of phosphoenolpyruvate carboxykinase from *Panicum maximum*. - Plant Physiol. *58* : 603 - 607, 1976.

27599 - **REBANE, K., AVARMAA, R.** : Fine-structured spectra of chlorophyll and effect of inhomogeneous broadening. - In : Molecular Spectroscopy of Dense Phases. Pp. 459 - 462. Elsevier sci. Publ. Co., Amsterdam 1976.

*27600 - **REES, A.R.** : Proportion of assimilates supplied to roots. - J. Sci. Food Agr. *25* : 232 - 233, 1974.

27601 - **REFFYE, P. de, SNOECK, J.** : Modèle mathématique de base pour l'étude et la simulation de la croissance et de l'architecture du *Coffea robusta*. - Café Cacao Thé *20* : 11 - 32, 1976.

27602 - **REGEHR, D.L., BAZZAZ, F.A.** : Low temperature photosynthesis in successional winter annuals. - Ecology *57* : 1297 - 1303, 1976.

27603 - **REGITZ, G., OHAD, I.** : Trypsin-sensitive photosynthetic activities in chloroplast membranes from *Chlamydomonas reinhardi*, y-1. - J. biol. Chem. *251* : 247 - 252, 1976.

27604 - **REHDER, H.** : Nutrient turnover studies in alpine ecosystems. II. Phytomass and nutrient relations in the *Caricetum firmae*. - Oecologia *23* : 49 - 62, 1976. [Primary production.]

27605 - **REIBACH, P.H., BENEDICT, C.R.** : Active species of "CO_2" utilized by phosphoenol pyruvate carboxylase in corn leaves. - Plant Physiol. *57* (Suppl.) : 6, 1976.

27606 - **REICH, R., SCHEERER, R.** : Effect of electric fields on the absorption spectrum of dye molecules in lipid layers. IV. Electrochromism of oriented chlorophyll *b*. - Ber. Bunsen-Ges. phys. Chem. *80* : 542 - 547, 1976.

27607 - **REICH, R., SCHEERER, R., SEWE, K.-U.** : Electrochromic spectra of dyes in lipid layers. - Ber. Bunsen-Ges. phys. Chem. *80* : 245 - 246, 1976. [Chl.]

27608 - **REICH, R., SCHEERER, R., SEWE, K.-U., WITT, H.T.** : Effect of electric fields on the absorption spectrum of dye molecules in lipid layers. V. Refined analysis of the field-indicating absorption changes in photosynthetic membranes by comparison with electrochromic measurements *in vitro*. - Biochim. biophys. Acta *449* : 285 - 294, 1976.

27609 - **REICOSKY, D.A., HANOVER, J.W.** : Seasonal changes in leaf surface waxes of *Picea pungens*. - Amer. J. Bot. *63* : 449 - 456, 1976. [Resistances.]

27610 - **REIMER, S., TREBST, A.** : Reversal of dibromothymoquinone inhibition of photosynthetic electron flow by thiol compounds. - Z. Naturforsch. *31 C* : 103, 1976.

27611 - **REISEN, W.K.** : The ecology of Honey Creek : Temporal patterns of the travertine periphyton and selected physico-chemical parameters, and *Myriophyllum* community productivity. - Proc. Okla. Acad. Sci. *56* : 69 - 74, 1976. [Ps.]

27612 - **REISSER, W.** : Die stoffwechselphysiologischen Beziehungen zwischen *Paramecium bursaria* EHRBG. und *Chlorella* spec. in der *Paramecium bursaria*-Symbiose. I. Der Stickstoff- und der Kohlenstoff-Stoffwechsel. - Arch. Microbiol. *107* : 357 - 360, 1976.

27613 - **REISSER, W.** : Die stoffwechselphysiologischen Beziehungen zwischen *Paramecium bursaria* EHRBG. und *Chlorella* spec. in der *Paramecium bursaria*-Symbiose. II. Symbiose-spezifische Merkmale der Stoffwechselphysiologie und der Cytologie des Symbioseverbandes und ihre Regulation. - Arch. Microbiol. *111* : 161 - 170, 1976. [Ps.]

27614 - **REITZEL, L., NIELSEN, N.C.** : Acetyl-CoA carboxylase during development of plastids in wild-type and mutant barley seedlings. - Europe. J. Biochem. *65* : 131 - 138, 1976.

27615 - **REMENNIKOV, S.M., CHAMAROVSKY, S.K., KONONENKO, A.A., RUBIN, A.B.** : Low potential oxidation-reduction titration of absorbance changes induced by pulsed laser in chromatophores of photosynthesizing bacteria *Ectothiorhodospira shaposhnikovii*. - Stud. biophys. *60* : 15 - 33, 1976.

27616 - **REMY, R., BEBEE, G., MOYSE, A.** : Electrophoretic analysis of pigment protein complexes from *Porphyridium (Rhodophyta)* and *Spirulina (Cyanophyta)* thylakoTds. - Phycologia *15* : 321 - 327, 1976.

27617 - **RENGER, G.** : Studies on the structural and functional organization of System II of photosynthesis. The use of trypsin as a structurally selective inhibitor at the outer surface of the thylakoid membrane. - Biochim. biophys. Acta *440* : 287 - 300, 1976.

27618 - **RENGER, G.** : The induction of a high resistance to 3-(3,4-dichlorophenyl-1,1-dimethylurea (DCMU) of oxygen evolution in spinach chloroplasts by trypsin treatment. - FEBS Lett. *69* : 225 - 230, 1976.

27619 - **RENGER, G., ERIXON, K., DÖRING, G., WOLFF, C.** : Studies on the nature of the inhibitory effect of trypsin on the photosynthetic electron transport of system II in spinach chloroplasts. - Biochim. biophys. Acta *440* : 278 - 286, 1976.

27620 - **RENGER, G., WOLFF, C.** : The existence of a high photochemical turnover rate at the reaction centers of system II in Tris-washed chloroplasts. - Biochim. biophys. Acta *423* : 610 - 614, 1976.

27621 - **REPO, E., HATCH, M.D.** : Photosynthesis in *Gomphrena celosioides* and its classification amongst C_4-pathway plants. - Aust. J. Plant Physiol. *3* : 863 - 876, 1976.

27622 - **RETHY, R., FREDERICQ, H., MATON, J., DE GREEF, J.** : The effect of different light treatments on the chlorophyll content of *Marchantia polymorpha* L. thalli. - Biol. Jaarb. Dodonaea *44* : 269 - 279, 1976.

27623 - **RETZLAFF, G., HAMM, R.** : The relationship between CO_2 assimilation and the metabolism of bentazone in wheat plants. - Weed Res. *16* : 263 - 266, 1976.

27624 - **RICHMOND, P., WARDLAW, I.F.** : On the translocation of sugar : van der Waals' forces and surface flow. - Aust. J. Plant Physiol. *3* : 545 - 549, 1976. [Photosynthates.]

27625 - **RIJKS, D.A.** : Water use by irrigated cotton in the Sudan. IV. Water use potential evaporation and yield. - J. appl. Ecol. *13* : 491 - 506, 1976. [Yield formation.]

27626 - **RIKHIREVA, G.T., PULATOVA, M.K., BALASHOV, S.P., NAZAROVA, N.M., CHEKULAEVA, L.N., DRUZHKO, A.B., SINESHCHEKOV, V.A., LITVIN, F.F.** : Izuchenie produktov fotoprevrashcheniya bakteriorodopsina metodami ÉPR i nizkotemperaturnoĭ spektrofotometrii. [Products of bacteriorhodopsin phototransformation by ESR methods and low-temperature spectrophotometry.] - Biofizika *21* : 1038 - 1045, 1976. [In R, ab : E.]

27627 - **RINEHART, C.A., HUBBARD, J.S.** : Energy coupling in the active transport of proline and glutamate by the photosynthetic halophile *Ectothiorhodospira halophila*. - J. Bacteriol. *127* : 1255 - 1264, 1976.

27628 - **RIOV, J., BROWN, G.** : Ferredoxin-$NADP^+$ reductase from *Tsuga canadensis*. A procedure for isolation and properties. - Physiol. Plant. *38* : 147 - 152, 1976.

27629 - **RIOV, J., BROWN, G.N.** : Comparative studies of activity and properties of ferredoxin-$NADP^+$ reductase during cold hardening of wheat. - Can. J. Bot. *54* : 1896 - 1902, 1976.

27630 - **RIPLEY, E.A., REDMANN, R.E.** : Grassland. - In : MONTEITH, J.L. (ed.) : Vegetation and the Atmosphere. Vol.2. Case Studies. Pp. 351 - 398. Academic Press, London - New York - San Francisco 1976. [Ps.]

27631 - **RISCH, N., BROCKMANN, H. Jr** : Zur absoluten Konfiguration der Chlorophylle, VIII. Die absolute Konfiguration der Bacteriochlorophylle *c, d* und *e* an C-2'. - Liebigs Ann. Chem. *1976* : 578 - 583, 1976.

27632 - **RITT, E., WALZ, D.** : Pigment containing lipid vesicles. I. Preparation and characterization of chlorophyll *a*-lecithin vesicles. - J. Membrane Biol. *27* : 41 - 54, 1976.

27633 - **ROBARTS, R.D.** : Primary productivity of the upper reaches of a South African estuary (Swartvlei). - J. exp. mar. Biol. Ecol. *24* : 93 - 102, 1976.

27634 - **ROBERTIE, P., GOLDSTEIN, L., BLACK, C.C.** : Hydrodynamic properties and carboxylase-oxygenase activities of homogenously purified ribulose-1,5-diphosphate carboxylase. - Plant Physiol. *57* (Suppl.) : 5, 1976.

27635 - **ROBERTS, E.H.** : The efficiency of photosynthesis : the hierarchy of cell, leaf and crop canopy. - In : DUCKHAM, A.N., JONES, J.G.W., ROBERTS, E.H. (ed.) : Food Production and Consumption. The Efficiency of Human Food Chains and Nutrient Cycles. Pp. 85 - 105. North-Holland, Amsterdam - Oxford 1976.

27636 - **ROBERTS, J.** : An examination of the quantity of water stored in mature *Pinus sylvestris* L. trees. - J. exp. Bot. *27* : 473 - 479, 1976. [Stomatal resistance.]

27637 - **ROBERTSON, A., DODGE, A.D., KERR, M.W.** : Electron transport capabilities of chloroplasts isolated from non-photosynthesizing leaves. - Planta *129* : 95 - 96, 1976.

27638 - **ROBERTSON, P.A.** : Photosynthetic and respiratory responses of natural populations of *Koeleria cristata* grown in three environmental regimes. - Bot. Gaz. *137* : 94 - 98, 1976.

27639 - **ROBINSON, J.M., GIBBS, M.** : Influence of pH upon the oxygen inhibition of two Calvin cycle enzymes from intact spinach chloroplasts. - Plant Physiol. *57* (Suppl.) : 60, 1976.

27640 - **ROBINSON, S.P., WISKICH, J.T.** : Factors affecting the ADP/O ratio in isolated chloroplasts. - Biochim. biophys. Acta *440* : 131 - 146, 1976.

27641 - **ROBINSON, S.P., WISKICH, J.T.** : Stimulation of carbon dioxide fixation in isolated pea chloroplasts by catalytic amounts of adenine nucleotides. - Plant Physiol. *59* : 156 - 162, 1976.

27642 - **RODGERS, J.H. Jr, HARVEY, R.S.** : The effect of current on periphytic productivity as determined using carbon-14. - Water Resour. Bull. *12* : 1109 - 1118, 1976.

27643 - **RODRIGUEZ, B.P., LAMBETH, V.N.** : Photosynthetic rate-yield component relationships in winter greenhouse tomatoes. - HortScience *11* : 430 - 431, 1976.

27644 - **RODRIGUEZ, D.B., RAYMUNDO, L.C., LEE, T.-C., SIMPSON, K.L., CHICHESTER, C.O.:** Carotenoid pigment changes in ripening *Momordica charantia* fruits. - Ann. Bot. *40* : 615 - 624, 1976.

27645 - **RODRIGUEZ, D.B., TANAKA, Y., KATAYAMA, T., SIMPSON, K.L., LEE, T.-C., CHICHESTER, C.O.** : Hydroxylation of β-carotene on Micro-Cel C. - J. agr. Food Chem. *24* : 819 - 822, 1976.

27646 - **ROGATYKH, N.P., YASINOVSKIĬ, V.G., ZUBAREV, T.N.** : Urovni membrannogo potentsiala u kletki atsetabulyarii. [Levels of the membrane potential in *Acetabularia* cell.] - Biofizika *21* : 656 - 660, 1976. [In R, ab : E.]

27647 - **ROGERS, I.S.** : The effect of plant density on the yield of three varieties of French beans (*Phaseolus vulgaris* L.). - J. hort. Sci. *51* : 481 - 488, 1976.

27648 - **ROMAGOUX, J.-C.** : Corrélations au niveau d'un sédiment lacustre entre énergie lumineuse. - Compt. rend. Acad. Sci. Paris, Sér. D *283* : 1049 - 1052, 1976. [Chl.]

27649 - **ROMAGOUX, J.-C.** : Observation sur les teneurs en pigments chlorophylliens d'un sédiment lacustre, en fonction de la profondeur et du temps. - Compt. rend. Acad. Sci. Paris, Sér. D *283* : 45 - 48, 1976.

27650 - **ROMAGOUX, J.C.** : Production primaire des algues unicellulaires benthiques. - An. Hydrobiol.   *7* : 61, 1976.

27651 - **ROMANKO, E.G., SELIVANKINA, S.Yu., OMANN, É.É.** : Vliyanie tsitokininov na aktivnost' ryada khloroplastnykh i tsitoplazmaticheskikh fermentov v étiolirovannykh prorostkakh rzhi. [Effect of cytokinins on the activity of cytoplasmic and chloroplast enzymes in etiolated rye seedlings.] - Fiziol. Rast. *23* : 543 - 549, 1976. [In R, ab : E.]

27652 - **ROMANO, J.-C.** : Contribution à l'étude de la phase de latence dans des cultures d'algues planctoniques. II. Modifications dans les pigments, les nucléotides adényliques (ATP, ADP, AMO) et les acides ribonucléiques. - Compt. rend. Séances Soc. Biol. *170* : 1069 - 1074, 1976.

27653 - **ROMIJN, J.C., AMESZ, J.** : Photochemical activities of reaction centers from *Rhodopseudomonas sphaeroides* at low temperature and in the presence of chaotropic agents. - Biochim. biophys. Acta *423* : 164 - 173, 1976.

27654 - **ROOK, D.A., HOBBS, J.F.F.** : Soil temperatures and growth of rooted cuttings of radiata pine. - New Zeal. J. Forest Sci. *5* : 296 - 305, 1976. [Ps.]

27655 - **ROSA, L., HALL, D.O.** : Phosphorylation in isolated chloroplasts coupled to dichlorophenyldimethylurea-insensitive silicomolybdate reduction. - Biochim. biophys. Acta *449* : 23 - 36, 1976.

27656 - **ROSE, C.W., BEGG, J.E., TORSSELL, B.W.R.** : Townsville stylo (*Stylosanthes humilis* H.B.K.). - In : **MONTEITH, J.L.** (ed.) : Vegetation and the Atmosphere. Vol.2. Case Studies. Pp. 151 - 169. Academic Press, London - New York - San Francisco 1976. [Ps.]

27657 - **ROSE, R., POSSINGHAM, J.** : Chloroplast growth and replication in germinating spinach cotyledons following massive γ-irradiation of the seed. - Plant Physiol. *57* : 41 - 46, 1976.

27658 - **ROSENBERG, N.J., VERMA, S.B.** : A system and program for monitoring CO_2 concentration, gradient, and flux in an agricultural region. - Agron. J. *68* : 414 - 418, 1976.

27659 - **ROSHCHINA, V.V., AKULOVA, E.A.** : Razvitie aktivnosti tsitokhroma f v protsesse zeleneniya étiolirovannykh prorostkov gorokha. [Activity of cytochrome f during greening of etiolated pea seedlings.] - Fiziol. Rast. *23* : 50 - 57, 1976. [In R, ab : E.]

27660 - **ROSHCHINA, V.V., AKULOVA, E.A., SHUBIN, L.M.** : Svetoindutsiruemye izmeneniya absorbtsii pri 590 nm v izolirovannykh khloroplastakh gorokha i ikh svyaz' s plastotsianinom. [Light-induced absorption changes at 590 nm in isolated pea chloroplasts and their relation with plastocyanin.] - Mol. Biol. (Moskva) *10* : 1311 - 1319, 1976. [In R, ab : E.]

27661 - **ROSING, J., SMITH, D.J., KAYALAR, C., BOYER, P.D.** : Medium ADP and not ADP already tightly bound to thylakoid membranes forms the initial ATP in chloroplast phosphorylation. - Biochem. biophys. Res. Commun. *72* : 1 - 8, 1976.

*27662 - **ROSS, P.E., KALFF, J.** : Phytoplankton production in Lake Memphremagog, Québec (Canada) - Vermont (U.S.A.). - Verh. int. Ver. Limnol. *19* (Part 2) : 760 - 769, 1975. [Ps, Chl.]

27663 - **ROSS, R.T., ANDERSON, R.J., HSIAO, T.-L.** : Stochastic modeling of light energy conversion in photosynthesis. - Photochem. Photobiol. *24* : 267 - 278, 1976.

27664 - **ROSSITER, R.C.** : The influence of defoliation on vegetation growth in swards of three strains of subterranean clover. - Aust. J. agr. Res. *27* : 197 - 206, 1976. [Growth analysis.]

27665 - **ROUGHAN, P.G., SLACK, C.R., HOLLAND, R.** : High rates of [1-^{14}C] acetate incorporation into the lipid of isolated spinach chloroplasts. - Biochem. J. *158* : 593 - 601, 1976.

27666 - **ROUHANI, I., BASSIRI, A.** : Changes in the physical and chemical characteristics of Shahani dates during development and maturity. - J. hort. Sci. *51* : 489 - 494, 1976. [Chl.]

27667 - **ROUSSAUX, J., HOFFELT, M., FARINEAU, N.** : Evolution des RNA ribosomaux au cours du verdissement de cotylédons de concombre en présence de 6-benzylaminopurine. - Can. J. Bot. *54* : 2328 - 2336, 1976.

27668 - **ROWELL, P., POWLS, R.** : A mutant strain of *Scenedesmus obliquus* deficient in ribulose diphosphate carboxylase, cytochrome f and photosystem II activity. - Biochim. biophys. Acta *423* : 65 - 79, 1976.

27669 - **ROY, H., PATTERSON, R., JAGENDORF, A.T.** : Identification of the small subunit of ribulose 1,5-bis phosphate carboxylase as a product of wheat leaf cytoplasmic ribosomes. - Arch. Biochem. Biophys. *172* : 64 - 73, 1976.

27670 - **ROY, H., PATTERSON, R., JAGENDORF, A.T.** : The identification of the small subunit of ribulose bisphosphate carboxylase as product of cytoplasmic polyribosomes of greening wheat. - Plant Physiol. *57* (Suppl.) : 35, 1976.

27671 - **ROZEMA, J.** : An ecophysiological study on the response to salt of four halophytic and glycophytic *Juncus* species. - Flora *165* : 197 - 209, 1976. [Chl.]

27672 - **ROZHDESTVENSKIĬ, V.I.** : Avtomaticheskoe upravlenie usloviyami fotosinteza i mineral'nogo pitaniya rasteniĭ kak osnova budushcheĭ tekhnologii promyshlennogo proizvodstva sel'skokhozyaĭstvennoĭ produktsii. [Automatic control of conditions for photosynthesis and mineral nutrition of plants as the basis for future technology in industrial production of agricultural products.] - Tr. agron. Fiz. agrofiz. nauch.-issled. Inst. *39* (Potentsial'naya Produktivnost' Rasteniĭ) : 159 - 164, 1976. [In R.]

27673 - **ROZOV, N.F.** : Biologiya i produktivnost' ovoshchnykh rasteniĭ v fitotrone pri trekhletnem bessmennom kul'tivirovanii. [Biology and productivity of fruit plants in a phytotron under three-year continuous cultivation.] - Izv. TSKhA *1976* (4) : 141 - 151, 1976. [In R, ab : E.]

27674 - **RUBIN, B.A., MERZLYAK, M.N., YUFEROVA, S.G.** : Okislenie lipidnykh komponentov v izolirovannykh khloroplastakh pod deĭstviem sveta. Substraty i produkty pereokisleniya lipidov. [Oxidation of lipid components in isolated chloroplasts by the action of light . Substrates and products of lipid peroxidation.] - Fiziol. Rast. *23* : 254 - 261, 1976. [In R, ab : E.]

27675 - **RUBIN, B.A., PEROVA, I.A., VORONKOV, L.A., GULYAEV, B.A., STADNICHUK, I.N.** : Spektral'nye kharakteristiki fotosistem, vydelennykh iz khloroplastov zdorovykh i porazhennykh vertitsilleznym viltom rasteniĭ khlopchatnika. [Spectral characteristics of photosystems isolated from chloroplasts of healthy and *Verticillium* wilt-diseased cotton plants.] - Nauch. Dokl. vyssh. Shkoly, biol. Nauki *19* (11) : 35 - 41, 1976. [In R.]

27676 - **RUBTSOV, P.M., EFREMOVICH, N.V., KULAEV, I.S.** : Svetozavisimyĭ sintez pirofosfata v khloroplastakh. [Light-dependent synthesis of pyrophosphate in chloroplasts.] - Dokl. Akad. Nauk SSSR *230* : 1236 - 1237, 1976. [In R.]

27677 - **RÜDIGER, W., BENZ, J., LEMPERT, U., SCHOCH, S., STEFFENS, D.** : Hemmung der Phytol-Akkumulation mit Herbiziden. Geranylgeraniol- und Dihydrogeranylgeraniol-haltiges Chlorophyll aus Weizenkeimlingen. - Z. Pflanzenphysiol. *80* : 131 - 143, 1976.

27678 - **RUDOĬ, A.B., VEZITSKIĬ, A.Yu.** : Biosintez khlorofillov *a* i *b* v temnote v ètiolirovannykh list'yakh, infil'trirovannykh ékzogennym khlorofillidom *a*. [Dark biosynthesis of chlorophylls *a* and *b* in etiolated leaves infiltrated with exogeneous chlorophyllide *a*.] - Biokhimiya *41* : 91 - 97, 1976. [In R, ab : E.]

27679 - **RULE, D.E., STABY, G.L.** : Growth and development of *Lycopersicon esculentum* MILL. and *Brassia actinophylla* under reduced pressures and modified atmospheric composition. - HortScience *11* (3, Sect. 2) : 299, 1976. [Chl.]

27680 - **RUMBERG, B., MUHLE, H.** : Investigation of the kinetics of proton translocation across the thylakoid membrane. - Bioelectrochem. Bioenerg. *3* : 373 - 403, 1976.

27681 - **RURAINSKI, H.J., HOPPE, H.J.** : Light-dependent development of photosynthetic competence in *Scenedesmus* mutant No. 8. - Biochim. biophys. Acta *430* : 105 - 112, 1976.

27682 - **RUSSO, J.W., KNAPP, W.W.** : A numerical simulation of plant growth. - Int. J. Biometeorol. *20* : 276 - 285, 1976.

27683 - **RUTMAN, G.I., SAAKOV, V.S., DRAPKIN, V.Z., MAKAROV, Yu.A., BARANOV, A.A., SHIRYAEVA, G.A.** : Proizvodnaya spektrofotometriya v biologicheskikh issledovaniyakh (prakticheskie skhemy i rekomendatsii). [Derivative spectrophotometry in biological studies (practical schemes and recommendations).] - Byull. vses. nauch. issled. Inst. Rastenievod. N.I. Vavilova *63* : 70 - 79, 1976. [Chl; in R.]

27684 - **RUTMAN, G.I., SAAKOV, V.S., DRAPKIN, V.Z., MAKAROV, Yu.A., UDOVENKO, G.V.** : Metody molekulyarnoĭ spektrofotometrii pri izuchenii ustoĭchivosti plastidnogo apparata. [Methods of molecular spectrofotometry for studying the resistance of plastid apparatus.] - Tr. prikl. Bot., Genet. Selektsii *57* (2) : 130 - 147, 1976. [In R, ab : E.]

27685 - **RYBERG, M., SUNDQVIST, C.** : The influence of 8-hydroxyquinoline on the accumulation of porphyrins in dark-grown wheat leaves treated with δ-aminolevulinic acid. - Physiol. Plant. *36* : 356 - 361, 1976.

27686 - **RYCHNOVSKÁ, M.** : Alluvial grassland hydrosere : Primary production and plant processes. - Pol. ecol. Stud. *2* (2) : 103 - 112, 1976.

27687 - **RYLE, G.J.A., COBBY, J.M., POWELL, C.E.:** Synthetic and maintenance respiratory losses of $^{14}CO_2$ in uniculm barley and maize. - Ann. Bot. *40* : 571 - 586, 1976. [Ps.]

27688 - **RYLE, G.J.A., POWELL, C.E.** : Effect of rate of photosynthesis on the pattern of assimilate distribution in the graminaceous plant. - J. exp. Bot. *27* : 189 - 199, 1976.

27689 - **RYSSOV-NIELSEN, H.** : Application of the *Pye Unicam SP 1800* and accessories in automated chlorophyll *a, b, c* and phaeophytin analysis. - SCAN 7 : 14 - 16, 1976.

27690 - **SAAKOV, V.S.** : Issledovanie tsentrov lokalizatsii povrezhdayushchikh vozdeĭstviĭ v membranakh khloroplastov metodami molekulyarnoĭ spektroskopii. [Investigation of centres of harmful influences in chloroplast membranes by means of molecular spectroscopy.] - Tr. prikl. Bot., Genet. Selektsii *57* (2): 17 - 34, 1976. [In R, ab : E.]

27691 - **SADLER, D.M.** : X-ray diffraction from chloroplast membranes oriented in a magnetic field. - FEBS Lett. *67* : 289 - 293, 1976.

27692 - **SAFIR, G.R., SCHNEIDER, C.L.** : Diffusive resistances of two sugarbeet cultivars in relation to their black root disease reaction. - Phytopathology *66* : 277 - 280, 1976.

27693 - **SAGE, J.** : Vapor Gard antitranspirant sprays for size increase on 5 apple cultivars. - HortScience *11* : 301, 1976. [Ps.]

*27694 - **SAGER, R.** : Patterns of inheritance of organelle genomes : molecular basis and evolutionary significance. - In : BIRKY, C.W. Jr., PERLMAN, P.S., BYERS, T.J. (ed.) : Genetics and Biogenesis of Mitochondria and Chloroplasts. Pp. 252 - 267. Ohio State University Press, Columbus 1975. [Chloroplast.]

27695 - **SAGER, R.** : The circular diploid model of chloroplast DNA in *Chlamydomonas*. - In : BÜCHER, T., NEUPERT, W., SEBALD, W., WERNER, W. (ed.) : Genetics and Biogenesis of Chloroplasts and Mitochondria. Pp. 295 - 303. North-Holland Publ. Co., Amsterdam - New York - Oxford 1976.

27696 - **SAGER, R., RAMANIS, Z.** : Chloroplast genetics of *Chlamydomonas*. I. Allelic segregation ratios. - Genetics *83* : 303 - 321, 1976.

27697 - **SAGER, R., RAMANIS, Z.** : Chloroplast genetics of *Chlamydomonas*. II. Mapping by cosegregation frequency analysis. - Genetics *83* : 323 - 340, 1976.

27698 - **SAGHER, D., GROSFELD, H., EDELMAN, M.** : Large subunit ribulosebisphosphate carboxylase messenger RNA from *Euglena* chloroplasts. - Proc. nat. Acad. Sci. USA *73* : 722 - 726, 1976.

27699 - **SAINIS, J.K., SANE, P.V.** : Photosynthetic and biochemical potential of the developing pods of red kidney beans. - Plant biochem. J. *3* : 97 - 104, 1976.

27700 - **SAKS, N.M., STONE, R.J., LEE, J.J.** : Autotrophic and heterotrophic nutritional budget of salt marsh epiphytic algae. - J. Phycol. *12* : 443 - 448, 1976.

27701 - **SALE, P.J.M.** : Effect of shading at different times on the growth and yield of the potato. - Aust. J. agr. Res. *27* : 557 - 566, 1976. [Growth analysis.]

27702 - **SALEM, L.** : Theory of photochemical reactions. - Science *191* : 822 - 830, 1976.

27703 - **SALVADOR, G.F., BENEY, G., NIGON, V.** : Control of δ-aminolevulinic acid synthesis during greening of dark-grown *Euglena gracilis*. - Plant Sci. Lett. *6* : 197 - 202, 1976.

27704 - **SAMOĬLOVA, O.P., SOLOV'EV, I.S., BLYUMENFEL'D, L.A.** : Zavisimost' kinetiki narastaniya i spada signala mikrovolnovoĭ fotoprovodimosti khlorelly ot sostoyaniya kul'tury. [Dependence of rise and decay kinetics of microwave photoconductivity signals of *Chlorella* cells on physiological state of the culture.] - Biofizika *21* : 1031 - 1034, 1976. [In R, ab : E.]

27705 - **SANADZE, G.A., DZHAIANI, G.I., BAAZOV, D.I., KHAKHUBIYA, Ts.G., EBRALIDZE, Sh.S., GVANTSELADZE, L.G.** : Veroyatnostno-statisticheskaya model' raspredeleniya ugleroda $C^{13}O_2$ v molekule izoprena pri fotosinteze. [Probability-statistical model for distribution of carbon from $^{13}CO_2$ in isoprene molecule during photosynthesis.] - Fiziol. Rast. *23* : 690 - 696, 1976. [In R, ab : E.]

27706 - **SANDERS, J.K.M., WATERTON, J.C.** : The chlorophyll-*a* radical cation : Determination of hyperfine coupling constants by nuclear magnetic resonance spectroscopy. - J. chem. Soc. chem. Commun. *1976* (7) : 247 - 248, 1976.

27707 - **SANDERS, J.L., BROWN, D.A.** : Effect of variations in the shoot : root ratio
upon the chemical composition and growth of soybeans. - Agron. J. *68* : 713 -
717, 1976.

27708 - **SANDERSON, S., HASLETT, B.G., BOULTER, D.** : The N-terminal amino acid sequence of cytochrome *f* from *Spirulina platensis*. - Phytochemistry *15* : 815 - 816,
1976.

27709 - **SANE, P.V., BHAGWAT, A.S.** : An analysis of the C_4 pathway - a possible mechanism for the regulation of CO_2 fixation. - Plant biochem. J. *3* : 1 - 10, 1976.

27710 - **SAN JOSE, J.J., MEDINA, Y.E.** : Organic matter production in the *Trachypogon*
savanna at Calabozo, Venezuela. - Trop. Ecol. *17* : 113 - 124, 1976. [Growth
analysis.]

27711 - **SAPOZHNIKOV, D.I., GABR, M.A., MASLOVA, T.G.** : Deĭstvie anaerobioza i nekotorykh metabolitov na reaktsii violaksantinovogo tsikla v list'yakh rasteniĭ.
[Effect of anaerobiosis and some metabolites on the reactions of violaxanthin
cycle in plant leaves.] - Fiziol. Rast. *23* : 31 - 35, 1976. [In R, ab : E.]

27712 - **SARANCHA, Yu.P., MAL'YAN, A.N., MAKAROV, A.D.** : Kolebatel'nyĭ kharakter adsorbtsii ADF i obrazovaniya ATF v khloroplastakh. [Oscillatory pattern of
ADP adsorption and ATP formation in chloroplasts.] - Fiziol. Rast. *23* : 615 -
616, 1976. [In R.]

27713 - **SARANCHA, Yu.P., MAL'YAN, A.N., MAKAROV, A.D.** : Vzaimodeĭstvie proizvodnykh
adenina s membranami khloroplastov. [Interaction of adenine derivatives with
chloroplast membranes.] - Biokhimiya *41* : 898 - 902, 1976. [In R, ab : E.]

*27714 - **SARDA, C., PRIOUL, J.-L., BOURDU, R.** : L'infrastructure du *Lolium multiflorum* sous des photopériodes induisant la floraison : modifications de l'équi-
libre entre chloroplastes, mitochondries et peroxysomes. - Compt. rend. Acad.
Sci. Paris, Sér. D *278* : 723 - 726, 1974.

27715 - **SARIĆ, M., MILIVOJEVIĆ, D.** : The effect of some ions on pigment content and
chloroplast structure. - Arh. poljoprivred. Nauke *29* : 103 - 116, 1976.

27716 - **SARKISYAN, S.A., ZAKARYAN, N.E., ABRAMYAN, L.Kh.** : Vliyanie gibberellovoĭ
kisloty na ul'trastrukturu khloroplastov list'ev kartofelya. [Effect of gibberellic acid on the ultrastructure of potato leaf chloroplasts.] - Uchen.
Zapiski erevansk. gos. Univ., estestv. Nauki *1976* [3(133)] : 132 - 138, 1976.
[In R, ab : Arm.]

27717 - **SÁRVÁRI, É., HALÁSZ, G., NYITRAI, P., LÁNG, F.** : Effect of lincocin treatment
on the greening process in bean (*Phaseolus vulgaris*) leaves. - Physiol.
Plant. *36* : 187 - 192, 1976.

27718 - **SASA, T., SUGAHARA, K.** : Photoconversion of protochlorophyll to chlorophyll *a*
in a mutant of *Chlorella regularis*. - Plant Cell Physiol. *17* : 273 - 279,
1976.

27719 - **SATO, K.** : [The growth responses of soybean plant to photoperiod and temperature I. Responses in vegetative growth.] - Proc. Crop Sci. Soc. Jap. *45* :
443 - 449, 1976. [Growth analysis; in Jap., ab : E.]

27720 - **SATO, K., ANDO, A.** : [Growth of reed canarygrass (*Phalaris arundinacea* L.)
as affected by environment and cutting II. The influence of cutting height
on the regrowth processes.] - Proc. Crop Sci. Soc. Jap. *45* : 464 - 470, 1976.
[Growth analysis; in Jap., ab : E.]

27721 - **SATO, K., INABA, K.** : [High temperature injury of ripening in rice plant. IV.
Effect of high temperature on ^{14}C-assimilation and translocation at an early
ripening period.] - Proc. Crop Sci. Soc. Jap. *45* : 151 - 155, 1976. [In Jap.,
ab : E.]

27722 - **SATO, K., INABA, K.** : [High temperature injury of ripening in rice plant. V.
On the early decline of assimilate storing ability of grains at high temperature.] - Proc. Crop Sci. Soc. Jap. *45* : 156 - 161, 1976. [In Jap., ab : E.]

27723 - **SATÔ, M., HASEGAWA, M.** : The latency of spinach chloroplast phenolase. - Phytochemistry *15* : 61 - 65, 1976.

27724 - **SATOH, K., STRASSER, R., BUTLER, W.L.** : A demonstration of energy transfer from Photosystem II to Photosystem I in chloroplasts. - Biochim. biophys. Acta *440* : 337 - 345, 1976.

27725 - **SATOH, M., OHYAMA, K.** : Studies on photosynthesis and translocation of photo-synthate in mulberry tree. V. Utilization of reserve substance in the process of regrowth after shoot pruning in a growing season. - Proc. Crop Sci. Soc. Jap. *45* : 51 - 56, 1976.

27726 - **SATOH, T., HOSHINO, Y., KITAMURA, H.** : *Rhodopseudomonas sphaeroides* forma sp. *denitrificans*, a denitrifying strain as a subspecies of *Rhodopseudomonas sphaeroides*. - Arch. Microbiol. *108* : 265 - 269, 1976.

27727 - **SATTER, R.L., GALSTON, A.W.** : The physiological functions of phytochrome. - In : GOODWIN, T.W. (ed.) : Chemistry and Biochemistry of Plant Pigments. 2nd Ed. Vol. 1. Pp. 680 - 735. Academic Press, London - New York - San Francisco 1976. [Chl, chloroplast.]

27728 - **SAUGIER, B.** : Sunflower. - In : MONTEITH, J.L. (ed.) : Vegetation and the Atmosphere. Vol. 2. Case Studies. Pp. 87 - 119. Academic Press, London - New York - San Francisco 1976.

27729 - **SAUNDERS, J.A., McCLURE, J.W.** : The occurrence and photoregulation of flavono-ids in barley plastids. - Phytochemistry *15* : 805 - 807, 1976.

27730 - **SAUNDERS, J.A., McCLURE, J.W.** : The distribution of flavonoids in chloroplasts of twenty five species of vascular plants. - Phytochemistry *15* : 809 - 810, 1976.

27731 - **SAUNDERS, V.A.** : Photosynthetic prokaryotes. - Nature *263* : 461 - 462, 1976.

27732 - **SAVIDGE, G.** : A preliminary study of the distribution of chlorophyll *a* in the vicinity of fronts in the Celtic and western Irish seas. - Estuar. coast. mar. Sci. *4* : 617 - 625, 1976.

27733 - **SAVITSKIĬ, I.L.** : O fotosinteticheskoĭ deyatel'nosti organov yabloni. [Pho-tosynthetic activity in apple tree organs.] - Fiziol. Biokhim. kul't. Rast. *8* : 53 - 56, 1976. [In R, ab : E.]

*27734 - **SAWADA, S.** : Net assimilation rates of wheat plants grown in two fields with different temperature conditions during winter. - Jap. J. Ecol. *23* : 243 - 250, 1973.

27735 - **SCAIFE, M.A., JONES, D.** : The relationship between crop yield (or mean plant weight) of lettuce and plant density, length of growing period, and initial plant weight. - J. agr. Sci. *86* : 83 - 91, 1976.

27736 - **SCARISBRICK, D.H., CARR, M.K.V., WILKES, J.M.** : The effect of sowing date and season on the development and yield of Navy beans (*Phaseolus vulgaris*) in south-east England. - J. agr. Sci. *86* : 65 - 76, 1976.

*27737 - **SCAWEN, M.D., RAMSHAW, J.A.M., BROWN, R.H., BOULTER, D.** : The amino-acid se-quence of plastocyanin from *Sambucus nigra* L. (elder). - Europe. J. Biochem. *44* : 299 - 303, 1974.

27738 - **SCHADLER, D.L., STEELE, J.A., DURBIN, R.D.** : Some effects of tentoxin on ma-ture and developing chloroplasts. - Mycopathologia *58* : 101 - 105, 1976.

27739 - **SCHÄFERS, H.-A., FEIERABEND, J.** : Ultrastructural differentiation of plastids and other organelles in rye leaves with a high-temperature-induced deficiency of plastid ribosomes. - Cytobiologie *14* : 75 - 90, 1976.

27740 - **SCHEPERS, A., SIBMA, L.** : Yield and dry matter content of early and late po-tatoes, as affected by monoculture and mixed cultures. - Potato Res. *19* : 73 - 90, 1976. [Primary production.]

27741 - **SCHIEMER, F., PROSSER, M.** : Distribution and biomass of submerged macrophytes in Neusiedlersee. - Aquat. Bot. *2* : 289 - 307, 1976.

27742 - **SCHILLING, N., SCHEIBE, R., BECK, E., KANDLER, O.** : Maltose phosphate in iso-lated spinach chloroplasts. - FEBS Lett. *61* : 192 - 193, 1976.

27743 - **SCHLETZ, K.** : Phototaxis bei *Volvox* - Pigmentsysteme der Lichtrichtungsperzep-tion. - Z. Pflanzenphysiol. *77* : 189 - 211, 1976.

27744 - SCHMID, G.H., JANKOWICZ, M., MENKE, W. : Cyclic photophosphorylation and
chloroplast structure in the labellum of the orchid *Aceras anthropophorum*. -
J. Microsc. Biol. cell. *26* : 25 - 28, 1976.

27745 - SCHMID, R., SHAVIT, N., JUNGE, W. : The coupling factor of photophosphorylation and the electric properties of the thylakoid membrane. - Biochim. biophys. Acta *430* : 145 - 153, 1976.

27746 - SCHMIDT, B. : Interaction of oxidized and reduced N-methylphenazonium methosulfate with Photosystem II. - Biochim. biophys. Acta *449* : 516 - 524, 1976.

27747 - SCHMIDT, B., RURAINSKI, H.J. : Light-dependent interactions of phenazine methosulfate with 3-(3,4-dichlorphenyl)-1,1-dimethylurea-poisoned chloroplasts.-
Z. Naturforsch. *31 C* : 722 - 729, 1976.

27748 - SCHMIDT, L., STEHLÍK, V. : Vysoké výnosy chrástu a jejich příčiny. [High yields of sugar beet leaves and the reason of it.] - Listy cukrov. (Praha) *92* :
1 - 9, 1976. [Photosynthates; in Czech, ab : R, E, G.]

27749 - SCHMIDT, W. : Experimental ecology. - Progr. Bot. *38* : 352 - 366, 1976. [Ps.]

27750 - SCHMIDT-VOGT, H., GROSS, K. : Untersuchungen zum winterlichen Gaswechsel der
Fichte (*Picea abies* [L.] KARST.) unter Freilandbedingungen. Ein Beitrag zum
Anbau der Fichte im Westen Europas. - Allg. Forst- Jagdz. *147* : 189 - 192,
1976.

27751 - SCHMIEDEKNECHT, M.,SCHLEGEL, H. : Chlorophyllgehalt und Anfälligkeit für
Puccinia striiformis WEST. - Tagungsber. Akad. Landwirtschaftswiss. DDR
143 : 363 - 371, 1976.

27752 - SCHMITZ, K., LOBBAN, C.S. : A survey of translocation in *Laminariales*
(Phaeophyceae). - Mar. Biol. *36* : 207 - 216, 1976. [Photosynthates.]

27753 - SCHNABL, H., MAYER, I. : Dark fixation of CO_2 by flowers of cut roses. -
Planta *131* : 51 - 55, 1976.

27754 - SCHNARRENBERGER, C., FOCK, H. : Interactions among organelles involved in
photorespiration. - In : STOCKING, C.R., HEBER, U. (ed.) : Transport in
Plants III. Intracellular Interactions and Transport Processes. Pp. 185 -
234. Springer-Verlag, Berlin - Heidelberg - New York 1976.

27755 - SCHNEIDER, H.A.W. : Enzymic capacities for chlorophyll biosynthesis. Activation and *de novo* synthesis of enzymes. - Z. Naturforsch. *31 C* : 55 - 63, 1976.

27756 - SCHOLZ, A. : Lichtorientierte Chloroplastenbewegung bei *Hormidium flaccidum* :
Perception der Lichtrichtung mittels Sammellinseneffekt. - Z. Pflanzenphysiol. *77* : 406 - 421, 1976.

27757 - SCHOLZ, A. : Lichtorientierte Chloroplastenbewegung bei *Hormidium flaccidum* :
Verschiedene Methoden der Lichtrichtungsperception und die wirksamen Pigmente.
- Z. Pflanzenphysiol. *77* : 422 - 436, 1976.

27758 - SCHOOLEY, R.E., GOVINDJEE : Cation-induced changes in the circular dichroism
spectrum of chloroplasts. - FEBS Lett. *65* : 123 - 125, 1976.

27759 - SCHOPFER, P., BAJRACHARYA, D., FALK, H. : Photocontrol of microbody and mito-
chondrion development : the involvement of phytochrome. - In : SMITH, H.
(ed.) : Light and Plant Development. Pp. 193 - 212. Butterworths, London -
Boston - Sydney - Wellington - Durban - Toronto 1976.

27760 - SCHRADER, L.E. : CO_2 metabolism and productivity in C_3 plants: An assessment.
- In : BURRIS, R.H., BLACK, C.C. (ed.) : CO_2 Metabolism and Plant Productivity. Pp. 385 - 396. University Park Press, Baltimore - London - Tokyo 1976.

*27761 - SCHREIBER, K. : Möglichkeiten zur Steuerung pflanzlicher Prozesse. - Wiss.
Fortschr. *24* : 407 - 411, 1974. [Ps.]

27762 - SCHREIBER, U. : Correlation between growth temperature and heat-induced chlorophyll fluorescence changes in *Scenedesmus obliquus*. - Carnegie Inst. Year
Book *75* : 472 - 477, 1976.

27763 - **SCHREIBER, U., COLBOW, K., VIDAVER, W.** : Analysis of temperature-jump chlorophyll fluorescence induction in plants. - Biochim. biophys. Acta *423* : 249 - 263, 1976.

27764 - **SCHREIBER, U., VIDAVER, W.** : The I-D fluorescence transient. An Indicator of rapid energy distribution changes in photosynthesis. - Biochim. biophys. Acta *440* : 205 - 214, 1976.

27765 - **SCHREIBER, U., VIDAVER, W.** : Rapid light-induced changes of energy distribution between photosystems I and II. - FEBS Lett. *62* : 194 - 197, 1976.

27766 - **SCHULTS, D.W., MALUEG, K.W., SMITH, P.D.** : Limnological comparison of culturally eutrophic Shagawa Lake and adjacent oligotrophic Burntside Lake, Minnesota. - Amer. Midland Nat. *96* : 160 - 178, 1976. [Chl.]

27767 - **SCHULZ-BALDES, M., LEWIN, R.A.** : Fine structure of *Synechocystis didemni* (*Cyanophyta : Chroccoccales*). - Phycologia *15* : 1 - 6, 1976. [Chromatophores.]

27768 - **SCHULZE, E.-D., LANGE, O.L., EVENARI, M., KAPPEN, L., BUSCHBOM, U.** : An empirical model of net photosynthesis for the desert plant *Hammada scoparia* (POMEL) ILJIN. I. Description and test of the model. - Oecologia *22* : 355 - 372, 1976.

27769 - **SCHULZE, E.-D., SCHULZE, I.** : Distribution and control of photosynthetic pathways in plants growing in the Namib Desert, with special regard to *Welwitschia mirabilis* HOOK. fil. - Madoqua *9* (3) : 5 - 13, 1976.

27770 - **SCHULZE, R.E.** : A physically based method of estimating solar radiation from suncards. - Agr. Meteorol. *16* : 85 - 101, 1976.

27771 - **SCHUMANN, B., BÖRNER, T.** : Über den Zusammenhang zwischen Lamellarproteinen und Thylakoidmorphologie bei der Biosynthese grüner, etiolierter und mutierter Plastiden von *Hordeum, Pelargonium* und *Lycopersicon*. I. Charakterisierung der plastidalen Lamellarproteine. - Acta histochem. *17* (Suppl.) : 153 - 155, 1976.

27772 - **SCHÜRMANN, P., WOLOSIUK, R.A., BREAZEALE, V.D., BUCHANAN, B.B.** : Two proteins function in the regulation of photosynthetic CO_2 assimilation in chloroplasts. - Nature *263* : 257 - 258, 1976.

27773 - **SCHUSTER, H., KOHLER, A., KREEB, K.** : Eine neue Methode zur Beurteilung der Belastbarkeit von submersen Makrophyten. - Verhandl. Ges. Ökol. *1976* : 335 - 345, 1976. [Ps.]

27774 - **SCHUSTER, H., KREEB, K.** : Indikationen von Schwermetallschädigungen an höheren Wasserpflanzen über den CO_2-Gaswechsel. - In : Daten und Dokumente zum Umweltschutz. Vol.19 (Vorträge der Tagung über Umweltforschung der Universität Hohenheim). Pp. 133 - 139. Dokumentationsstelle der Universität Hohenheim, Stuttgart-Hohenheim 1976.

27775 - **SCHWARTZBACH, S.D., SCHIFF, J.A., KLEIN, S.** : Biosynthetic events required for lag elimination in chlorophyll synthesis in *Euglena*. - Planta *131* : 1 - 9, 1976.

27776 - **SCHWENN, J.D., DEPKA, B., HENNIES, H.H.** : Assimilatory sulfate reduction in chloroplasts : Evidence for the participation of both stromal and membrane-bound enzymes. - Plant Cell Physiol. *17* : 165 - 176, 1976.

27777 - **SCHWENN, J.D., TREBST, A.** : Photosynthetic sulfate reduction by chloroplasts. - In : BARBER, J. (ed.) : The Intact Chloroplasts. Pp. 315 - 334. Elsevier, Amsterdam - New York - Oxford 1976.

27778 - Scientific Research Station, Hsiaotsum Brigade, Takou People's Commune, Yian Shih County, Honan and Specialty of Plant Physiology and Biochemistry, Practice Group, Class of 1973, Department of Biology, Peking University : [Studies on photorespiration in several different variaties of wheat by biochemical methods.] - Acta bot. sin. *18* : 293 - 299, 1976. [In Chin., ab : E.]

27779 - **SCURTU, D.** : Influenţa îngrăşămintelor minerale asupra unor caracteristici ale suprafeţei foliare la sfeclă şi grîu de toamnă. [The effects of mineral fertilizers on some characteristics of the foliar surface in beet and winter wheat.] - Lucrări științ. Acad.Ști. agr. silv. Sta. Cerc. agr. (Suceava) Vol. omag. Pp. 273 - 280. 1976. [Growth analysis; in Roum., ab : E, F, R.]

27780 - SEELY, G.R. : Chlorophyll-poly(vinylpyridine) complexes. V. Energy transfer from chlorophyll to bacteriochlorophyll. - J. phys. Chem. *80* : 441 - 446, 1976.

27781 - SEELY, G.R. : Chlorophyll-poly(vinylpyridine) complexes. VI. Sensitized fluorescence in chlorophyll *b*-chlorophyll *a* systems. - J. phys. Chem. *80* : 447 - 451, 1976.

27782 - SEIDEL, D. : Experimentell-ökologische Untersuchungen an Waldbodenmoosen als Grundlage einer kausalanalytischen Interpretation synökologischer Aspekte. - Flora *165* : 163 - 196, 1976.

27783 - SELLAMI, A. : Évolution des adénosine phosphates et de la charge énergétique dans les compartiments chloroplastique et non-chloroplastique des feuilles de blé. - Biochim. biophys. Acta *423* : 524 - 539, 1976.

27784 - SELLDÉN, G., SELSTAM, E. : Changes in chloroplast lipids during the development of photosynthetic activity in barley etio-chloroplasts. - Physiol. Plant. *37* : 35 - 41, 1976.

27785 - SELLNER, K.G., ZINGMARK, R.G., MILLER, T.G. : Interpretations of ^{14}C method of measuring total annual production of phytoplankton in a South Carolina estuary. - Bot. mar. *19* : 119 - 125, 1976.

27786 - SELMAN, B.R. : Phenylenediamine restoration of photosynthetic electron flux in DBMIB-inhibited chloroplasts. - J. Bioenerg. Biomembranes *8* : 143 - 156, 1976.

27787 - SELMAN, B.R., HAUSKA, G.A. : The energization of chloroplasts in the dark, induced by reduction-oxidation reactions across the thylakoid membrane. - FEBS Lett. *71* : 79 - 82, 1976.

27788 - SELMAN, B.R., PSCZOLLA, G. : Dark oxidation-reduction coupled phosphorylation in sonicated chloroplast vesicles. - FEBS Lett. *61* : 135 - 139, 1976.

27789 - SELVAM, R., GNANAM, A. : Acid phosphatase activity in the thylakoid membranes of chloroplasts of *Sorghum vulgare*. - Plant Physiol. *57* (Suppl.) : 74, 1976.

27790 - SEMENENKO, V.E., KASATKINA, T.I., RUDOVA, T.S. : Obratimoe podavlenie sinteza belka fraktsii I pod vliyaniem 2-dezoksi-D-glyukozy. [Reversible suppression of fraction I protein synthesis by 2-deoxy-D-glucose.] - Fiziol. Rast. *23* : 1225 - 1231, 1976. [In R, ab : E.]

27791 - SEMENENKO, V.E., ZVEREVA, M.G., KLIMOVA, L.A. : Éndogennaya regulyatsiya fotosinteza na urovne sinteza nukleinovykh kislot i belkov khloroplasta. [Endogenous regulation of photosynthesis at the level of synthesis of nucleic acids and proteins in the chloroplast.] - In : Itogi Issledovaniya Mekhanizma Fotosinteza. Pp. 65 - 66. Pushchino 1976. [In R.]

27792 - SEMIKHATOVA, O.A., EGOROVA, L.I. : Deĭstvie sveta na protsess reaktivatsii fotosinteza list'ev kukuruzy posle vozdeĭstviya vysokoĭ temperatury. [Effect of light on the recovery of photosynthesis in maize leaves after heat stress.] - Bot. Zh. *61* : 313 - 323, 1976. [In R, ab : E.]

27793 - SEMPIO, di C., MONTALBINI, P., TORRE, G. DELLA, CAPPELLI, C., TAMBURI, F., TAMBURI, G., BARBERINI, B., FERRANTI, F. : Variazioni metaboliche che accompagnano la comparsa dei sintomi di marciume apicale, per squilibrio idrio, in frutti di Pomodoro "San Marzano". [Metabolic changes associated with the appearance of blossom end rot symptoms, caused by water shortage, in "San Marzano" tomato fruits.] - Phytopathol. mediterr. *15* : 29 - 39, 1976. [Ps; in Ital., ab : E, F.]

27794 - SEN, D.N., BHANDARI, D.C., BANSAL, R.P. : Antholysis of *Sesamum indicum* L. (TIL). - Curr. Sci. *45* : 248 - 249, 1976. [Chl, Car.]

27795 - SENGER, H., BORNMAN, C.H. : *Welwitschia mirabilis:* Formation and distribution of chlorophylls. - Z. Pflanzenphysiol. *80* : 261 - 270, 1976.

27796 - SENGER, H., OH-HAMA, T. : Quantum yield and conformational changes during greening and bleaching of *Chlorella protothecoides*. - Plant Cell Physiol. *17* : 551 - 556, 1976.

27797 - SENGUPTA, C., SEN, S.P. : Photosynthetic $^{14}CO_2$ fixation in *Oxalis corniculata*. - Plant biochem. J. *3* : 44 - 53, 1976.

27798 - SEO, T., OHTAKI, E. : Infrared device for measurement of carbon dioxide fluctuations under field conditions. III. Adaptation to infrared hygrometry. - Ber. Ôhara Inst. landwirtsch. Biol. Okayama Univ. *16* : 191 - 198, 1976.

27799 - SERDYUK, A.L., MAĬDUROVA, V.E., CHERNENKO, L.N. : Matematicheskaya formula rascheta ploshchadi listovoĭ poverkhnosti u pertsev i baklazhanov. [Mathematical formula for calculating the area of leaf surface in pepper and eggplant.] - Fiziol. Rast. *23* : 840 - 841, 1976. [In R.]

27800 - SERVAITES, J.C., OGREN, W.L. : Effect of pH on photosynthesis, oxygen inhibition, and photorespiration in soybean leaf cells. - Plant Physiol. *57* (Suppl.) : 59, 1976.

*B27801 - ŠESŤÁK, Z., ČATSKÝ, J. (ed.) : Photosynthesis Bibliography. Vol. 1 (2 parts) - 1966/1970.-Dr. W. Junk b. v.-Publ., The Hague 1974.

*B27802 - ŠESŤÁK, Z., ČATSKÝ, J. (ed.) : Photosynthesis Bibliography. Vol. 2 - 1971. - Dr. W. Junk b. v.-Publ., The Hague 1975.

27803 - ŠESŤÁK, Z., ČATSKÝ, J. : Bibliography of reviews and methods of photosynthesis - 35, 36, 37, 38. - Photosynthetica *10* : 93 - 105, 208 - 227, 348 - 354, 466 - 476, 1976.

27804 - ŠESŤÁK, Z., DEMETER, S. : Changes in circular dichroism and *P*700 content of chloroplasts during leaf ontogenesis. - Photosynthetica *10* : 182 - 187, 1976.

27805 - SHABEL'SKAYA, Ė.F., GVARDIYAN, V.N. : O fuktsional'noĭ aktivnosti khloroplastov kartofelya na raznykh ėtapakh ikh degradatsii, vyzvannoĭ deĭstviem temnoty. [Functional activity of potato chloroplasts at different stages of their degradation caused by darkness.] - In : Voprosy Estestvoznaniya. Pp. 130 - 133. Izd. minsk. gos. pedagog. Inst. im. A.M. Gor'kogo, Minsk 1976. [In R.]

27806 - SHADI, A., MANSUROVA, S.Ė., TSYDENDAMBAEV, V.D., KULAEV, I.S. : O biosinteze polifosfatov v khromatoforakh *Rhodospirillum rubrum*. [Polyphosphate biosynthesis in *Rhodospirillum rubrum* chromatophores.]-Mikrobiologiya *45* : 333 - 336, 1976. [In R, ab : E.]

27807 - SHADRIKOV, O.A., POLYAKOV, M.A. : Ustanovka dlya issledovaniya fotosinteza v lazernykh luchakh s pomoshch'yu infrakrasnogo gazoanalizatora (IKG). [Apparatus for studying photosynthesis in laser beams using an infra-red analyser.] - Ėlektron. Obrab. Mater. *1976* (5) : 72 - 74, 1976. [In R.]

27808 - SHAH, F.H., MALIK, S.K., SHEIKH, A.S. : Degradation of chlorophyll during drying and storage of leaf protein concentrate. - Pak. J. sci. ind. Res. *19* : 193 - 194, 1976.

27809 - SHAHAK, Y., POSNER, H.B., AVRON, M. : Evidence for a block between plastoquinone and cytochrome f in a photosynthetic mutant of *Lemna* with abnormal flowering behavior. - Plant Physiol. *57* : 577 - 579, 1976.

27810 - SHAMSI, S.R.A. : Effect of a light-break on the growth and development of *Epilobium hirsutum* and *Lythrum salicaria* in short photoperiods. - Ann. Bot. *40* : 153 - 162, 1976. [Chl.]

27811 - SHAR, A.O. : Stability in a plankton model. -   Int.Rev.ges.Hydrobiol. *61*:841- 845, 1976.

27812 - SHARKEY, P.J., PATE, J.S. : Translocation from leaves to fruits of a legume, studied by a phloem bleeding technique : diurnal changes and effects of continuous darkness. - Planta *128* : 63 - 72, 1976.

*27813 - SHATILOV, I.S., MAZEIN, V.L. : Fotosinteticheskiĭ potentsial i produktivnost' fotosinteza klevera krasnogo. [Photosynthetic potential and productivity of photosynthesis in red clover.] - Izv. TSKhA (Moskva) *1974* (6) : 28 - 37, 1974. [In R.]

27814 - **SHAVER, G.R., BILLINGS, W.D.** : Carbohydrate accumulation in tundra graminoid plants as a function of season and tissue age. - Flora *165* : 247 - 267, 1976.

27815 - **SHAVIT, N., LIEN, S., SAN PIETRO, A.** : Kinetic studies on the phosphorylation of membrane-bound and free ADP in chloroplast membranes. - Plant Physiol. *57* (Suppl.) : 23, 1976.

27816 - **SHAW, A.B., ANDERSON, M.M., McCARTY, R.E.** : Role of galactolipids in spinach chloroplast lamellar membranes. II. Effects of galactolipid depletion on phosphorylation and electron flow. - Plant Physiol. *57* : 724 - 729, 1976.

27817 - **SHCHERBAKOVA, I.Yu., GILLER, Yu.E.** : O děĭstvii ingibitorov biosinteza RNK i belka na shibatovskiĭ sdvig. [Action of inhibitors of RNA and protein biosynthesis on the Shibata shift.] - Dokl. Akad. Nauk tadzh. SSR *19* (12) : 45 - 48, 1976. [In R, ab : Tajik.]

27818 - **SHEATH, R.G.** : Changes in plastid structure and photosynthetic rate of *Porphyra leucosticta* conchocelis phase in response to dark incubation. - J. Phycol. *12* (Suppl.) : 22, 1976.

27819 - **SHEEHY, J.E., CHAPAS, L.C.** : The measurement and distribution of irradiance in clear and overcast conditions in four temperate forage grass canopies. - J. appl. Ecol. *13* : 831 - 840, 1976.

27820 - **SHELL, G.S.G., LANG, A.R.G.** : Movements of sunflower leaves over a 24-h period. - Agr. Meteorol. *16* : 161 - 170, 1976. [Ps.]

27821 - **SHELP, B., URSINO, D.J.** : Radiation-induced changes in the translocation of photoassimilates in soybean. - Plant Physiol. *57* (Suppl.) : 28, 1976.

27822 - **SHEMIN, D.** : 5-aminolaevulinic acid dehydratase : structure, function, and mechanism. - Phil. Trans. roy. Soc. London *B 273* : 109 - 115, 1976.

27823 - **SHERIDAN, R.P.** : Sun-shade ecotypes of a bluegreen alga in a hot spring. - J. Phycol. *12* : 279 - 285, 1976. [Ps, Chl.]

27824 - **SHERIDAN, R.P., SANDERSON, C., KERR, R.** : Effects of pulp mill emissions on lichens in the Missoula Valley, Montana. - Bryologist *79* : 248 - 252, 1976. [Ps, Chl.]

27825 - **SHERIDAN, R.P., ULIK, T.** : Adaptive photosynthesis responses to temperature extremes by the thermophilic cyanophyte *Synechococcus lividus*. - J. Phycol. *12* : 255 - 261, 1976.

27826 - **SHERIFF, D.W., McGRUDDY, E.** : New apparatus for control of temperature and humidity in leaf and small plant chambers. - J. exp. Bot. *27* : 1376 - 1381, 1976.

27827 - **SHERMAN, L.A.** : Infection of *Synechococcus cedrorum* by the cyanophage AS-1M III. Cellular metabolism and phage development. - Virology *71* : 199 - 206, 1976. [Ps.]

27828 - **SHERMAN, W.V., KORENSTEIN, R., CAPLAN, S.R.** : Energetics and chronology of phototransients in the light response of the purple membrane of *Halobacterium halobium*. - Biochim. biophys. Acta *430* : 454 - 458, 1976.

27829 - **SHERMAN, W.V., SLIFKIN, M.A., CAPLAN, S.R.** : Kinetic studies of phototransients in bacteriorhodopsin. - Biochim. biophys. Acta *423* : 238 - 248, 1976.

27830 - **SHEVELUKHA, V.S., DOVNAR, V.S.** : Fotosinteticheskie aspekty modeli sortov zernovykh kul'tur intensivnogo tipa. [Photosynthetic aspects of a model for intensive type cereal cultivars.] - Sel'skokhoz. Biol. *11* : 218 - 225, 1976. [In R, ab : E.]

27831 - **SHIBATA, H., OCHIAI, H.** : Studies on δ-amino levulinic acid dehydratase in radish cotyledons during chloroplast development. - Plant Cell Physiol. *17* : 281 - 288, 1976.

27832 - **SHIBATA, H., SUEKANE, T., OCHIAI, H.** : [Thiol content and photoreductive activity of developing chloroplast.] - J. agr. chem. Soc. Japan *50* : 49 - 54, 1976. [In Jap., ab : E.]

27833 - **SHIBLES, R.** : Terminology pertaining to photosynthesis. - Crop Sci. *16* : 437 - 439, 1976.

27834 - **SHIDA, S., MATSUNAGA, R., TSUZUKI, E.** : [Monogenic heterosis due to the gene
regulating *chlorina* mutant in yellow lupine.] - Bull. Fac. Agr., Miyazaki
Univ. 23 : 533 - 538, 1976. [In Jap., ab : E.]

27835 - **SHIDA, S., MATSUNAGA, R., TSUZUKI, E.** : [Inheritance of two virescent mutants
in rice plants and variation of chlorophyll content during growth period in
those plants.] - Bull. Fac. Agr. Miyazaki Univ. 23 : 539 - 544, 1976.
[In Jap., ab : E.]

27836 - **SHIEH, P., PACKER, L.** : Photo-induced potentials across a polymer stabilized
planar membrane, in the presence of bacteriorhodopsin. - Biochem. biophys.
Res. Commun. 71 : 603 - 609, 1976.

27837 - **SHIKHOBALOV, V.V., ZABOTIN, A.I.** : Dinamika komponentov uglekislotnogo gazoobmena zeleneyushchikh prorostkov pshenitsy i kukuruzy. [Dynamics of the
components of CO_2 exchange in greening seedlings of wheat and maize.] -
In : Gazometricheskoe Issledovanie Fotosinteza i Dykhaniya Rasteniĭ. Pp.
148 - 151. Akad. Nauk SSSR, Tartu 1976. [In R.]

27838 - **SHIL'KROT, G.S.:** Litoral'nyĭ prirodnyĭ kompleks i ego rol' v evtrofirovanii
vodoemov. [Litoral natural complex and its role in eutrofication of water
reservoirs.] - In : Antropogennoe Evtrofirovanie Ozer. Pp. 82 - 116, 119.
Nauka, Moskva 1976. [In R.]

27839 - **SHINN, J.H., CLEGG, B.R., STUART, M.L., THOMPSON, S.E.** : Exposures of field-
-grown lettuce to geothermal air pollution - photosynthetic and stomatal responses. - J. environm. Sci. Health A 11 : 603 - 612, 1976.

27840 - **SHINOHARA, T., KITANO, H., FUKUDA, M.** : [Effects of inhibitors of photosynthesis on the ozone injury to tobacco plants.] - Bull. Okayama Tobacco exp.
Sta. 36 : 83 - 86, 1976. [In Jap., ab : E.]

27841 - **SHIOI, Y., TAKAMIYA, K., NISHIMURA, M.** : Isolation and some properties of
NAD^+ reductase of the green photosynthetic bacterium *Prosthecochloris aestuarii*. - J. Biochem. (Tokyo) 79 : 361 - 371, 1976.

27842 - **SHIOI, Y., TAKAMIYA, K., NISHIMURA, M.** : Light-induced oxidation-reduction
reactions of cytochromes in the green sulfur photosynthetic bacterium *Prosthecochloris aestuarii*. - J. Biochem. (Tokyo) 80 : 811 - 820, 1976.

27843 - **SHIPMAN, L.L., COTTON, T.M., NORRIS, J.R., KATZ, J.J.** : New proposal for
structure of special-pair chlorophyll. - Proc. nat. Acad. Sci. USA 73 :
1791 - 1794, 1976.

27844 - **SHIPMAN, L.L., NORRIS, J.R., KATZ, J.J.** : Quantum mechanical formalism for
computation of the electronic spectral properties of chlorophyll aggregates.-
J. phys. Chem. 80 : 877 - 882, 1976.

27845 - **SHIRAZI, G.A., STONE, J.F., BACON, C.M.** : Oscillatory transpiration in a cotton plant. II A model. - J. exp. Bot. 27 : 619 - 633, 1976. [Resistances.]

27846 - **SHIRAZI, G.A., STONE, J.F., BACON, C.M.** : Oscillatory transpiration in a cotton plant : An IBM/CSMP computer simulation. - Oklahoma State Univ. techn.
Bull. 143 : 1 - 27, 1976. [Resistances.]

27847 - **SHIRAZI, G.A., STONE, J.F., TODD, G.W.** : Oscillatory transpiration in a cotton plant. I. Experimental characterization. - J. exp. Bot. 27 : 610 - 618,
1976. [Resistances.]

*27848 - **SHISHCHENKO, S.V.** : Vliyanie ul'trafioletovogo izlucheniya na strukturu i
funktsiyu ènergeticheskikh organoidov list'ev gorokha. [Effect of ultraviolet radiation on the structure and function of energetic organoids of pea
leaves.] - Èlektron. Obrabotka Materialov 1973 (1) : 66 - 69,95, 1973. [Chl;
in R.]

27849 - **SHIVASHANKAR, K., VLASSAK, K., LIVENS, J.** : A comparison of the effect of
straw incorporation and CO_2 enrichment on the growth, nitrogen fixation and
yield of soya beans. - J. agr. Sci. 87 : 181 - 185, 1976. [Chl.]

27850 - **SHKLYAEV, Yu.N., SHVETSOVA, T.D.** : Deĭstvie magniya i kobal'ta na temnoustoĭ-
chivost' pigmentov list'ev pshenitsy. [Effect of magnesium and cobalt on the

dark resistance of wheat leaf pigments.] - Agrokhimiya *1976* (9) : 121 - 123, 1976. In R.]

27851 - SHLYK, A.A., CHAÏKA, M.T., KLYUCHAREVA, A.N. : Sopryazhenie lokalizatsii novykh molekul khlorofillov i belkov v membrannoĭ sisteme khloroplastov. [Relationship of localization of new chlorophyll molecules and proteins in the membrane system of chloroplasts.] - Dokl. Akad. Nauk SSSR *226* : 1232 - 1235, 1976. [In R.]

27852 - SHLYK, A.A., CHKANIKOVA, R.A., VLASENOK, L.I., KURILO, E.S., VRUBEL', S.V. : Izbiratel'noe raspredelenie svezheobrazovannykh molekul khlorofillov *a* i *b* pri raznoĭ glubine fragmentatsii fotosinteticheskikh membran digitoninom. [Selective separation of newly formed molecules of chlorophylls *a* and *b* by different fragmentation level of photosynthetic membranes by digitonin.] - Dokl. Akad. Nauk SSSR *228* : 970 - 973, 1976. [In R.]

27853 - SHLYK, A.A., PRUDNIKOVA, I.V., MITSUK, Z.I., SUKHOVER, L.K. : Aktivatsiya temnovogo biosinteza khlorofilla *b* v postétiolirovannykh list'yakh pod vliyaniem khloramfenikola. [Activation of the chlorophyll *b* dark biosynthesis in postetiolated leaves under the effect of chloramphenicol.] - Dokl. Akad. Nauk SSSR *230* : 244 - 247, 1976. [In R.]

27854 - SHLYK, A.A., PRUDNIKOVA, I.V., SAVCHENKO, G.E., AVERINA, N.G., KOSTYUK, N.N., PARAMONOVA, T.K., SHEVCHUK, S.N. : Tsentry biosinteza khlorofilla i regulirovanie protsessa formirovaniya pigmentnogo apparata fotosinteza. [Centers of chlorophyll biosynthesis and the regulation of the process of formation of the pigment apparatus of photosynthesis.] - Izv. Akad. Nauk SSSR, Ser. biol. *1976* : 101 - 120, 1976. [In R, ab : E.]

27855 - SHMAT'KO, I.G., SHAPOVAL, A.I., SHEVCHUK, N.V. : Ustoĭchivost' zelenykh pigmentov k vodnomu defitsitu i povyshennym temperaturam. [Resistance of green pigments to water deficit and elevated temperatures.] - In : Metody Otsenki Ustoĭchivosti Rasteniĭ k Neblagopriyatnym Usloviyam Sredy. Pp. 48 - 54. Kolos, Leningrad 1976. [In R.]

27856 - SHMELEVA, V.A., IVANOVA, B.N., AKULOVA, E.A. : Fotofosforilirovanie i élektronnyĭ transport v khloroplastakh gorokha, vyrashchennogo pri raznoĭ intensivnosti sveta. [Photophosphorylation and electron transport in chloroplasts of pea grown under different illuminance.] - Fiziol. Rast. *23* : 869 - 876, 1976. [In R, ab : E.]

27857 - SHMELEVA, V.L., RUZIEVA, R.Kh., IVANOV, B.N., MUZAFAROV, E.N., AKULOVA, E.A.: Funktsional'naya aktivnost' khloroplastov gorokha, vyrashchennogo pri razlichnoĭ osveshchennosti. [Functional activity of chloroplasts in pea grown at various illumination.] - Fiziol. Biokhim. kul't. Rast. *8* : 612 - 618, 1976. [In R, ab : E.]

27858 - SHOAF, W.T., LIUM, B.W. : Improved extraction of chlorophyll *a* and *b* from algae using dimethyl sulfoxide. - Limnol. Oceanogr. *21* : 926 - 928, 1976.

27859 - SHOAF, W.T., LIUM, B.W. : The measurement of adenosine triphosphate in pure algal cultures and natural aquatic samples. - J. Res. US geol. Survey *4* : 241 - 245, 1976.

27860 - SHOMER-ILAN, A., WAISEL, Y. : Further comments on the effects of NaCl on photosynthesis in *Aeluropus litoralis*. - Z. Pflanzenphysiol. *77* : 272 - 273, 1976.

27861 - SHUBIN, V.V., SINESHCHEKOV, V.A., LITVIN, F.F. : Peremennaya fluorestsentsiya nativnykh form khlorofilla pri -196 °C. [Variable fluorescence of native forms of chlorophyll at -196 °C.]-Biofizika *21* : 760 - 762, 1976. [In R, ab : E.]

27862 - SHUMILOVA, A.A., FEDOSEENKO, A.A., STEPANOVA, A.M. : Vliyanie sveta na funktsionirovanie tsikla Krebsa v list'yakh kukuruzy. [Effect of light on the functioning of the Krebs Cycle in maize leaves.] - Nauch. Dokl. vyssh. Shkoly, biol. Nauki *19* (9) : 87 - 91, 1976. [Ps; in R.]

27863 - SHUTILOVA, N.I., KUTYURIN, V.M. : Vydelenie i issledovanie trekh vidov pigment-belkovolipidnykh kompleksov (PBLK) khloroplastov gorokha : PBLK reaktsionnogo tsentra fotosistemy 1, PBLK reaktsionnogo tsentra fotosistemy 2 i vspomogatel'nogo svetosobirayushchego kompleksa. [Isolation and study of

three types of pigment-lipoprotein complexes (PLC) from pea chloroplasts :
PLC of the reaction centre of photosystem 1, PLC of the reaction centre of
photosystem 2, and auxiliary antenna complex.]-Fiziol. Rast. *23* : 42 - 49,
1976. [In R, ab : E.]

27864 - **SHUTILOVA, N.I., ZHIGAL'TSOVA, Z.V., KUTYURIN, V.M.** : Issledovanie fotoindutsiruemykh izmenenii karotinoidnogo sostava u izolirovannykh pigment-bel-
kovolipoidnykh kompleksov fotosistem 1 i 2 khloroplastov. [Photo-induced
changes in carotenoid composition of pigment-protein-lipid complexes of photosystems 1 and 2 isolated from chloroplasts.] - Fiziol. Rast. *23* : 452 -
459, 1976. [In R, ab : E.]

27865 - **SHUTTLEWORTH, W.J.** : Experimental evidence for the failure of the Penman-Mon-
teith equation in partially wet conditions. - Boundary-Layer Meteorol. *10* :
91 - 94, 1976. [Stomatal resistance.]

27866 - **SHUVALOV, V.A.** : The study of the primary photoprocesses in photosystem I of
chloroplasts. Recombination luminescence, chlorophyll triplet state and tri-
plet-triplet annihilation. - Biochim. biophys. Acta *430* : 113 - 121, 1976.

27867 - **SHUVALOV, V.A., KLIMOV, V.V.** : Pervichnoe razdelenie zaryadov mezhdu moleku-
lami pigmentov v bakterial'nykh reaktsionnykh tsentrakh. [Primary distribution of charge among pigments in the bacterial reaction centres.] - In :
Itogi Issledovaniya Mekhanizma Fotosinteza. Pp. 8 - 12. Pushchino 1976.
[In R.]

27868 - **SHUVALOV, V.A., KLIMOV, V.V.** : The primary photoreactions in the complex cy-
tochrome-P-890.P-760 (bacteriopheophytin$_{760}$) of *Chromatium minutissimum*
at low redox potentials. - Biochim. biophys. Acta *440* : 587 - 599, 1976.

27869 - **SHUVALOV, V.A., KLIMOV, V.V., KRAKHMALEVA, I.N., MOSKALENKO, A.A.,
KRASNOVSKII, A.A.** : Fotoprevrashcheniya bakteriofeofitina v reaktsionnykh
tsentrakh *Rhodospirillum rubrum* i *Chromatium minutissimum*. [Phototransforma-
tions of bacteriopheophytin in reaction centers of *Rhodospirillum rubrum* and
Chromatium minutissimum.] - Dokl. Akad. Nauk SSSR *227* : 984 - 987, 1976.
[In R.]

27870 - **SHUVALOV, V.A., KLIMOV, V.V., KRASNOVSKII, A.A.** : Issledovanie pervichnykh
fotoprotsessov v legkikh fragmentakh khloroplastov. [Primary photoprocesses
in light fragments of chloroplasts.] - Mol. Biol. (Moskva) *10* : 326 - 339,
1976. [In R, ab : E.]

27871 - **SHUVALOV, V.A., KRAKHMALEVA, I.N., KLIMOV, V.V.** : Photooxidation of *P*-960
and photoreduction of *P*-800 (bacteriopheophytin *b*-800) in reaction centers
from *Rhodopseudomonas viridis*. Exciton interaction between the pigment molecules. - Biochim. biophys. Acta *449* : 597 - 601, 1976.

27872 - **SIDERER, Y., HARDT, H., MALKIN, S.** : Effect of uncouplers and ADRY reagents
on delayed and triggered emission from isolated chloroplasts. - FEBS Lett.
69 : 19 - 22, 1976.

27873 - **SID'KO, F.Ya., CHUCHALIN, A.I., TIKHOMIROV, A.A.** : Dykhanie poseva pshenitsy
v zavisimosti ot urovnya FAR i BIKR. [Respiration of a wheat stand as dependent on PhAR and near-IR radiation.] - In : Gazometricheskoe Issledovanie Fotosinteza i Dykhaniya Rastenii. Pp. 129. Akad. Nauk SSSR, Tartu 1976. [Opti-
mization of Ps; in R.]

27874 - **SIEFERMANN, D., YAMAMOTO,H.Y.**: Light-induced de-epoxidation in lettuce chloroplasts. VI. De-epoxidation in grana and in stroma lamellae. - Plant Physiol. *57* : 939 - 940, 1976.

27875 - **SIEGELMAN, M.H., RASCHED, I.R., KUNERT, K.-J., KRONECK, P., BÖGER, P.** :
Plastocyanin : Possible significance of quaternary structure. - Europe. J.
Biochem. *64* : 131 - 140, 1976.

27876 - **SIEGENTHALER, P.-A., DEPÉRY, F.** : Influence of unsaturated fatty acids in
chloroplasts. Shift of the pH optimum of electron flow and relations to ΔpH,
thylakoid internal pH and proton uptake. - Europe. J. Biochem. *61* : 573 -
580, 1976.

27877 - **SIGEE, D.C., EPTON, H.A.S.** : Ultrastructural changes in resistant and susceptible varieties of *Phaseolus vulgaris* following artificial inoculation with *Pseudomonas phaseolica*. - Physiol. Plant Pathol. *9* : 1 - 8, 1976. [Chloroplast.]

27878 - **SIGGEL, U.** : The function of plastoquinone as electron and proton carrier in photosynthesis. - Bioelectrochem. Bioenerg. *3* : 302 - 318, 1976.

27879 - **SILBERSTEIN, B.R., MALKIN, S., HAAS, E.** : A novel short-lived emission from the photosynthetic bacterium *Rhodospirillum rubrum*. - FEBS Lett. *63* : 299 - 303, 1976.

27880 - **SILVA, J.F. DA, FADAYOMI, R.O., WARREN, G.F.** : Cotyledon disc bioassay for certain herbicides. - Weed Sci. *24* : 250 - 252, 1976. [Ps inhibitors.]

27881 - **SILVANOVICH, M.P., HILL, R.D.** : Subunit studies of coupling factor 1 of bean chloroplasts. - Can. J. Biochem. *54* : 481 - 487, 1976.

27882 - **SIMMELSGAARD, S.E.** : Adaptation to water stress in wheat. - Physiol. Plant. *37* : 167 - 174, 1976.

27883 - **SIMPSON, D.I., LEE, T.H.** : Plastoglobules of leaf chloroplasts of two cultivars of *Capsicum annuum*. - Cytobios *15* : 139 - 148, 1976.

27884 - **SIMPSON, D.J., BAQAR, M.R., McGLASSON, W.B., LEE, T.H.** : Changes in ultrastructure and pigment content during development and senescence of fruits of normal and *rin* and *nor* mutant tomatoes. - Aust. J. Plant Physiol. *3* : 575 - 587, 1976.

27885 - **SIMPSON, K.L., LEE, T.-C., RODRIGUEZ, D.B., CHICHESTER, C.O.** : Metabolism in senescent and stored tissues. - In : GOODWIN, T.W. (ed.) : Chemistry and Biochemistry of Plant Pigments. 2nd Ed. Vol. 1. Pp. 779 - 842. Academic Press, London - New York - San Francisco 1976. [Chl, Car.]

27886 - **SINCLAIR, T.R., BUZZARD, G.H., KNOERR, K.R.** : A pressure method for frequent differential calibration of CO_2 infrared analyzer. - Photosynthetica *10* : 188 - 192, 1976.

27887 - **SINCLAIR, T.R., DE WIT, C.T.** : Analysis of the carbon and nitrogen limitations to soybean yield. - Agron. J. *68* : 319 - 324, 1976.

27888 - **SINCLAIR, T.R., MURPHY, C.E. Jr., KNOERR, K.R.** : Development and evaluation of simplified models for simulating canopy photosynthesis and transpiration. - J. appl. Ecol. *13* : 813 - 829, 1976.

27889 - **SINESHCHEKOV, O.A., ANDRIANOV, V.K., KURELLA, G.A., LITVIN, F.F.** : Bioèlektricheskie yavleniya odnokletochnoĭ zhgutikovoĭ vodorosli, ikh svyaz' s fototaksisom i fotosintezom. [Bioelectric phenomena in unicellular flagellar alga and their relation to phototaxis and photosynthesis.] - Fiziol. Rast. *23* : 229 - 237, 1976. [In R, ab : E.]

27890 - **SINESHCHEKOV, V.A., LITVIN, F.F.** : Lyuminestsentsiya bakteriorodopsina purpurnykh membran iz kletok *Halobacterium halobium*. [Luminescence of bacteriorhodopsin in purple membranes from *Halobacterium halobium* cells.] - Biofizika *21* : 313 - 320, 1976. [In R, ab : E.]

27891 - **SINGER, B., SAGER, R., RAMANIS, Z.** : Chloroplast genetics of *Chlamydomonas*. III. Closing the circle. - Genetics *83* : 341 - 354, 1976.

27892 - **SINGH, M.** : The theory of production methods. - Trop. Ecol. *16* : 14 - 27, 1976.

*27893 - **SINGH, U.N., AMBASHT, R.S.** : Biotic stress and variability in structure and organic (net primary) production of grassland communities at Varanasi, India. - Trop. Ecol. *16* (2) : 86 - 95, 1975.

*27894 - **SINGH, U.N., AMBASHT, R.S.** : Energy conserving efficiency of a forest grassland at Varanasi. - Acta bot. indica *3* (2) : 132 - 135, 1975.

27895 - **SINGH, V.P., GUPTA, K.C.** : Effect of mosaic virus on productivity of Bermuda grass. - Photosynthetica *10* : 201 - 203, 1976.

27896 - **SINHA, S.K.** : Genetical aspects of photosynthetic potential in crop plants. - Plant biochem. J. *3* : 81 - 90, 1976.

27897 - **SIONIT, N., KRAMER, P.J.** : Water potential and stomatal resistance of sunflower and soybean subjected to water stress during various growth stages. -
Plant Physiol. *58* : 537 - 540, 1976.

27898 - **SIRENKO, L.A., MYSLOVICH, V.O., GORYUSHIN, V.A., MIKHAĬLYUK, D.P.** :
Vliyanie infektsii tsianofaga AM-1 na metabolizm sine-zelenoĭ vodorosli.
[Effect of infection with cyanophage AM-1 on metabolism of the blue-green
alga.] - Fiziol. Rast. *23* : 1214 - 1218, 1976. [In R, ab : E.]

27899 - **SISSON, W.B., CALDWELL, M.M.** : Photosynthesis, dark respiration, and growth
of *Rumex patientia* L. exposed to ultraviolet irradiance (288 to 315 nanome-
ters) simulating a reduced atmospheric ozone column. - Plant Physiol. *58* :
563 - 568, 1976.

27900 - **SISSONS, C.H.** : Improved technique for accurate and convenient assay of biological reactions liberating $^{14}CO_2$. - Anal. Biochem. *70* : 454 - 462, 1976.

27901 - **SIVALINGAM, P.M., IKAWA, T., NISIZAWA, K.** : Physiological roles of a substance 334 in algae. - Bot. Mar. *19* : 9 - 21, 1976. [Ps, Chl.]

27902 - **SIZOVA, L.I.** : Vliyanie postradiatsionnogo khraneniya semyan na strukturnye
mutatsii khromosom u khlorofil'nykh mutantov podsolnechnika. [Effect of post-
irradiation storage of seeds on structural chromosome mutations in chlorophyll mutants of sunflower.] - Genetika *12* (1) : 12 - 17, 1976. [In R, ab :E.]

27903 - **SIZOVA, L.I.** : Vliyanie stareniya semyan na strukturnye mutatsii khromosom,
indutsirovannye gamma-luchami, u khlorofil'nykh mutantov podsolnechnika.
[Effect of seed ageing on structural chromosome mutations induced by gamma-
-irradiation in chlorophyll mutants of sunflower.] - Genetika *12*(7) : 24 - 30,
1976. [In R, ab : E.]

27904 - **SKULACHEV, V.P.** : A hypothesis of the evolution of biological energy trans-
ducers. - Origins Life *7* : 145 - 160, 1976. [Ps.]

27905 - **SKULACHEV, V.P.** : Conversion of light energy into electric energy by bacteriorhodopsin. - FEBS Lett. *64* : 23 - 25, 1976.

27906 - **SLABAS, A.R., WALKER, D.A.** : Enzymic reconstitution of photosynthetic carbon
assimilation. Pentose phosphate-dependent O_2 evolution by illuminated enve-
lope-free chloroplasts from *Spinacia oleracea*. - Arch. Biochem. Biophys.
175 : 590 - 597, 1976.

27907 - **SLABAS, A.R., WALKER, D.A.** : Localization of inhibition by adenosine diphosphate of phosphoglycerate-dependent oxygen evolution in a reconstituted chloroplast system. - Biochem. J. *154* : 185 - 192, 1976.

27908 - **SLABAS, A.R., WALKER, D.A.** : Transient inhibition by ribose 5-phosphate of
photosynthetic O_2 evolution in a reconstituted chloroplast system. - Biochim. biophys. Acta *430* : 154 - 164, 1976.

27909 - **SLABAS, A.R., WALKER, D.A.** : Inhibition of spinach phosphoribulokinase by
DL-glyceraldehyde. - Biochem. J. *153* : 613 - 619, 1976.

27910 - **SLATER, E.C.** : The role of mitochondrial and chloroplast ATPases in oxidative
and photosynthetic phosphorylation. - J. Biochem. (Tokyo) *79* : 45p, 1976.

*27911 - **SLATER, J.H.** : The control of carbon dioxide assimilation and ribulose 1,5-
-diphosphate carboxylase activity in *Anacystis nidulans* grown in a light-li-
mited chemostat. - Arch. Mikrobiol. *103* : 45 - 49, 1975.

27912 - **SLATYER, R.O.** : Water deficits in timberline trees in the Snowy Mountains of
south-eastern Australia. - Oecologia *24* : 357 - 366, 1976. [Ps.]

27913 - **SLAWYK, G., MINAS, H.J., MINAS, M., PACKARD, T.T.** : A further investigation
on the primary productivity in the divergence zone near the French Mediter-
ranean coast. - Int. Rev. ges. Hydrobiol. *61* : 373 - 381, 1976.

27914 - **SLOCUM, R.D., LAWREY, J.D.** : Viability of the epizoic lichen flora carried
and dispersed by green lacewing (*Nodita pavida*) larvae. - Can. J. Bot.
54 : 1827 - 1831, 1976. [Ps.]

B27915 - **SLONOV, L.Kh.** : Fotosintez i Produktivnost' Yuzhnoĭ Konopli. [Photosynthesis
and Productivity of Southern Hemp.] - El'brus, Nal'chik 1976. [In R.]

27916 - **SLOOTEN, L., SYBESMA, C.** : Photoinactivation of photophosphorylation and
dark ATP-ase in *Rhodospirillum rubrum* chromatophores. - Biochim. biophys.
Acta *449* : 565 - 581, 1976.

27917 - **SLOVACEK, R.E., BANNISTER, T.T.** : Evidence against proton gradient formation
being the cause of chlorophyll fluorescence quenching by N-methylphenazonium
methosulfate. - Biochim. biophys. Acta *430* : 165 - 181, 1976.

27918 - **SLOVTSOV, R.I., ARABI, A.K., DOROZHKINA, L.A.** : Vliyanie gerbitsidov na nakoplenie sakharov v kornyakh svekly.[Effect of herbicides on the accumulation
of sugars in beet roots.] - Khim. sel'sk. Khoz. *14* (11) : 55 - 60, 1976.
[In R.]

27919 - **SMERAGE, G.H.** : Matter and energy flows in biological and ecological systems.
- J. theor. Biol. *57* : 203 - 223, 1976.

27920 - **SMID, A.E., PEASLEE, D.E.** : Growth and CO_2 assimilation by corn as related
to potassium nutrition and simulated canopy shading. - Agron. J. *68* : 904 -
908, 1976.

27921 - **SMILLIE, R.M.** : Photosystem II in symbiotic algae. - Aust . J. Plant Physiol.
3 : 133 - 139, 1976.

27922 - **SMILLIE, R.M.** : Temperature control of chloroplast development. - In :
BÜCHER, T., NEUPERT, W., SEBALD, W., WERNER, S. (ed.) : Genetics and Bioge-
nesis of Chloroplasts and Mitochondria. Pp. 103 - 110. North-Holland Publ.
Co., Amsterdam - New York - Oxford 1976.

27923 - **SMILLIE, R.M., HENNINGSEN, K.W., NIELSEN,N.C., von WETTSTEIN, D.** :
The influence of cations and methylamine on structure and function of thylakoid membranes from barley chloroplasts. - Carlsberg. Res. Commun. *41* : 27 -
56, 1976.

27924 - **SMILLIE, R.M., NIELSEN, N.C., HENNINGSEN, K.W., von WETTSTEIN, D.** :
Using photosynthetic mutants of barley to study development of photochemical
capacity in chloroplast lamellae. - Proc. aust. biochem. Soc. *9* : 50, 1976.

27925 - **SMIRNOVA, L.F.** : Vliyanie urovnya gruntovykh vod osushchennykh torfyanikov
na nakopleniya khlorofilla i karotina v list'yakh kukuruzy. [Effect of the
level of subsurface waters of reclaimed peat bogs on the accumulation of
chlorophyll and carotene in the leaves of maize plants.] - In : **KAKHNOVICH,
L.V.** (ed.) : Optimizatsiya Fotosinteticheskogo Apparata Vozdeĭstviem Razlichnykh Faktorov. Pp. 124 - 134. Izd. BGU, Minsk 1976. [In R.]

27926 - **SMIRNOVA, N.N.** : Vliyanie aminokislot na soderzhanie fotoassimiliruyushchikh
pigmentov i fotosintez u *Phragmites communis* TRIN. [The effect of amino acids
on the content of photoassimilating pigments and photosynthesis in *Phragmites
communis* TRIN.] - Gidrobiol. Zh. *12* (4) : 89 - 92, 1976. [In R.]

27927 - **SMITH, B.B., REBEIZ, C.A.** : Reconstitution of membrane-bound Mg-protoporphy-
rin monoester (MPE) *in vitro*. - Plant Physiol. *57* (Suppl.) : 72, 1976.

27928 - **SMITH, B.N.** : Evolution of C_4 photosynthesis in response to changes in carbon
and oxygen concentrations in the atmosphere through time. - BioSystems *8* :
24 - 32, 1976.

27929 - **SMITH, B.N., OLIVER, J., McMILLAN, C.** : Influence of carbon source, oxygen
concentration, light intensity and temperature on $^{13}C/^{12}C$ ratios in plant
tissues. - Bot. Gaz. *137* : 99 - 104, 1976.

27930 - **SMITH, C.J.** : Adaptations to environmental change in *Picea sitchensis* seed-
ling-changes in photosynthesis, growth and substrate availability. - Ann.
Bot. *40* : 1003 - 1015, 1976.

*27931 - **SMITH, D.** : Influence of temperature on growth of Froker oats for forage.
I. Dry matter yields and growth rates. - Can. J. Plant Sci. *54* : 725 - 730,
1974.

27932 - **SMITH, D.F., WIEBE, W.J.** : Constant release of photosynthate from marine phytoplankton. - Appl. environm. Microbiol. *32* : 75 - 79, 1976.

27933 - **SMITH, D.J., BOYER, P.D.** : Demonstration of a transitory tight binding of
ATP and of committed Pi and ADP during ATP synthesis by chloroplasts. -
Proc. nat. Acad. Sci. USA *73* : 4314 - 4318, 1976.

27934 - **SMITH, D.J., STOKES, B.O., BOYER, P.D.** : Probes of initial phosphorylation events in ATP synthesis by chloroplasts. - J. biol. Chem. *251* : 4165 - 4171, 1976.

27935 - **SMITH, E.W., TOLBERT, N.E., KU, H.S.** : Variables affecting the CO_2 compensation point. - Plant Physiol. *58* : 143 - 146, 1976.

27936 - **SMITH, F.A., RAVEN, J.A.** : Transport and regulation of cell pH. - In : LÜTTGE, U., PITMAN, M.G. (ed.) : Transport in Plants II. Part A. Cells. Pp. 317 - 346. Springer-Verlag, Berlin - Heidelberg - New York 1976. [Ps.]

27937 - **SMITH, I.K.** : Sulfate transport in tobacco cells. - Plant Physiol. *57* (Suppl.) : 95, 1976. [Ps inhibitors.]

27938 - **SMITH, R.C., TYLER, J.E.** : Transmission of solar radiation into natural waters. - In : SMITH, K.C. (ed.) : Photochemical and Photobiological Reviews. Vol. 1. Pp. 117 - 155. Plenum Press, New York - London 1976.

27939 - **SMITH, W.K., NOBEL, P.S.** : Temperature and water relationships for sun and shade leaves of a desert broadleaf (*Hyptis emoryi* TORR.). - Plant Physiol. *57* (Suppl.) : 104, 1976.

27940 - **SMRCHEK, J.C., CAIRNS, J. Jr., DICKSON, K.L., KING, P.H., RANDALL, C.W., CROWE, J., HUBER, D., OLVER, J.W.** : The effects of various tertiary treatment nutrient removal schemes on periphyton communities in model laboratory streams. - Virginia polytech. Inst. State Univ., Virginia Water Resources Res. Center Bull. *86* : 1 - 124, 1976. [Chl.]

27941 - **SO, M.L., THROWER, L.B.** : The host-parasite relationship between *Vigna sesquipedalis* and *Uromyces appendiculatus*. I. Development of parasitic colonies and the pattern of photosynthesis. - Phytopathol. Z. *85* : 320 - 332, 1976.

27942 - **SO, M.L., THROWER, L.B.** : The host-parasite relationship between *Vigna sesquipedalis* and *Uromyces appendiculatus*. II. Movement of photosynthate and levels of growth substances. - Phytopathol. Z. *86* : 252 - 265, 1976.

27943 - **SOFROVÁ, D., VILÍM, V., LEBLOVÁ, S.** : Effect of trypsin on photochemical reactions of the blue-green alga *Plectonema boryanum*. - Photosynthetica *10* : 40 - 46, 1976.

27944 - **SOFROVÁ, D., VILÍM, V., LEBLOVÁ, S.** : Non-reactivation of the trypsin-inhibited Hill reaction activity in the blue-green alga *Plectonema boryanum*. - Photosynthetica *10* : 198 - 200, 1976.

27945 - **SOJKA, G.A., GEST, H., MARRS, B.** : Molecular biology of photosynthetic bacteria. - BioScience *26* : 119 - 120, 1976.

27946 - **SOKOLOVE, P.M., MARSHO, T.V.** : Ascorbate-independent carotenoid de-epoxidation in intact spinach chloroplasts. - Biochim. biophys. Acta *430* : 321 - 326, 1976.

27947 - **SOLAZZI, A., TOLOMIO, C.** : Effetti della profondita' sull'apparato fotosintetico delle alghe. I - I pigmenti clorofilliani in *Halimeda tuna* LAM. (*Chlorophyceae Siphonales*). [Sea depth effects on the algal photosynthetic apparatus. I - Chlorophyll pigments of *Halimeda tuna* LAM. (*Chlorophyceae Siphonales*).] - Mem. Biol. mar. Oceanogr. N.S. *6* (1) : 21 - 27, 1976. [In Ital., ab : E.]

27948 - **SOLDATENKOV, S.V., KOVALEVA, L.B., YUZBEKOV, A.K.** : Mekhanizm temnovogo i svetozavisimogo obrazovaniya yablochnoĭ kisloty v list'yakh *Bryophyllum daigremontianum*. [Mechanism of dark- and light- dependent formation of malic acid in the leaves of *Bryophyllum daigremontianum*.] - Vestn. leningr. Univ., Biol. *1976* (3) : 113 - 117, 1976. [In R, ab : E.]

27949 - **SOLOV'EV, E.V., KARMANOV, V.G.** : Avtomaticheskiĭ perenosnyĭ pribor dlya izucheniya uglekislotnogo gazoobmena rasteniĭ. [Automatic transportable apparatus for studying CO_2 exchange in plants.] - In : Gazometricheskoe Issledovanie Fotosinteza i Dykhaniya Rasteniĭ. Pp. 130 - 132. Akad. Nauk SSSR, Tartu 1976. [In R.]

27950 - **SOLOV'EV, I.S., SAMOĬLOVA, O.P., BLYUMENFEL'D, L.A.** : Vliyanie dlinnovolnovoĭ podsvetki na kineticheskie kharakteristiki signalov mikrovolnovoĭ fotoprovodimosti kletok khlorelly i ėffekt Ėmersona. [The effect of long-wave

illumination on the kinetic characteristics of microwave photoconductivity
signals and Emerson effect for *Chlorella* cells.] - Biofizika *21* : 1035 -
1037, 1976. [In R, ab : E.]

27951 - **SONDAHL, M.R., CROCOMO, O.J., SODEK, L.** : Measurements of ^{14}C incorporation
by illuminated intact leaves of coffee plants from gas mixtures containing
$^{14}CO_2$. - J. exp. Bot. *27* : 1187 - 1195, 1976.

27952 - **SONG, P.-S., KOKA, P., PRÉZELIN, B.B., HAXO, F.T.** : Molecular topology of
the photosynthetic light-harvesting pigment complex, peridinin-chlorophyll
a-protein, from marine dinoflagellates. - Biochemistry *15* : 4422 - 4427, 1976.

27953 - **SOROKIN, E.M.** : Kinetika lyuminestsentsii khlorofilla i perenosa èlektronov
v fotosisteme II fotosinteticheskogo apparata rasteniĭ. [Kinetics of chlorophyll luminescence and electron transmission in photosystem II of plant photosynthetic apparatus.] - Izv. Akad. Nauk SSSR,Ser. biol. *1976* : 894 - 902,
1976. [In R, ab : E.]

27954 - **SOROKIN, E.M.** : Kinetika zamedlennoĭ fluorestsentsii khlorofilla *a in vivo*
v otsutstvie transporta èlektronov na aktseptornoĭ storone fotosistemy 2.
[Kinetics of retarded chlorophyll *a* fluorescence *in vivo* in absence of electron transport on the acceptor side of photosystem 2.] - Biofizika *21* :
665 - 668, 1976. [In R, ab : E.]

27955 - **SOROKIN, E.M.** : Zavisimost' dlitel'nosti fluorestsentsii khlorofilla *a* v khloroplastakh ot kontsentratsii ferritsianida i vitamina K_3. [Dependence of
fluorescence duration of chlorophyll *a* in chloroplasts on the concentration
of ferricyanide and vitamin K_3.] - Biofizika *21* : 181 - 182, 1976. [In R,
ab : E.]

27956 - **SOURNIA, A.** : Abondance du phytoplancton et absence de récifs coralliens sur
les côtes des Îles Marquises. - Compt. rend. Acad. Sci. Paris, Sér.D *282* :
553 - 555, 1976. [Primary production, Chl.]

27957 - **SOURNIA, A.** : Primary production of sands in the lagoon of an atoll and the
role of foraminiferan symbionts. - Mar. Biol. *37* : 29 - 32, 1976.

27958 - **SOURNIA, A.** : Ecologie et productivité d'une Cyanophycée en milieu coralli-
en : *Oscillatoria limosa* AGARDH. - Phycologia *15* : 363 - 366, 1976.

27959 - **SOURNIA, A.** : Oxygen metabolism of a fringing reef in French Polynesia. -
Helgoländer wiss. Meeresuntersuch. *28* : 401 - 410, 1976.

27960 - **SOURNIA, A., RICARD, M.** : Données sur l'hydrologie et la productivité du la-
gon d'un atoll fermé (Takapoto,Iles Tuamotu). - Vie Milieu, Sér. B *26* :
243 - 279, 1976.

27961 - **SOURNIA, A., RICARD, M.** : Phytoplankton and its contribution to primary productivity in two coral reef areas of French Polynesia. - J. exp. mar. Biol.
Ecol. *2* : 129 - 140, 1976.

27962 - **SPIESS, H., ARNOLD, C.G.** : Ribosomal proteins from the chloroplast of *Chlamydomonas reinhardii* and two streptomycin-resistant mutants. - Ber. deut.
bot. Ges. *88* : 391 - 398, 1976.

27963 - **SPILLER, H., BOOKJANS, G., BÖGER, P.** : The influence of oxygen on nitrite
reduction in a reconstituted system. - Z. Naturforsch. *31 C* : 565 - 568,
1976. [Ferredoxin.]

27964 - **SPODNIEWSKA, I.** : Changes in the structure and production of phytoplankton
in Mikołajskie Lake 1963 - 1972. - Limnologica *10* : 299 - 306, 1976.

27965 - **SPONHOLTZ, D.K., BRAUTIGAN, D.L., LOACH, P.A., MARGOLIASH, E.** : Preparation
of cytochrome c_2 from *Rhodospirillum rubrum*. - Anal. Biochem. *72* : 255 - 260,
1976.

27966 - **SPREY, B.** : Intrathylakoidal occurrence of ribulose 1,5-diphosphate carboxylase in spinach chloroplasts. - Z. Pflanzenphysiol. *78* : 85 - 89, 1976.

27967 - **SPREY, B., GLIEM, G., JÁNOSSY, A.G.S.** : Iron containing inclusions in chloroplasts of *Nicotiana clevelandii* and *Nicotiana glutinosa*. I. X-ray microana-
lysis and ultrastructure. - Z. Pflanzenphysiol. *79* : 165 - 176, 1976.

27968 - **SPREY, B., LAETSCH, W.M.** : Chloroplast envelopes of *Spinacia oleracea* L. II. Ultrastructure of chloroplast envelopes and lamellae. - Z. Pflanzenphysiol. *78* : 146 - 163, 1976.

27969 - **SPREY, B., LAETSCH, W.M.** : Chloroplast envelopes of *Spinacia oleracea* L. III. Freeze-fracturing of chloroplast envelopes. - Z. Pflanzenphysiol. *78* : 360 - 371, 1976.

27970 - **SREENIVASAN, A.** : Limnological studies of and primary production in temple pond ecosystems. - Hydrobiology *48* : 117 - 123, 1976.

27971 - **SRIDHAR, R., REDDY, P.R., ANJANEYULU, A.** : Physiology of rice tungro virus disease : changes in chlorophyll, carbohydrates, amino acids and phenol contents. - Phytopathol. Z. *86* : 136 - 143, 1976.

27972 - **SRIVASTAVA, H.S.** : Inhibition of maize seedling growth by chloramphenicol. - Indian J. Plant Physiol. *19* : 53 - 59, 1976. [Chl.]

27973 - **SRIVASTAVA, H.S.** : Some aspects of nitrate assimilation in the seedlings of normal and opaque-2 mutant of maize. - J. exp. Bot. *27* : 1215 - 1222, 1976. [Chl.]

27974 - **SSYMANK, V., STEUP, M., SENGER, H.** : Studies on nucleic acids in plastids of the pigment mutant C-2A' of *Scenedesmus obliquus*. - Plant Cell Physiol. *17* : 787 - 798, 1976.

27975 - **STABENAU, H.** : Microbodies from *Spirogyra*. Organelles of a filamentous alga similar to leaf peroxisomes. - Plant Physiol. *58* : 693 - 695, 1976.

27976 - **STABENAU, H.** : Peroxisomes in the alga *Spirogyra*. - Plant Physiol. *57* (Suppl.) : 94, 1976.

*27977 - **STAEHELIN, L.A.** : Chloroplast membrane structure. Intramembranous particles of different sizes make contact in stacked membrane regions. - Biochim. biophys. Acta *408* : 1 - 11, 1975.

27978 - **STAEHELIN, L.A.** : Reversible particle movements associated with unstacking and restacking of chloroplast membranes *in vitro*. - J. Cell Biol. *71* : 136 - 158, 1976.

27979 - **STAKHOV, L.F., MAKAROV, A.D., IVANOV, B.N.** : Pterin-belkovyĭ kompleks i ferredoksin kak vozmozhnye komponenty molekulyarnogo kompleksa v fotosisteme I. [Pterin-protein complex and ferredoxin as possible components of a molecular complex in photosystem I.] - Biokhimiya *41* : 655 - 659, 1976. [In R, ab : E.]

27980 - **STANEV, V.P.** : Izmeneniya na fotosintetichnite pokazateli na sl"nchogleda v zavisimost ot toreneto i g"stotata na poseva. [Changes in sunflower photosynthetic parameters in dependence on fertilizer application and seeding density.] - Fiziol. Rast. (Sofia) *2* (2) : 28 - 38, 1976. [In Bulg., ab : E, R.]

27981 - **STANHILL, G.** : Cotton. - In : **MONTEITH, J.L.** (ed.) : Vegetation and the Atmosphere. Vol. 2. Case Studies. Pp. 121 - 150. Academic Press, London - New York - San Francisco 1976. [Production.]

27982 - **STANHILL, G.** : Trends and deviations in the yield of the English wheat crop during the last 750 years. - Agro-Ecosystems *3* : 1 - 10, 1976. [Primary production.]

27983 - **STANHILL, G., MORESHET, S., FUCHS, M.** : Effect of increasing foliage and soil reflectivity on the yield and water use efficiency of grain sorghum. - Agron. J. *68* : 329 - 332, 1976.

27984 - **STANKOVA, P.G.** : Vliyanie na pochvenoto zasushavane v"rkhu intenzivnostta na rastezha pri nyakoi sortove zimna meka pshenitsa. [Effect of water stress on the growth rate of some winter soft wheat cultivars.] - Fiziol. Rast. (Sofia) *2* (1) : 13 - 21, 1976. [In Bulg., ab : E, R.]

27985 - **STANKOVIĆ, Ž.** : Medijum kao faktor za izolaciju intaktnih etioplasta. [Medium as isolation factor for intact etioplasts.] - Acta bot. croat. *35* : 65 - 70, 1976. [In Croat., ab : E.]

27986 - **STANLEY, D.W.** : Productivity of epipelic algae in tundra ponds and a lake near Barrow, Alaska. - Ecology *57* : 1015 - 1024, 1976. [Ps, Chl.]

27987 - **STANLEY, D.W.** : A carbon flow model of epipelic algal productivity in Alaskan tundra ponds. - Ecology *57* : 1034 - 1042, 1976.

27988 - **STANLEY, D.W., DALEY, R.J.** : Environmental control of primary productivity in Alaskan tundra ponds. - Ecology *57* : 1025 - 1033, 1976.

*27989 - **STANLEY, R.A., NAYLOR, A.W.** : Glycolate metabolism in Eurasian watermilfoil (*Myriophyllum spicatum*). - Plant Physiol. *29* : 60 - 63, 1973. [Photorespiration.]

27990 - **STARCK, Z.** : Różne aspekty wpływu regulatorów wzrostu na fotosyntezę i przemieszczanie metabolitów. [Various aspects of the effect of growth regulators on photosynthesis and metabolite transport.] - Wiadom. bot. *20* : 81 - 95, 1976. [In Pol.]

27991 - **STARCK, Z., UBYSZ, L.** : Source-sink relationships in radish plant. - Acta Soc. Bot. Pol. *45* : 477 - 493, 1976. [Ps.]

27992 - **STEELE, J.A., UCHYTIL, T.F., DURBIN, R.D., BHATNAGAR, P., RICH, D.H.** : Chloroplast coupling factor 1 : A species-specific receptor for tentoxin. - Proc. nat. Acad. Sci. USA *73* : 2245 - 2248, 1976.

27993 - **STEEMANN NIELSEN, E., BRUUN LAURSEN, H.** : Effect of $CuSO_4$ on the photosynthetic rate of phytoplankton in four Danish lakes. - Oikos *27* : 239 - 242, 1976.

27994 - **STEEMANN NIELSEN, E., ROCHON, T.** : The influence of extremely high concentrations of inorganic P at varying pH on the growth and photosynthesis of unicellular algae. - Int. Rev. ges. Hydrobiol. *61* : 407 - 415, 1976.

27995 - **STEER, B.T., PEARSON, C.J.** : Photosynthate translocation in *Capsicum annuum*. - Planta *128* : 155 - 162, 1976.

27996 - **STEFANOVIĆ, L., PLESNIČAR, M.** : Delovanje atrazina na fotosintezu nekih linija i hibrida kukuruza. [The effect of atrazine on photosynthesis in some inbred lines and hybrids of maize.] - Acta bot. croat. *35* : 77 - 85, 1976. [In Croat., ab : E.]

27997 - **STEFFENS, D., BLOS, I., SCHOCH, S., RÜDIGER, W.** : Lichtabhängigkeit der Phytolakkumulation. Ein Beitrag zur Frage der Chlorophyll-Biosynthese. - Planta *130* : 151 - 158, 1976.

27998 - **STEGMAN, E.C., SCHIELE, L.H., BAUER, A.** : Plant water stress criteria for irrigation scheduling.- Trans.ASAE *19*: 850 - 855, 1976. [Ps.]

27999 - **STEIN, M., WILLENBRINK, J.** : Zur Speicherung von Saccharose in der wachsenden Zuckerrübe. - Z. Pflanzenphysiol. *79* : 310 - 322, 1976. [Photosynthates.]

28000 - **STEMLER, A.** : Dark uptake of HCO_3^- by washed chloroplast grana. - Carnegie Inst. Year Book *75* : 477 - 479, 1976.

28001 - **STEMLER, A.** : Binding of bicarbonate ions to chloroplast grana. - Plant Physiol. *57* (Suppl.) : 60, 1976.

28002 - **STEPANOVA, A.M., NIKIFOROVA, L.F.** : Issledovanie kharaktera zavisimosti skorosti fotofosforilirovaniya khloroplastov gorokha ot nekotorykh faktorov vneshneĭ sredy. [Characteristics of the effect of some environmental factors of photophosphorylation rate in pea chloroplasts.] - Vest. leningr. Univ.,Biol. *1976* (3) : 111 - 117, 155, 1976. [In R, ab : E.]

28003 - **STEPANOVA, A.M., SHUMILOVA, A.A.** : Ingibirovanie fotosinteza transakonitovoĭ kislotoĭ. [Inhibition of photosynthesis by transaconitic acid.] - Vestn. leningr. Univ.,Biol. *1976* (3) : 118 - 120, 1976. [In R, ab : E.]

28004 - **STEPHENS, D.W., GILLESPIE, D.M.** : Phytoplankton production in the Great Salt Lake, Utah, and a laboratory study of algal response to enrichment. - Limnol. Oceanogr. *21* : 74 - 87, 1976.

28005 - **STEPHENSON, R.A., BROWN, R.H., ASHLEY, D.A.** : Translocation of ^{14}C-labeled assimilate and photosynthesis in C_3 and C_4 species. - Crop Sci. *16* : 285 - 288, 1976.

28006 - **STEUP, M., PEAVEY, D.G., GIBBS, M.** : The regulation of starch metabolism by inorganic phosphate. - Biochem. biophys. Res. Commun. *72* : 1554 - 1561, 1976.

28007 - **STEVENS, B.J., CURGY, J.-J., LEDOIGT, G., ANDRÉ, J.** : Comparative studies of ribosomes from mitochondria, chloroplasts and cytoplasm. Morphology and electrophoretic behavior. - In : BÜCHER, T., NEUPERT, W., SEBALD, W., WERNER, S. (ed.) : Genetics and Biogenesis of Chloroplasts and Mitochondria. Pp. 731 - 740. North-Holland Publ. Co., Amsterdam - New York - Oxford 1976.

28008 - **STEVENS, C.L.R., MYERS, J.** : Characterization of pigment mutants in a blue-green alga, *Anacystis nidulans*. - J. Phycol. *12* : 99 - 105, 1976.

28009 - **STEWART, R., CODD, G.A.** : The purification and photoregulation of ribulose diphosphate carboxylase from *Scenedesmus quadricauda*. - Plant Physiol. *57* (Suppl.) : 6, 1976.

28010 - **STIGTER, C.J., BIRNIE, J., JANSEN, P.** : Multi-point temperature measuring equipment for crop environment, with some results on horizontal homogeneity in a maize crop. 1. Field results. - Neth. J. agr. Sci. *24* : 223 - 237, 1976.

28011 - **STIGTER, C.J., LENGKEEK, J.G., KODIJMAN, J.** : A simple worst case analysis for estimation of correct scanning rate in a micrometeorological experiment. - Neth. J. agr. Sci. *24* : 3 - 16, 1976.

28012 - **ST.JOHN, J.B., HILTON, J.L.** : Structure versus activity of substituted pyridazinones as related to mechanism of action. - Weed Sci. *24* : 579 - 582, 1976. [Chl, Car.]

28013 - **STOBBS, T.H., IMRIE, B.C.** : Variation in yield, canopy structure, chemical composition and *in vitro* digestibility within and between two *Desmodium* species and interspecific hybrids. - Trop. Grasslands *10* : 99 - 106, 1976.

28014 - **STOCKER, O.** : The water-photosynthesis syndrome and the geographical plant distribution in the Saharan deserts. - In : LANGE, O.L., KAPPEN, L., SCHULZE, E.-D. (ed.) : Water and Plant Life. Pp. 506 - 521. Springer-Verlag, Berlin - Heidelberg - New York 1976.

B28015 - **STOCKING, C.R., HEBER, N.** (ed.) : Transport in Plants III. Intracellular Interactions and Transport Processes. (Ecycl. Plant Physiol. N.S. Vol. 3). Springer-Verlag, Berlin - Heidelberg - New York 1976. [Ps.]

28016 - **STOCKNER, J.G., COSTELLA, A.C.** : Field and laboratory studies on effects of pulp mill effluent on growth of marine phytoplankton in coastal waters of British Columbia. - Environmental Protect. Serv. Rep. *5-PR-76-9* : 1 - 60, 1976. [Production.]

28017 - **STOCKNER, J.G., SHORTREED, K.R.S.** : Autotrophic production in Carnation Creek, a coastal rainforest stream on Vancouver Island, British Columbia. - J. Fish. Res. Board Can. *33* : 1553 - 1563, 1976.

28018 - **STOEV, K., DOBREVA, S.** : Influence du mode de conduite de la vigne sur la photosynthèse et la distribution des substances elaborées. - Connais. Vigne Vin *10* : 125 - 139, 1976.

*28019 - **STOEV, K., DOBREVA, S., SLAVCHEVA, T., GADEVSKA, A.** : Prouchvane na fotosintezata pri st"bleno otglezhdane na lozata. [Photosynthesis of bole cultivated grape vine.] - Gradin. lozar. Nauka *10* (4) : 89 - 97, 1973. [In Bulg., ab : F, R.]

28020 - **STOEV, K.D., DOBREVA, S.I.** : Fotosintez vinogradnoĭ lozy i raspredelenie assimilyatov v zavisimosti ot formirovaniya kustov. [Photosynthesis in grapevines and the distribution of assimilates as affected by shrub farming.] - Sel'skokhoz. Biol. *11* : 622 - 626, 1976. [In R, ab : E.]

28021 - **STOLOVITSKIĬ, Yu.M., EVSTIGNEEV, V.B.** : Perenos ėlektrona pri osveshchenii i razdelenie zaryadov v khlorofillovykh sistemakh. [Electron transport during irradiation and charge separation in chlorophyll systems.] - In : Itogi Issledovaniya Mekhanizma Fotosinteza. Pp. 20 - 25. Pushchino 1976. [In R.]

28022 - **STONER, E.R., BAUMGARDNER, M.F., SWAIN, P.H.** : Determining density of maize canopy from digitized photographic data. - Agron. J. *68* : 55 - 59, 1976.

*28023 - **STOY, V.** : Assimilatbildung und -verteilung als Komponenten der Ertragsbildung beim Getreide. - Angew. Bot. *47* : 17 - 26, 1973.

28024 - STRAIN, B.R., HIGGINBOTHAM, K.O., MULROY, J.C. : Temperature preconditioning
and photosynthetic capacity of *Pinus taeda* L. - Photosynthetica *10* : 47 - 53,
1976.

28025 - STRAŠKRABA, M. : Empirical and analytical models of eutrophication. - In :
EUTROSYM '76 - Proceedings of the International Symposium on Eutrophication
and Rehabilitation of Surface Waters. Vol. III. Pp. 352 - 371. Karl-Marx-
-Stadt 1976. [Chl.]

28026 - STRAŠKRABA, M. : Development of an analytical phytoplankton model with parameters empirically related to dominant controlling variables. - Abhandl.
Akad. Wiss. DDR *1974* (Umweltbiophysik) : 33 - 65, 1976. [Ps, Chl.]

28027 - STRASSER, R.J., BIFANO, T., BUTLER, W.L. : Absorbance changes associated
with thylakoid fusion. - Plant Physiol. *57* (Suppl.) : 74, 1976.

28028 - STRASSER, R.J., BUTLER, W.L. : Energy transfer in the photochemical apparatus
of flashed bean leaves. - Biochim. biophys. Acta *449* : 412 - 419, 1976.

28029 - STRASSER, R.J., BUTLER, W.L. : Correlation of absorbance changes and thylakoid fusion with the induction of oxygen evolution in bean leaves greened by
brief flashes. - Plant Physiol. *58* : 371 - 376, 1976.

28030 - STRAUSS, G. : Optical spectroscopy of bilayer membranes. - Photochem. Photobiol. *24* : 141 - 153, 1976. [Chl.]

28031 - STRITZKE, J.F., CROY, L.I., McMURPHY, W.E. : Effect of shade and fertility
on NO_3-N accumulation, carbohydrate content, and dry matter production of
tall fescue. - Agron. J. *68* : 387 - 389, 1976.

28032 - STRØBAEK, S., GIBBONS, G.C. : Ribulose-1,5-diphosphate carboxylase from barley (*Hordeum vulgare*). Isolation, characterization, and peptide mapping
studies of the subunits. - Carlsberg Res. Commun. *41* : 57 - 72, 1976.

28033 - STRØBAEK, S., GIBBONS, G.C., HASLETT, B., BOULTER, D., WILDMAN, S.G. :
On the nature of the polymorphism of the small subunit of ribulose-1,5-di-
phosphate carboxylase in the amphidiploid *Nicotiana tabacum*. - Carlsberg
Res. Commun. *41* : 335 - 343, 1976.

28034 - STROTMANN, H., BICKEL, S., HUCHZERMEYER, B. : Energy-dependent release of
adenine nucleotides tightly bound to chloroplast coupling factor CF_1. -
FEBS Lett. *61* : 194 - 198, 1976.

28035 - STROTMANN, H., MURAKAMI, S. : Energy transfer between cell compartments. -
In : STOCKING, C.R., HEBER, U. (ed.) : Transport in Plants III. Intracellu-
lar Interactions and Transport Processes. Pp. 398 - 416. Springer-Verlag,
Berlin - Heidelberg - New York 1976.

28036 - STROUSE, C.E. : Structural studies related to photosynthesis : a model for
chlorophyll aggregates in photosynthetic organisms. - In : LIPPARD, S.J.
(ed.) : Chemistry. Vol. 21. Pp. 159 - 177. John Wiley & Sons, New York -
- London - Sydney - Toronto 1976.

28037 - SUBCHINSKI, V.K., RUUGE, É.K., TIKHONOV, A.N. : Issledovanie strukturnykh
perestroek membran khloroplastov metodom paramagnitnykh zondov. [Structural
rearrangement of chloroplast membranes studied by paramagnetic probes.] -
Fiziol. Rast. *23* : 660 - 665, 1976. [In R, ab : E.]

28038 - SUBRAHMANYAM, P., GOPAL, G.R., MALAKONDAIAH, N., REDDY, M.N. : Physiological
changes in rust infected groundnut leaves. - Phytopathol. Z. *87* : 107 - 113,
1976. [Chl.]

28039 - SUD'INA, O.G., GOLOD, M.G., DOVBYSH, K.P., BAÏDULOVA-BABKO, T.Yu. :
Dynamika vmistu pigmentiv ta khlorofilaznoï aktyvnosti riznykh bilkovykh
fraktsiï v ontogenezi lystka. [Dynamics of pigment content and chlorophyllase activity of different protein fractions in leaf ontogenesis.] - Ukr. bot.
Zh. *33* : 132 - 136, 1976. [In Ukr., ab : E.]

28040 - SUGIYAMA, T., SHIRAHASHI, K. : [C_4 photosynthesis from the viewpoint of enzymes.] - Kagaku To Seibutsu[Chemistry and Life] *14* : 23 - 27, 1976. [In Jap.]

28041 - SUMMERFIELD, R.J. : Use of glasshouses as an adjunct to field research on
soybeans. - World Soybean Res. *1976* : 358 - 370, 1976. [Models of growth.]

28042 - **SUMPER, M., HERRMANN, G.** : Biogenesis of purple membrane : regulation of bacterio-opsin synthesis. - FEBS Lett. *69* : 149 - 152, 1976.

28043 - **SUMPER, M., HERRMANN, G.** : Biosynthesis of purple membrane : Control of retinal synthesis by bacterio-opsin. - FEBS Lett. *71* : 333 - 336, 1976.

28044 - **SUN-FU SHIH** : Methods of computing area. - Agron. J. *68* : 827 - 829, 1976.

28045 - **SÜSS, K.-H.** : Identification of chloroplast thylakoid membrane polypeptides : coupling factor of photophosphorylation (CF_1) and cytochrome *f*. - FEBS Lett. *70* : 191 - 196, 1976.

28046 - **SÜSS, K.-H., SCHMIDT, O., MACHOLD, O.** : The action of proteolytic enzymes on chloroplast thylakoid membranes. - Biochim. biophys. Acta *448* : 103 - 113, 1976.

28047 - **SUTCLIFFE, W.H. Jr., ORR, E.A., HOLM-HANSEN, O.** : Difficulties with ATP measurements in inshore waters. - Limnol. Oceanogr. *21* : 145 - 149, 1976.

28048 - **SUTER, W., LUTZ, H.U., BACHOFEN, R.** : Phosphate binding to chromatophores of *Rhodospirillum rubrum*. - Europe.J. Biochem. *67* : 57 - 60, 1976.

28049 - **SUTTON, B.G., TING, I.P., TROUGHTON, J.H.** : Seasonal effects on carbon isotope composition of cactus in a desert environment. - Nature *261* : 42 - 43, 1976.

28050 - **SUZUKI, S.** : The role of cotyledons in growth and photosynthesis of radish plants. - J. jap. Soc. hort. Sci. *45* : 275 - 282, 1976.

28051 - **SVOBODA, J., HOŠEK, P.** : Arctic sun simulator for ecophysiological studies. - Arctic alpine Res. *8* : 393 - 398, 1976.

28052 - **SWAN, F.R. Jr., LIETH, H.** : Measurement of solar radiation under forest canopies by use of chlorophyll extracts. - J. Biogeogr. *3* : 237 - 247, 1976.

28053 - **SWANSON, C.A., HODDINOTT, J., SIJ, J.W.** : The effect of selected sink leaf parameters on translocation rates. - In : **WARDLAW, I.F., PASSIOURA, J.B.** (ed.) : Transport and Transfer Processes in Plants. Pp. 347 - 356. Academic Press, New York - San Francisco - London 1976. [Ps.]

28054 - **SWANSON, C.D., BACHMANN, R.W.** : A model of algal exports in some Iowa streams. - Ecology *57* : 1076 - 1080, 1976. [Chl.]

28055 - **SWENBERG, C.E., DOMINIJANNI, R., GEACINTOV, N.E.** : Effects of pigment heterogeneity on fluorescence in photosynthetic units. - Photochem. Photobiol. *24* : 601 - 604, 1976.

28056 - **SWENBERG, C.E., GEACINTOV, N.E., POPE, M.** : Bimolecular quenching of excitons and fluorescence in the photosynthetic unit. - Biophys. J. *16* : 1447 - 1452, 1976.

28057 - **ŚWIEBODA, M.** : Chlorophyll content in pine (*Pinus silvestris* L.) needles exposed to flue dust from lead and zinc works. - Acta Soc. Bot. Pol. *45* : 411 - 419, 1976.

28058 - **SÝKOROVÁ, M.** : Intensity of photosynthesis of the leaves and of the bracts of maize (*Zea mays* L.) when measured by the gravimetric method. - Acta Fac. Rerum nat. Univ. comen., Physiol. Plant *11* : 61 - 67, 1976.

28059 - **SYVERTSEN, J.P., NICKELL, G.L., SPELLENBERG, R.W., CUNNINGHAM, G.L.** : Carbon reduction pathways and standing crop in three Chihuahuan desert plant communities. - Southwest. Natur. *21* : 311 - 320, 1976.

28060 - **SZABOLCS, J.** : Some studies on the stereochemistry of carotenoids. - Pure appl. Chem. *47* : 147 - 159, 1976.

28061 - **SZABUNIEWICZ, B.** : Inicjalne mechanizmy rozwoju fotosyntezy. [Initial mechanisms of the development of photosynthesis.] - Kosmos (Warszawa), Ser. A *25* (1) : 7 - 34, 1976. [In Pol.]

28062 - **SZANIAWSKI, R.K.** : Charakterystyka niektórych procesów fizjologicznych w trakcie rozwoju i starzenia aparatu asymilacyjnego roślin iglastych. [Characteristics of some physiological processes during development and aging of the assimilatory apparatus of coniferous plants.] - Wiad. bot. *20* (3) : 147 - 154, 1976. [Ps; In Pol.]

28063 - **SZANIAWSKI, R.K., WIERZBICKI, B., ŻELAWSKI, W.** : Calibration of differential
infra-red CO_2 analysers by use of dilute bicarbonate solution. - Photosynthetica *10* : 86 - 88, 1976.

*28064 - **SZAREK, S.R., JOHNSON, H.P., TING, I.P.** : Seasonal patterns of acid metabolism and gas exchange in *Opuntia basilaris*. - Amer. J. Bot. *61* (5,Suppl.) :
31, 1974. [Ps.]

28065 - **SZAREK, S.R., TROUGHTON, J.H.** : Carbon isotope ratios in Crassulacean Acid
Metabolism plants. Seasonal patterns from plants in natural stands. - Plant
Physiol. *58* : 367 - 370, 1976.

28066 - **SZAREK, S.R., TROUGHTON, J.H.** : Seasonal variations of carbon isotope fractionation in CAM plants. - Plant Physiol. *57* (Suppl.) : 32, 1976.

28067 - **SZAREK, S.R., WOODHOUSE, R.M.** : Ecophysiological studies of Sonoran Desert
plants. I. Diurnal photosynthesis patterns of *Ambrosia deltoidea* and *Olne-
ya tesota*.-Oecologia *26* : 225 - 234, 1976.

*28068 - **SZUJKÓ-LACZA, J., FEKETE, G.** : Examination of development and growth of
Brachypodium silvaticum and *Euphorbia cyparissias* in oakwoods. - Acta bot.
Acad. Sci. hung. *20* : 147 - 158, 1974. [Growth analysis.]

*28069 - **SZUJKÓ-LACZA, J., SZŐCS, Z., HORNOK, L.** : LAR, RGR and NAR parameters inter-
nodially, and crop volatile oil yield in *Pimpinella anisum* L. sown periodically. - Acta bot. Acad. Sci. hung. *21* : 175 - 188, 1975.

28070 - **SZWARCBAUM, I., SHAVIV, G.** : A Monte-Carlo model for the radiation field in
plant canopies. - Agr. Meteorol. *17* : 333 - 352, 1976.

28071 - **TABITA, F.R., McFADDEN, B.A.** : Molecular and catalytic properties of ribulose
1,5-bisphosphate from the photosynthetic extreme halophile *Ectothiorhodospira
halophila*. - J. Bacteriol. *126* : 1271 - 1277, 1976.

28072 - **TABITA, F.R., STEVENS, S.E., Jr., GIBSON, J.L.** : Carbon dioxide assimilation
in blue-green algae : initial studies on the structure of ribulose 1,5-bis-
phosphate carboxylase. - J. Bacteriol. *125* : 531 - 539, 1976.

*28073 - **TADZHIEVA, F.N.** : Soderzhanie pigmentov v assimiliruyushchikh organakh kus-
tarnikov, polukustarnikov i derev'ev Yugo-Zapadnogo Kyzylkuma. [Pigment content in assimilatory organs of shrubs, semi-shrubs, and trees of South-West-
ern Kyzylkum.] - Uzb. biol. Zh. *1975* (4) : 15 - 17, 68 - 69, 1975. [In R,
ab : Uz.]

28074 - **TAĞEEVA, S.V., POPOV, V.I.** : Razvitie strukturnoĭ organizatsii membran khlo-
roplastov. [Development of structural organization of chloroplast membranes.]
- In : Tezisy Dokladov X Vsesoyuznoĭ Konferentsii po Elektronnoĭ Mikroskopii.
Vol. 2. Pp. 342 - 343. Moskva 1976. [In R.]

28075 - **TAĞEEVA, S.V., POPOV, V.I.** : Interpretatsiya nadmolekulyarnoĭ organizatsii
fotosinteticheskikh membran. [Interpretation of supermolecular organization
of photosynthetic membranes.] - Dokl. Akad. Nauk SSSR *228* : 476 - 479, 1976.
[In R.]

28076 - **TAGUCHI, S.** : Short-term variability of photosynthesis in natural marine
phytoplankton populations. - Mar. Biol. *37* : 197 - 207, 1976.

28077 - **TAGUCHI, S.** : Relationship between photosynthesis and cell size of marine
diatoms. - J. Phycol. *12* : 185 - 189, 1976.

28078 - **TAĬLAKOV, N., SAPARGEL'DYEV, G.** : Vliyanie zasoleniya na intenzivnost' fotosinteza kul'turnykh rasteniĭ. [Effect of salination on photosynthetic rate of
cultivated plants.] - Izv. Akad. Nauk turkm. SSR, Ser. biol. Nauk *1976*(2): 39-
- 44, 1976. [In R, ab : E, Turk.]

28079 - **TAKABE, T., NISHIMURA, M., AKAZAWA, T.** : Presence of two subunit types in ri-
bulose-1,5-bisphosphate carboxylase from blue-green algae. - Biochem. biophys.
Res. Commun. *68* : 537 - 544, 1976.

28080 - **TAKAHAMA, U., NISHIMURA, M.:** Effects of electron donor and acceptors, electron
transfer mediators, and superoxide dismutase on lipid peroxidation in illuminated chloroplast fragments. - Plant Cell Physiol. *17* : 111 - 118, 1976.

28081 - **TAKAHAMA, U., SHIMIZU, M., NISHIMURA, M.** : Temperature-jump-induced release of hydrogen ions from chloroplasts and its relaxation characteristics in the presence of ionophores. - Biochim. biophys. Acta *440* : 261 - 265, 1976.

28082 - **TAKAHASHI, F., KIKUCHI, R.** : Photoelectrolysis using chlorophyll electrodes. - Biochim. biophys. Acta *430* : 490 - 500, 1976.

28083 - **TAKAHASHI, M.** : [Carbon dioxide and metabolic regulation.] - Yuki Gosei Kagaku Kyokai Shi *34* : 294 - 299, 1976. [In Jap., ab : E.]

28084 - **TAKAHASHI, M., ASADA, K.** : Removal of Mn from spinach chloroplasts by sodium cyanide and the binding of Mn^{2+} to Mn-depleted chloroplasts. - Europe. J. Biochem. *64* : 445 - 452, 1976.

28085 - **TAKEDA, G.** : [Ecological studies on the photosynthesis of winter cereals. I. Diurnal changes in the photosynthesis of two-rowed barley under field conditions in winter.] - Proc. Crop Sci. Soc. Jap. *45* : 17 - 24, 1976. [In Jap., ab : E.]

28086 - **TAKEDA, G., UDAGAWA, T.** :[Ecological studies on the photosynthesis of winter cereals III. Changes of the photosynthetic ability of various organs with growth.] - Proc. Crop Sci. Soc. Jap. *45* : 357 - 368, 1976. [In Jap., ab : E.]

28087 - **TAKEDA, T., YAJIMA, M., AOKI, M., HAKOYAMA, S., SAITO, H., ONO, H.** : [Chamber method for estimating the primary productivity of rice plant population.] - Proc. Crop Sci. Soc. Jap. *45* : 139 - 150, 1976. [In Jap., ab : E.]

28088 - **TALLING, J.F.** : The depletion of carbon dioxide from lake water by phytoplankton. - J. Ecol. *64* : 79 - 121, 1976. [Ps, Chl.]

28089 - **TAN, C.S., BLACK, T.A.** : Factors affecting the canopy resistance of a Douglas--fir forest. - Boundary-Layer Meteorol. *10* : 475 - 488, 1976.

28090 - **TANAKA, A., KIKUCHI, K.** : [Nutrio-physiological studies on field beans (*Phaseolus vulgaris* L.). 3. Changes in the photosynthetic rate of individual leaves during growth of the determinate and the semi-determinate.] - Nippon Dojo-Hiryogaku Zasshi *47* : 506 - 510, 1976. [In Jap.]

28091 - **TANAKA, A., KIKUCHI, K.** : [Nutrio-physiological studies on field beans (*Phaseolus vulgaris* L.). 4. Effects of the removal of pods or leaves.] - Nippon Dojo-Hiryogaku Zasshi *47* : 537 - 541, 1976. [In Jap.]

28092 - **TANAKA, I.** : Climatic influence on photosynthesis and respiration of rice. - In : Proceedings of the Symposium on Climate & Rice. Pp. 223 - 247. Int. Rice Research Institute, Manila 1976.

28093 - **TANAKA, T.** : Regulation of plant type and carbon assimilation of rice. - Jap. agr. Res. Quart. *10* (4) : 161 - 167, 1976.

28094 - **TANAKA, Y., KATAYAMA, T.** : Comparative biochemistry of carotenoids in algae - VI. Carotenoids in *Phylloderma sacrum*, *Lyngbya* sp. and *Spirogyra* sp. - Kagoshima Daigaku Suisangakubu Kiyo [Mem. Fac. Fish., Kagoshima Univ.] *25* (1) : 27 - 32, 1976.

28095 - **TANAS'EV, V.K.** : Soderzhanie khlorofilla i stepen' aktivnosti katalazy v list'yakh sortov yabloni v raznykh uchastkakh krupnoob"emnoĭ krony. [Chlorophyll content and rate of catalase activity in leaves of apple tree cultivars in various parts of the voluminous crown.] - Tr. kishinev. sel'skokhoz. Inst. *154* : 35 - 38, 1976. [In R.]

28096 - **TANIYAMA, T., YAMASHITA, K., KOIKE, T.** : [Studies on the mechanism of injurious effects of toxic gases on crop plants XIII. Effects of ozone in the air on the apparent photosynthesis of corn, rice and peanut plants.] - Proc. Crop Sci. Soc. Jap. *45* : 9 - 16, 1976. [In Jap., ab : E.]

28097 - **TANNER, C.B., JURY, W.A.** : Estimating evaporation and transpiration from a row crop during incomplete cover. - Agron. J. *68* : 239 - 243, 1976. [Growth analysis.]

28098 - **TATENO, K., OJIMA, M.** : [Effects of temperature and soil water content during grain filling period on the yields of grain sorghum.] - Proc. Crop Sci. Soc. Jap. *45* : 63 - 68, 1976. [Primary production; in Jap., ab : E.]

28099 - **TAYLOR, A.O., ROWLEY, J.A., HUNT, B.J.** : Potential of new summer grasses in
Northland. 1. Warm-season yields under dryland and irrigation. - New Zeal. J.
agr. Res. *19* : 127 - 133, 1976. [Primary production.]

28100 - **TAYLOR, A.O., ROWLEY, J.A., HUNT, B.J.** : Potential of new summer grasses in
Northland. II. A further range of grasses. - New Zeal. J. agr. Res. *19* : 477 -
- 481, 1976. [Primary production.]

28101 - **TAYLOR, B.K., GOUBRAN, F.H.** : Effects of phosphate and pH stress on the
growth and function of apple roots. - Plant Soil *44* : 149 - 162, 1976. [Growth
analysis.]

28102 - **TAYLOR, R.J., PEARCY, R.W.** : Seasonal patterns of the CO_2 exchange characteristics of understory plants from a deciduous forest. - Can. J. Bot. *54* :
1094 - 1103, 1976.

28103 - **TEDRO, S.M., MEYER, T.E., KAMEN, M.D.** : Primary structure of a high potential
iron-sulfur protein from the purple non-sulfur photosynthetic bacterium *Rhodopseudomonas gelatinosa*. - J. biol. Chem. *251* : 129 - 136, 1976.

28104 - **TELFER, A., NICOLSON, J., BARBER, J.** : Cation control of chloroplast structure and chlorophyll *a* fluorescence yield and its relevance to the intact chloroplast. - FEBS Lett. *65* : 77 - 83, 1976.

28105 - **TEL-OR, E., STEWART, W.D.P.** : Photosynthetic electron transport, ATP synthesis and nitrogenase activity in isolated heterocysts of *Anabaena cylindrica*.
- Biochim. biophys. Acta *423* : 189 - 195, 1976.

28106 - **TENHUNEN, J.D., WEBER, J.A., YOCUM, C.S., GATES, D.M.** : Development of a photosynthesis model with an emphasis on ecological applications. II. Analysis
of a data set describing the P_M surface. - Oecologia *26* : 101 - 119, 1976.

28107 - **TENHUNEN, J.D., YOCUM, C.S., GATES, D.M.** : Development of a photosynthesis
model with an emphasis on ecological applications. I. Theory. - Oecologia
26 : 89 - 100, 1976.

28108 - **TERASAKI, W.L., BROOKER, G.** : Automated method for the quantitation of ortho-
phosphate. - Anal. Biochem. *75* : 447 - 453, 1976.

28109 - **TERJUNG, W.H., LOUIE, S.S-F., O'ROURKE, P.A.** : Seasonally based photosynthesis
model, predicting world food productivity. - Int. J. Biometeorol. *20* : 267 -
- 270, 1976.

28110 - **TERJUNG, W.H., LOUIE, S.S.F., O'ROURKE, P.A.** : Toward an energy budget model
of photosynthesis predicting world productivity. - Vegetatio *32* : 31 - 53,
1976.

28111 - **TERPSTRA, W.** : Chlorophyllase and lamellar structure in *Phaeodactylum tricornutum*. III. Situation of chlorophyllase in pigmented membranes. - Z. Pflanzenphysiol. *80* : 177 - 188, 1976.

28112 - **TERRY, N.** : Effects of sulfur on the photosynthesis of intact leaves and isolated chloroplasts of sugar beets. - Plant Physiol. *57* : 477 - 479, 1976.

28113 - **TERRY, N.** : Photosynthesis, growth, and the role of chloride. - Plant Physiol.
57 (Suppl.) : 95, 1976.

28114 - **TERSKOV, I.A., SPIROV, V.V., KHARUK, V.I.** : Dvukhvolnovoĭ skaniruyushchiĭ
mikrospektroreflektometr. [Dual wavelength scanning microspectroreflectometer.]
- Izv. sibir. Otd. Akad. Nauk SSSR, Ser. biol. Nauk *1976* (10) : 121 - 128,
1976. [Chl; in R, ab : E.]

28115 - **TERSKOV, I.A., TRENKENSHU, A.P., SID'KO, F.Ya., BELYANIN, V.N.** : Rost i éffektivnost' fotosinteza khlorelly pri preryvistom obluchenii. [Growth and effectiveness of *Chlorella* photosynthesis during intermittent irradiation.] - Dokl.
Akad. Nauk SSSR *230* : 998 - 1001, 1976. [In R.]

28116 - **TETLEY, R.M., KRIVAK, B.M.** : Carbohydrate metabolism in oat leaves. I. Photosynthetic accumulation of fructans in detached leaf segments. - Plant Physiol. *57* (Suppl.) : 48, 1976.

28117 - **TEVINI, M.** : Veränderungen der Glyko- und Phospholipidgehalte während der
Blattvergilbung. - Planta *128* : 167 - 171, 1976. [Chl.]

28118 - THERRIAULT, J.-C., LACROIX, G. : Nutrients, chlorophyll, and internal tides
        in the St. Lawrence estuary. - J. Fish. Res. Board Can. *33* : 2747 - 2757,
        1976.

28119 - THIEDE, B. : Die funktionelle Regulation der Chloroplastenstruktur von *Chla-
        mydobotrys stellata*. - Protoplasma *87* : 361 - 385, 1976.

28120 - THINH, L.V., GRIFFITHS, D.J. : Amino acid composition of autotrophic and he-
        terotrophic cultures of the Emerson strain of *Chlorella*. - Plant Cell Physiol.
        17 : 193 - 196, 1976.

28121 - THOMAS, H. : Delayed senescence in leaves treated with protein synthesis in-
        hibitor MDMP. - Plant Sci. Lett. *6* : 369 - 377, 1976. [Chl, Fraction 1 pro-
        tein.]

28122 - THOMAS, J.B., BOLLEN, M.H.M., KLIJN, W.J. : Photobleaching and dark-bleach-
        ing of *Euglena gracilis* chloroplast fragments. - Acta bot. neerl. *25* : 361 -
        - 369, 1976.

28123 - THOMAS, J.C., BROWN, K.W., JORDAN, W.R. : Stomatal response to leaf water
        potential as affected by preconditioning water stress in the field. - Agron.
        J. *68* : 706 - 708, 1976. [Growth analysis.]

28124 - THOMAS, R.J., HIPKIN, C.R., SYRETT, P.J. : The interaction of nitrogen assi-
        milation with photosynthesis in nitrogen deficient cells of *Chlorella*. - Plan-
        ta *133* : 9 - 13, 1976.

28125 - THOMPSON, D.R., HINCKLEY, T.M. : Modeling tree water status: simulating the
        diurnal course of xylem pressure potential and leaf resistance. - Plant Phy-
        siol. *57* (Suppl.) : 28, 1976.

28126 - THORESON, B., HAERTEL, L., MOORE, D.G. : Landsat imagery as an indicator of
        prairie lake algal blooms and water transparency. - Proc. S.D. Acad. Sci. *55*:
        56 - 72, 1976. [Chl.]

28127 - THORNBER, J.P., ALBERTE, R.S. : Chlorophyll-proteins : Membrane-bound photo-
        receptor complexes in plants. - In : MARTONOSI, A. (ed.) : The Enzymes of
        Biological Membranes. Vol. 3. Pp. 163 - 190. J.Wiley & Sons, London - New
        York - Sydney - Toronto 1976.

26128 - THRONDSEN, J. : Occurrence and productivity of small marine flagellates. -
        Norw. J. Bot. *23* : 269 - 293, 1976. [Chl.]

28129 - THRONDSEN, J., HEIMDAL, B.R. : Primary production, phytoplankton and light
        in Straumsbukta near Tromsø. - Astarte *9* (2) : 51 - 60, 1976.

28130 - THROWER, S.L., THROWER, L.B. : Translocation of labelled assimilate in potas-
        sium-deficient plants. - New Phytol. *77* : 541 - 545, 1976.

28131 - THURNAUER, M.C., NORRIS, J.R. : Magnetophotoselection applied to the triplet
        state observed by EPR in photosynthetic bacteria. - Biochem. biophys. Res.
        Commun. *73* : 501 - 506, 1976.

28132 - TIBONI, O., DI PASQUALE, G., CIFERRI, O. : Ribosomes and translation factors
        from isolated spinach chloroplasts. - Plant Sci. Lett. *6* : 419 - 429, 1976.

28133 - TICHÁ, I. : Der Beitrag einzelner Blätter an der Pflanze zur Photosynthese
        des gesamten Blattapparates im Laufe der Pflanzenentwicklung. - Biol. Plant.
        18 : 237 - 240, 1976.

28134 - TICHÁ, I. : Photosynthesis of plants in two controlled environments. - In :
        Doklady i Tezisy Dokladov Koordinatsionnogo Soveshchaniya SĚV, Fotosintez
        I-18.3. Pp. 32 - 46. Szeged 1976.

28135 - TIEDE, D.M., PRINCE, R.C., DUTTON, P.L. : EPR and optical spectroscopic pro-
        perties of the electron carrier intermediate between the reaction center bac-
        teriochlorophylls and the primary acceptor in *Chromatium vinosum*. - Biochim.
        biophys. Acta *449* : 447 - 467, 1976.

28136 - TIEDE, D.M., PRINCE, R.C., REED, G.H., DUTTON, P.L. : EPR properties of the
        electron carrier intermediate between the reaction center bacteriochlorophylls
        and the primary acceptor in *Chromatium vinosum*. - FEBS Lett. *65* : 301 - 304,
        1976.

28137 - **TIEN, H.T.** : Electronic processes and photoelectric aspects of bilayer lipid membranes. - Photochem. Photobiol. *24* : 97 - 116, 1976.

28138 - **TIKU, B.L.** : Effect of salinity on the photosynthesis of the halophyte *Salicornia rubra* and *Distichlis stricta*. - Physiol. Plant. *37* : 23 - 28, 1976.

*28139 - **TILNEY-BASSETT, R.A.E.** : Genetics of variegated plants. - In : BIRKY, C.W., Jr., PERLMAN, P.S., BYERS, T.J. (ed.) : Genetics and Biogenesis of Mitochondria and Chloroplasts. Pp. 268 - 308. Ohio State Univ. Press, Columbus 1975. [Chloroplast.]

28140 - **TILZER, M.M., SCHWARZ, K.** : Seasonal and vertical patterns of phytoplankton light adaptation in a high mountain lake. - Arch. Hydrobiol. *77* : 488 - 504, 1976. [Ps, Chl.]

28141 - **TIMMIS, R.** : Methods of screening tree seedlings for frost hardiness. - In : CANNELL, M.G.R., LAST, F.T. (ed.) : Tree Physiology and Yield Improvement. Pp. 421 - 435. Academic Press, London - New York - San Francisco 1976. [Ps, Chl, leaf reflectance.]

28142 - **TING, I.P.** : Malate dehydrogenase and other enzymes of C_4 acid metabolism in marine plants. - Aust. J. Plant Physiol. *3* : 121 - 127, 1976.

28143 - **TING, I.P.** : Crassulacean acid metabolism in natural ecosystems in relation to annual CO_2 uptake patterns and water utilization. - In : BURRIS, R.H., BLACK, C.C. (ed.) ; CO_2 Metabolism and Plant Productivity. Pp. 251 - 268. Univ. Park Press, Baltimore - London - Tokyo 1976.

28144 - **TINGEY, D.T., FITES, R.C., WICKLIFF, C.** : Differential foliar sensitivity of soybean cultivars to ozone associated with differential enzyme activities. - Physiol. Plant. *37* : 69 - 72, 1976. [Resistances.]

28145 - **TINUS, R.W.** : Photoperiod and atmospheric CO_2 level interact to control black walnut (*Juglans nigra* L.) seedling growth. - Plant Physiol. *57* (Suppl.) : 106, 1976.

*28146 - **TISHCHENKO, N.N.** : Izmenenie sostava belkov khloroplastov i aktivnosti aminotransferaz u rasteniĭ s razlichnym tipom fiksatsii CO_2 v zavisimosti ot azotnogo pitaniya. [Changes in the composition of chloroplast lipids and aminotransferase activity in plants of different type of CO_2 fixation in dependence on nitrogen fixation.] - Vestn. leningrad. Univ., Ser. biol. *1975* [21 (4)]: 115 - 122, 1975. [In R, ab : E.]

28147 - **TITLYANOV, É.A.** : Adaptatsiya benticheskikh rasteniĭ k svetu. I. Znachenie sveta v raspredelenii morskikh prikreplennykh vodorosleĭ. [Adaptation of benthic plants to light. I. Importance of light in the distribution of attached marine algae.] - Biol. Morya *1* : 3 - 12, 1976. [Ps; in R, ab : E.]

28148 - **TITOV, A.F., OLIMPIENKO, G.S.** : Chastota khlorofilldefektnykh prorostkov v selektsionnykh potomstvakh rasteniĭ ovsyanitsy lugovoĭ (*Festuca pratensis* HUDS.). [Frequency of chlorophyll-deficient seedlings in breeding progenies of meadow fescue (*Festuca pratensis* HUDS.).] - Genetika *12* (2) : 162 - 164, 1976. [In R, ab : E.]

28149 - **TIŢU, H., DUMITRESCU, M.** : Chloroplast ultrastructure and peroxidase isoenzymes in *Spinacia oleracea* L. seedlings derived from seeds treated with X-rays. - Rev. roum. Biol., Sér. Biol. vég. *21* : 61 - 65, 1976.

28150 - **TOKUNAGA, F., IWASA, T., YOSHIZAWA, T.** : Photochemical reaction of bacteriorhodopsin. - FEBS Lett. *72* : 33 - 38, 1976.

28151 - **TOLBERT, N.E.** : Glycollate oxidase and glycollate dehydrogenase in marine algae and plants. - Aust. J. Plant Physiol. *3* : 129 - 132, 1976.

28152 - **TOLBERT, N.E., GAREY, W.** : Apparent total CO_2 equilibrium point in marine algae during photosynthesis in sea water. - Aust. J. Plant Physiol. *3* : 69 - 72, 1976.

28153 - **TOLBERT, N.E., OSMOND, C.B.** : The Great Barrier Reef photorespiration expedition : introduction. - Aust. J. Plant Physiol. *3* : 1 - 8, 1976.

28154 - **TOLBERT, N.E., RYAN, F.J.** : Glycolate biosynthesis and metabolism during photorespiration. - In : BURRIS, R.H., BLACK, C.C. (ed.) : CO_2 Metabolism and

Plant Productivity. Pp. 141 - 159. Univ. Park Press, Baltimore - London - Tokyo 1976.

28155 - **TOLLIN, G.** : Model systems for photosynthetic energy conversion. - J. phys. Chem. *80* : 2274 - 2277, 1976.

28156 - **TOLSTOGUZOVA, B.G.** : Dinamika nakopleniya khlorofilla v list'yakh yabloni s razlichnoĭ formoĭ krony. [Dynamics of chlorophyll content in leaves of apple-trees with different crown framing.] - Sel'skokhoz. Biol. *11* : 559 - 564, 1976. [In R, ab : E.]

28157 - **TOMBESI, L., DE ROSSI, C., FRANCAVIGLIA, R., FRATICELLI, A.** : Influenza della temperatura e degli elementi fertilizzanti sulla produttività in ambiente controllato. Nota II - Azione esercitata dalla concimazione azotata. [Influence of temperature and fertilization on yield in a controlled environment. Note II - Effect of nitrogenous fertilizer.] - Ann. Ist. sperim. Nutr. Piante *7* (2) : 1 - 26, 1976. [Dry-matter production; in Ital., ab : E.]

28158 - **TÖRMÄLÄ, T., RAATIKAINEN, M.** : Primary production and seasonal dynamics of the flora and fauna of the field stratum in a reserved field in Middle Finland. - J. sci. agr. Soc. Finland *48* : 363 - 385, 1976.

28159 - **TORRES, A.M.R., O'FLAHERTY, L.M.** : Influence of pesticides on *Chlorella, Chlorococcum, Stigeoclonium (Chlorophyceae), Tribonema, Vaucheria (Xanthophyceae)* and *Oscillatoria (Cyanophyceae)*. - Phycologia *15* : 25 - 36, 1976.

28160 - **TOWPASZ, K.** : Primary production of the herb layer in the forest association *Tilio-carpinetum* of the Pogórze Wielickie Region (near Bochnia). - Bull. Acad. pol. Sci., Sér. Sci. biol. Cl. II *24* : 205 - 211, 1976.

28161 - **TOYOSHIMA, Y.** : [Photochemical reaction in bimolecular lipid membrane systems containing chlorophyll.] - Hyomen *14* : 317 - 326, 1976. [In Jap.]

28162 - **TRAVERS, M.** : Le microplancton du Golfe de Marseille : Pigments phytoplanctoniques, estimations de production. - Tethys *7* : 137 - 168, 1975 (1976).

28163 - **TREBST, A.** : Artificial energy conservation in bacterial photosynthetic electron transport. - Z. Naturforsch. *31 c* : 152 - 156, 1976.

28164 - **TREBST, A.** : Coupling sites, native and artificial, in photophosphorylation by isolated chloroplasts. - Trends biochem. Sci. *1* (3) : 60 - 62, 1976.

28165 - **TREBST, A., REIMER, S., DALLACKER, F.** : Properties of photoreductions by photosystem II. - Plant Sci. Lett. *6* : 21 - 24, 1976.

28166 - **TREFFRY, T.** : Changes in the surface properties of chloroplast membranes from dark-grown *Pinus silvestris* following illumination. - Plant Sci. Lett. *6* : 193 - 196, 1976.

28167 - **TREGUBENKO, M.Ya., FILIPPOV, G.L., VISHNEVSKIĬ, N.V.** : Fiziologicheskie osnovy vysokoproduktivnogo ispol'zovaniya vody kukuruzoĭ v usloviyakh orosheniya. [Physiological basis for high water use efficiency in irrigated maize.] - In: Biologicheskie Osnovy Povysheniya Urozhaev Kukuruzy i Drugikh Polevykh Kul'tur v Severnoĭ Stepi USSR. Pp. 10 - 15. Sinel'nikov. selek.-opyt. Sta., Dnepropetrovsk 1976. [In R.]

28168 - **TREGUBENKO, M.Ya., FILIPPOV, G.L., VISHNEVSKIĬ, N.V., MAKSIMOVA, L.A.** : Fotosinteticheskaya deyatel'nost' razlichnykh po skorospelosti gibridov kukuruzy v usloviyakh orosheniya. [Photosynthetic activity of irrigated maize hybrids differing in earliness.] - In : Biologicheskie Osnovy Povysheniya Urozhaev Kukuruzy i Drugikh Polevykh Kul'tur v Severnoĭ Stepi USSR. Pp. 21 - 25. Sinel'nikov. selek.-opyt. Sta., Dnepropetrovsk 1976. [In R.]

28169 - **TREGUBENKO, M.Ya., FILIPPOV, G.L., VISHNEVSKIĬ, N.V., MAKSIMOVA, L.A.** : Intensivnost' i produktivnost' fotosinteza oroshaemoĭ kukuruzy. [Photosynthetic rate and productivity of photosynthesis in irrigated maize.] - In : Biologicheskie Osnovy Povysheniya Urozhaev Kukuruzy i Drugikh Polevykh Kul'tur v Severnoĭ Stepi USSR. Pp. 16 - 21. Sinel'nikov. selek.-opyt. Sta., Dnepropetrovsk 1976. [In R.]

28170 - **TRENCH, R.K., OHLHORST, S.** : The stability of chloroplasts from siphonaceous algae in symbiosis with sacoglossan molluscs. - New Phytol. *76* : 99 - 109, 1976.

28171 - **TRENKENSHU, A.P., SID'KO, F.Ya., BELYANIN, V.N.** : Ob êffektivnosti fotosinteza khlorelly pri razlichnykh dlitel'nostyakh temnovykh periodov v svetoimpul's-
nom rezhime. [Effectivity of photosynthesis in *Chlorella* exposed to various
dark periods in light-impulse regime.] - In : Gazometricheskoe issledovanie
Fotosinteza i Dykhaniya RasteniĬ. Pp. 133 - 134. Akad. Nauk SSSR, Tartu 1976.
[In R.]

28172 - **TRENKENSHU, A.P., SID'KO, F.Ya., BELYANIN, V.N.** : Svetoimpul'snye kharakteristiki fotosinteza *Chlorella vulgaris*. [Light-impulse characteristics of photosynthesis of *Chlorella vulgaris*.] - Fiziol. Rast. *23* : 702 - 709, 1976. [In
R, ab : E.]

28173 - **TRENKENŠU, A.P., BELJANIN, V.N., SIDKO, F.Ja.** : Photobiosynthesis of the mic-
roalga *Synechococcus elongatus* upon exposure to intermittent light. - Arch.
Hydrobiol. Suppl. *49* (Algol. Stud. *15*) : 176 - 184, 1976. [Chl.]

28174 - **TRIFONOVA, I.S.** : Fitoplankton i ego produktsiya. [Phytoplankton and its production.] - In : Biologicheskaya Produktivnost' Ozera Krasnogo i Usloviya Ee
Formirovaniya. Pp. 69 - 104. Nauka, Leningrad 1976. [In R.]

28175 - **TRINKLER, Yu.G.** : Izuchenie chistoĬ produktivnosti fotosinteza v bol'shom
tsikle razvitiya kartofelya. [Study of net productivity of photosynthesis in
a big cycle of potato growth.] - Fiziol. Rast. *23* : 397 - 400, 1976. [In R.]

28176 - **TRIOLO, L., BASSANELLI, C.** : Carbonic anhydrase net photosynthesis and photorespiration in Zn-deficient leaves of *Triticum durum*. A hypothesis of çarbonic
anhydrase function. - Agrochimica *20* : 457 - 465, 1976.

28177 - **TROUGHTON, J.H.** : Translocation in *Zea mays* leaves. - In : WARDLAW, I.F.,
PASSIOURA, J.B. (ed.) : Transport and Transfer Processes in Plants. Pp. 339 -
- 345. Academic Press, New York - San Francisco - London 1976.

28178 - **TRÜPER, H.G.** : Higher taxa of the phototrophic bacteria : *Chloroflexaceae*
fam. nov., a family for the gliding, filamentous, phototrophic "green" bacteria. - Int. J. syst. Bact. *26* : 74 - 75, 1976.

28179 - **TSCHAKALOVA, E.** : Struktur-Funktionsbeziehungen bei *Phaseolus vulgaris* L. unter besonderer Berücksichtigung der Photosynthese. I. Untersuchung der Spaltöff-
nungen und des Interzellularvolumens von *Phaseolus vulgaris*-Blättern im Ver-a
lauf der Ontogenese. - Godishnik sofiĬ. Univ., biol. Fak., Kn. 2, *68* : 1 - 10,
1976.

28180 - **TSCHAKALOVA, E., HOFFMANN, P.** : Strukturelle und funktionelle Grundlagen des
photosynthetischen Gaswechsels bei *Triticum aestivum* L. - Wiss. Z. Humboldt-
-Univ. Berlin, math.-naturwiss. Reihe *25* : 723 - 736, 1976.

28181 - **TSCHAKALOWA, E.** : Struktur-Funktionsbeziehungen bei *Phaseolus vulgaris* L. unter besonderer Berücksichtigung der Photosynthese II. Struktur des Photosyn-
theseapparates. - Godishnik sofiĬ. Univ., biol. Fak., Kn. 2, *69* : 5 - 19,
1975/1976.

28182 - **TSCHÄPE, M., PEISKER, M.** : Über die Beziehung zwischen CO_2-Aufnahme und Sto-
mataweite. - Kulturpflanze *24* : 119 - 132, 1976.

28183 - **TSENOVA, E.N., VASSILEVA, V.S., FEDINA, I.S., VUNKOVA, R.N., GUSHCHINA, L.M.,
VAKLINOVA, S.G.** : Influence of nitrate and ammonia nitrogen of the activity
of some enzymes of photosynthesis and photorespiration in greening plants of
maize. - Dokl. bolg. Akad. Nauk *29* : 701 - 704, 1976.

28184 - **TSUJIMOTO, H.Y., McSWAIN, B.D., HIYAMA, T., ARNON, D.I.** : Effect of $NADP^+$ on
light-induced cytochrome changes in membrane fragments from a blue-green alga.
- Biochim. biophys. Acta *423* : 303 - 312, 1976.

28185 - **TSYARÊNTS'EŬ, V.M., KOSHALEVA, L.L., BAKHNOVA, K.V., IVANSKAYA, G.A., KAZARA-
VA, R.A.** : Uplyŭ fosfarnaga zhyŭlennya na fotasintêtychnuyu dzeĬnasts' raslin
il'nu-daŭguntsu. [Effect of phosphorus nutrition on the photosynthetic activity
of fibre-flax leaves.] - Vestsi Akad. Navuk belarus. SSR, Ser. biyal. Navuk
1976 (5) : 19 - 24, 137, 1976. [In Belorus., ab : R.]

28186 - **TSYGANKOVA, T.A.** : Sezonnaya i dnevnaya dinamika karotina i askorbinovoĬ kisloty u *Festuca sulcata* HACK. [Seasonal and diurnal dynamics of carotene and
ascorbic acid in *Festuca sulcata* HACK.] - Nauch. Dokl. vyssh. Shkoly, biol.
Nauki *19* (7) : 142, 1976. [In R.]

28187 - **TUCKER, C.J., GARRATT, M.W.** : Three-dimensional chlorophyll concentrations in a high biomass blue grama canopy. - J. Range Management *29* : 170 - 171, 1976.

28188 - **TUCKER, C.J., MAXWELL, E.L.** : Sensor design for monitoring vegetation canopies. - Photogram. Eng. remote Sensing *42* : 1399 - 1410, 1976. [Chl.]

28189 - **TUQUET, C., GUILLOT-SALOMON, T., FARINEAU, J., SIGNOL, M.** : Biogenèse des membranes plastidiales dans les feuilles étiolées d'Orge soumises à des éclairs répétés développement d'accolements de thylacoïdes, synthèse de phosphatidylglycérol et apparition de cytochrome b_{559} (forme haut potentiel). - Physiol. vég. *14* : 11 - 30, 1976.

28190 - **TURCZYŃSKA, J., WIŚNIEWSKI, R.J.** : Effect of heated waters on biocenosis of the moderately polluted Narew River. Phytoplankton. - Pol. Arch. hydrobiol. *23* : 507 - 517, 1976. [Chl.]

28191 - **TURGEON, A.J., LESTER, G.** : Xanthophyll levels in turfgrass clippings. - Agron. J. *68* : 946 - 948, 1976.

28192 - **TURNER, R.E.** : Geographic variations in salt marsh macrophyte production : A review. - Contrib. mar. Sci. *20* : 47 - 68, 1976. [Solar energy conversion.]

28193 - **TYLDUM, M.K., NILSEN, S.** : Dusk and dawn effect on spruce photosynthesis. Comparative studies on long term $IR-CO_2$-analyses of whole plants and short term $^{14}CO_2$-incorporation in detached leaves. - Z. Pflanzenphysiol. *79* : 121 - - 131, 1976.

28194 - **TYZNIK, D.J., PARKINSON, C.C.** : The light compensation point as an index for foliage plant lighting. - HortScience *11* (3, Sect. 2) : 301, 1976.

28195 - **UCHIJIMA, Z.** : Microclimate of the rice crop. - In : Climate and Rice. Pp. 115 - 140. Int. Rice Res. Inst., Los Baños 1976. [Ps.]

28196 - **UCHIJIMA, Z.** : Water consumption in crop production. - In : Science for Better Environment. Pp. 184 - 193. HESC, Kyoto 1976. [Ps.]

28197 - **UCHIJIMA, Z.** : Maize and rice. - In : MONTEITH, J.L. (ed.) : Vegetation and the Atmosphere. Vol. 2. Case Studies. Pp. 33 - 64. Academic Press, London - - New York - San Francisco 1976. [Ps.]

28198 - **UDEL'NOVA, T.M., BOĬCHENKO, E.A., KARYAKIN, A.V.** : Polivalentnye metally v khloroplastakh. [Polyvalent metals in chloroplasts.] - Fiziol. Rast. *23* : 1154 - 1159, 1976. [In R, ab : E.]

28199 - **UDOVENKO, G.V., SINEL'NIKOVA, V.N., SEMUSHINA, L.A., EVDOKIMOV, V.M.** : Deĭstvie zasoleniya na fotosinteziruyushchuyu deyatel'nost' rasteniĭ i otlozhenie zapasnykh veshchestv. [Effect of salinity on the photosynthetic activity of plants and storage of storage substances.] - Byull. vses. nauch. issled. Inst. Rastenievod. Im. N.I. Vavilova *63* : 40 - 44, 1976. [In R.]

28200 - **UDOVENKO, G.W., SAAKOV, V.S.** : Resistenz der Getreidepflanzen gegen ungünstige Bedingungen des Milieus : Physiologische und genetische Aspekte. - Wiss. Z. Humboldt-Univ. Berlin, math.-naturwiss. Reihe *25* : 776 - 786, 1976. [Ps.]

28201 - **UEDAN, K., SUGIYAMA, T.** : Purification and characterization of phosphoenolpyruvate carboxylase from maize leaves. - Plant Physiol. *57* : 906 - 910, 1976.

28202 - **UEKI, T., KATAOKA, M., MITSUI, T.** : Structural order in chromatophore membranes of *Rhodospirillum rubrum*. - Nature *262* : 809 - 810, 1976.

28203 - **UHRIG, H., TEVINI, M.** : Effekte der Phospholipase *D* auf den Elektronentransport und die Lipidzusammensetzung isolierter Spinatchloroplasten. - Planta *128* : 173 - 178, 1976.

28204 - **UKELES, R., ROSE, W.E.** : Observations on organic carbon utilization by photosynthetic marine microalgae. - Mar. Biol. *37* : 11 - 28, 1976.

28205 - **ULLRICH-EBERIUS, C.I., LÜTTGE, U., NEHER, L.** : CO_2 uptake by barley leaf slices as measured by photosynthetic O_2 evolution. - Z. Pflanzenphysiol. *79* : 336 - 346, 1976.

28206 - ULLRICH-EBERIUS, C.I., LÜTTGE, U., NEHER, L. : Energy relations of phosphate
        uptake and distribution in barley leaf slices as affected by cutting and
        adaptive ageing. - Z. Pflanzenphysiol. *79* : 347 - 359, 1976. [Ps, Chl.]

28207 - UNGER, J. : Einsatzmöglichkeiten für Wachstumsregulatoren zur pflanzlichen
        Prozeßsteuerung bei ausgewählten Körnerhülsenfrüchten (*Phaseolus* spec., *Pi-
        sum* spec.). - Beitr. trop. Landwirtsch. Veterinärmed. *14* : 61 - 69, 1976.

28208 - UNGER, K., MEIXNER, P. : Zur Methodik der Strahlungsmessung im photosynthe-
        tisch aktiven Bereich. - Z. Meteorol. *26* : 257 - 267, 1976.

28209 - UNSWORTH, M.H., BISCOE, P.V., BLACK, V. : Analysis of gas exchange between
        plants and polluted atmospheres. - In : MANSFIELD, T.A. (ed.) : Effects of
        Air Pollutants on Plants. Pp. 5 - 16. Cambridge Univ. Press, Cambridge - Lon-
        don - New York - Melbourne 1976.

28210 - UPHOFF, G.D., HERGENRADER, G.L. : A portable quantum meter-spectroradiometer
        for use in aquatic studies. - Freshwater Biol. *6* : 215 - 219, 1976.

28211 - URBÁNOVICH, T.A., IVANCHENKO, V.M. : Ob ingibiruyushchem êffekte aprofena na
        fosforiliruyushchuyu aktivnost' khloroplastov. [Inhibitory effect of aprophe-
        ne on phosphorylating activity of chloroplasts.] - Vestsi Akad. Navuk belarus.
        SSR, Ser. biyal. Navuk *1976* (4) : 109 - 111, 1976. [In R.]

28212 - USHAROVA, G.P. : Vliyanie norm vyseva i sposobov poseva na strukturu urozhaya
        ozimoĭ pshenitsy v usloviyakh oroshaemoĭ zony yugo-vostoka Kazakhstana. [The
        effect of sowing standards and types of canopy on yield structure of winter
        wheat in irrigated zone of South-Eastern Kazakhstan.] - In : Fotosintez i
        Produktivnost' Ozimoĭ Pshenitsy na Yugo-Vostoke Kazakhstana. Pp. 57 - 70, 131.
        Nauka kazakh. SSR, Alma-Ata 1976. [In R.]

28213 - UTKILEN, H.C. : Thiosulphate as electron donor in the blue-green alga *Anacys-
        tis nidulans.* - J. gen. Microbiol. *95* : 177 - 180, 1976.

28214 - VACEK, K., NAUŠ, J., ŠVÁBOVÁ, M., VAVŘINEC, E., KAPLANOVÁ, M., HÁLA, J. :
        Excitation energy transfer in chlorophyll *a* molecules in model systems. - In :
        Molecular Spectroscopy of Dense Phases. Pp. 463 - 466. Elsevier sci. Publ.
        Comp., Amsterdam 1976.

28215 - VACEK, K., VAVŘINEC, E., KAPLANOVÁ, M. : Mirror symmetry of absorption and
        fluorescence spectra of chlorophyll *a* solutions measured at room temperature.
        - Acta Univ. Carol. - Math. Phys. *17* : 45 - 50, 1976.

28216 - VAGERA, J., NOVÁK, F.J., VYSKOT, B. : Anther cultures of *Nicotiana tabacum* L.
        mutants. - Theor. appl. Genet. *47* (3) : 109 - 114, 1976. [Chl.]

28217 - VAISBERG, A.J., SCHIFF, J.A. : Events surrounding the early development of
        Euglena chloroplasts. 7. Inhibition of carotenoid biosynthesis by the herbi-
        cide SAN 9789 (4-chloro-5-(methylamino)-2-(α,α,α-trifluoro-*m*-tolyl)-3-(2H)py-
        ridazinone) and its developmental consequences. - Plant Physiol. *57* : 260 -
        - 269, 1976.

28218 - VAISBERG, A.J., SCHIFF, J.A., LI, L., FREEDMAN, Z. :Events surrounding the
        early development of *Euglena* chloroplasts. VII. Photocontrol of the source of
        reducing power for chloramphenicol reduction by the ferredoxin-NADP reductase
        system. - Plant Physiol. *57* : 594 - 601, 1976.

28219 - VAKLINOVA, S., POPOVA, L., DIMITROVA, O. : On the C-4 pathway in *Zea* maize,
        the participation of ferredoxine in the CO_2 reduciion and activity of malic
        enzyme. - Dokl. bolg. Akad. Nauk *29* : 705 - 708, 1976.

28220 - VAKLINOVA, S., VASILEVA, V., FEDINA, I., TSENOVA, E., VUNKOVA, R., GUSHCHINA,
        L. : Changes in the activity of certain enzymes of photosynthesis and photo-
        respiration under the influence of ammonia and nitrate nitrogen in barley
        seedlings. - Dokl. bolg. Akad. Nauk *29* : 869 - 872, 1976.

28221 - VALANNE, N. : Development of chloroplast structure and photosynthetic compe-
        tence in dark-adapted moss protonemata after exposure to light. - Protoplasma
        89 : 359 - 369, 1976.

28222 - VALANNE, N., ARO, E.-M. : Incorporation of 5-aminolevulinic acid in the chlo-
        rophyll-protein complexes of the moss *Ceratodon purpureus.* - Physiol. Plant.
        37 : 218 - 222, 1976.

28223 - VALENTINE, J.P., BINGHAM, S.W. : Influence of algae on amitrole and atrazine
residues in water. - Can. J. Bot. *54* : 2100 - 2107, 1976. [Ps.]

28224 - VALLÉE, J.-C., MARTIN, C., VANSUYT, G. : Influence d'un apport exogène de
proline sur le métabolisme aminé de *Nicotiana tabacum* var. Xanthi n.c.; rôle
de la lumière et du gaz carbonique. - Compt. rend. Acad. Sci. Paris, Sér. D
282 : 1861 - 1863, 1976. [Photorespiration.]

28225 - VALLEJOS, R.H., ANDREO, C.S. : Sulphydryl groups in photosynthetic energy conservation : further evidence of vicinal dithiols involvement shown by light-
-dependent effects of *o*-iodosobenzoate. - FEBS Lett. *61* : 95 - 99, 1976.

25226 - VALLESPINÓS RIERA, F. : Comunidades bentónicas de sustrato duro del litoral
NE. español. III. Pigmentos y producción. [Benthic communities in hard substrates from NE Spanish coast. III. Pigments and primary productivity.] - Invest. Pesq. *40* : 515 - 532, 1976. [In Span., ab : E.]

28227 - VAMBUTAS, V., BERTSCH, W. : Does AMP participate in photosynthetic phosphorylation ? - Biochem. biophys. Res. Commun. *73* :686 - 693, 1976.

28228 - VAN, T.K., GARRARD, L.A., WEST, S.H. : Effects of UV-B radiation on net photosynthesis of some crop plants. - Crop Sci. *16* : 715 - 718, 1976.

28229 - VAN, T.K., HALLER, W.T., BOWES, G. : Photosynthesis of three submerged aquatic macrophytes. - Plant Physiol. *57* (Suppl.) : 6, 1976.

28230 - VAN, T.K., HALLER, W.T., BOWES, G. : Comparison of the photosynthetic characteristics of three submersed aquatic plants. - Plant Physiol. *58* : 761 - 768,
1976.

28231 - Van BAVEL, C.H.M., AHMED, J. : Dynamic simulation of water depletion in the
root zone. - Ecol. Model. *2* (3) : 189 - 212, 1976. [Ps.]

28232 - Van BEEUMEN, J., AMBLER, R.P., MEYER, T.E., KAMEN, M.D., OLSON, J.M., SHAW,
E.K. : The amino acid sequences of the cytochromes *c*-555 from two green sulphur bacteria of the genus *Chlorobium*. - Biochem. J. *159* : 757 - 774, 1976.

28233 - Van BENTHEM, N.J., MOED, J.R. : Installation of additional counters in the
(Kipp) solarimeter integrator CC 1. - Freshwater Biol. *6* : 381 - 382, 1976.

28234 - Van der BENT, S.J., SCHAAFSMA, T.J., GOEDHEER, J.C. : Detection of triplet
states in algae by zero-field resonance. - Biochem. biophys. Res. Commun.
71 : 1147 - 1152, 1976.

28235 - VanderMEULEN, D.L., GOVINDJEE : Anthroyl stearate as a fluorescent probe of
chloroplast membranes. - Biochim. biophys. Acta *449* : 340 - 356, 1976.

28236 - Van GORKOM, H.J., PULLES, M.P.J., HAVEMAN, J., den HAAN, G.A. : Primary reactions of photosystem II at low pH. I. Prompt and delayed fluorescence. -
Biochim. biophys. Acta *423* : 217 - 226, 1976.

28237 - Van GRONDELLE, R., DUYSENS, L.N.M., Van der WAL, H.N. : Function of three
cytochromes in photosynthesis of whole cells of *Rhodospirillum rubrum* as studied by flash spectroscopy. Evidence for two types of reaction center. - Biochim. biophys. Acta *449*: 169 - 187, 1976.

28238 - Van GRONDELLE, R., ROMIJN, J.C., HOLMES, N.G. : Photoreduction of the long
wavelength bacteriopheophytin in reaction centers and chromatophores of the
photosynthetic bacterium *Chromatium vinosum*. - FEBS Lett. *72* : 187 - 192,
1976.

28239 - Van HASSELT, P.R. : Protection of *Cucumis* leaf pigments against photo-oxida-
tive degradation during chilling. - Acta bot. neer. *25* : 41 - 50, 1976.

28240 - Van HASSELT, P.R., STRIKWERDA, J.T. : Pigment degradation in discs of the
thermophilic *Cucumis sativus* as affected by light, temperature, sugar application and inhibitors. - Physiol. Plant. *37* : 253 - 257, 1976.

28241 - Van KEULEN, H., de WIT, C.T., LOF, H. : The use of simulation models for pro-
ductivity studies in arid regions. - In : LANGE, O.L., KAPPEN, L., SCHULZE,
E.-D. (ed.) : Water and Plant Life. Pp. 408 - 420. Springer-Verlag, Berlin -
- Heidelberg - New York 1976.

28242 - VAN LEEUWEN, P.H., VAN OORSCHOT, J.L.P. : Effects of some phenylurea herbicides on photosynthesis of two wheat varieties. - Weed Res. 16 : 11 - 14, 1976.

28243 - VANNINI, G.L., FASULO, M.P., BRUNI, A. : Effetto dell'acido 2,3,5-triiodoben-
zoico sul processo d'inverdimento di Euglena gracilis eziolata. [Effect of
2,3,5-triiodobenzoic acid on the greening process in etiolated Euglena gracilis.] - G. bot. ital. 110 : 65 - 75, 1976. [In Ital., ab : E.]

28244 - Van OORSCHOT, J.L.P. : Effects in relation to water and carbon dioxide exchange of plants. - In : AUDUS, L.J. (ed.) : Herbicides. Physiology, Biochemistry, Ecology. Vol. 1. Pp. 305 - 333. Academic Press, London - New York - Sar
Francisco 1976.

28245 - Van SAMBEEK, J.W., PICKARD, B.G. : Mediation of rapid electrical, metabolic,
transpirational, and photosynthetic changes by factors released from wounds.
III. Measurements of CO_2 and H_2O flux. - Can. J. Bot. 54 : 2662 - 2671, 1976.

26246 - VAN STEVENINCK, R.F.M. : Effect of hormones and related substances on ion
transport. - In : LÜTTGE, U., PITMAN, M.G. (ed.) : Transport in Plants II.
Part B. Tissues and Organs. Pp. 307 - 342. Springer-Verlag, Berlin - Göttingen - Heidelberg 1976. [Ps.]

28247 - VARLET-GRANCHER, C., BONHOMME, R., CASTANEDA, P.L. : Rendimiento energético
de un cultivo de caña de azúcar. [Energy yield of a sugar cane crop.] - Turrialba 26 : 139 - 143, 1976. [In Span., ab : F.]

28248 - VASCONCELOS, A.C. : Synthesis of chloroplast proteins by Euglena chloroplasts.
- Plant Physiol. 57 (Suppl.) : 37, 1976.

28249 - VASCONCELOS, A.C. : Synthesis of proteins by isolated Euglena gracilis chloroplasts. - Plant Physiol. 58 : 719 - 721, 1976.

28250 - VASCONCELOS, A.C., MENDIOLA-MORGENTHALER, L.R., FLOYD, G.L., SALISBURY, J.L.:
Fractionation and analysis of polypeptides of Euglena gracilis chloroplasts.
- Plant Physiol. 58 : 87 - 90, 1976.

28251 - VAVŘINEC, E. : Difference absorption spectrum of chlorophyll a in polar and
non-polar solvents. - Acta Univ. Carolin. - Math. Phys. 17 : 51 - 54, 1976.

28252 - VECHER, A.S., BARDYSHEV, M.A., KLINGER, Yu.E., NIKOL'SKIĬ, Yu.K. : Osobennosti raspredeleniya nekotorykh mikroélementov v yadrakh i khloroplastakh poli-
ploidnykh form. [Peculiarities of distribution of some microelements in nuc-
lei and chloroplasts of polyploid forms.] - Dokl. Akad. Nauk belorus. SSR
20 : 1042 - 1044, 1056, 1976. [In R.]

28253 - VECHER, A.S., KALER, V.L., PREDKEL', K.I., ADAMCHIK, G.G. : Soderzhanie tsitokhromov b_6 i f v razlichnykh plastidakh, vydelennykh v vodnoĭ i nevodnoĭ
sredakh. [Content of cytochromes b_6 and f in plastids isolated in aqueous and
nonaqueous media.] - Dokl. Akad. Nauk belorus.SSR 20 : 274 - 277, 1976. [In
R.]

28254 - VEERASEKARAN, P., KIRKWOOD, R.C., FLETCHER, W.W. : The mode of action of asu-
lam [menthyl(4-aminobenzenesulphonyl) carbamate] in bracken. - Bot. J. linn.
Soc. 73 : 247 - 268, 1976. [Photosynthates.]

28255 - VENKATESH, C.S., EMMANUEL, C.J.S.K. : Spontaneous chlorophyll mutations in
Bombax L. - Silvae Genet. 25 : 137 - 139, 1976.

28256 - VERMA, S.B., ROSENBERG, N.J. : Vertical profiles of carbon dioxide concentration in stable stratification. - Agr. Meteorol. 16 : 359 - 369, 1976.

28257 - VERMA, S.B., ROSENBERG, N.J., BLAD, B.L., BARADAS, M.W. : Resistance-energy
balance method for predicting evapotranspiration : determination of boundary
layer resistance and evaluation of error effects. - Agron. J. 68 : 776 - 782,
1976.

28258 - VERMEGLIO, A., BRETON, J., MATHIS, P. : Trapping at low temperature of oriented chloroplasts : Application to the study of antenna pigments and of the trap
of Photosystem 1. - J. supramol. Struct. 5 : 109 - 117, 1976.

28259 - VERMEGLIO, A., CLAYTON, R.K. : Orientation of chromophores in reaction centers
of Rhodopseudomonas sphaeroides. Evidence for two absorption bands of the dimeric primary electron donor. - Biochim. biophys. Acta 449 : 500 - 515, 1976.

28260 - **VERNOTTE, C., BRIANTAIS, J.-M., REMY, R.** : Light harvesting pigment protein complex requirement for spill-over changes induced by cations. - Plant Sci. Lett. *6* : 135 - 141, 1976.

28261 - **VERSHININ, A.V., SOKOLOV, V.A., SHUMNYĬ, V.K.** : Fiziologo-biokhimicheskie aspekty monogibridnogo geterozisa, poluchennogo na osnove khlorofil'nykh mutantov u gorokha. Soobshchenie I. Pigmentnyĭ sostav i belki khloroplastov. [Physiological and biochemical aspects of monohybrid heterosis derived from pea chlorophyll mutant. I. Pigment composition and proteins of chloroplasts.] - Genetika *12* (2) : 52 - 58, 1976. [In R, ab : E.]

28262 - **VERSHININA, E.I.** : Fotosinteticheskaya deyatel'nost' rasteniĭ yarovoĭ pshenitsy i urozhaĭ zerna v svyazi s priemami vozdelyvaniya. [Photosynthetic activity of spring wheat plants and grain yield in connection with agrotechnics.] - Tr. gor'kov. sel'skokhoz. Inst. *69* (Botanika i Fiziologiya Rasteniĭ) : 25 - - 28, 1976. [In R.]

28263 - **VICENTE, C., ESTÉVEZ, M.P.** : Inhibition of photolysis by chloroatranorine in isolated chloroplasts from *Evernia prunastri*'s phycobiont. - Bol. real. Soc. españ. Hist. nat. (Secc. Biol.) *74* : 17 - 23, 1976.

28264 - **VIDAL, J., CAVALIE, G., GADAL, P.** : Etude de la phosphoenol-pyruvate carboxylase du Haricot et du Sorgho par electrophorese sur gel de polyacrylamide. - Plant Sci. Lett. *7* : 265 - 270, 1976.

28265 - **VIDOVIČ, J., PIOVARČI, A.** : Spatial distribution of leaf area in the stand and the productivity of stand-spatial distribution of leaf area as a factor of radiation distribution in the maize stand and its productivity under the conditions of southern Slovakia. - Rostl. Výroba (Praha) *22* : 1053 - 1064, 1976.

28266 - **VIEIRA da SILVA, J.** : Water stress, ultrastructure and enzymatic activity. - In : LANGE, O.L., KAPPEN, L., SCHULZE, E.-D. (ed.) : Water and Plant Life. Pp. 207 - 224. Springer-Verlag, Berlin - Heidelberg - New York 1976. [Chl, chloroplast.]

*28267 - **VIGIL, E.L.** : Structure and function of plant microbodies. - Sub-cell. Biochem. *2* : 237 - 285, 1973.

28268 - **VIĬL, Yu., PYARNIK, T.** : Osobennosti funktsionirovaniya vosstanovitel'nogo pentozofosfatnogo tsikla pri razlichnykh kontsentratsiyakh kisloroda. [Characteristics of functioning of the reductive pentose phosphate pathway at different oxygen concentrations.] - In : Gazometricheskoe Issledovanie Fotosinteza i Dykhaniya Rasteniĭ. Pp. 23 - 25. Akad. Nauk SSSR, Tartu 1976. [In R.]

28269 - **VIĬL', Yu.A.** : Deĭstvie kisloroda na fotosinteticheskiĭ metabolizm ugleroda v list'yakh fasoli pri limitiruyushchikh i nasyshchayushchikh kontsentratsiyakh CO_2. [Effect of oxygen on the photosynthetic metabolism of carbon in French bean leaves under limiting and saturating CO_2 concentrations.] - In : Regulyatsiya Rosta i Pitanie Rasteniĭ. Pp. 171 - 176. Zinatne, Riga 1976. [In R.]

28270 - **VIKMANE, M.Ya., MAURINYA, Kh.A.** : Izmenenie nekotorykh fiziologicheskikh protsessov u rasteniĭ tomatov pod vliyaniem okisliteleĭ i vosstanoviteleĭ. [Changes in some physiological processes in tomato plants influenced by oxidants and reductants.] - In : Regulyatsiya Rosta i Pitanie Rasteniĭ. Pp. 91 - 96. Zinatne, Riga 1976. [Chl; in R.]

28271 - **VILLAR, J.-G.** : Photoelectrochemical effects in the electrolyte-pigment-metal system. I. Metal-free phthalocyanine film description of short-circuit photocurrents for thin films of pigment. - J. Bioenerg. Biomembranes *8* : 173 - 187, 1976. [Ps model systems.]

28272 - **VILLAR, J.-G.** : Photoelectrochemical effects in the electrolyte-pigment-metal system. II. Metal-free phthalocyanine film action spectra of short-circuit photocurrents with increase of the film thickness. - J. Bioenerg. Biomembranes *8* : 189 - 198, 1976. [Ps model systems.]

28273 - **VILLAR, J.-G.** : Photoelectrochemical effects in the electrolyte-pigment-metal system. III. Chlorophyll films short-circuit photocurrent transients light energy conversion efficiency. - J. Bioenerg. Biomembranes *8* : 199 - 208, 1976.

28274 - VITOLA, A.K., KRISTKALNE, S.Kh., SELGA, M.P., GUBAR', G.D., KREĬTSBERG, O.É.:
Fiziologo-biokhimicheskie parametry lista v zavisimosti ot svetovogo rezhima
i mineral'nogo pitaniya. [Leaf physiological and biochemical properties in
relation to light regime and mineral nutrition.] - In : Regulyatsiya Rosta i
Pitanie Rasteniĭ. Pp. 161 - 170. Zinatne, Riga 1976. [In R.]

28275 - VLASOV, B.E., DADYKIN, V.P. : Opyt postroeniya matematicheskoĭ modeli raboty
ust'ichnogo apparata list'ev. [Attempt to construct a mathematical model of
stomata functioning.] - In : Gazometricheskoe Issledovanie Fotosinteza i Dykhaniya Rasteniĭ. Pp. 26 - 28. Akad. Nauk SSSR, Tartu 1976. [In R.]

28276 - VLASYUK, P.A., KHMARA, L.A., KLIMOVITSKAYA, Z.M., SEMICHAEVSKIĬ, V.D. : Spektral'nye svoĭstva khlorofilla v khloroplastakh i subkhloroplastnykh fraktsi-
yakh v zavisimosti ot urovnya margantsa v rasteniyakh. [Spectral properties
of chlorophyll in chloroplasts and subchloroplast fractions depending on manganese level in plants.] - Fiziol. Biokhim. kul't. Rast. *8* : 6 - 11, 1976.
[In R, ab : E.]

28277 - VLCEK, L., GASSMAN, M. : Reversal of α,α'-dipyridyl-induced porphyrin and
phorbin synthesis in etiolated red kidney bean leaves. - Plant Physiol. *57*
(Suppl.) : 45, 1976.

28278 - VOGEL, S. : Der Einfluß von Phenol und von phenologischen Verbindungen auf
den Gaswechsel autotropher und heterotropher Planktonten. - Verhandl. Ges.
Ökol. *1976* : 361 - 369, 1976.

28279 - VOĬNOVSKAYA, K.K. : K voprosu o sostoyanii khlorofilla v list'yakh rasteniĭ
i metodakh ego izucheniya. [State of chlorophyll in plant leaves and methods
for its study.] - In : Fotosintez i Produktivnost' Ozimoĭ Pshenitsy na Yugo-
-Vostoke Kazakhstana. Pp. 70 - 76, 131. Nauka kaz. SSR, Alma-Ata 1976. [In
R.]

28280 - VOLK, G.M., DUDECK, A.M. : Abnormal color response of turf ryegrass to top-
-dressed isobutylidene diurea. - Agron. J. *68* : 534 - 536, 1976. [Chl.]

28281 - VOLODARSKIĬ, A.D., CHAĬKA, M.T., ABRAMCHIK, L.M., CHAYANOVA, S.S., TIKHONOV-
SKAYA, N.G., SAVCHENKO, G.E. : Immunokhimicheskaya kharakteristika pigment-
-belkovykh komponentov membran ètioplastov i khloroplastov. [Immunochemical
characteristics of pigment-protein components from membranes of etioplasts
and chloroplasts.] - Fiziol. Rast. *23* : 1207 - 1213, 1976. [In R, ab : E.]

28282 - VOLODARSKIĬ, N.I., BYSTRYKH, E.E. : Nekotorye osobennosti fotosinteticheskoĭ
deyatel'nosti vysokoproduktivnykh sortov pshenitsy (obzor). [Some characteristics of photosynthetic activity of high-productive wheat cultivars (review).] - Sel'skokhoz. Biol. *11* : 328 - 336, 1976. [In R, ab : E.]

28283 - VOOKOVÁ, B. : Caloric values of leaves of *Cornus mas* L., *Crataegus oxyacan-
tha* L. and *Ligustrum vulgare* L. - Biológia (Bratislava) *31* : 737 - 744, 1976.

28284 - VOORN, G., MITCHELL, P. : A coulometric device for measuring total oxygen demand. - Int. Lab. *1976* : 25 - 26, 29, 1976.

28285 - VOSKRESENSKAYA, N.P. : Regulyatornaya rol' sinego sveta v formirovanii aktivnosti fotosinteticheskogo apparata. [Regulatory role of blue light in formation of photosynthetic apparatus activity.] - Fiziol. Biokhim. kul't. Rast.
8 : 339 - 348, 1976. [In R, ab : E.]

28286 - VOSKRESENSKAYA, N.P., MAZHUL', M.M., POLYAKOV, M.A. : Fotoregulyatsiya gazoobmena CO_2 i metabolizma ugleroda. [Photoregulation of CO_2 gas exchange and
carbon metabolism.] - In : Itogi Issledovaniya Mekhanizma Fotosinteza. Pp.
67 - 71. Pushchino 1976. [In R.]

28287 - VOSKRESENSKAYA, N.P., POLYAKOV, A.M. : Regulyatornoe deĭstvie sinego sveta na
fotosinteticheskiĭ gazoobmen : spektr deĭstviya na svetovoe nasyshchenie gazoobmena CO_2 u list'ev landysha. [Regulatory effect of blue light on photosynthetic gas exchange : action spectrum and light saturation of carbon dioxide gas exchange in the leaves of lily of the valley.] - Fiziol. Rast. *23* :
10 - 16, 1976. [In R, ab : E.]

28288 - VOZNESENSKAYA, E.V. : Ul'trastruktura assimiliruyushchikh organov nekotorykh
vidov sem. *Chenopodiaceae*. 1. [The ultrastructure of assimilating organs in

some species of *Chenopodiaceae* family. I.] - Bot. Zh. *61* : 342 - 351, 1976.
[In R, ab : E.]

28289 - **VOZNESENSKAYA, E.V.** : Ul'trastruktura assimiliruyushchikh organov nekotorykh
vidov sem.*Chenopodiaceae*. II. [Ultrastructure of assimilating organs of some
species of the family *Chenopodiaceae*. II.]-Bot. Zh. *61* : 1546 - 1557, 1976.
[In R, ab : E.]

28290 - **VOZNYAK, V.M., PROSKURYAKOV, I.I., ELFIMOV, E.I., KIM, V.A., EVSTIGNEEV, V.B.:**
Svobodno-radikal'nye sostoyaniya v pervichnykh fotokhimicheskikh i okislitel'-
no-vosstanovitel'nykh reaktsiyakh fotosinteticheskikh pigmentov. [Free-radical
states in primary photochemical and redox reactions of photosynthetic pigments.] - In : Itogi Issledovaniya Mekhanizma Fotosinteza. Pp. 25 - 29. Pushchino 1976. [In R.]

28291 - **VREDENBERG, W.J.** : Electrical interactions and gradients between chloroplast
compartments and cytoplasm. - In : **BARBER, J.** (ed.) : The Intact Chloroplast.
Pp. 53 - 88. Elsevier, Amsterdam - New York - Oxford 1976.

28292 - **VREDENBERG, W.J., BULYCHEV, A.A.** : Changes in the electrical potential across
the thylakoid membranes of illuminated intact chloroplasts in the presence of
membrane-modifying agents. - Plant Sci. Lett. *7* : 101 - 107, 1976.

28293 - **VSEVOLODOV, N.N., KAYUSHIN, L.P.** : Spektral'nye prevrashcheniya v purpurnykh
membranakh iz *H. halobium*. [Spectral transformations in purple membranes from
H. halobium.] - Stud. biophys. *59* : 81 - 87, 1976. [In R, ab : E.]

28294 - **VYAS, N.L., GARG, R.K., VYAS, L.N.** : Plant biomass and net production relations of *Terminalia tomentosa* WIGHT et ARN. at deciduous forest near Udaipur
(Rajasthan), India. - Flora *165* : 381 - 387, 1976.

28295 - **WAALAND, J.R.** : Growth of the red alga *Iridaea cordata* (TURNER) BORY in semi-
-closed culture. - J. exp. mar. Biol. Ecol. *23* : 45 - 53, 1976.

28296 - **WAFFORD, J.D., WHITBREAD, R.** : Effects of leaf infections by *Septoria nodorum*
BERK. on the translocation of ^{14}C-labelled assimilates in spring wheat. - Ann.
Bot. *40* : 83 - 90, 1976.

28297 - **WAGNER, E.** : The nature of photoperiodic time measurement : energy transduction and phytochrome action in seedlings of *Chenopodium rubrum*. - In : **SMITH,
H.** (ed.) : Light and Plant Development. Pp. 419 - 443. Butterworth, London
1976. [Ps.]

28298 - **WAGNER, G., HOPE, A.B.** : Proton transport in *Halobacterium halobium*. - Aust.
J. Plant Physiol. *3* : 665 - 676, 1976.

28299 - **WAGNER, G., OESTERHELT, D.** : Lichtabhängiger K^+-Transport und seine Kopplung
an die elektrogene H^+-Pumpe Bakteriorhodopsin bei *Halobacterium halobium*. -
Ber. deut. bot. Ges. *89* : 289 - 292, 1976.

28300 - **WALCOTT, J.J., LAING, D.R.** : Some physiological aspects of growth and yield
in wheat crops : a comparison of a semidwarf and a standard height cultivar.
- Aust. J. exp. Agr. anim. Husb. *16* : 578 - 587, 1976. [Ps.]

28301 - **WALDRON, J.C.** : Nitrogen compounds transported in the xylem of sugar cane. -
Aust. J. Plant Physiol. *3* : 415 - 419, 1976. [Ps.]

28302 - **WALKER, A.J., HO, L.C.** : Young tomato fruits induced to export carbon by cooling. - Nature *261* : 410 - 411, 1976. [Photosynthates.]

28303 - **WALKER, D.A.** : CO_2 fixation by intact chloroplasts : Photosynthetic induction
and its relation to transport phenomena and control mechanisms. - In : **BARBER,
J.** (ed.) : The Intact Chloroplast. Pp. 235 - 278. Elsevier, Amsterdam - New
York - Oxford 1976.

28304 - **WALKER, D.A.** : Plastids and intracellular transport. - In : **STOCKING, C.R.,
HEBER, U.** (ed.) : Transport in Plants III. Intracellular Interactions and
Transport Processes. Pp. 85 - 136. Springer-Verlag, Berlin - Heidelberg - New
York 1976.

28305 - **WALKER, D.A., LILLEY, R. McC.** : Ribulose bisphosphate carboxylase - an enigma
resolved ? - In : **SUNDERLAND, N.** (ed.) : Perspectives in Experimental Biology.

Vol. 2. Botany. Pp. 189 - 198. Pergamon Press, Oxford - New York - Toronto -
- Sydney - Paris - Braunschweig 1976.

28306 - WALKER, D.A., SLABAS, A.R. : Stepwise generation of the natural oxidant in a
reconstituted chloroplast system. - Plant Physiol. 57 : 203 - 208, 1976.

28307 - WALKER, D.A., SLABAS, A.R., FITZGERALD, M.P. : Photosynthesis in a reconstituted chloroplast system from spinach. Some factors affecting CO_2-dependent
oxygen evolution with fructose-1,6-bisphosphate as substrate. - Biochim. biophys. Acta 440 : 147 - 162, 1976.

28308 - WALKER, J.R.L., McWHA, J.A. : A simple demonstration of CO_2-fixation and acid
production in CAM plants. - J. biol. Educ. 10 : 169 - 172, 1976.

28309 - WALLACE, A., MUELLER, R.T. : Behavior of iron-inefficient plants when grown
in combinations of calcareous and noncalcareous soils. - Commun. Soil Sci.
Plant Analysis 7 (1) : 107 - 110, 1976. ·

28310 - WALLACE, A., MUELLER, R.T., KAAZ, H.W. : Iron oxide as an iron source in pott-
ing-soil mixtures. - Commun. Soil Sci. Plant Analysis 7 (1) : 125 - 127, 1976.

28311 - WALLACE, A., PATEL, P.M., ROMNEY, E.M., ALEXANDER, G.V. : Iron chlorosis
caused by $MgCO_3$. - Commun. Soil Sci. Plant Analysis 7 (1) : 27 - 35, 1976.

28312 - WALLACE, A., ROMNEY, E.M., ALEXANDER, G.V. : Lime-induced chlorosis caused by
excess irrigation water. - Commun. Soil Sci. Plant Analysis 7 (1) : 47 - 49,
1976. [Chl.]

28313 - WALLACE, A., WOOD, R.A., SOUFI, S.M. : Cation-anion balance in lime-induced
chlorosis. - Commun. Soil Sci. Plant Analysis 7 (1) : 15 - 26, 1976.

28314 - WALLACE, D.G. : Prediction of the secondary and tertiary structure of plastocyanin. - Biophys. Chem. 4 : 123 - 130, 1976.

28315 - WALLACE, D.G., BOULTER, D. : Immunological comparison of higher plant plasto-
cyanins. - Phytochemistry 15 : 137 - 141, 1976.

28316 - WALLACE, D.H., PEET, M.M., OZBUN, J.L. : Studies of CO_2 metabolism in Phaseo-
lus vulgaris L. and applications in breeding. - In : BURRIS, R.H., BLACK, C.
C. (ed.) : CO_2 Metabolism and Plant Productivity. Pp. 43 - 58. Univ. Park
Press, Baltimore - London - Tokyo 1976.

28317 - WALLENTINUS, I. : Productivity studies on Cladophora glomerata (L.) KÜTZING
in the northern Baltic proper. - In : PERSOONE, G., JASPERS, E. (ed.) :
Proceedings of the 10th European Symposium on Marine Biology. Vol. 2. Pp. 631-
- 651. Universa Press, Wettern 1976.

28318 - WALLIHAN, E.F., SHARPLESS, R.G., PRINTY, W. : Chlorophyll concentration in
navel orange leaves in relation to iron status, leaf age, and season. - J.
amer. Soc. hort. Sci. 101 : 425 - 427, 1976.

28319 - WALTER, G. : Zur Regulation des Chlorophyllstoffwechsels. - Wiss. Z. Humboldt-
-Univ. Berlin, math.-naturwiss. Reihe 25 : 759. - 768, 1976.

28320 - WALTER, G., PESTER, A. : Die hydrolytische Aktivität der Chlorophyllase in
Keimpflanzen von Triticum aestivum L. verschiedener Entwicklungsstadien. -
Wiss. Z. Humboldt-Univ. Berlin, math.-naturwiss. Reihe 25 : 796 - 802, 1976.

28321 - WALTON, D., HARRISON, M., GALSON, E. : ABA levels and metabolism in water-
-stressed bean plants. - Plant Physiol. 57 (Suppl.) : 62, 1976.

28322 - WALTON, D.W.H. : Dry matter production in Acaena (Rosaceae) on a subantarctic
island. - J. Ecol. 64 : 399 - 415, 1976. [Chl.]

28323 - WALTON, D.W.H., SMITH, R.I.L. : Some limitations on plant growth and development in tundra regions - an investigation using phytometers. - New Phytol.
76 : 501 - 510, 1976. [Growth analysis.]

28324 - WALZ, D. : Pigment containing lipid vesicles. II. Interaction of valinomycin
with lecithin as sensed by chlorophyll a. - J. Membrane Biol. 27 : 55 - 81,
1976.

28325 - WANDERS, J.B.W. : The role of benthic algae in the shallow reef of Curaçao
(Netherlands Antilles) II. Primary productivity of the Sargassum beds on the

north-east coast submarine plateau. - Aquat. Bot. *2* : 327 - 335, 1976.

28326 - WANG HSI, SHIH YI-PING : [Physiological analysis of the ripening-promoting
         effect of ethrel in rice.] - Acta bot. sin. *18* : 150 - 155, 1976. [Ps, Chl;
         in Chin., ab : E.]

28327 - WANG, J.H., YANG, M. : Generation of ATP by chloroplasts through solvent per-
         turbation. - Biochem. biophys. Res. Commun. *73* : 673 - 678, 1976.

28328 - WANG, R.T., MYERS, J. : On the distribution of excitation energy to two photo·
         reactions of photosynthesis. - Photochem. Photobiol. *23* : 405 - 410, 1976.

28329 - WANG, R.T., MYERS, J. : Simultaneous measurement of action spectra for photo-
         reactions I and II of photosynthesis. - Photochem. Photobiol. *23* : 411 - 414,
         1976.

28330 - WARA-ASWAPATI, O., BRADBEER, J.W. : The inhibition of the development of pho-
         tosystem II-mediated electron transport in greening bean leaves by D-*threo*-
         -chloramphenicol. - Plant Sci. Lett. *7* : 95 - 100, 1976.

28331 - WARD, B. : On choice of hydrogen-ion buffer for photosynthesis studies with
         cyanobacteria. - Microbios Lett. *3* : 101 - 106, 1976.

28332 - WARDEN, J.T. : Experimental examination of the "energy upconversion" theory
         for green plant photosynthesis. - Proc. nat. Acad. Sci. USA *73* : 2773 - 2775,
         1976.

28333 - WARDEN, J.T. : Flash photolysis-electron spin resonance studies of photosys-
         tem I. A fast reduction component of P-700$^+$. - Biochim. biophys. Acta *440* :
         89 - 97, 1976.

28334 - WARDEN, J.T., BLANKENSHIP, R.E., SAUER, K. : A flash photolysis ESR study of
         photosystem II. Signal II$_{vf}$, the physiological donor to P-680$^+$. - Biochim.
         biophys. Acta *423* : 462 - 478, 1976.

28335 - WARDEN, J.T., BOLTON, J.R. : Combined optical and electron spin resonance
         kinetic spectrometer. - Rev. sci. Instrum. *47* : 201- 204, 1976. [Measurement
         of ESR kinetics in Ps bacteria.]

28336 - WARDLAW, I.F. : Assimilate partitioning : cause and effect. - In : WARDLAW,
         I.F., PASSIOURA, J.B. (ed.) : Transport and Transfer Processes in Plants.
         Pp. 381 - 391. Academic Press, New York - San Francisco - London 1976.

28337 - WARDLAW, I.F. : Assimilate movement in *Lolium* and *Sorghum* leaves. I. Irradian-
         ce effects on photosynthesis, export and the distribution of assimilates. -
         Aust. J. Plant Physiol. *3* : 377 - 387, 1976.

28338 - WARDLAW, I.F., MARSHALL, C. : Assimilate movement in *Lolium* and *Sorghum* lea-
         ves. II. Irradiance effects on the products of photosynthesis. - Aust. J.
         Plant Physiol. *3* : 389 - 400, 1976.

28339 - WARDLAW, I.F., MONCUR, L. : Source, sink and hormonal control of transloca-
         tion in wheat. - Planta *128* : 93 - 100, 1976.

B28340 - WARDLAW, I.F., PASSIOURA, J.B. (ed.) : Transport and Transfer Processes in
         Plants. - Academic Press, New York - San Francisco - London 1976. [Photosyn-
         thates, resistances.]

28341 - WARING, R.H., RUNNING, S.W. : Water uptake, storage and transpiration by coni-
         fers : a physiological model. - In : LANGE, O.L., KAPPEN, L., SCHULZE, E.-D.
         (ed.) : Water and Plant Life. Pp. 189 - 202. Springer-Verlag, Berlin - Heidel-
         berg - New York 1976. [Resistances.]

28342 - WARRINGTON, I., PEET, M., PATTERSON, D., BUNCE, J., HELLMERS, H. : Physiolo-
         gical response of soybean to thermoperiod in relation to growth. - Plant Phy-
         siol. *57* (Suppl.) : 105, 1976.

28343 - WARRINGTON, I.J., MITCHELL, K.J. : The influence of blue- and red-biased
         light spectra on the growth and development of plants. - Agr. Meteorol. *16* :
         247 - 262, 1976. [Growth analysis.]

28344 - WARRINGTON, I.J., MITCHELL, K.J., HALLIGAN, G. : Comparisons of plant growth
         under four different lamp combinations and various temperature and irradiance
         levels. - Agr. Meteorol. *16* : 231 - 245, 1976.

28345 - **WASIELEWSKI, M.R., STUDIER, M.H., KATZ, J.J.** : Covalently linked chlorophyll
a dimer : a biomimetic model of special pair chlorophyll. - Proc. nat. Acad.
Sci. USA *73* : 4282 - 4286, 1976.

28346 - **WATADA, A.E., NORRIS, K.H., WORTHINGTON, J.T., MASSIE, D.R.** : Estimation of
chlorophyll and carotenoid contents of whole tomato by light absorbance technique. - J. Food Sci. *41* : 329 - 332, 1976.

28347 - **WATANABE, I.** : Transformation factor from CO_2 net assimilation to dry matter
in crop plants. - JARQ - Jap. agr. Res. Quart. *10* : 114 - 118, 1976.

28348 - **WATANABE, I., OGIHARA, H., KONNO, S., TABUCHI, K.** :[Revised standard of examination for leaf colour of soybean.]- Proc. Crop Sci. Soc. Jap. *45* : 173 -
- 174, 1976. [In Jap.]

28349 - **WATANABE, M., MIYOSHI, Y., FURUYA, M.** : Phototaxis in *Cryptomonas* sp. under
condition suppressing photosynthesis. - Plant Cell Physiol. *17* : 683 - 690,
1976.

28350 - **WATTS, W.R., NEILSON, R.E., JARVIS, P.G.** : Photosynthesis in Sitka spruce
(*Picea sitchensis* (BONG.) CARR.) VII. Measurements of stomatal conductance
and $^{14}CO_2$ uptake in a forest canopy. - J. appl. Ecol. *13* : 623 - 638, 1976.

28351 - **WAYGOOD, E.R., LAW, G.** : CO_2 fixation in leaves and chloroplasts of maize. -
Plant Physiol. *57* (Suppl.) : 33, 1976.

28352 - **WEBB, D.P.** : Root growth in *Acer saccharum* MARSH. seedlings : Effects of light
intensity and photoperiod on root elongation rates. - Bot. Gaz. *137* : 211 -
- 217, 1976. [Production.]

28353 - **WEBB, W.L., ZAERR, J.B.** : Carbon dioxide efflux of Douglas-fir seedlings in
light and dark. - Photosynthetica *10* : 388 - 393, 1976.

*28354 - **WEBER, C.I.** : Recent developments in the measurement of the response of plank-
ton and periphyton to changes in their environment. - In : GLASS, G. (ed.) :
Bioassay Techniques and Environmental Chemistry. Pp. 119 - 138. Ann Arbor
Science Publishers Inc., Ann Arbor, Mich. 1973.

28355 - **WEDDING, R.T., BLACK, M.K., PAP, D.** : Malate dehydrogenase and NAD malic enzyme in the oxidation of malate by sweet potato mitochondria. - Plant Physiol. *58* : 740 - 743, 1976.

28356 - **WEEDON, B.C.L.** : Synthesis of carotenoids and related polyenes. - Pure appl.
Chem. *47* : 161 - 171, 1976.

28357 - **WEIDNER, M., STEINBISS, H., KREMER, B.P.** : Correlations between photosynthetic enzymes, CO_2-fixation and plastid structure in an albino mutant of *Zea
mays* L. - Planta *131* : 263 - 270, 1976.

28358 - **WEIN, R.W., RENCZ, A.N.** : Plant cover and standing crop sampling procedures
for the Canadian High Arctic. - Arct. alp. Res. *8* : 139 - 150, 1976.

28359 - **WEISE, G., HORBACH, W., HORNIG, L., GNAUCK, A.H.** : Der Kohlendioxidumsatz
submerser Makrophyten als Indikationskriterium des Wassergütezustandes. - Acta hydrochim. hydrobiol. *4* : 95 - 101, 1976.

28360 - **WEISSENBÖCK, G.** : Accumulation of flavonoids in the leaves and the plastids of
oat seedlings (*Avena sativa* L.). - Nova Acta leopoldina - Suppl. *7* (Secondary
Metabolism and Coevolution) : 97 - 101, 1976.

28361 - **WEISSENBÖCK, G., PLESSER, A., TRINKS, K.** : Flavonoidgehalt und Enzymaktivi-
täten isolierter Haferchloroplasten (*Avena sativa* L.). - Ber. deut. bot. Ges.
89 : 457 - 472, 1976.

28362 - **WELLBURN, A.R.** : Evidence for chlorophyll esterified with geranyl-geraniol
in newly greened leaves. - Biochem. Physiol. Pflanzen *169* : 265 - 271, 1976.

28363 - **WELLBURN, A.R., HAMPP, R.** : Movement of labelled metabolites from mitochondria to plastids during development. - Planta *131* : 17 - 20, 1976.

28364 - **WELLBURN, A.R., HAMPP, R.** : Fluxes of gibberellic and abscisic acids, toget-
her with that of adenosine 3',5'-cyclic phosphate, across plastid envelopes
during development. - Planta *131* : 95 - 96, 1976.

28365 - **WELLBURN, F.A.M., WELLBURN, A.R.** : Novel chloroplasts and unusual cellular ultrastructure in the "resurrection" plant *Myrothamnus flabellifolia* WELW. (*Myrothamnaceae*). - Bot. J. linn. Soc. *72* : 51 - 54, 1976.

28366 - **WEST, D.W., GAFF, D.F.** : A controlled-environment leaf chamber to allow measurement of gas exchange by leaves undergoing rapid fluctuations in temperature. - J. exp. Bot. *27* : 205 - 213, 1976.

28367 - **WEST, D.W., GAFF, D.F.** : The effect of leaf water potential, leaf temperature and light intensity on leaf diffusion resistance and the transpiration of leaves of *Malus sylvestris*. - Physiol. Plant. *38* : 98 - 104, 1976.

28368 - **WEST, L.D., MUZIK, T.J., WITTERS, R.E.** : Differential gas exchange responses of two biotypes of redroot pigweed to atrazine. - Weed Sci. *24* : 68 - 72, 1976.

28369 - **WESTERMAN, P.W., BARFIELD, B.J., LOEWER, O.J., WALKER, J.N.** : Evaporative cooling of a partially-wet and transpiring leaf - I. Computer model and its evaluation using wind-tunnel experiments. - Trans. ASAE *19* : 881 - 888, 1976. [Resistances.]

28370 - **WESTERMAN, P.W., BARFIELD, B.J., LOEWER, O.J., WALKER, J.N.** : Evaporative cooling of a partially-wet and transpiring leaf - II. Simulated effect of variations in environmental conditions, leaf properties, and surface water characteristics. - Trans. ASAE *19* : 889 - 893, 896, 1976. [Resistances.]

28371 - **WESTRIN, H., ALBERTSSON, P.-Å., JOHANSSON, G.** : Hydrophobic affinity partition of spinach chloroplasts in aqueous two-phase systems. - Biochim. biophys. Acta *436* : 696 - 706, 1976.

28372 - **WETHEY, D.S., PORTER, J.W.** : Sun and shade differences in productivity of reef corals. - Nature *262* : 281 - 282, 1976.

28373 - **WETTSTEIN, D. von** : Genetic regulation of membrane synthesis in chloroplasts as studied with lethal gene mutants. - In : **BOLIS, L., HOFFMAN, J.F., LEAF, A.** (ed.) : Membranes and Disease. Pp. 123 - 130. Raven Press, New York 1976.

28374 - **WHITBREAD, R., BHATTI, M.A.R.** : The translocation of ^{14}C-labelled assimilates in dwarf bean plants infected with *Xanthomonas phaseoli* (E.F.SM.) DOWSON. - Ann. Bot. *40* : 499 - 509, 1976.

28375 - **WHITEAKER, G., GERLOFF, G.C., GABELMAN, W.H., LINDGREN, D.** : Intraspecific differences in growth of beans at stress levels of phosphorus. - J. amer. Soc. hort. Sci. *101* : 472 - 475, 1976. [Ps.]

28376 - **WHITTINGHAM, C.P.** : Function in photosynthesis. - In : **GOODWIN, T.W.** (ed.): Chemistry and Biochemistry of Plant Pigments. 2nd Ed. Vol. 1. Pp. 624 - 654. Academic Press, London - New York - San Francisco 1976.

28377 - **WHITTLE, S.J.** : The major chloroplast pigments of *Chlorobotrys regularis* (WEST) BOHLIN (*Eustigmatophyceae*) and *Ophiocytium majus* NAEGELI (*Xanthophyceae*). - Brit. phycol. J. *11* : 111 - 114, 1976.

28378 - **WICKRAMASINGHE, R.H.** : Model role for bacteriorhodopsin for solar energy utilization by primordial organisms. - Cytobios *17* : 31 - 33, 1976.

28379 - **WIEDENROTH, E.-M.** : Chlorophyllbildung und Gaswechsel von Weizenkeimpflanzen (*Triticum aestivum* L.) unter dem Einfluß variierter Temperatur- und Sauerstoffbedingungen im Wurzelbereich. - Tagungsber. Akad. Landwirtschaftwiss. DDR *143* : 297 - 309, 1976.

28380 - **WIEDENROTH, E.-M.** : Methodik der Erfassung des Gaswechsels in Sproß- und Wurzelsystemen intakter Jungpflanzen. - Wiss. Z. Humboldt-Univ. Berlin, math.-naturwiss. Reihe *25* : 737 - 741, 1976.

28381 - **WIELGOLASKI, F.E.** : The effect of herbage intake by sheep on primary production, ratios top-root and dead-live aboveground parts (Hardangervidda, Norway). - Pol. ecol. Stud. *2* : 67 - 76, 1976.

28382 - **WIEN, H.C., ALTSCHULER, S.L., OZBUN, J.L., WALLACE, D.H.** : ^{14}C-assimilate distribution in *Phaseolus vulgaris* L. during the reproductive period. - J. amer. Soc. hort. Sci. *101* : 510 - 513, 1976.

28383 - **WIESBERG, L.H.G.** : Natürliche ^{13}C-Isotopenmarkierung in Bäumen. - Z. Pflanzenphysiol. *79* : 292 - 299, 1976.

28384 - **WILDMAN, R.B., HUNT, P.** : Phytoferritin associated with yellowing in leaves
of *Cocos nucifera (Arecaceae)*. - Protoplasma *87* : 121 - 134, 1976.

*28385 - **WILDMAN, S.G., CHEN, K., GRAY, J.C., KUNG, S.D., KWANYUEN, P., SAKANO, K.** :
Evolution of ferredoxin and Fraction I protein in the genus *Nicotiana*. - In :
BIRKY, C.W., Jr., PERLMAN, P.S., BYERS, T.J. (ed.) : Genetics and Biogenesis
of Mitochondria and Chloroplasts. Pp. 309 - 329. Ohio State Univ. Press, Columbus 1975.

28386 - **WILDNER, G.F.** : The role of ribulose-1,5-bisphosphate carboxylase and its
oxygenase activity in the events of photorespiration. - Ber. deut. bot. Ges.
89 : 349 - 360, 1976.

28387 - **WILDNER, G.F., HENKEL, J.** : Specific inhibition of the oxygenase activity of
ribulose-1,5-bisphosphate carboxylase. - Biochem. biophys. Res. Commun. *69* :
268 - 275, 1976.

28388 - **WILKINSON, M.J., SMITH, H.** : Properties of phosphoenol pyruvate carboxylase
from *Bryophyllum fedtschenkoi* leaves and fluctuations in carboxylase activity
during the endogenous rhythm of carbon dioxide output. - Plant Sci. Lett. *6* :
319 - 324, 1976.

28389 - **WILLERT, D.J. von, KIRST, G.O., TREICHEL, S., WILLERT, K. von** : The effect
of leaf age and salt stress on malate accumulation and phosphoenolpyruvate
carboxylase activity in *Mesembryanthemum crystallinum*. - Plant Sci. Lett.
7 : 341 - 346, 1976.

28390 - **WILLERT, D.J. von, TREICHEL, S., KIRST, G.O., CURDTS, E.** : Environmentally
controlled changes of phosphoenolpyruvate carboxylases in *Mesembryanthemum*.
- Phytochemistry *15* : 1435 - 1436, 1976.

28391 - **WILLIAMS, G.J. III, KEMP, P.R.** : Temperature relations of photosynthetic response in populations of *Verbascum thapsus* L. - Oecologia *25* : 47 - 54, 1976.

28392 - **WILLIAMS, J.H., WILSON, J.H.H., BATE, G.C.** : The influence of defoliation
and pod removal on growth and dry matter distribution in groundnuts (*Arachis
hypogaea* L. cv. Makulu Red). - Rhod. J. agr. Res. *14* : 111 - 117, 1976.
[Growth analysis.]

28393 - **WILLIAMS, L.E., KENNEDY, R.A.** : Relationship between primary photosynthetic
products, photorespiration and stage of leaf development in *Zea mays*. - Plant
Physiol. *57* (Suppl.) : 33, 1976.

28394 - **WILLIAMS, P.J. LeB., YENTSCH, C.S.** : An examination of photosynthetic production, excretion of photosynthetic products, and heterotrophic utilization of
dissolved organic compounds with reference to results from a coastal subtropical sea. - Mar. Biol. *35* : 31 - 40, 1976.

28395 - **WILLIAMS, W.P., SALAMON, Z.** : Enhancement studies on algae and isolated chloroplasts. Part I. Variability of photosynthetic enhancement in *Chlorella pyrenoidosa*. - Biochim. biophys. Acta *430* : 282 - 299, 1976.

28396 - **WILLIAMS, W.P., SALAMON, Z., MUALLEM, A., BARBER, J., MILLS, J.** : Enhancement
studies on algae and isolated chloroplasts. Part II. Enhancement of oxygen
evolution in intact chloroplasts. - Biochim. biophys. Acta *430* : 300 - 311,
1976.

28397 - **WILLIAMSON, P.** : Above-ground primary production of chalk grassland allowing
for leaf death. - J. Ecol. *64* : 1059 - 1075, 1976.

28398 - **WILLISON, J.H.M., DAVEY, M.R.** : Fraction 1 protein crystals in chloroplasts
of isolated tobacco leaf protoplasts : A thin-section and freeze-etch morphological study. - J. Ultrastruct. Res. *55* : 303 - 311, 1976.

28399 - **WILLMER, C.M., JOHNSTON, W.R.** : Carbon dioxide assimilation in some aerial
plant organs and tissues. - Planta *130* : 33 - 37, 1976.

28400 - **WILSON, J.R.** : Variation of leaf characteristics with level of insertion on
a grass tiller. I Development rate, chemical composition and dry matter digestibility. - Aust. J. agr. Res. *27* : 343 - 354, 1976.

28401 - WINKENBACH, F., FALK, H., LIEDVOGEL, B., SITTE, P. : Chromoplasts of *Tropaeolum majus* L. : Isolation and characterization of lipoprotein elements. - Planta *128* : 23 - 28, 1976.

28402 - WINOGRAD, N., SHEPARD, A., KARWEIK, D.H., KOESTER, V.J., FONG, F.K. : X-ray photoelectron spectroscopic studies of the thermal stability of chlorophyll *a* monohydrate. - J. amer. chem. Soc. *98* : 2369 - 2370, 1976.

28403 - WINTER, J., KANDLER, O. : Misleading data on isotope distribution in malate--^{14}C from CAM plants caused by fumarase activity of *Lactobacillus plantarum*. - Z. Pflanzenphysiol. *78* : 103 - 112, 1976.

28404 - WINTER, K., LÜTTGE, U. : Balance between C_3 and CAM pathway of photosynthesis. - In : LANGE, O.L., KAPPEN, L., SCHULZE, E.-D.(ed.) : Water and Plant Life. Pp. 323 - 334. Springer-Verlag, Berlin - Heidelberg - New York 1976.

28405 - WINTER, K., LÜTTGE, U. : Malate accumulation in leaf slices of *Mesembryanthemum crystallinum* in relation to osmotic gradients between the cells and the medium. - Aust. J. Plant Physiol. *3* : 653 - 663, 1976.

28406 - WINTER, K., TROUGHTON, J.H., CARD, K.A. : δ^{13}C values of grass species collected in the northern Sahara desert. - Oecologia *25* : 115 - 123, 1976.

28407 - WINTER, K., TROUGHTON, J.H., EVENARI, M., LÄUCHLI, A., LÜTTGE, U. : Mineral ion composition and occurrence of CAM-like diurnal malate fluctuations in plants of coastal and desert habitats of Israel and the Sinai. - Oecologia *25* : 125 - 143, 1976.

28408 - WINZELER, H., HUNT, L.A., MAHON, J.D. : Ontogenetic changes in respiration and photosynthesis in a uniculm barley. - Crop Sci. *16* : 786 - 790, 1976.

28409 - WITT, H.T. : Primary actions of excited molecules in the functional membrane of photosynthesis. - In : BIRKS, J.B. (ed.) : Excited States of Biological Molecules. Pp. 245 - 261. Wiley-Interscience, London - New York 1976.

28410 - WITT, H.T. : Biophysikalische Primärvorgänge in der Photosynthesemembran. Ergebnisse mit pulsspektroskopischen Methoden. - Naturwissenschaften *63* : 23 - - 27, 1976.

28411 - WITT, H.T., SCHLODDER, E., GRÄBER, P. : Membrane-bound ATP synthesis generated by an external electrical field. - FEBS Lett. *69* : 272 - 276, 1976.

28412 - WOJCIECHOWSKA, W. : Biomass dynamics of dominant species in the phytoplankton of two lakes varying in trophy. - Ekol. pol. *24* : 447 - 459, 1976.

28413 - WOLEDGE, J., LEAFE, E.L. : Single leaf and canopy photosynthesis in a ryegrass sward. - Ann. Bot. *40* : 773 - 783, 1976.

*28414 - WOLF, H., SCHEER, H. : Stereochemistry and chiroptic properties of pheophorbides and related compounds. - Ann. New York Acad. Sci. *206* : 549 - 567, 1973. [Chl.]

28415 - WOLF, H.U., ZANDER, R., LANG, W. : An automated continuous determination of oxygen with high sensitivity. - Anal. Biochem. *74* : 585 - 591, 1976.

28416 - WOLIŃSKA, D. : Functional and structural changes in chloroplasts of senescent tobacco leaves. - Acta Soc. Bot. Pol. *45* : 341 - 352, 1976.

28417 - WOLLGIEHN, R., LERBS, S., MUNSCHE, D. : Synthesis of ribosomal RNA in chloroplasts from tobacco leaves of different age. - Biochem. Physiol. Pflanzen *170* : 381 - 387, 1976.

28418 - WOLLMAN, F.-A., THOREZ, D. : Oscillation du niveau de fluorescence initiale chez *Chlorella pyrenoïdosa* après préillumination, puis addition de DCMU. - Compt. rend. Acad. Sci. Paris, Sér. D *283* : 1345 - 1348, 1976.

28419 - WOLOSIUK, R.A., BUCHANAN, B.B. : Studies on the regulation of chloroplast NADP-linked glyceraldehyde-3-phosphate dehydrogenase. - J. biol. Chem. *251* : 6456 - 6461, 1976.

28420 - WONG, D., GOVINDJEE : Effects of lead ions on photosystem I in isolated chloroplasts : Studies on the reaction center *P*700. - Photosynthetica *10* : 241 - - 254, 1976.

28421 - WONG, S.L., CLARK, B., PAINTER, D.S. : Application of underwater light measurements in nutrient and production studies in shallow rivers. - Freshwater
Biol. 6 : 543 - 550, 1976. [Ps.]

28422 - WOO, K.C., OSMOND, C.B. : Glycine decarboxylation in mitochondria isolated
from spinach leaves. - Aust. J. Plant Physiol. 3 : 771 - 785, 1976. [Chl.]

28423 - WOOD, P.M. : Electron transport between plastoquinone and cytochrome c-552
in Euglena chloroplasts. - FEBS Lett. 65 : 111 - 116, 1976.

28424 - WOOD, P.M., BENDALL, D.S. : The reduction of plastocyanin by plastoquinol-1
in the presence of chloroplasts. A dark electron transfer reaction involving
components between the two photosystems. - Europe. J. Biochem. 61 : 337 -
- 344, 1976.

28425 - WOODLEY, S.J., FENSOM, D.S., THOMPSON, R.G. : Biopotentials along the stem
Helianthus in association with short-term translocation of ^{14}C and chilling.
- Can. J. Bot. 54 : 1246 - 1256, 1976.

28426 - WOODWARD, R.G. : Photosynthesis and expansion of leaves of soybean grown in
two environments. - Photosynthetica 10 : 274 - 279, 1976.

28427 - WOODWARD, R.G., BEGG, J.E. : The effect of atmospheric humidity on the yield
and quality of soya bean. - Aust. J. agr. Res. 27 : 501 - 508, 1976. [Stoma-
tal resistance.]

28428 - WOODWARD, R.G., RAWSON, H.M. : Photosynthesis and transpiration in dicotyledonous plants. II. Expanding and senescing leaves of soybean. - Aust. J. Plant
Physiol. 3 : 257 - 267, 1976.

28429 - WOOLERY, M.L., LEWIN, R.A. : The effects of lead on algae. IV. Effects of Pb
on respiration and photosynthesis of Phaeodactylum tricornutum (Bacillario-
phyceae). - Water Air Soil Pollut. 6 : 25 - 31, 1976.

28430 - WOOLHOUSE, H.W., BATT, T. : The nature and regulation of senescence in plastids. - In : SUNDERLAND, N. (ed.) : Perspectives in Experimental Biology. Vol.
2. Botany. Pp. 163 - 175. Pergamon Press, Oxford - New York 1976.

28431 - WORLEY, S., Jr., RAMEY, H.H., Jr., HARRELL, D.C., CULP, T.W. : Ontogenetic
model of cotton yield. - Crop Sci. 16 : 30 - 34, 1976.

28432 - WORT, D.J. : Mechanism of plant growth stimulation by naphthenic acid. II.
Enzymes of CO_2 fixation, CO_2 compensation point, bean embryo respiration. -
Plant Physiol. 58 : 82 - 86, 1976.

28433 - WRISCHER, M., LJUBEŠIĆ, N., DEVIDÉ, Z. : Ultrastructural and functional characteristics of plastids in the leaves of Ligustrum ovalifolium HASSK. var. au-
reum. - Acta bot. croat. 35 : 57 - 64, 1976.

28434 - WRÓBEL, D., SALAMON, Z., FRĄCKOWIAK, D. : Quenching of chlorophyll c fluorescence by chlorophyll a. - Acta phys. pol. A49 : 269 - 274, 1976.

28435 - WU, P.-H.L. : Effects of illumination and temperature on metabolic patterns
of the ^{14}C-incorporation by the moss, Dicranum scoparium. - Ohio J. Sci.
76 : 103 - 109, 1976. [Photosynthates.]

28436 - WUTTKE, H.-G. : Chromoplasts in Rosa rugosa : Development and chemical characterization of tubular elements. - Z. Naturforsch. 31C : 456 - 460, 1976.
[Chloroplast.]

28437 - WYDRZYNSKI, T., GOVINDJEE, ZUMBULYADIS, N., SCHMIDT, P.G., GUTOWSKY, H.S. :
NMR studies on chloroplast membranes. - In : RESING, H.A., WADE, C.G. (ed.):
Magnetic Resonance in Colloid and Interface Science. ACS Symposium Series No.
34. Pp. 471 - 482. Amer. chem. Soc., Washington, D.C. 1976.

28438 - WYDRZYNSKI, T., ZUMBULYADIS, N., SCHMIDT, P.G., GOVINDJEE : NMR studies on
photosynthesis : proton relaxation as a monitor of membrane-bound manganese
and of the charge accumulating states during oxygen production. - Biophys. J.
16 (2, Part 2): 161a, 1976.

28439 - WYDRZYNSKI, T., ZUMBULYADIS, N., SCHMIDT, P.G., GUTOWSKY, H.S., GOVINDJEE :
Proton relaxation and charge accumulation during oxygen evolution in photosynthesis. - Proc. nat. Acad. Sci. USA 73 : 1196 - 1198, 1976.

28440 - **YABUKI, K.** : [Growth environment and photosynthesis of plants. 1.] - Nogyo
Oyobi Engei [Agriculture Horticulture] *51* : 1083 - 1087, 1976. [In Jap.]

28441 - **YABUKI, K.** : [Growth environment and photosynthesis of plants. 2.] - Nogyo
Oyobi Engei [Agriculture Horticulture] *51* : 1215 - 1220, 1976. [In Jap.]

28442 - **YABUKI, K.** : [Growth environment and photosynthesis of plants. 3.] - Nogyo
Oyobi Engei [Agriculture Horticulture] *51* : 1329 - 1333, 1976. [In Jap.]

28443 - **YABUKI, K.** : [Growth environment and photosynthesis of plants. 4.] - Nogyo
Oyobi Engei [Agriculture Horticulture] *51* : 1461 - 1464, 1976. [In Jap.]

28444 - **YAGI, T.** : Separation of hydrogenase-catalyzed hydrogen-evolution system from
electron-donating system by means of enzymic electric cell technique. - Proc.
nat. Acad. Sci. USA *73* : 2947 - 2949, 1976. [Inorganic model system.]

28445 - **YAGI, T., MUKOHATA, Y.** : Adenylate regulation of photosynthetic electron transport and the coupling sites of phosphorylation in spinach chloroplasts. - J.
Bioenerg. Biomembranes *8* : 247 - 255, 1976.

28446 - **YAMAMOTO, Y.** : Effect of some physical and chemical factors on the germination
of akinetes of *Anabaena cylindrica*. - J. gen. appl. Microbiol. *22* : 311 - 323,
1976. [Ps.]

28447 - **YAMAMOTO, Y., NISHIMURA, M.** : Characteristics of light-induced H$^+$ transport
in spinach chloroplast at lower temperatures I. Relationship between H$^+$ transport and physical changes of the microenvironment in chloroplast membranes. -
Plant Cell Physiol. *17* : 11 - 16, 1976.

28448 - **YAMAMOTO, Y., NISHIMURA, M.** : Characteristics of light-induced H$^+$ transport
in spinach chloroplast at lower temperatures II. A possible role of structu-
red water in H$^+$ transport of chloroplast near 0 $^\circ$C.- Plant Cell Physiol. *17* :
17 - 21, 1976.

28449 - **YAMASHITA, T., TOMITA, G.** : Light-reactivation of (Tris-washed)-DPIP-treated
chloroplasts : Manganese incorporation, chlorophyll fluorescence, action spectrum and oxygen evolution. - Plant Cell Physiol. *17* : 571 - 582, 1976.

28450 - **YAMAZAKI, S., TAKISAWA, H., TAMAURA, Y., HIROSE, S., INADA, Y.** : Effect of
troponin component TN-C on the inhibition of adenosine triphosphatase activities of mitochondria and chloroplasts by troponin component TN-I. - FEBS
Lett. *66* : 23 - 26, 1976.

28451 - **YANG, H.S., LEUNG, K.-H., ARCHER, M.C.** : Preparation and properties of bacterial chromatophores entrapped in polyacrylamide. - Biotechnol. Bioeng. *18* :
1425 - 1432, 1976.

28452 - **YANNAI, Y., EPEL, B.L., NEUMANN, J.** : Photophosphorylation in stable chloroplast fragments from the alga *Chlamydomonas reinhardi*. - Plant Sci. Lett. *7* :
295 - 304, 1976.

28453 - **YASINOVSKII, V.G., ROGATYKH, N.P., ZUBAREV, T.N.** : Bioélektricheskie kharakteristiki *Acetabularia mediterranea*. [Bioelectric characteristics of *Acetabularia mediterranea*.] - Fiziol. Rast. *23* : 180 - 186, 1976. [In R, ab : E.]

28454 - **YASUE, T., KAWASE, Y.** : [Studies on the Japanese barnyard millet as soiling
crop. III. Mesocotyl elongation in Japanese barnyard millet seedling.] - Proc.
Crop Sci. Jap. *45* : 91 - 98, 1976. [Leaf growth; in Jap., ab : E.]

28455 - **YEN, H.-C., MARRS, B.** : Map of genes for carotenoid and bacteriochlorophyll
biosynthesis in *Rhodopseudomonas capsulata*. - J. Bacteriol. *126* : 619 - 629,
1976.

B28456 - **YEOMAN, M.M.** (ed.) : Cell Division in Higher Plants. - Academic Press, London - New York - San Francisco 1976. [Chloroplast.]

28457 - **YOCUM, C.F.** : Photosystem II-mediated cyclic photophosphorylation. - Biochem.
biophys. Res. Commun. *68* : 828 - 835, 1976.

28458 - **YOCUM, C.F.** : Photosystem II cyclic photophosphorylation. - Plant Physiol.
57 (Suppl.) : 23, 1976.

28459 - **YOKOI, Y.** : Growth and reproduction in higher plants. I. Theoretical analysis by mathematical models. – Bot. Mag. (Tokyo) *89* : 1 – 14, 1976. [Biomass production.]

28460 – **YOKOI, Y.** : Growth and reproduction in higher plants. II. Analytical study of growth and reproduction of *Erythronium japonicum.* – Bot. Mag. (Tokyo) *89* : 15 – 31, 1976. [Biomass production.]

28461 – **YOSHIDA, S., CORONEL, V.** : Nitrogen nutrition, leaf resistance, and leaf photosynthetic rate of the rice plant. – Soil Sci. Plant Nutr. *22* : 207 – 211, 1976.

28462 – **YOSHIDA, S., SHIOYA, M.** : Photosynthesis of the rice plant under water stress. – Soil Sci. Plant Nutr. *22* : 169 – 180, 1976.

28463 – **YOSHIDA, T.** : [On the stomatal frequency in barley. I. The relationship between stomatal frequency and photosynthesis.] – Jap. J. Breed. *26* : 130 – 136, 1976. [In Jap., ab : E.]

28464 – **YURKOVSKIĬ, A.K., BRAMANE, A.Ė., YURKOVSKA, V. A.** : Tsikl izmeneniĬ pigmentov fitoplanktona i bakterioplanktona v BaltiĬskom more v 1974 g. [A cycle of variation of phytoplankton pigments and bacterioplankton in the Baltic Sea in 1974.] – Okeanologiya *16* : 830 – 838, 1976. [In R, ab : E.]

28465 – **YURKOVSKIS, A., LINE, R., BRAMANE, A., SIDREVITS, L., YURKOVSKA, V., VITINYA, M.** : Phosphorus and the seasonal dynamics of phyto-, bacterio-, and zooplankton in the Baltic Sea in 1974. – Ann. biol. *31* : 75 – 79, 1976.

28466 – **ZAGALSKY, P.F.** : Carotenoid-protein complexes. – Pure appl. Chem. *47* : 103 – – 120, 1976.

28467 – **ZAKHAROVA, N.I., GRIGOROVICH, V.I., KUTYURIN, V.M.** : O soderzhanii margantsa v khlorofill-belkovykh kompleksakh vysshikh rasteniĬ. [Manganese content in chlorophyll-protein complexes of higher plants.] – Biokhimiya *41* : 14 – 19, 1976. [In R, ab : E.]

28468 – **ZAKRZHEVSKIĬ, D.A., ROZONOVA, L.N., KALASHNIKOV, Yu.E.** : O fotosinteticheskom razlozhenii vody. [Photosynthetic water splitting.] – In : Itogi Issledovaniya Mekhanizma Fotosinteza. Pp. 47 – 51. Pushchino 1976. [In R.]

28469 – **ZANNONI, D., MELANDRI, B.A., BACCARINI-MELANDRI, A.** : Energy transduction in photosynthetic bacteria. XI. Further resolution of cytochromes of *b* type and the nature of the sensitive oxidase present in the respiratory chain of *Rhodopseudomonas capsulata.* – Biochim. biophys. Acta *449* : 386 – 400, 1976.

28470 – **ZAVITKOVSKI, J.** : Ground vegetation biomass, production, and efficiency of energy utilization in some northern Wisconsin forest ecosystems. – Ecology *57* : 694 – 706, 1976.

28471 – **ZDANOWSKI, B.** : Wieloletnie zmiany w produkcji pierwotnej pelagialu jezior podgrzanych. [Long-term changes of the pelagic primary production in heated lakes.] – Rocz. Nauk roln. H *97* (3) : 123 – 139, 1976. [In Pol., ab : E, R.]

28472 – **ZDANOWSKI, B.** : The influence of mineral fertilization of phytoplankton production in lakes of various trophic types. – Ekol. pol. *24* : 167 – 195, 1976.

28473 – **ZEINALOV, Yu.** : A simple equation allowing calculation of an average value of the misses (α) after very short flashes ($\beta_i = 0$). – Photosynthetica *10* : 83 – 85, 1976.

28474 – **ŻELAWSKI, W.** : Variation in the photosynthetic capacity of *Pinus sylvestris.* – In : CANNELL, M.G.R., LAST, F.T. (ed.) : Tree Physiology and Yield Improvement. Pp. 99 – 109. Academic Press, London – New York – San Francisco 1976.

28475 – **ŻELAWSKI, W., WALKER, R.B.** : Photosynthesis, respiration, and dry matter production. – In : MIKSCHE, J.P. (ed.) : Modern Methods in Forest Genetics. Pp. 89 – 119. Springer-Verlag, Berlin – Heidelberg – New York 1976.

28476 – **ZELDIN, M.H., SCHIFF, J.A.** : Blue light induced absorption changes in intact cells of *Euglena gracilis* var. *bacillaris* mutant W_3BUL. – Plant Physiol. *57* (Suppl.) : 22, 1976.

28477 – **ZELITCH, I.** : Biochemical and genetic control of photorespiration. – In :

BURRIS, R.H., BLACK, C.C. (ed.) : CO_2 Metabolism and Plant Productivity. Pp. 343 - 358. Univ. Park Press, Baltimore - London - Tokyo 1976.

28478 - ZELITCH, I. : Biokhimicheskaya i geneticheskaya regulyatsiya fotodykhaniya. [Biochemical and genetic regulation of photorespiration.] - Fiziol. Biokhim. kul't. Rast. 8 : 483 - 492, 1976. [In R, ab : E.]

28479 - ZELITCH, I. : The biochemistry of photorespiration. - In : SMITH, H. (ed.) : Commentaries in Plant Science. Pp. 51 - 61. Pergamon Press, Oxford - New York - Toronto - Sydney - Paris - Frankfurt 1976.

28480 - ZELITCH, I. : The mechanism of action of glycidate, an inhibitor of glycolate synthesis and photorespiration. - Plant Physiol. 57 (Suppl.) : 54, 1976.

28481 - ZEN'KEVICH, E.I., LOSEV, A.P. : Osobennosti agregatsii 4-vinil-protokhlorofilla i protokhlorofilla v rastvorakh. [Features of 4-vinyl-protochlorophyll and protochlorophyll aggregation in solutions.] - Mol. Biol. (Moskva) 10 : 294 - - 304, 1976. [In R, ab : E.]

B28482 - ZHARKIKH, A.A. : Zhizn' Rastitel'nogo Organizma i Voda. [Plant Life and Water.] - Khabarovskoe Knizhnoe Izd., Blagoveshchensk 1976. [Photosynthates; in R.]

28483 - ZHESTKOVA, I.M., MOLOTKOVSKII, Yu.G. : Predpolagaemyĭ mekhanizm aktivatsii i regulirovaniya Mg^{2+}-ATFazy khloroplastov. [Presumed mechanism of activation and regulation of Mg^{2+}-ATPase in chloroplasts.] - Fiziol. Rast. 23 : 1188 - - 1196, 1976. [In R, ab : E.]

28484 - ZHUKOV, V.G., FIRSOV, N.N. : Fotoassimilyatsiya organicheskikh soedineniĭ Thiocapsa roseopersicina. [Photoassimilation of organic compounds by Thiocapsa persicina.] - Mikrobiologiya 45 : 946 - 950, 1976. [Ps; in R, ab : E.]

*28485 - ZHUKOVA, G.Ya. : Plastidnyĭ apparat zarodysha khloro- i leĭkoěmbriofitov i ego funktsional'noe znachenie. [Plastid apparatus of the germ of chloro- and leucoembryophytes and its functional significance.] - Dokl. Akad. Nauk SSSR 224 : 1425 - 1427, 1975. [In R.]

28486 - ZHUKOVA, G.Ya., YAKOVLEV, M.S. : Khloroplasty pochechki zarodysha iskopaemogo plodika lotosa (ělektronnomikroskopicheskoe issledovanie). [Chloroplasts of embryo plumule of fossil lotus seeds (an electron microscopy study).] - Bot. Zh. 61 : 869 - 872, 1976. [In R, ab : E.]

*28487 - ZHURBITSKIĬ, Z.I. : Vliyanie postoyannogo ělektricheskogo polya na absorbtsiyu CO_2 list'yami rasteniĭ. [Effect of a constant electric field on the absorption of CO_2 by plant leaves.] - Dokl. Akad. Nauk SSSR 223 : 1273 - 1275, 1975. [In R.]

28488 - ZICKLER, A., WITT, H.T., BOHEIM, G. : Estimation of the light-induced electrical potential at the functional membrane of photosynthesis using a voltage-dependent ionophore. - FEBS Lett. 66 : 142 - 148, 1976.

28489 - ZIEGLER, H., OSMOND, C.B., STICHLER, W., TRIMBORN, P. : Hydrogen isotope discrimination in higher plants : correlations with photosynthetic pathway and environment. - Planta 128 : 85 - 92, 1976.

28490 - ZIEGLER, I., MAREWA, A., SCHOEPE, E. : Action of sulphite on the substrate kinetics of chloroplastic NADP-dependent glyceraldehyde-3-phosphate dehydrogenase. - Phytochemistry 15 : 1627 - 1632, 1976.

28491 - ZILINSKAS, B.A., GOVINDJEE : Stabilization by glutaraldehyde fixation of chloroplast membranes against inhibitors of oxygen evolution. - Z. Pflanzenphysiol. 77 : 302 - 314, 1976.

28492 - ZIMMERMANN, G., KELLY, G.J., LATZKO, E. : Efficient purification and molecular properties of spinach chloroplast fructose 1,6-bisphosphatase. - Europe. J. Biochem. 70 : 361 - 367, 1976.

28493 - ZINDLER-FRANK, E. : Oxalate biosynthesis in relation to photosynthetic pathway and plant productivity - a survey. - Z. Pflanzenphysiol. 80 : 1 - 13, 1976.

B28494 - ZLOBIN, V.S. : Pervichnaya Produktsiya i Kul'tivirovanie Morskogo Fitoplanktona. [Primary Production and Cultivation of Marine Phytoplankton.] - Pishchevaya Promyshlennost', Moskva 1976. [In R.]

28495 - ZVEREVA, E.G., BARTKOV, B.I. : Raspredelenie assimilyatov u soi v period na-
         liva plodov pri pereuvlazhnenii pochvy. [Distribution of assimilates in soy-
         bean during fruit growth with soil excessive moistening.] - Fiziol. Biokhim.
         kul't. Rast. 8 : 204 - 208, 1976. [In R, ab : E.]

Authors' names are presented in the form in which they appear in the respective publication. The names from papers published in Cyrillic characters are transcribed as
shown on p. III od this volume. Alternative spellings and forms of the name of the
same author are usually cross-indexed. The numbers in *italics* refer to publications
in which the respective author acts as an editor.

A

ABARSUA, A.Z. 26709
ABDEL-RAHMAN, M. 25162
ABELIOVICH, A. 25163
ABERNETHY, R.H. 25164
ABRAMCHIK, L.M. 28281
ABRAMYAN, L.Kh. 27716
ABREU, I. 27380
ACEVEDO, E. 26437-8
ACKEFORS, H. 25165
ACKER, S. 25166
ACKERMAN, T.L. 25303
ACKERSON, L.C. 27326
ACOCK, B. 25167-8
ADAMCHIK, G.G. 28253
ADAMOVA, N.P. 26731, 27273
ADAMS, G.M.W. 25524
ADAMS, J.E. 25169
ADAMS, M.S. 26754
ADEDIPE, N.O. 25170-1
ADELANA, B.O. 25172
ADJEI-TWUM, D.C. 25173
ADLER, K. 25174
ADYGEZALOV, V.F. 25175
AÉROV, I.L. 25176
AFGAN, N.H. *27253*
AFLAFO, C. 25177
AGADZHANYAN, Zh.G. 27110
AGRIKOVA, I.M. 25957
AGUILAR, M.I. 26029-30
AGUILAR-MÁRTINEZ, M. 25442
AGZAMOV, A. 25381-2
AHMED, J. 28231
AHMED, M.B.
    see BAKR AHMED, M.
AIGA, I. 27078
AKAGI, J.M. 26422
AKANOV, É.N. 25178
AKAZAWA, T. 25179, 25249, 26959, 27263,
    28079
ÅKERLUND, H.-E. 25180, 25217
AKERS, C.P. 25181
AKOYUNOGLOU, G. 25237, 25636
AKOYUNOGLOU, G.A. 27043
AKSENOV, V.P. 26822
AKULOVA, E.A. 25182-3, 26500, 26653-4,
    27659-60, 27856-7
ALBERTE, R.S. 25184-6, 26990, 27033,
    28127
ALBERTINE, K.H. 25187
ALBERTSSON, P.-Å. 25180, 25188, 25217,
    28371
AL'BITSKAYA, O.N. 25699, 25875, 26725
ALBU, M. 27480

ALDERFER, R.G. 25189
ALEXANDER, A.G. 25190
ALEXANDER, G.V. 28311-2
ALEXANDER, V. 25738-9
ALFANO, R.R. 25191
ALIEV, É.A. 27150
ALIEV, K.A. 25192
ALLAKHVERDOV, B.L. 27481
ALLAWAY, W.G. 25193-4
ALLEN, L.H.Jr. 25195
ALLEN, M.D.B. 26111
ALLEN, M.J. 25196
ALLEN, M.M. 25197
ALLEN, R.J.Jr. 25198
AL-MASHHADANI, Y. 27162
ALMASSY, R.J. 25855
ALMODÓVAR, L.R. 25598
ALSCHER, R. 25199
ALSCHER, R.G. 25200
ALTSCHULER, S.L. 28382
ALVAGER, T. 25201
AMBASHT, R.S. 27893-4
AMBLER, R.P. 28232
AMESZ, J. 25202, 27653
AMEZAGA, A.de 26179
AMIRDZHANOV, A.G. 25203
AMIRI, Z. 25204
AMPOFO, S.T. 25205-7
AMUNDSON, R.G. 26895
ANAGNOSTAKIS, S. 26336
ANDERSAG, R. 25208
ANDERSEN, A.S. 25209
ANDERSEN, F.Ø. 25210
ANDERSON, L.E. 25211-4
ANDERSON, L.L. 25713
ANDERSON, M.M. 27816
ANDERSON, O.R. 25215
ANDERSON, R.J. 27663
ANDERSON, W.R. 25216
ANDERSSON, B. 25180, 25217
ANDERSSON, L. 25918
ANDO, A. 27720
ANDRE, C. 25218
ANDRÉ, M. 28007
ANDREENKO, T.I. 27535
ANDREEVA, N.E. 25219-20
ANDREEVA, T.F. 25221
ANDREO, C.S. 25222, 26967, 28225
ANDREWS, A.K. 25223
ANDREWS, R. 25753
ANDREWS, R.A. 26713
ANDREWS, T.J. 26333-4, 26958
ANDRIANOV, V.K. 25589, 27889

ANGUS, J.F.   25224
ANIKEENKO, A.P.   25225
ANISIMOV, A.A.   25226
ANISIMOVA, I.N.   26822
ANJANEYULU, A.   27971
ANTON, J.A.   25227
AOKI, M.   28087
APEL, K.   25228
APEL, P.   25229-32, 27425
APELBAUM, A.   25233
APELL, G.S.   26159
APONASENKO, A.D.   25234
ap REES, T.   25235
ARABEĬ, N.M.   25472
ARABI, A.K.   27918
ARAD (MALIS), S.   25236
ARBOIT, D.   27577
ARCHER, M.C.   27374, 28451
ARENTS, J.C.   26347
ARGYROUDI-AKOYUNOGLOU, J.H.   25237
ARKHANGEL'SKAYA, M.A.   26853
ARKIN, G.F.   25169, 25238
ARMENTANO, T.V.   25239
ARMITAGE, T.L.   25574
ARMOND, P.A.   25240-1, 25245, 25819
ARMSON, K.A.   27094
ARMSTRONG, W.   25780
ARNASON, T.   25242-3
ARNOLD, C.G.   27962
ARNOLD, W.   25244
ARNON, D.I.   26380, 27104, 28184
ARNTZEN, C.J.   25240-1, 25245-6, 25819
ARO, E.-M.   28222
ARPIN, N.   25247
ARRON, G.P.   25529
ARTYUKH, A.D.   25480
ASADA, K.   25248, 26730, 28084
ASAMI, S.   25249
ASCENSO, J.C.   25250
ASHLEY, D.A.   26093, 28005
ASHOUR, N.I.   25297
ASPINALL, D.   25460
ASTON, M.J.   25251
ATABEKYAN, E.A.   27110
ATANASIU, L.   25252
AUCLAIR, A.N.D.   25253-4
AUCLAIR, D.   25255
AUDUS, L.J.   *27184, 28244*
AUFHAMMER, W.   25256
AUGUSTIN, P.   25257-8
AUGUSTINE, J.J.   25259-61
AUNG, L.H.   27283
AUSTENFELD, F.-A.   25262
AUSTIN, D.J.   26138
AUSTIN, L.A.   25263
AUSTIN, R.B.   25264, 27343-4
AVAKYAN, L.M.   26221
AVARMAA, R.   27599
AVDEEVA, T.A.   25221
AVERINA, N.G.   27854
AVETISOV, V.A.   25265
AVIRAM, I.   25266
AVIVI-BLEISER, N.   26562
AVRAMEAS, S.   *25421*

AVRON, M.   25211-2, 25267, 25425, 27454-
      -6, 27809
AVSIEVICH, N.A.   26118
AXELSSON, L.   25268-9
AYALA, R.P.   26527
AYLES, G.B.   25270
AYRES, P.G.   25271, 26369
AZAM, F.   25272, 26391
AZAMATOV, M.A.   26723
AZOV, Y.   25163
AZZI, A.   25273

B

BAARS, J.A.   25274-5
BAAZOV, D.I.   27705
BABCOCK, G.T.   25276
BABENKO, V.I.   25277-8
BACCARINI-MELANDRI, A.   28469
BACHE, D.H.   25279
BACHMANN, P.   25280
BACHMANN, R.W.   26543, 28054
BACHOFEN, R.   25281, 28048
BACKMAN, T.W.   25282
BACON, C.M.   27845-6
BACONE, J.   25283
BADGER, M.   25395
BADGER, M.R.   25284-5, 26958
BADINA, G.V.   25311
BAGDASARYAN, E.G.   26232
BAGGE, P.   25286
BAHL, J.   25287-9
BAHR, J.T.   25290, 26526
BAĬDULOVA-BABKO, T.Yu.   28039
BAIER, W.   25291
BAIRD, B.A.   25292
BAJRACHARYA, D.   27759
BAKER, N.R.   25293-5
BAKER, R.A.   25461
BAKER, R.H.   27337
BAKHNOVA, K.V.   28185
BAKHRAMDZHANOVA, N.A.   25381
BAKKER, E.P.   25296, 25927
BAKR AHMED, M.   25297
BALAKRISHNAN, T.K.   27220
BALASHOV, S.P.   25298, 27626
BALASUBRAMANIAN, V.   25299
BALDING, F.R.   25300
BALDRY, C.W.   25769
BALL, E.   26980
BALLESTER, A.   27067
BALNOKIN, Yu.V.   25301
BALTSCHEFFSKY, M.   25302
BAMBERG, S.A.   25303
BANASZAK, J.   25304
BANCROFT, K.   25305
BANKS, M.S.   25306
BANNISTER, T.T.   27917
BANSAL, R.P.   27794
BANSE, K.   25307
BAQAR, M.R.   27884

CATHEY, G.W.   27102
CATLIN, P.B.   27514-5
ÇATOGLIO, J.A.   25648
CATSKÝ, J.   25654, *B27801-2*, 27803
CATTOLICO, R.A.   25655
CAUBERGS, R.   25829
CAVALIÉ, G.   25619, 28264
CAVELL, S.   25656
CEDEÑO-MALDONADO, A.   25658
CERNUSCA, A.   25659-61
CESARENI, G.   25662
CEULEMANS, R.   25663
CHABOT, B.F.   25664
CHADEFAUD, M.   25665
CHAIKA, M.T.   27851, 28281
CHALLA, H.   25666
CHAMAROVSKY, S.K.   27615
      see CHAMOROVSKIĬ, S.K.
      see CHAMOROVSKY, S.K.
CHAMBROY, Y.   25667
CHAMOROVSKIĬ, S.K.   25668
      see CHAMAROVSKY, S.K.
      see CHAMOROVSKY, S.K.
CHAMOROVSKY, S.K.   25669
      see CHAMAROVSKY, S.K.
      see CHAMOROVSKIĬ, S.K.
CHAMPAGNOL, F.   25670
CHAMPIGNY, M.-L.   25671-2
CHAND, P.   25673
CHANG, N.K.   25674
CHAPADOS, C.   26871
CHAPARRO, A.   25675
CHAPAS, L.C.   27819
CHAPMAN, A.R.O.   25676-7
CHAPMAN, D.J.   25678
CHAPMAN, G.W.Jr.   25679
CHAPRA, S.C.   25680
CHARLES-EDWARDS, D.A.   25167, 25681-3
CHARNETSKI, W.A.   27016
CHARRA, R.   27266
CHARTIER, M.   27187
CHARTIER, P.   25514, 25684, 27187
CHATTAR, M.S.   25685
CHATTERJEE, A.   25686
CHATTERTON, N.J.   25687
CHAYANOVA, S.S.   28281
CHEBOTAR', A.A.   26666
CHECCUCCI, A.   25688
CHEKULAEVA, L.N.   25369, 27626
CHELM, B.K.   25689
CHEMARIN, N.G.   26799, 26801
CHEMERILOVA, V.I.   25690
CHEN, C.-H.   25394, 25691
CHEN, K.   25692-3, 28385
CHENG, D.M.H.   25694
CHENG, K.H.   25759
CHENIAE, G.   25695
CHERMNYKH, L.N.   25696
CHERNAVINA, I.A.   25697
CHERNENKO, L.N.   27799
CHERNETSKIĬ, S.S.   26927
CHERNISHEVA, S.   26101
CHERNOMORSKIĬ, S.A.   25698
CHERNYAD'EV, I.I.   25699, 27007
      see CHERNYADYEV, I.I.

CHERNYADYEV, I.I.   25875
      see CHERNYAD'EV, I.I.
CHERNYSHENKO, T.I.   26240
CHERNYSHEVA, S.V.   26757
CHERRY, J.H.   27159
CHEVALLIER, D.   25700-1
CHEVROU, R.B.   25702
CHIANG I-HWA   25984-5
CHIBISOV, A.K.   25219-20, 25703
CHICHESTER, C.O.   27644-5, 27885
CHIEN YUE-CHIN   25984-5
CHIN, P.   25704
CHIRANJEEVI, V.   25705
CHISHOLM, S.W.   25272, 25706
CHITTENDEN, J.   26509
CHKANIKOVA, R.A.   27852
CHLOUPEK, O.   25707
CHMORA, S.N.   25568, 25708-10
CHOLLET, R.   25711-3, 27539
CHOPRA, N.M.   25714
CHOUDHARY, D.K.   25715
CHOUDHURI, M.A.   25435
CHOUDHURY, N.K.   25716, 25810
CHOUSSY, M.   26127
CHOW, C.T.   25717-8
CHOW, W.S.   25719-20
CHRISTELLER, J.T.   25721, 26834
CHRISTY, A.L.   25722-3
CHRÔST, R.J.   25724
CHU, A.C.P.   25725
CHUA, N.-H.   25378, 25726
CHUB, A.I.   25727
CHUCHALIN, A.I.   27873
CHUGUNOV, V.A.   25957, 26102
CHUGUNOVA, N.G.   25696
CHURCH, J.M.F.   25728
CHUROVÁ, K.   25729
ÇIFERRI, O.   25730, 28132
ČIRKOVA-GEORGIEVA, M.   25731
ČIRKOVA-GEORGIEVSKA, M.   25732
CLABAULT, G.   25453
CLARK, B.   28421
CLARK, J.A.   26531
CLARK, J.G.   26884
CLARK, R.A.   27255
CLARKE, R.H.   25733-6
CLARKSON, N.M.   25737
CLASBY, R.   25738
CLASBY, R.C.   25739, 27250
CLAUSSEN, W.   25740
CLAYTON, R.K.   25741, 28259
CLEGG, B.R.   27839
CLÉMENT-METRAL, J.   27504
CLÉMENT-METRAL, J.D.   27505
CLERC, M.H.le
      see Le CLERC, M.H.
CLINE, R.G.   25742
CLINGENPEEL, W.J.   25513
CLOERN, J.E.   25743
CLOSS, G.L.   25517
COATS, G.E.   25744
COBB, A.H.   25745
COBBY, J.M.   27687
COCKING, E.C.   *25912*

MICHEL-VILLAZ, M. 25761-2, 27128
MIFLIN, B.J. 25449, 27090
MIKHAÏLOVA, L.M. 25507
MIKHAÏLOVA, S. 27129
MIKHAÏLOVA, S.P. 26853
MIKHAÏLOVA, T.P. 27560
MIKHAÏLUK, D.P. 27898
MIKHOV, A. 26125
MIKSCHE, J.P. *28475*
MIKULSKA, E. 25802, 27130
MIKULSKA, M. 27494
MIKUL'SKAYA, S.A. 26096, 26497
MILBOURN, G.M. 26964-5
MILES, C.D. 27131-2
MILES, D. 26887
MILFORD, G.F.J. 27133
MILICA, C.I. 27398
MILIVOJEVIČ, D. 27715
MILKUS, B.N. 27134
MILLAR, B.D. 25843
MILLER, G.J. 25228, 27137
MILLER, K.R. 25228, 27135-8
MILLER, L.N. 25592
MILLER, P.C. 27139
MILLER, R.F. 25538
MILLER, T.G. 27785
MILLER, W.E. 26212
MILLERIOUX, G. 25851
MILLS, J. 25316, 28396
MILLS, J.D. 27140
MILOHNIĆ, J. 26552
MILOSAVLJEVIĆ, M. 27141
MILTHORPE, F.L. 25194, 25604, 27142
MINAS, H.J. 27913
MINAS, M. 27143-4, 27913
MINDICH, L. 26681
MISHRA, D. 26581
MIŠOVIĆ, M. 27145
MISRA, H.P. 27146
MISRA, J.B. 27147
MITCHELL, C.A. 27148
MITCHELL, K.J. 28343-4
MITCHELL, P. 28284
MITROFANOV, B.A. 27149-50
MITROFANOVA, S.V. 27150
MITSUF, A. 26805
MITSUI, A. 27151-3
MITSUI, T. 28202
MITSUK, Z.I. 27853
MITSULOV, N. 27154
MIYACHI, S. 26402, 27155-6
MIYAKE, H. 26613, 27157-8
MIYOSHI, Y. 28349
MIZRAHI, Y. 27159
MŁODZIANOWSKI, F. 27160
MOCK, J.J. 27421
MOED, J.R. 28233
MOEREELS, E. 25829-30
MOERMANS, R. 25663
MOFFATT, J.D. 25793
MOGA, I. 27396
MOGILEVA, G.A. 27161
MOHAMMAD, A.M.S. 27162
MOHANTY, P. 25432-3, 27163

MOHAPATRA, P.K. 25810
MOHAPATRA, S.C. 27164
MOHR, H. 26077, 26603
MOK, M.C. 27165
MOKRONOSOV, A.T. 26155
MOLCHANOV, A.G. 27166
MOLCHANOV, M.I. 27167-70
MOLDAU, Kh.A. 26593, 27171
MOLNÁR, E. 27507
MOLOTKOVSKY, Yu.G. 28483
MONAGHAN, J.L. 27172
MONCUR, L. 28339
MONÉGER, R. 25287-9, 26292, 26855
MONGER, T.G. 27173
MONICA, R.F.la 27174
MONOSÓV, E.Z. 26727
MONSI, M. 27227
MONTALBINI, P. 27793
MONTALVO, M.R.de
    see DE MONTALVO, M.R.
MONTECUCCO, C. 25273
MONTEITH, J.L. *25195, 25571, 26515,*
    26912, 26923, 27587, 27630, 27656,
    27728, 27981, 28197
MONTENY, B. 27175
MONTES, G. 25530, 27176
MONTGOMERY, R.E. 26137
MOON, R. 25825
    see MOON, R.E.
MOON, R.E. 25824, 27177-8
    see MOON, R.
MOONEY, H.A. 25902, 25924, 27179-80
MOORE, D.G. 28126
MOORE, K.G. 25205-7, 27430
MOORE, P.H. 27055
MOORE, T.C. 27181
MORADSHAHI, A. 27182
MORDACHEVA, G.S. 26063
MOREL, A. 26530
MOREL, C. 27183
MORELAND, D.E. 27184
MORESHET, S. 26080, 27983
MORGAN, K. 27185
MORGAN, N.L. 26215
MORGAN, W.R. 26119
MORGENTHALER, J.-J. 27113, 27186
MORI, Y. 26134
MOROT-GAUDRY, J.-F. 27187
MOROZOV, V.L. 27188
MORRÉ, D.J. 26463
MORRIS, I. 25346
MORRIS, W.J. 26387
MORRISON, I.N. 27189
MORTON, I. *28766*
MOSKALENKO, A.A. 25957, 27869
MOSS, D.N. 26068, 26537, 27190
MOSS, G.I. 27191
MOSS, G.P. 27192-3
MOUDRIANAKIS, E.N. 27194
MOURSI, M.A. 27195
MOUSSEAU, M. 27196
MOUTOUNET, M. 27197
MOVSESYAN, G.M. 25822
MOYSE, A. 27198, 27452, 27616
MUALLEM, A. 28396
MUCHOW, R.C. 27199

ROSSITER, R.C.   26213, 27664
ROTH, H.   26708
ROTTENBERG, H.   25296
ROUGHAN, P.G.   27665
ROUHANI, I.   27666
ROUSSAUX, J.   27667
ROUSSEAU, B.   25904
ROWELL, P.   27668
ROWLEY, J.A.   28099-100
ROY, H.   27669-70
ROZEMA, J.   27671
ROZHDESTVENSKII, V.I.   27672
ROZHKO, I.I.   26221
ROZONOVA, L.N.   28468
ROZOV, N.F.   27673
RUBIN, A.B.   25668-9, 26731, 26773,
    26974, 27075, 27260, 27273-4,
    27535-7, 27615
RUBIN, B.A.   25362, 26118, 27674-5
RUBTSOV, P.M.   27676
RUDENKO, T.I.   27018
RÜDIGER, W.   27677, 27997
RUDOI, A.B.   27678
RUDOVA, T.S.   27790
RUIZ, J.B.
    see BONILLA RUIZ, J.
RULE, D.E.   27679
RUMBERG, B.   27680
RUMYANTSEVA, V.B.   26735
RUNNING, S.W.   28341
RUPPEL, H.G.   25874
RURAINSKI, H.J.   27681, 27747
RUSSELL, D.J.   26242
RUSSELL, J.S.   25737
RUSSO, J.W.   27682
RUSZKOWSKA, M.   26991
RUTKOVSKAYA, V.A.   27534
RUTMAN, G.I.   25310, 26758, 27683-4
RUTSKAYA, L.A.   27288
RUUGE, É.K.   28037
RUYTERS, G.   26755
RUZIEVA, R.Kh.   27857
RYAN, F.J.   27420, 28154
RYBERG, M.   27685
RYCHNOVSKÁ, M.   27686
RYLE, H.J.A.   27687-8
RYSSOV-NIELSEN, H.   27689
RYZHOVA, E.F.   26977

S

SAAKOV, V.S.   25310, 27683-4, 27690,
    28200
SADLER, D.M.   27691
SADOVNIKOVA, L.G.   26751
SAFIR, G.R.   27692
SAGALAEVA, A.P.   27124
SAGE, J.   27693
SAGER, R.   27694-7, 27891
SAGHER, D.   27698
SAGROMSKY, H.   25838

SAHU, G.   27210
SAHU, S.   26615-6
SAÏ, P.K.   26063
SAIJO, Y.   26530
SAINIS, J.K.   27699
SAITO, H.   28087
SAKANO, K.   28385
SAKATA, K.   25953
SAKS, N.M.   27700
SALAMON, Z.   26062, 28395-6, 28434
SALCHEVA, G.   26131
SALE, P.J.M.   27701
SALEM, L.   27702
SALISBURY, J.L.   28250
SALONTAI, A.   25312
SALVADOR, G.F.   27703
SAMBEEK, J.W.van
    see VAN SAMBEEK, J.W.
SAMOILOVA, O.P.   27704, 27950
SAMOKHVAL, E.G.   26710
SAMSUDDIN, Z.   25663
SAMUILOV, V.D.   25323, 25884
SANADI, D.R.   26720, 27390
SANADZE, G.A.   27125-6, 27705
SANDERS, J.K.M.   25844, 27706
SANDERS, J.L.   27707
SANDERSON, C.   27824
SANDERSON, S.   27708
SANE, P.V.   25408-9, 27699, 27709
SAN JOSE, J.J.   27710
SANKHLA, N.   25338, 26446
SAN PIETRO, A.   26172-4, 27815
SANTARIUS, K.A.   26335
SANTOS, M.F.   27120-1
SAPARGEL'DYEV, G.   28078
SAPOZHNIKOV, D.I.   27711
SARANCHA, Yu.P.   27482, 27712-3
SARDA, C.   27714
SARIĆ, M.   27715
SÁRKÁNY, S.   26201
SARKISYAN, S.A.   27716
SARPE, N.   27397
SÁRVÁRI, É.   27717
SASA, T.   27718
SASSA, S.   26205
SASTRY, P.S.N.   25613
SATO, K.   27719-22
SATÔ, M.   27723
SATOH, K.   27724
SATOH, M.   26775, 27725
SATOH, T.   27726
SATTER, R.L.   27727
SAŬCHANKA, G.Ya.   26594
SAUER, K.   25200, 25276, 25932, 28334
SAUGIER, B.   27728
SAUNDERS, J.A.   27729-30
SAUNDERS, V.A.   27731
SAUVEZON, R.   25911
SAVCHENKO, G.E.   27854, 28281
SAVIDGE, G.   27732
SAVITSKII, I.L.   27733
SAWADA, S.   27734
SAXENA, M.C.   27389
SAYED, A.M.   25297

SCAIFE, M.A. 27735
SCARISBRICK, D.H. 27736
SCAWEN, M.D. 27566, 27737
SCHAAFSMA, T.J. 25735, 26679-80, 28234
SCHADLER, D.L. 27738
SCHAEDLE, M. 26044-6
SCHÄFERS, H.-A. 27739
SCHAFFNER, J.C. 25504
SCHASTNYĬ, A.K. 25472
SCHÄTZLER, H.P. 26789
SCHAUB, H. 26403
SCHEER, H. 28414
SCHEERER, R. 27606-8
SCHEIBE, R. 27742
SCHEIL, I. 26756
SCHEJTER, A. 25266, 26178
SCHEPERS, A. 27740
SCHEPERS, J.S. 25407
SCHIELE, L.H. 27998
SCHIEMER, F. 27741
SCHIFF, J.A. 25427-8, 25577, 25748-9,
    25936, 26842, 27775, 28217-8, 28476
SCHILLING, N. 27742
SCHINDLER, D.W. 27231
SCHLEGEL, H. 27751
SCHLESER, G. 26049
SCHLETZ, K. 27743
SCHLIMME, E. 25487
SCHLODDER, E. 28411
SCHLUE, U. 25459
SCHMID, E. *26985*
SCHMID, G.H. 26711, 27114, 27744
SCHMID, R. 27745
SCHMIDT, B. 27746-7
SCHMIDT, H.W. 26276
SCHMIDT, L. 27748
SCHMIDT, O. 28046
SCHMIDT, P.G. 28437-9
SCHMIDT, W. 27749
SCHMIDT-VOGT, H. 27750
SCHMIEDEKŃECHT, M. 27751
SCHMITZ, K. 26772, 27752
SCHMUCK, C.L. 27364
SCHNABL, H. 27753
SCHNARRENBERGER, C. 27754
SCHNEIDER, C.L. 27692
SCHNEIDER, H.A.W. 27755
SCHNEPF, E. 26984
SCHNETTER, M.-L. 25644
SCHOCH, S. 27677, 27997
SCHOEPE, E. 28490
SCHOLZ, A. 27756-7
SCHONHORST, M.H. 25546
SCHOOLEY, R.E. 27758
SCHOPFER, P. 27759
SCHRADER, L.E. 27760
SCHRADER-REICHHARDT, U. 26004
SCHRAMM, W. 26661-2
SCHREIBER, K. 27761
SCHREIBER, M.M. 26409
SCHREIBER, U. 27762-5
SCHRÖER, P. 25399
SCHULTS, D.W. 27766
SCHULTZ, G. 25416

SCHULZ-BALDES, M. 27767
SCHULZE, E.-D. *25419, 25885, 25902,*
    25970, 25970, 26114, 26265, *26265,*
    26296, 26335, 26438, 26446, 26494,
    26580, *26689-90, B26846, 26908,*
    26920, 26970, 27768-9, *28014, 28241,*
    28266, 28341, 28404
SCHULZE, I. 27769
SCHULZE, R.E. 27770
SCHUMACHER, A. 25889
SCHUMANN, B. 25497, 26362, 27771
SCHUPHAN, W. 27205
SCHÜRMANN, P. 25583, 27010, 27772
SCHUSTER, H. 27773-4
SCHWARTZBACH, S.D. 27775
SCHWARZ, K. 28140
SCHWEIGER, H.G. 26583-5
SCHWENN, J.D. 27776-7
SCOPES, R.K. 25656
SCOTT, J.R. 25374
SCOTT, R.K. 26092
SCURTU, D. 27779
SEARS, B.B. 25786
SEBALD, W. *25378, 25427, 25463, 25497,*
    25524, 25692, 25726, 26003, 26094,
    26144, 26208, 26215, 26253, 26260,
    26414, 26430, 26562, 26774, 26885,
    27056, 27498, 27503, 27922, 28007
SEELY, G.R. 27780-1
SEIDEL, D. 27782
SEIPEL, T.M. 27148
SELGA, M.P. 28274
SELIVANKINA, S.Yu. 27651
SELLAMI, A. 27783
SELLDÉN, G. 25918, 27784
SELLNER, K.G. 27785
SELMAN, B.R. 25941, 26946, 27786-8
SELSTAM, E. 27784
SELVAM, R. 27789
SEMENENKO, V.E. 27790-1
SEMENOV, A.Yu. 25884
SEMENOVA, A.N. 26693
SEMENOVA, G.A. 26831
SEMICHAEVSKIĬ, V.D. 28276
SEMIKHATOVA, O.A. 27792
SEMPIO, C.di 27793
SEMUSHINA, L.A. 28199
SEN, D.N. 27794
SEN, S.P. 27797
SENGER, H. 26793-6, 27795-6, 27974
SENGUPTA, C. 27797
SEN GUPTA, P. 25809
SEO, T. 27322-3, 27798
SERDYUK, A.L. 27799
SERDYUK, O.P. 26666-7
SEREBRYANYĬ, A. 27346
SERGEEVA, E.A. 26591
SERVAITES, J.C. 27800
ŠESTÁK, Z. 27496, *B27801-2,* 27803-4
SEWE, K.-U. 27607-8
SHABEL'SKAYA, È.F. 27805
SHADI, A. 27806
SHADRIKOV, O.A. 27807
SHAGADAEVA, L.M. 27023

SHAH, F.H.   27808
SHAH, R.C.   25412
SHAHAK, Y.   27809
SHAHEEN, A.M.   25930, 27195
SHALTIEL, S.   *25212*
SHAMSI, S.R.A.   27810
SHAPIRO, S.L.   25630-2
SHAPOSHNIKOV, G.L.   27169-70
SHAPOVAL, A.I.   27855
SHAR, A.O.   27811
SHARKEY, P.J.   27812
SHARMA, A.N.   25412
SHARMA, N.N.   25673
SHARMA, R.   25434
SHARPLESS, R.G.   28318
SHATILOV, I.S.   27813
SHAVER, G.R.   27814
SHAVIT, N.   25177, 27745,.27815
SHAVIV, G.   ·28070
SHAW, A.B.   27816
SHAW, E.K.   27340, 27342, 28232
SHCHERBAKOVA, I.Yu.   26143, 27817
SHEATH, R.G.   27818
SHEEHY, J.E.   27819
SHEIKH, A.S.   27808
SHEIKH, K.H.   *25526, 25653*
SHEININ, D.M.   25823
SHELDON, R.B.   25523
SHELL, G.S.G.   26844, 27820
SHELP, B.   27821
SHEMIN, D.   26248, 27822
SHEN, T.C.   27506
SHEPARD, A.   28402
SHERIDAN, R.P.   27823-5
SHERIFF, D.W.   27826
SHERMAN, L.A.   27827
SHERMAN, W.V.   27828-9
SHEVCHUK, N.V.   27855
SHEVCHUK, S.N.   27854
SHEVELUKHA, V.S.   27830
SHIBATA, H.   27831-2
SHIBATA, K.   26480-2, 26703-4, 27222
SHIBLES, R.   25881, 27833
SHIDA, S.   27834-5
SHIEH, P.   27836
SHIH, S.-F.
     see SUN-FU SHIH
SHIH YI-PING   28326
SHIKHOBALOV, V.V.   27837
SHIL'KROT, G.S.   27838
SHILO, M.   26148, 27544
SHIMABUKURO, R.H.   25545
SHIMIZU, M.   28081
SHINN, J.H.   27839
SHINOHARA, T.   '27840
SHIOI, Y.   27841-2
SHIOYA, M.   28462
SHIPMAN, L.L.   27843-4
SHIRAHASHI, K.   28040
SHIRAZI, G.A.   27845-7
SHIROYAMA, T.   26212
SHIRYAEVA, G.A.   27683
SHISHCHENKO, S.V.   27848
SHITOVA, I.P.   27161

SHIVASHANKAR, K.   27849
SHIVE, J.   26169
SHKLYAEV, Yu.N.   27850
SHLYK, A.A.   26594, 27394, 27851-4
SHLYKOVA, I.M.   26674
SHMAT'KO, I.G.   27855
SHMELEVA, V.A.   27856
SHMELEVA, V.L.   27857
SHOAF, W.T.   27858-9
SHOCHAT, S.   25616
SHOMER-ILAN, A.   27860
SHORTREED, K.R.S.   28017
SHROPSHIRE, F.M.   25902
SHUBIN, L.M.   27660
SHUBIN, V.V.   26938, 27861
SHUGAEVA, E.V.   26853
SHUMILOVA, A.A.   27862, 28003
SHUMNYI, V.K.   28261
SHUTILOVA, N.I.   26176, 27863-4
SHUTOV, M.D.   25823
SHUTTLEWORTH, W.J.   27865
SHUVALOV, V.A.   25363, 26687, 26830,
     27866-71
SHUVALOVA, N.P.   25366-7, 26781
SHVETSOVA, T.D.   27850
SIBERT, J.   27400
SIBMA, L.   27740
SICKO, L.M.   26528
SIDERER, Y.   27872
SIDKO, F.Ja.   28173
     see SID'KO, F.Ya.
SID'KO, F.Ya.   25234, 25383, 26151,
     27873, 28115, 28171-2
SIDREVITS, L.   28465
SIEFERMANN, D.   27874
SIEGELMAN, H.W.   25402, 25586, 26326
SIEGELMAN, M.H.   27875
SIEGENTHALER, P.-A.   27876
SIGEE, D.C.   27877
SIGGEL, U.   27878
SIGNOL, M.   28189
SIHRA, C.K.   25968
SIJ, J.W.   28053
SIKORSKA, U.   25724
SILBERSTEIN, B.R.   27879
SILSBURY, J.H.   26087
SILVA, J.F.da   27880
SILVANOVICH, M.P.   27881
SIM, S.L.   26688
SIMKINS, J.   26138
SIMMELSGAARD, S.E.   27882
SIMPSON, D.I.   27883
SIMPSON, D.J.   27884
SIMPSON, J.R.   25842
SIMPSON, K.L.   26157, 27644-5, 27885
SINCLAIR, J.   25242-3, 27208
SINCLAIR, T.R.   27886-8
SINEL'NIKOVA, V.N.   28199
SINESHCHEKOV, O.A.   27889
SINESHCHEKOV, V.A.   26938, 27391,
     27626, 27861, 27890
SINGER, B.   27891
SINGH, H.G.   27074
SINGH, J.S.   25753

SZAREK, S.R.   28064-7
SZÖCS, Z.   27507, 28069
SZUJKÓ-LACZA, J.   28068
SZWARCBAUM, I.   28070
SZWEYKOWSKA, A.   27160

T

TABITA, F.R.   28071-2
TABUCHI, K.   28348
TADZHIEVA, F.N.   27558, 28073
TAGEEVA, S.V.   26830-1, 27481-3,
    28074-5
TAGUCHI, S.   28076-7
TAILAKOV, N.   28078
TAKABE, T.   28079
TAKAHAMA, U.   28080-1
TAKAHASHI, F.   28082
TAKAHASHI, M.   28083-4
TAKAHASHI, M.-A.   26730
TAKAMIYA, K.   27841-2
TAKEDA, G.   26506, 28085-6
TAKEDA, T.   28087
TAKISAWA, H.   28450
TAKO, T.   26091
TALANAVA, K.S.   26096
TALBERT, R.E.   27336
TALLING, J.F.   28088
TALPASAYI, E.R.S.   27572
TĂMAȘ, V.   27480
TAMAURA, Y.   28450
TAMBURI, F.   27793
TAMBURI, G.   27793
TAN, C.S.   28089
TAN, P.Y.   26922
TAN, S.F.   26922
TAN, T.Q.   26812
TANABE, Y.   26505
TANAKA, A.   28090-1
TANAKA, I.   28092
TANAKA, M.   25953
TANAKA, T.   28093
TANAKA, Y.   27645, 28094
TANAS'EV, V.K.   28095
TANIYAMA, T.   28096
TANNENBAUM, S.R.   27374
TANNER, C.B.   28097
TARAPCHAK, S.J.   25680
TARKHNISHVILI, G.M.   27126
TATARINTSEV, N.P.   27019
TATENO, K.   28098
TATTAR, T.A.   26579
TAYLOR, A.O.   28099-100
TAYLOR, B.K.   28101
TAYLOR, R.F.   25816
TAYLOR, R.J.   28102
TAYLOR, S.   25556
TEDRO, S.M.   28103
TELEGDY KOVÁTS, L.   25392, 26766
TELFER, A.   27140, 28104
TELLYAEV, A.Yu.   25507
TEL-OR, E.   25425, 25629, 28105
TEMPEL, N.R.   25549

TENHUNEN, J.D.   28106-7
TEPASKE, E.R.   27349-50
TERASAKI, W.L.   28108
TEREKHOVA, I.V.   25699, 25875, 26725
TERJUNG, W.H.   28109-10
TERPSTRA, W.   28111
TERRY, N.   28112-3
TERSKOV, I.A.   26151, 28114-5
TETLEY, R.M.   28116
TEVINI, M.   28117, 28203
THERRIAULT, J.-C.   28118
THERRIEN, H.P.   25349
THIEDE, B.   28119
THINH, L.V.   28120
THOM, M.   27055
THOMAS, H.   28121
THOMAS, J.B.   26683, 28122
THOMAS, J.C.   25572, 26916, 28123
THOMAS, R.J.   28124
THOMPSON, D.R.   28125
THOMPSON, K.H.   27341
THOMPSON, R.G.   28425
THOMPSON, S.E.   27839
THORESON, B.   28126
THOREZ, D.   28418
THORNBER, J.P.   25185-6, 26457, 26573,
    26990, 27033, 28127
THORNE, S.W.   25456
THORNLEY, J.H.M.   25168, 25771
THORPE, M.R.   25683
THRONDSEN, J.   28128-9
THROWER, L.B.   27941-2, 28130
THROWER, S.L.   28130
THUNIS, M.   27417
THURNAUER, M.C.   28131
TIBONI, O.   25730, 28132
TICHÁ, I.   25654, 28133-4
TIEDE, D.M.   28135-6
TIEFERT, M.A.   27194
TIEN, H.T.   28137
TIESZEN, L.L.   26533, 27139
TIKHOMIROV, A.A.   27873
TIKHONOV, A.N.   28037
TIKHONOVSKAYA, N.G.   28281
TIKU, B.L.   28138
TILLMAN, R.F.   25725
TILNEY-BASSETT, R.A.E.   26649, 28139
TILZNER, M.M.   27376, 28140
TIME, I.V.   25388
TIMKOVICH, R.   25855
TIMMIS, R.   28141
TIMOFEEV, K.N.   25893, 26731, 26974,
    27273
TING, I.P.   26282, 26535, 28049, 28064,
    28142-3
TINGEY, D.T.   28144
TINGLE, C.L.   25524
TINUS, R.W.   28145
TIRIMANNA, A.S.L.   26956
TISHCHENKO, N.N.   27008, 28146
TITLYANOV, É.A.   27438, 28147
TITOV, A.F.   28148
TIȚU, H.   28149
TKACHUK, K.S.   26241
TODD, G.W.   27847

VON WILLERT, K.
    see WILLERT, K.von
VOOKOVÁ, B. 28283
VOORN, G. 28284
VORONKOV, L.A. 27675
VORONOVA, E.A. 27435
VOSKRESENSKAYA, N.P. 25892-3, 28285-7
VOZNESENSKAYA, E.V. 28288-9
VOZNYAK, V.M. 26657-8, 28290
VREDENBERG, W.J. 25590-1, 28291-2
VRIES, D.A.de
    see DE VRIES, D.A.
VRIES, H.G.de
    see GORTER de VRIES, H.
VRUBEL', S.V. 27852
VSEVOLODOV, N.N. 28293
VUJIČIĆ, R. 27141.
VUNKOVA, R. 28220
VUNKOVA, R.N. 28183
VYARK, É. 26620
VYAS, L.N. 28294
VYAS, N.L. 28294
VYSKOT, B. 28216

W

WAALAND, J.R. 28295
WAALS, J.H.van der
    VAN DER WAALS, J.H.
WADA, K. 26305-6
WADE, C.G. *28437*
WADSWORTH, R.M. 25943
WAFFORD, J.D. 28296
WAGNER, E. 28297
WAGNER, G. 25720, 28298-9
WAGNER, J. 26854
WAISEL, Y. 27860
WAL, H.N.van der
    see VAN DER WAL, H.N.
WALCOTT, J.J. 28300
WALDRON, J.C. 28301
WALI, M.K. 26724
WALKER, A.J. 28302
WALKER, D.A. 25833, 26360, 26376,
    27906-9, 28303-7
WALKER, I.D. 25547
WALKER, J.N. 28369-70
WALKER, J.R.L. 28308
WALKER, L.L. 26431
WALKER, R.B. 26895, 28475
WALL, B.H. 26502
WALLACE, A. 26684-5, 28309-13
WALLACE, D.G. 28314-5
WALLACE, D.H. 28316, 28382
WALLACE, J.S. 25430
WALLENTINUS, I. 28317
WALLIHAN, E.F. 28318
WALNE, P.L. 27383
WALTER, G. 28319-20
WALTON, D. 28321
WALTON, D.W.H. 28322-3

WALZ, D. 27632, 28324
WANDERS, J.B.W. 28325
WANG, J.H. 28327
WANG, R.T. 28328-9
WANG HSI 28326
WARA-ASWAPATI, O. 25529, 28330
WARD, B. 25548, 28331
WARDEN, J. 26559
WARDEN, J.T. 25477, 28332-5
WARDLAW, I.F. *25193, 25722-3, 25765,*
    25966, 25967, 26122, 26320, 26349,
    26349, 26522, 26612, 26775, 27410,
    27573, 27624, *28053, 28177,* 28336-9,
    28338, B28340
WAREING, P.F. 27412
WARING, R.H. 26133, 28341
WARREN, G.F. 27880
WARREN WILSON, J. 25168
WARRINGTON, I. 28342
WARRINGTON, I.J. 28343-4
WARRIT, B. 26843
WASIELEWSKI, M.R. 28345
WATADA, A.E. 28346
WATANABE, I. 28347-8
WATANABE, M. 28349
WATERS, R. 27406
WATERTON, J.C. 25844, 27706
WATSON, L. 26319-20
WATSON, R.L. 27521
WATTS, W.R. 28350
WAYGOOD, E.R. 28351
WEBB, D.P. 28352
WEBB, W.L. 28353
WEBER, C.I. 28354
WEBER, J.A. 28106
WEBSTER, G.R. 26331
WEDDING, J.B. 25646
WEDDING, R.T. 28355
WEEDON, B.C.L. 27193, 28356
WEGFAHRT, P. 26678
WĘGLEŃSKA, T. 25455
WEIDNER, M. 27279, 28357
WEIN, R.W. 28358
WEINER, B. 26049
WEISE, G. 26166, 28359
WEISSENBÖCK, G. 28360-1
WEISTROP, J. 27056
WELLBURN, A.R. 25745, 26008, 26277-80,
    28362-5
WELLBURN, F.A.M. 28365
WERDAN, K. 25980-1
WERNER, S. *25378, 25427, 25463, 25497,*
    25524, 25692, 25726, 26003, 26094,
    26144, 26208, 26215, 26253, 26260,
    26414, 26430, 26562, 26774, 26885,
    27056, 27498, 27503, 27922, 28007
WEST, D.W. 28366-7
WEST, L.D. 28368
WEST, S.H. 28228
WESTERMAN, P.W. 28369-70
WESTRIN, H. 28371
WETHEY, D.S. 28372
WETTSTEIN, D.von 26053, 26562, 27923-4,
    28373
WETZEL, R.G. 26428

This index contains a selection of primary items chosen according to their importance in photosynthesis research and to their relevance and occurrence. The word "Photosynthesis" is not regarded as a main theme, but partial processes, photosynthetic parameters and the factors affecting photosynthesis are listed. The processes and other characteristics are summarized into several main themes when presented in combination with individual factors, e.g. carbon fixation pathways, electron transport chain, chlorophyll, gas exchange, ecosystem productivity and canopy functioning (including photosynthate distribution and translocation), photorespiration, resistances to CO_2 and water vapour transfer, etc.

Several items from branches related to photosynthesis research were also chosen for convenience, e.g. dealing with respiration, plant growth and development, water relations, anatomy, etc. These items contain only references to papers within the scope of this bibliography.

A

Abscisic acid   see   Growth regulators ...

Absorbance in canopy   see   Canopy, radiation profile ...

Accumulation of dry matter   see   Biomass distribution ...; Dry-matter production ...; Ecosystem production ...

Achlorophyllous cells and organs, respiration   see   Respiration of achlorophyllous tissues in light, light inhibition of respiration
Action spectra   see   Irradiance, spectral composition ...

Adenosine triphosphate   see   ATP

Aerodynamic methods, bioclimatological methods (sampling, measurement of wind, rain, dew, etc.)
     25195, 25279, 27434, 27912, 28010-1, 28475

Age of algae, leaf, plant   see   Ontogeny ...; Canopy, leaf age

Agrotechnics and chlorophyll   27980, 28156

Agrotechnics and ecosystem productivity and canopy functioning   25454, 25613, 25997, 26030, 26874, 27094, 27740, 27982, 28212, 28381

Agrotechnics and gas exchange   25613, 26014, 26182, 26373, 26591, 28020

Air-conditioning in photosynthesis measurement   see   Gasometric system, conditioning of air

Air-flow rate   see   Wind ...

Albedo, canopy   see   Canopy, radiation distribution

Algae and photosynthetic bacteria, cultivation (cf. also Algae-mass cultures productivity)   26357, B28494

Algae and secondary production of reservoirs
     25743, 26130, 26242, 26371, 26953, 27281, 27392, 27766, 28076, 28394, 28465

Algae, blue green, chromatophores in   see   Chromatophore ...

Algae carotenoids   see   Xanthophylls of algae

Algae chlorophylls see Chlorophylls a,d

Algae, CO_2 and O_2 exchange see Gas exchange in algae

Algae, depth distribution in reservoirs
25210, 25282, 25390-1, 25426, 25455, 25485, 25525, 25588, 25617, 25633, 25694,
25841, 25849-50, 25894, 26002, 26051, 26163, 26179, 26217, 26242, 26273,
26330, 26358, 26371, 26407, 26420, 26518, 26529, 26540, 26571, 26582, 26656,
26823, 26849, 26953, 27057, 27143-4, 27225, 27227, 27376, 27400, 27459, 27470,
27534, 27567-8, 27633, 27648, 27741, 27838, 27913, 27938, 27956-8, 27960-1,
27970, 27986, 27988, 28004, 28118, 28129, 28140, 28162, 28174, 28394, 28471-2,
B28494

Algae in sewage cleaning 26267, 26420

Algae life cycles see Ontogeny of algae...

Algae mass cultures productivity ($cf.$ also Algae and photosynthetic bacteria, cultivation) 26702, 26882, 27652, 28295

Algae photosynthesis and production
25210, 25272, 25286, 25367, 25426, 25452, 25525, 25541, 25588, 25633, 25739,
25763, 25806, 25847, 25849-50, 26002, 26089, 26110, 26163, 26179, 26214,
26217, 26242, 26287, 26304, 26344, 26371, 26458, 26516, 26529, 26540, 26563,
26568, 26570-1, 26752, 26838, 27227, 27231-2, 27377, 27392-3, 27418, 27463,
27470, 27534, 27567, 27633, 27650, 27662, 27859, 27913, 27956, 27959-60,
27970, 28026, 28076, 28088, 28129, 28152, 28162, 28174, 28317, 28325, 28394

Algae, primary productivity in reservoirs ($cf.$ also Chlorophyll as measure of production of algae and water reservoirs)
25341-3, 25788, 25848, 26072, 26344, 26705, 26875, 27371, 27811, 27940

Algae, primary productivity, methods ($cf.$ also O_2 determination (other than O_2 electrode); O_2 electrode)
25389, 25647, 25724, 25805, 25847, 25851, 26108, 26129, 26195, 26330, 26387,
26393, 26563, 26697-8, 27067, 27112, 27144, 27266, 27376, 27400, 27418,
27642, 27785, 27859, 28076, 28126, 28284, 28354

Algae synchronous cultures see Algae and photosynthetic bacteria, cultivation;
Ontogeny of algae ...

Altitude see Pressure, altitude ...

Amino acids see Proteins, amino acids, nucleic acids ...

δ-Aminolaevulinic acid see Chlorophyll biosynthesis ...

Anaerobic atmosphere see N_2, anaerobic atmosphere ...

Antibiotics and carbon fixation pathways 25199, 25672, 25730, 26144, 27790, 28303

Antibiotics and carotenoids 27972

Antibiotics and chlorophyll
25200, 25433, 25504, 25730, 26094, 26143, 26253, 26310, 26415-6, 26594, 26885,
27025, 27041, 27076, 27081, 27140, 27394, 27775, 27817, 27831, 27853, 27972

Antibiotics and chloroplast (chromatophore)
25463, 25558, 25730, 25853, 26144, 26252, 26950, 27076, 27214, 27249, 27407,
28292

Antibiotics and electron transport chain
25590, 25804, 26194, 26253, 26310, 26415, 26442, 26968, 27023, 27076, 27262,
27275, 27353-5, 27394, 27448, 27491, 27856, 28307, 28330, 28483

Antibiotics and gas exchange 25853, 25913, 26583, 26585, 27972

Antigens see Electron transport chain, serological analysis

Antitranspirants B25775, 25813, 27693, 27983

Architecture of canopy see Canopy ...

Assimilates see Photosynthates

Assimilation chamber
        25251, 25257, 25264, 25331, 25418, 25502, 25549, 26027, 26044, B26120, 26164,
        26372, 26375, 26506, 26743, 26745, 26895, 27280, 27421, 27461, 27826, 28085,
        28087, 28366, 28380, 28475

ATP   25177, 25184, 25226, 25244, 25261, 25305, 25317, 25320, 25401, 25487, 25508-
        -9, 25522, 25528, 25671, 25913, 25949-50, 25950, 25983-6, 26148-9, 26193,
        26198, 26238, 26268, 26291, 26322, 26332-3, 26341-2, 26441, 26472-3, 26500,
        26517, 26566, 26574, 26714, 26726, 26756, 26767, 26773, 26835, 26872, 26890,
        26980, 26997, 27006, 27017-9, 27037, 27039, 27064, 27091-2, 27112, 27131,
        27194, 27204, 27244, 27255, 27353, 27374, 27376, 27504, 27553, 27589-91,
        27592, 27597-8, 27640-1, 27652, 27661, 27665, 27712-3, 27745, 27783, 27788-9,
        27809, 27815, 27859, 27898, 27907-9, 27933-4, 28035, 28037, 28047, 28227,
        28297, 28303, 28304, 28307, 28327, 28354, 28411, 28445

ATP, methods 25305, 25509, 25777, 26391, 26590, 28047

ATPase, coupling factor 1
        25177, 25292, 25317, 25387, 25399, 25417, 25420, 25427, 25461, 25465, 25467,
        25638-9, 25830, 25861, 25888, 25891, 25914, 26103-4, 26210, 26230, 26301,
        26352, 26354, 26366, 26374, 26388, 26418, 26523, 26554, 26631, 26665, 26968-
        -9, 27004-6, 27017, 27037-8, 27113, 27138, 27194, 27204, 27243-5, 27335,
        27338, 27390, 27482, 27486, 27495, 27554, 27661, 27712-3, 27745, 27788, 27816,
        27829, 27881, 27898, 27907, 27910, 27916, 27992, 28034-5, 28045, 28164, 28411,
        28450, 28483

ATPase, methods 25492, 26301, 26969, 27243, 27330

Autotrophy see Carbon metabolism types ...

B

Bacteria, photosynthetic see Photosynthetic bacteria ...

Bacteriochlorophylls (*cf*. also Chlorophyll, *Chlorobium*)
        25263, 25323, 25488, 25562, 25606, 25733-4, 25741, 25746, 25756-7, 25773,
        25791, 25884, 25889, 25903, 25957, B26185, 26205, B26220, 26247, 26395, 26401,
        26404, 26406, 26410, 26486, 26508, 26545, 26609-10, 26614, 26658, 26672,
        26681, 26687, 26727, 26731,   26761, 26763, 26808, 26942-3, 26974, 26985,
        27000, 27173, 27254, 27273, 27282, 27303-4, 27340-2, 27381, 27504-5, 27518-9,
        27536, 27615, 27631, 27653, 27726, 27731, 27780-1, 27822, 27867-9, 27871,
        27879, 27904, 27945, 28036, 28131, 28135-6, 28155, 28232, 28237-8, 28259,
        28290, 28335, 28409, 28455

Bacteriorhodopsin see *Halobacterium* photosynthesis

Bibliographies of photosynthesis, biographies B27801-2, 27803

Biliproteins see also Phycocyanins; Phycoerythrins

Biliproteins absorption spectra *in vitro* 25361, 25579, 26061, 26992, 27299

Biliproteins absorption spectra *in vivo*   25867, 25869, 25958, 26158, 26207, 26250,
      26992, 27299, 27615, 28008

Biliproteins biosynthesis, precursors   25439, B26185

Biliproteins chemical structure   25402, 26159, B26185, B26220, 27299

Biliproteins complexes *in vivo*   26158, B26185, 26207, 27299, 27615

Biliproteins delayed light emission, luminescence   25223

Biliproteins determination, column chromatography   25579, 25631, 26160

Biliproteins determination, electrophoresis and other methods   25402, 25579, 26098,
      26160, 26250

Biliproteins determination, spectral methods   27086, 27804

Biliproteins energetic states *in vitro*   25394, 26082

Biliproteins fluorescence *in vitro*   25431, 25973

Biliproteins fluorescence *in vivo*   25958, 26098, 26168, 26207, 27319

Biliproteins in model systems   25361, 25394, 26062

Biliproteins in mutants   see   Mutants, biliproteins in

Biliproteins in photosynthesis mechanism   25973, 26160, B26185, 26718, 27319

Biliproteins in physiology of photosynthesis   26608

Biliproteins luminescence *in vitro*   25431

Bioclimatological methods   see   Aerodynamic methods ...

Biomass distribution and redistribution in plant
      25164, 25171-2, 25230, 25232, 25264, 25299, 25319, 25435, 25469, 25473,
      25531, 25566, 25594, 25660, 25666, 25725, 25737, 25753, 25758, 25771, 25810,
      25845, 25857, 25917, 25970, 25997, 26017, 26028-30, 26055, 26073, 26085,
      26096, 26110, 26132, 26197, 26202, 26204, 26213, 26246, 26263, 26267, 26274,
      26361, 26383, 26400, 26427, 26438, 26461, 26465, 26474, 26484, 26501, 26506,
      26530, 26533, 26537, 26607, 26643, 26688, 26776, 26788, 26798, 26803-4,
      26812, 26827-8, 26843, 26847, 26904, 26906, 26912, 26952, 26954, 27013-4,
      27062, 27094, 27101, 27108, 27139, 27209, 27212, 27216, 27228, 27235, 27248,
      27253, 27327, 27333, 27343-4, 27362, 27409, 27430, 27437, 27457, 27466,
      27521, 27538, 27540, 27563, 27647, 27664, 27719-20, 27725, 27734-6, 27748,
      27812, 27882, 27931, 27983-4, 27991, 28023, 28050, 28086, 28099, 28134, 28157,
      28175, 28207, 28212, 28300, 28312; 28337, 28347, 28381, 28392, 28400, 28408,
      28427, 29470, 28477, 28495

Biopotentials   see   Chloroplast and chromatophore biopotentials

Biosphere production   see   Ecosystem production ...

Blinks effect   see   Emerson effect, Blinks effect

Boundary layer of air   see   Resistance, leaf boundary layer

Bundle sheaths   see   Carbon metabolism types ...

C

$^{13}C/^{12}C$ ratio, $\delta^{13}C$ 25437, 25446-7, 25599, 25787, 25878, 25926, 26261, 26368, 27769,
      28383, 28406-7

Canopy photosynthesis and PhAR profile (*cf.* also Canopy, radiation profile)
     25571, 25659, 26027, 26030, 26188, 26515, 26857, 27139, 27635, 28231

Canopy photosynthesis, direct measurement
     25167-8, 25195, 25264, 25967, 26028, 26213, 26372, 26515, 26537, 27139,
     27343, 27587, 27630, 27635, 27728, 27981, 28059, 28087, 28093, 28195, 28350,
     28413

Canopy photosynthesis, energy balance 25366, 25571, 26515, 26531, 27434, 27587,
     27656, 27981

Canopy photosynthesis, mass and momentum balance 25195, 26515, 26531, 27030, 27166,
     27434, 27587, 27728, 27981, 28195

Canopy photosynthesis, model see Model ...

Canopy, radiation distribution; reflection, transmission, absorption, albedo, *etc.*
     (*cf.* also Canopy, radiation profile)
     25308, 25484, 25660, 25662, 25792, 25885-6, 26034, 26080, 26086, 26372,
     26530, 26844, 26877, 26898, 26920, 27014, 27253, 27333, 27587, 27728, 27888,
     27939, 27981, 28051-2, 28070, 28195, 28197, 28265, 28413

Canopy, radiation profile (*cf.* also Canopy, radiation distribution; Canopy photosyn-
     thesis and PhAR profile)
     25195, 25203, 25238, 25264, 25484, 25539, 25659-61, 25683, 25909, 26027,
     26080, 26086, 26501, 26515, 26530-1, 26840, 26877, 26879, 26898, 26910,
     26947, 27139, 27175, 27220, 27253, 27270, 27325, 27333, 27587, 27630, 27728,
     27819, 27920, 28070, 28197, 28369

Canopy, resistances for CO_2 and water vapour transfer see Resistances for CO_2 and
     water vapour transfer at canopy level

Canopy structure, methods 25482-3, 26878, 28022

Canopy, sunlit and shaded leaf area 25813, 26844, 26848, 27362, 27820, 27888

Canopy, temperature profiles 25450-1, 25571, 25659, 25661, 26337, 26515, 26531,
     26947, 27052, 27630, 27728, 28010, 28197, 28256-7

Canopy, turbulence inside 25195, 25659, 26337, 26531, 27087, 27587, 27630, 27728,
     28087, 28195, 28197, 28256

Canopy, vertical distribution of leaf area
     25203, 25659-62, 25989, 26027-8, 26047, 26531, 27362, 27587, 27630, 28013,
     28051, 28187, 28195, 28197, 28265

Canopy, water vapour profiles 25195, 25571, 25661, 26515, 26531, 26848, 26947, 27630,
     28011, 28197

Carbon-14 see Carbon isotopes ...

Carbon balance, plant
     25207, 25319, 25411, 25664, 25666, 25674, 25753, 25771, 26073, 26383-5, 26591,
     26883-4, 27011, 27097, 27212, 27895, 28302, 28347

Carbon dioxide see CO_2

Carbon fixation in isolated chloroplasts and its products
     25671, 25870, 26314, 26772, 26900, 27284, 27399

Carbon fixation pathways, comparison in mesophyll and bundle sheaths cells
     25711, 25912, 26315-7, 26444, 26574, 26786, 27007, 27364, 27597, 27621, 28493

Carbon fixation pathways enzymes, methods   see Enzymes of carbon fixation pathways
...; Ribulose 1,5-bisphosphate carboxylase ...; Phosphoenolpyruvate carboxylase ...; Malic enzyme ...

Carbon fixation pathways in photosynthetic bacteria   see Photosynthetic bacteria,
carbon fixation pathways

Carbon fixation pathways, intermediary types of carbon fixation 25443, 25445

Carbon fixation pathways, intermediates
25280, 25346, 25870, 26078, 26314-7, 26341, 26549, B26551, 26565-6, 26598,
26721-2, 26770, 26864, 26900, 27247, 27255, 27359, 27461, 27524, 27621, 27797,
28303-4, 28307

Carbon fixation pathways, model   see Model ...

Carbon fixation pathways, primary acceptors of CO_2
25678, 25831, 26317, 26623, 26668, 26872, 27091, 27106, 27255, 27359, 28304,
28306

Carbon isotopes (^{14}C, ^{11}C, ^{13}C), use in photosynthesis measurement
25190, 25207, 25256, 25346, 25350, 25405, 25411, 25436, 25447, 25489, 25527,
25576, 25595, 25599, 25602, 25739, 25809, 25828, 25847, 25858, 25878, 26024,
26027-8, 26037, 26049, 26078, 26121, 26218, 26228, 26261, 26281, 26349,
26384, 26387, 26451-2, 26489, 26498, 26503-4, 26533, 26535, 26542, 26589,
26607, 26626, 26845, 26876, 26973, 26979, 27027, 27067, 27167, 27170, 27183,
27210, 27290, 27343-4, 27347, 27359-60, 27364, 27402, 27461, 27477, 27494,
27514-5, 27524, 27533, 27564, 27642, 27687, 27721, 27725, 27785, 27812, 27900,
27929, 27948, 27951, 27995, 27999, 28005, 28177, 28337-9, 28374, 28475, 28477

Carbon metabolism types and carbon fixation pathways
B25603, 25612, 25768-9, 26006, 26155, 26316, 26428, 26786, 27116, 27198,
27292, 27419, 27928, 28120, 28204, 28305

Carbon metabolism types and carotenoids 25225, 25895

Carbon metabolism types and chlorophyll 25925, 26786

Carbon metabolism types and chloroplast 25718, 25878, 27407, 27469, 27557, 28119

Carbon metabolism types and ecosystem productivity and canopy functioning
B25603, 26979, 27539, 28005

Carbon metabolism types and electron transport chain B25603, 26473, 26786

Carbon metabolism types and gas exchange B25603, 25760, 25768, 25967, 26091, 26240,
26475, 26641, 26644-5, 26880, 27089, 27142, 27427, 27469, 28005, 28146

Carbon metabolism types and photorespiration 26427, 26786, 27469, 27769

Carbon metabolism types and resistances to CO_2 and water vapour transfer 26091, 26240

Carbonic anhydrase
25329, 25503, 25650, 25657, 25670, 25699, 25768, 25875, 26023-4, 26118, 26203,
26478, 26539, 26725, 26748, 26858, 26858, 27263, 27317-8, 27329, 27368, 27605,
28176, 28205, 28220, 28229, 28304, 28368

Carbonic anhydrase, methods 25670, 26606, 27329

Carboxylation see Carbon fixation pathways ...

Carboxylation resistance see Resistance, carboxylation and excitation

Carotenes 25376, 25648, 25762, 25794, 25838, 25876, 26225-6, 26666-7, 26707, 27060,

Carotenes (continued)
        27066, 27110, 27165, 27237, 27405, 27480, 27527, 27586, 27863, 28094, 28186,
        28377

Carotenoids absorption spectra *in vitro*   25648, 25714, 25794, 26127, 26157, 26270,
        26491, 27526, 27608, 27677, 27684, 28094, 28377

Carotenoids absorption spectra *in vivo*   25600, 26225, 26808, 27193, 27608, 28027,
        28029, 28060

Carotenoids and production of algae and water reservoirs   25218, 26214, 27266, 27940,
        28162, 28317, 28464-5

Carotenoids and production of higher plants   28282

Carotenoids biosynthesis, precursors
        25416, 25553-4, 25816, 25852, 25876, 26069, 26153, B26185, 26226, 26328, 26820,
        27184, 27304, 27444, 27472, 27586, 27645, 27677, 27755, 27784, 27805, 27972,
        28217, 28356, 28379

Carotenoids chemical structure
        25553, 25555, 25586, 25815, B26185, 26186, 26678, 26915, 27192-3, 27885,
        28060, 28466

Carotenoids complexes *in vivo*   25746, B26185, 26326, 26573, 26682, 26818, 27952,
        28111, 28466

Carotenoids degradation
        25248, 25432, 25714, 25716, 26491, 26634, 26716, 26740, 27071, 27159, 27387,
        27527, 27711, 27850, 27855, 27864, 27885, 28239-40

Carotenoids determination *cf.*also Pigments determination, sampling and extraction

Carotenoids determination, column chromatography   25794, 25815, 26157, B26185, 27645,
        28094

Carotenoids determination, electrophoresis and other methods   25815, B26185

Carotenoids determination, paper chromatography and thin-layer chromatography
        25442, 25794, 25815, 25876, B26185, 26290, 26312, 26820, 26922, 26956, 27066,
        27197, 27511, 28191, 28217

Carotenoids determination, spectral methods   25794, 25815, B26185, 26311-2, 26556,
        27511

Carotenoids energetic states *in vitro*   25808, 25976, 26082, 26832, 26913, 27015

Carotenoids energetic states *in vivo*   25380, 26808, 27114, 27173, 27868, 27952

Carotenoids, enzymes of synthetic and degradation processes   25553, 26153

Carotenoids fluorescence *in vivo*   22101

Carotenoids in model systems   25393, 25691, 25808, 25975-6, 25832, 26913, 27015,
        27045

Carotenoids in photosynthesis mechanism
        25202, 25244, 25248, 25302, 25335, 25344, 25353, 25533, 25645, 25691, 25746,
        25761, 25969, B26185, B26220, 26225, 26808, 26872, 26962, 27015, 27114,
        27127, 27173, 27340, 27512, 27608, 27711, 27864, 27868, 27874, 27946, 27952,
        28027, 28029

Carotenoids in physiology of photosynthesis   25383, 27511, 27513, 27925

Carotenoids in seeds and fruits 25225, 25553, 25779, 26186, 26225, 27510, 27526, 27644, 28186, 28401

Carotenoids precursors see Carotenoids biosynthesis ...

Chamber, assimilation see Assimilation chamber

Chemiosmotic hypothesis, proton transport in chloroplast
       25240, 25313-5, 25317, 25401, 25461, 25466, 25522, 25591, 25719-20, 25789,
       25804, 25980, 26056, 26103-5, 26193-4, 26198, 26210, B26220, 26259, 26268,
       26301, 26332, 26342-3, 26388, 26411-2, 26500, 26523, 26574, 26718, 26720,
       26767, 26850, 26980, 27006, 27075, 27092-3, 27262, 27335, 27353-5, 27454,
       27456, 27491, 27680, 27772, 27783, 27917, 27936, 27945, 28035, 28163-4,
       28227, 28292, 28327, 28410, 28438-9, 28447-8

Chlorobium chlorophyll see Chlorophylls, *Chlorobium*

Chlorophyll absorption spectra *in vitro*
       25204, 25517, 25563, 25567, 25614, 25714, 25748, 25974, 26041, 26043, 26053,
       B26220, 26270, 26508, 26609, 26712-3, 26718, 26738, 26763, 26871, 26897,
       26942, 27000, 27082, 27114, 27178, 27303, 27309, 27381, 27394, 27608, 27684-5,
       27794-5, 28215, 28222, 28251, 28324, 28377, 28481

Chlorophyll absorption spectra *in vivo*
       25166, 25176, 25200, 25263, 25268, 25295, 25310, 25379, 25433, 25440, 25535,
       25557-8, 25570, 25668, 25688, 25703, 25741, 25746, 25749, 25757, 25782,
       25791, 25837-8, 25867, 25869, 25884, 25888, 25899, 25960, 26019, 26168,
       B26220, 26326, 26406, 26454, 26483, 26545, 26556, 26573, 26658, 26710, 26718,
       26727, 26757-8, 26796, 26822, 26829-30, 26873, 26935, 26937, 26942, 26985,
       27050, 27071-2, 27114, 27163, 27222, 27279, 27340-1, 27490, 27512-3, 27608,
       27615, 27675, 27726, 27795, 27804, 27817, 27848, 27863, 27870, 27946, 28008,
       28029, 28036, 28114, 28122, 28127, 28200, 28217, 28226, 28258, 28260, 28276,
       28281, 28467, 28476

Chlorophyll and its products determination *cf.* also Pigments determination, samp-
       ling and extraction

Chlorophyll and its products determination, column chromatography 25204, 25698,
       26053, B26185, 26404, 27685

Chlorophyll and its products determination, electrophoresis and other methods
       B26185, 26404, 28046

Chlorophyll and its products determination, *in vivo*
       25263, 25557, 25812, 25882, 26293, 26379, 26454, 26732, 26998, 27273, 27525,
       27863, 28279

Chlorophyll and its products determination, paper chromatography, thin-layer chroma-
       tography 25204, 25748, 25895, B26185, 26270-1, 26404, 26518, 26521, 26717,
       26842, 27000, 27511, 27632, 27677, 28377

Chlorophyll and its products determination, spectral methods
       25201, 25204, 25234, 25394, 25849, 25895, 26129, B26185, 26270, 26404, 26453,
       26519, 26556, 26562, 26986, 26998, 27000, 27027, 27251, 27511, 27683-4,
       27689, 28346, 28348

Chlorophyll and production of higher plants 25422, 25644, 27386, 28282

Chlorophyll as measure of production of algae and water reservoirs
       25163, 25165, 25210, 25218, 25234, 25270, 25390-1, 25426, 25509, 25541, 25580,
       25588, 25598, 25633, 25647, 25651, 25680, 25739, 25743, 25760, 25763, 25841,
       25849-50, 25894, 25952, 25960, 25994, 26002, 26012, 26089, 26130, 26163,
       26212, 26214, 26242, 26270-1, 26358, 26387, 26407, 26510, 26516, 26518, 26529,
       26540, 26543, 26582, 26655-6, 26823, 26838, 26849, 26953, 27177, 27206, 27227,

Chlorophyll as measure of production of algae and water reservoirs (continued)
27250, 27266, 27281, 27298, 27376, 27392-3, 27400, 27459, 27567, 27613,
27642, 27648, 27650, 27662, 27732, 27766, 27913, 27940, 27956-8, 27960-1,
27986, 27988, 28004, 28017, 28025-6, 28047, 28054, 28076, 28088, 28118, 28126,
28128, 28140, 28162, 28174, 28190, 28226, 28317, 28354, 28464-5

Chlorophyll biosynthesis and precursors
25166, 25200, 25204, 25237, 25268-9, 25293, 25295, 25334, 25344-5, 25440,
25462, 25498, 25504, 25517, 25564, 25567, 25596, 25616, 25689, 25700, 25747-8,
25852-3, 25899-900, 25904, 25918, 25933, 25936-8, 25988, 26053, 26057, 26069-
-70, 26094, 26097, 26099, 26113, 26139, 26143, 26153, B26185, 26191-2, 26205,
26215, 26248, 26276, 26278, 26310, 26340, 26351, 26377, 26399, 26415, 26437,
26481, 26545, 26558, 26562, 26594, 26603-4, 26630, 26667, 26681, 26706, 26737,
26768, 26842, 26890, 26928, 26936-7, 26949, 26980, 27043, 27076-8, 27081,
27083, 27260, 27303-4, 27314-5, 27329, 27380, 27397, 27407, 27444, 27452,
27505, 27564-5, 27637, 27659, 27677-8, 27685, 27703, 27718, 27727, 27755,
27775, 27784, 27795-6, 27805, 27822, 27831-2, 27837, 27851-4, 27927, 27972,
27997, 28039, 28156, 28166, 28170, 28217-8, 28277, 28319, 28362, 28373, 28379,
28476

Chlorophyll chemical structure
25562-3, 25735, 25844, 26083, B26185, B26220, 26398, 26508, 26521, 26555,
26717, 26935, 26986, 27536, 27631, 27706, 27843, 28036, 28222, 28414

Chlorophyll complexes *in vitro*
25614, 26041-2, 26083, 26712-3, 26763, 26871, 26942, 28036, 28324, 28481

Chlorophyll complexes *in vivo*
25166, 25174, 25184, 25186, 25200, 25228, 25241, 25246, 25263, 25268-9,
25297, 25311, 25323, 25379, 25440, 25456, 25462, 25488, 25564, 25570, 25606,
25616, 25650, 25668, 25703, 25746, 25773, 25791, 25820, 25822, 25837-8, 25881,
25888-9, 25899, 25957, 26048, 26053, 26063, 26099, 26101, 26143, 26160, 26168,
26172, B26185, B26220, 26252-3, 26293, 26326-7, 26352-3, 26362, 26414, 26457,
26545, 26556, 26573, 26594, 26604, 26608-9, 26632, 26672, 26674, 26682-3,
26700, 26709, 26718, 26757-8, 26761, 26774, 26818, 26822, 26890, 26927,
26930, 26935-8, 26977, 26985, 26995, 27043, 27078, 27128, 27137, 27222,
27341-2, 27407, 27489, 27512-3, 27615, 27619, 27668, 27683, 27690, 27717,
27771, 27795, 27817, 27823, 27852, 27863-4, 27867, 27943, 27952, 28021, 28028,
28036, 28045, 28055, 28111, 28122, 28127, 28222, 28232, 28234, 28238, 28258,
28260, 28279, 28281, 28345, 28376, 28410, 28416, 28467

Chlorophyll degradation
25204, 25233, 25432, 25434, 25490, 25550-1, 25620, 25714, 25716, 25849,
25852, 25938, 26019, 26097, 26099, 26407, 26511, 26581, 26586-7, 26602,
26617, 26634, 26716, 26763, 26814, 26929, 26934, 27025, 27042, 27044, 27071,
27159, 27251, 27257, 27291, 27381, 27383, 27387, 27393, 27423, 27500, 27520,
27527, 27572, 27649, 27666, 27683, 27808, 27850, 27855, 27864, 27869, 27885,
27997, 28116-7, 28121-2, 28155, 28190, 28217, 28239-40, 28312, 28319, 28414,
28421

Chlorophyll delayed light emission, luminescence *in vitro*
26686, 26763, 26868, 27599, 27953, 28214, 28481

Chlorophyll delayed light emission, luminescence *in vivo*
25244, 25313, 25339, 25378, 25645, 25703, 25836, 25932, 26479-82, 26559,
26830, 27059, 27075, 27264, 27866, 27868, 27870, 27872, 27953

Chlorophyll determination see Chlorophyll and its products determination...

Chlorophyll energetic states *in vitro* (*cf.* also Chlorophyll in model systems)
25219-20, 25394, 25735-6, 25800, 26082, B26220, 26459, 26556, 26670, 26673,
26679, 26832, 27015, 27780-1, 27844, 28021, 28036, 28161, 28214, 28273,
28290

Chlorophyll energetic states *in vivo*
    25201, 25244-5, 25355, 25544, 25631, 25703, 25733-4, 25751, 25846, 25881,
    26060, 26394-6, 26410, 26556, 26610, 26672, 26675-6, 26680, 26720, 26764,
    26942-3, 27100, 27114-5, 27128, 27173, 27282, 27341, 27378-9, 27866-7,
    27952, 28131, 28136, 28155, 28234, 28238, 28332, 28409

Chlorophyll energetics model   see   Model ...

Chlorophyll, enzymes of synthesis and degradation (other than chlorophyllase)
    25345, 25462, 25881, 25937, 26153, 26205, 26248, 26558, 26681, 26750, 26949,
    27509, 27775, 27822, 27831

Chlorophyll fluorescence *in vitro*
    25316, 25517, 25567, 25973, 26041-2, B26220, 26713, 27082, 27114, 27781,
    28030, 28214-5, 28434, 28481

Chlorophyll fluorescence *in vivo*
    25166, 25201, 25211, 25240, 25245-6, 25263, 25293, 25295, 25315, 25339,
    25344, 25353, 25356, 25378-9, 25385-6, 25396, 25456, 25476, 25544, 25570,
    25606, 25616, 25630, 25636, 25645, 25733, 25746, 25819-20, 25840, 25846,
    25861, 25868, 25892, 25899, 25994, 26048, 26063, 26100-1, 26143, 26168,
    26175, 26196, 26224, 26293-4, 26298, 26326, 26396, 26406, 26411-2, 26468,
    26480, 26523, 26538, 26559, 26582, 26588, 26608, 26615, 26632-3, 26675-6,
    26687, 26703-4, 26710, 26718, 26767, 26792, 26794, 26796-7, 26813, 26829-30,
    26899, 26935-8, 26955, 27064, 27067, 27084, 27111, 27114-5, 27128, 27140,
    27173, 27273, 27319, 27332, 27345, 27378-9, 27391, 27505, 27512, 27619,
    27675, 27678, 27717, 27746-7, 27762-5, 27780, 27809, 27817, 27861, 27863,
    27868, 27870, 27879, 27901, 27917, 27927, 27946, 27952, 27954-5, 28028,
    28055, 28104, 28111, 28122, 28234, 28236, 28260, 28276, 28418, 28449, 28491

Chlorophyll forms see Chlorophyll complexes ...

Chlorophyll in flowers B27046, 27118-9, 28436

Chlorophyll in model systems (*cf.* also Chlorophyll energetic states *in vitro*)
    25219-20, 25227, 25393-4, 25400, 25464, 25466, 25468, 25517, 25565, 25691,
    25704, 25790, 25800, 25974, 26041, 26043, 26267, 26395, 26413, 26657, 26679,
    26713, 26734, 26751, 26764, 26832, 26868, 26897, 26942, 27015, 27040, 27045,
    27082, 27308, 27606-8, 27632, 27781, 27904, 28021, 28030, 28082, 28161,
    28214, 28271-3, 28290, 28324, 28345, 28402

Chlorophyll in photosynthesis mechanism
    25166, 25185-6, 25202, 25241, 25244-5, 25263, 25323, 25344, 25348, 25353,
    25355, 25379, 25441, 25490, 25533, 25570, 25606, 25616, 25631, 25691, 25733,
    25751, 25761, 25773, 25834, 25846, 25879-81, 25888, 25918, 25957, 26043,
    26160, B26185, B26220, 26223-4, 26257-8, 26298, 26352, 26394-5, 26398, 26413,
    26456, 26479, 26523, 26538, 26556, 26561, 26573, 26609-10, 26614, 26672,
    26676, 26680, 26718, 26720, 26734, 26761, 26765, 26872, 26914, 26938, 26943,
    26974, 26990, 27015, 27064, 27072, 27083-4, 27100, 27114-5, 27127-8, 27137,
    27173, 27222, 27254, 27273, 27282, 27309, 27340, 27512, 27518-9, 27528,
    27608, 27615, 27619, 27668, 27764, 27864, 27867-8, 27878, 27952, 28021, 28029,
    28036, 28127, 28131, 28155, 28234, 28258, 28260, 28290, 28332, 28345, 28409-10

Chlorophyll in physiology of photosynthesis
    25184, 25259-60, 25293-4, 25383, 25404, 25406, 25441, 25616, 25810, 25853,
    25867, 26099, 26123, 26353, 26512, 26597, 26684, 26700, 26869, 27036, 27069,
    27129, 27317, 27350, 27460, 27463, 27508, 27511, 27513, 27567, 27925, 28008,
    28128, 28172-3, 28188, 28379, 28433

Chlorophyll in seeds and fruits 25634, 26511, 27265, 27520, 27526, 27561

Chlorophyll luminescence see Chlorophyll delayed light emission ...

Chlorophyll, methods see Chlorophyll and its products determination ...

Chlorophyll number see Chlorophyll in physiology of photosynthesis

Chlorophyll precursors see Chlorophyll biosynthesis ...

Chlorophyll unit see Photosynthetic (chlorophyll) unit

Chlorophyllase 25462, 25938, 26097, 26107, 26404, 26777, 27855, 27885, 28039, 28111,
        28320

Chlorophylls a,b content and their ratio
        25166, 25176, 25180, 25184-6, 25204, 25217, 25228, 25237, 25240, 25245-6,
        25261, 25287, 25293-4, 25297, 25335, 25364, 25368, 25379, 25423-4, 25445,
        25456, 25470, 25474, 25480, 25518, 25525, 25620, 25679, 25686, 25700, 25705,
        25716, 25795, 25819, 25838, 25862, 25867, 25872, 25887, 25896, 25898, 25905-6,
        25913, 25918, 25930, 25962, 25988, 25991-2, 26019, 26033, 26057, 26078, 26099,
        26101, 26125, 26131, 26139, 26148, 26172, 26181, B26185, 26219, B26220, 26246,
        26270-1, 26284, 26327, 26353, 26361, 26392, 26400-1, 26404, 26408, 26415,
        26440, 26470, 26493, 26495, 26508, 26511, 26518, B26551, 26554-5, 26573,
        26575, 26597, 26614, 26617, 26634, 26666-7, 26680, 26683, 26700, 26707, 26734,
        26749, 26757, 26777, 26779, 26786, 26795, 26800, 26812, 26830, 26855, 26872-
        -3, 26910, 26917, 26925, 26927-8, 26930, 26934, 26977, 26985-6, 26990, 26994,
        27027, 27035, 27052, 27057, 27060-1, 27066, 27110, 27189, 27216, 27222, 27237-
        -9, 27270, 27284, 27288, 27309, 27348, 27373, 27380-1, 27386, 27388, 27394,
        27405, 27443-4, 27488-9, 27494, 27508, 27527, 27547, 27552, 27559-60, 27568,
        27581, 27599, 27649, 27671, 27683, 27689, 27690, 27715, 27717, 27794-5, 27816,
        27824, 27834-5, 27848, 27852, 27854, 27863, 27874, 27884-5, 27924, 27926,
        27947, 27978, 27980, 27996, 28055, 28057, 28117, 28122, 28180, 28187, 28205,
        28221-2, 28240, 28258, 28261, 28276, 28290, 28297, 28317, 28357, 28361, 28422,
        28467, 28477

Chlorophylls c, d
        25442, 25648, 25651, 25849, 25895, 26025, 26116, 26127, B26185, B26220,
        26270-1, 26404, 26453, 26508, 26518, 26521, 27381, 27511, 27513, 27649,
        27689, 27921, 27940, 28162, 28317, 28377, 28434

Chlorophylls, *Chlorobium*
        25480, 25562-3, B26185, B26220, 26508, 26614, 26630, 26657, 26672, 27381,
        27631, 28036

Chloroplast see also Thylakoid; Stroma of chloroplast; Pyrenoid; Ribosome of chlo-
        roplast; Phycobilisome

Chloroplast and chromatophore biopotentials
        25175, 25202, 25296, 25313, 25393, 25589, 25884, 25961, 26198, 27075, 27208,
        27285, 27455, 27646, 27719, 27745, 27764, 27836, 27889, 28137, 28291-2,
        28453, 28488

Chloroplast and chromatophore chemical composition (*cf.* also Lipids, fatty acids, and
        chloroplast; Proteins, amino acids, nucleic acids, and chloroplast, *etc.*)
        25174, 25177, 25263, 25277, 25287-9, 25377-8, 25392, 25497, 25504, 25529,
        25585, 25730, 25757, 25802, 25874, 25888, 25891, 25953, 25979, 26033, 26059,
        26071, 26095, 26119, 26144, 26147, 26158, 26167, 26174, 26219, 26225, 26341,
        26353, 26378, 26415-6, 26432, 26455, 26463, 26522, 26553-4, 26665, 26769,
        26774, 26818, 26833, 26872, 26894, 26899, 26907, 26917, 26926, 26942, 26948,
        26995, 27004, 27025, 27057, 27090, 27092, 27130, 27167-70, 27181, 27186,
        27205, 27286, 27288, 27390, 27452, 27468, 27492, 27499, 27546, 27657, 27661,
        27665, 27667, 27674, 27676, 27690, 27712-3, 27729-30, 27738, 27742, 27771,
        27784, 27789-91, 27816, 27832, 27863, 27874, 27924, 27962, 28034-5, 28045-6,
        28048, 28084, 28127, 28132, 28189, 28198, 28203, 28248-50, 28252, 28281,
        28360-1, 28373, 28416-7, 28467

Chloroplast and chromatophore dimensions
        25530, 25770, 25878, 26077, 26095, 26674, 26715, 26818, 26842, 26926, 26928,
        27249, 27316, 27443, 27488-9, 27499, 27715, 28180-1, 28288

Chloroplast and chromatophore distribution in cell   see   Chloroplast and chromato-
        phore number and distribution

Chloroplast and chromatophore fragments
        25180, 25192, 25196, 25211, 25217, 25228, 25245-6, 25263, 25287, 25323,
        25347-8, 25470, 25504, 25703, 25745, 25837, 25884, 25927, 25968, 26022,
        26063, 26101, 26172-4, 26176, 26219, B26220, 26441, 26463, 26468, 26561,
        26602, 26615, 26677, 26694, 26710, 26793, 26829-30, 26910, 26917, 27075,
        27104, 27155, 27312, 27399, 27486, 27581, 27603, 27675-6, 27681, 27713,
        27784, 27789, 27816, 27829, 27831, 27851-2, 27863, 27870, 27934, 27968,
        27978, 28111, 28184, 28236, 28276, 28332, 28452

Chloroplast and chromatophore number and distribution
        25528, 25530, 25655, 25781, 25906, 26674, 26853, 26926, 26928, 26994, 27443,
        27488, 27497, 27499, 27657, 28288, 28357

Chloroplast and chromatophore replication, ontogeny
        25287-8, 25373, 25377, 25500, 25528, 25530, 25551, 25616, 25621, 25655,
        25689, 25745, 25837, 25853, 25904, 26077, 26142, 26201, 26279-80, 26340,
        26447-8, 26462, 26467, 26649, 26842, 26890, 27054, 27081, 27157, 27167,
        27217, 27249, 27316, 27351, 27407, 27443, 27482, 27497-9, 27577, 27614,
        27657, 27717, 27818, 27884, 27922, 27967, 27985, 28149, 28217, 28363-4,
        28430, 28436

Chloroplast and chromatophore volume changes
        25621, 26276, 26435, 26467, 26496, 27018, 27028, 27351, 27355, 27454, 27497,
        27556, 27576, 27796, 28449

Chloroplast immobilization   see   Photosystems stabilization ...

Chloroplast, isolated, carbon fixation in   see   Carbon fixation in isolated chloro-
        plasts ...

Chloroplast, isolated, gas exchange by
        25372, 25712, 25833, 26008, 26206, 26333-4, 26565-6, 26703, 27453, 27583-4,
        27618, 27641, 27655, 27665, 27906-8, 27943, 28000, 28112, 28198, 28227,
        28263, 28307, 28334, 28396, 28449, 28457

Chloroplast isolation
        25180, 25188, 25301, 25757, 25770, 26153, 26206, 26302, 26554, 26790, 26894,
        27263, 27279, 27394, 27562, 27583-4, 27615, 27968, 27985, 28253, 28304, 28331,
        28371, 28451

Chloroplast, localization of electron transport chain in thylakoid   see   Electron
        transport chain localization in thylakoid

Chloroplast movements
        25558-60, 26088, 26321-4, 26346, 26435, 26951, 27047, 27051, 27719, 27727,
        27756-7

Chloroplast outer membrane
        25500, 25745, 25770, 25833, 25979, 26036, 26090, 26126, 26234, 26276, 26279,
        26334-5, 26342, 26447, 26566, 26767, 26816-7, 26893, 27186, 27241, 27304,
        27468-9, 27562, 27640, 28249, 28364, 28456

Chloroplast proteins (and other photosynthetic proteins), methods
        25199, 25228, 25890, 26158, 26432, 26455, 27214

Chloroplast ultrastructure (cf. also Chloroplast outer membrane; Stroma of chloro-
        plast; Thylakoid, granum)
        25180-1, 25184, 25187-8, 25228, 25237, 25240-1, 25331, 25338, 25360, 25374,
        25421, 25425, 25431, 25486, 25500, 25512, 25528, 25530, 25532, 25564, 25579,
        25622, 25624-7, 25767, 25770, 25786, 25789, 25802, 25837, 25852, 25891, 25959,
        25977, 25988, 25995, 26007, 26025, 26057, 26071, 26077, 26094-5, 26101,

Chloroplast ultrastructure (continued)
26126-7, 26158, 26167, 26170, 26201, 26206, B26220, 26297, 26302, 26317, 26342, 26353, 26359, 26361, 26366, 26440, 26449, 26463, 26492, 26520, 26557, 26596-7, 26601, 26613, 26632-3, 26664, 26667, 26674, 26699-700, 26715, 26720, 26768, 26791, 26831, 26890, 26894, 26899, 26916, 26950, 27002, 27010, 27028-9, 27054, 27057-8, 27061, 27105-6, 27118-21, 27130, 27134-8, 27141, 27157-8, 27160, 27164, 27176, 27184, 27211, 27217, 27233, 27246, 27261, 27268, 27279, 27305-6, 27316, 27351-2, 27383, 27399, 27407, 27450, 27481, 27483, 27487, 27497-9, 27525, 27530, 27576-7, 27581, 27613, 27621, 27657, 27668, 27691, 27714-6, 27739, 27744, 27767, 27784, 27816, 27818, 27848, 27883-4, 27922-3, 27967-9, 27978, 27985, 28029, 28074-5, 28119, 28138, 28149, 28166, 28181, 28189, 28202, 28217, 28221, 28250, 28274, 28278, 28289, 28357, 28365, 28384, 28398, 28401, 28416, 28433, 28436, 28485-6

Chromatophore in photosynthetic bacteria and blue-green algae (*cf*. also Chloroplast and chromatophore ...)
25579, 25665, 25773, 26942, 27254, 27806, 27904, 27945, 28048, 28202, 28451

Circular dichroism see Dichroisms ...

Clark electrode see O_2 electrode

Clock, biological see Diurnal changes ...

CO_2 and biliproteins 26168

CO_2 and carbon fixation pathways
25216, 25284-5, 25290, 25331, 25395, 25713, 25721, 25787, 25868, 26052, 26402, 26428, B26551, 26650, 26911, 26958, 27182, 27524, 27605, 27760, 27928, 28083, 28269, 28303

CO_2 and carotenoids 25798

CO_2 and chlorophyll
25233, 25798, 25868, 26079, 26168, 26814, B27001, 27314, 27849, 28477

CO_2 and chloroplast 27314

CO_2 and ecosystem productivity and canopy functioning
25209, 25230-1, 25331, 25568, 25599, 25623, 25681, 25758, 26029, 26031, 26079, 26151, 26383, 26474, 26544, 26648, 26776, 26784, 27190, 27539-40, 27849, 28145, 28477

CO_2 and electron transport chain 25320, 26333, 28001, 28083

CO_2 and gas exchange (*cf*. also CO_2 and gas exchange, analysis of CO_2 curves)
25167-8, 25221, 25258, 25261, 25331-2, 25382, 25395, 25406, 25414-5, 25419, 25437, 25457, 25476, 25502-3, 25568, 25571, 25578, 25599, 25609, 25611, 25642, 25666, 25674, 25684, 25709-10, 25723, 25772, 25783, 25873, 25923, 25996, 26009, 26023-4, 26045, 26065, 26079, 26110, 26164, 26227, 26235, 26237, 26241, 26280, 26333, 26368, 26389, 26397, 26402, 26436, 26464, 26478, 26567, 26593, 26641, 26647, 26655, 26718, 26776, 26785, 26835-6, 26863, 26872, 26880, 26987, B27001, 27156, 27180, 27182, 27187, 27278, 27370, 27440, 27460, 27548, 27574, 27641, 27750, 27760, 27827, 27839, 27887, 27951, 28001, 28088, 28092, 28106, 28205-6, 28229-30, 28477

CO_2 and gas exchange, analysis of CO_2 curves
25476, 25709, 26164, 26227, 27574, 28107, 28193

CO_2 and photorespiration
25261, 25395, 25721, 25776, 26626, 26629, 26784, 26880, 27402, 27460, 27754, 27760, 27929, 28269

CO_2 and resistances to CO_2 and water vapour transfer
25604, 25611, 26065, 26237, 26241, 26524, 26776, 27356, 27574, 27839, 28367

CO_2 and respiration 25502

CO_2 compensation concentration
25168, 25261, 25271, 25283, 25406, 25414, 25445, 25528, 25653-4, 25684,
25708, 25712, 25755, 25858, 26009, 26052, 26055, 26078, 26162, 26164, 26227,
26402, 26446, 26464, 26475, 26618-9, 26719, 26767, 26776, 26784, 26863,
26987-8, 27187, 27198, 27215, 27289, 27313, 27356, 27363, 27414, 27422,
27425, 27431, 27460, 27597, 27760, 27951, 28229-30, 28399, 28428, 28432,
28479

CO_2 exchange see Gas exchange ...

CO_2 fixation, dark see Dark CO_2 fixation

CO_2 measurement, infra-red gas analyser see Infra-red gas analyser for CO_2

CO_2 measurement (other than with infra-red gas analyser)
25350, 25595, 25667, 25828, 25858, 26135, 26638, 26745, 27949, 28475, 28475

CO_2 transfer across membranes 26858, 26966, 27401, 27469, 27936, 28304

CO_2 transfer, theory 26239, 28304

Cold (hardiness) see Temperature, low ...

Combustion heat see Calorimetry

Compensation irradiance
25168, 25549, 26300, 26370, 26661, 26863, 27198, 27431, 27613, 28077, 28194,
28229, 28462

Compensation point, CO_2 see CO_2 compensation concentration

Compensation point, light see Compensation irradiance

Competition in ecosystem 25613, 26947, 27336, 27740

Conductance for transfer of gases see Resistance ...

"Contribution" of individual organs see Biomass distribution and redistribution;
Photosynthate translocation ...

Correlations within plant 25881

Cosmic radiation see Ionizing radiation ...

Coupling factor 1 see ATPase ...

Cover, vegetative see Canopy ...; Ecosystem ...

Crassulacean Acid Metabolism see CAM

Cultivar differences, carbon fixation pathways 26118, 26777, 27053, 27384, 28385

Cultivar differences, carotenoids 25479, 26187, 26777, 27110, 27925

Cultivar differences, chlorophyll
25310, 25364, 25404, 25479, 26050, 26187, 26552, 26757, 26777, 27405, 27460,
27520, 27690, 27925, 28200, 28282

Cultivar differences, chloroplast 25277, 27647, 27883

Cultivar differences, ecosystem productivity and canopy functioning
25232, 25264, 25357, 25364, 25405, 25412, 25576, 26050, 26134, 26263, 26466,
26778, 27210, 27384-5, 27386, 27389, 27778, 27819, 27982, 28041, 28262, 28300

Cultivar differences, electron transport chain 25392, 26001, 26766, 27161

Cultivar differences, gas exchange
      25261, 25264, 25312, 25328, 25404, 25415, 25615, 25708, 25828, 25967, 26118,
      26392, 26647, 26777, 27210, 27384, 27389, 27460, 27778, 27830, 27896, 27951,
      28018-9, 28092, 28282, 28300, 28375

Cultivar differences, photorespiration 27460, 27778, 27951, 28092

Cultivar differences, resistances to CO_2 and water vapour transfer
      25264, 25364, 26118

Cultivar differences, respiration 27389, 28092, 28300

Cultivation of algae and photosynthetic bacteria see Algae and photosynthetic bacte-
      ria, cultivation; Algae mass cultures productivity

Cuticular CO_2 and O_2 exchange 26826

Cuticular resistance see Resistance, cuticular

Cytochromes
      25183, 25266, 25385-7, 25393, 25427, 25470, 25511, 25565, 25606, 25669, 25701,
      25704, 25757, 25855, 25890, 25893, 25964, 25988, 26084, 26172, 26178, B26220,
      26334, 26382, 26423-5, 26442-3, 26456, 26486, 26608, 26660, 26694, 26794,
      26796, 26807, 26826, 26872, 26990, 27006, 27086, 27116, 27254, 27275, 27304,
      27309, 27340, 27345, 27447-9, 27518-9, 27535, 27537, 27615, 27617, 27659,
      27668, 27708, 27809, 27841-2, 27863, 27868, 28045, 28135-6, 28184, 28189,
      28232, 28237, 28253, 28333, 28376, 28423-4, 28469

Cytochromes, methods 25988, 26660, 27965

D

Dark CO_2 fixation
      25446, 25516, 25528, 25780, 25784, 25863, 26004, B26551, 26563, 26645, 27107,
      27427, 27542, 28000-1, 28064, 28077, 28390

Data recording and processing 25549, 25728, 26635

Decapitation see Defoliation ...

Defoliation, decapitation, ear and root removal, effect on carbon fixation pathways
      27384

Defoliation, decapitation, ear and root removal, effect on carotenoids 27239

Defoliation, decapitation, ear and root removal, effect on chlorophyll 26775, 27239,
      27384

Defoliation, decapitation, ear and root removal, effect on ecosystem productivity
      and canopy functioning
      25189, 25336, 25818, 25832, 25917, 26213, 26452, 26465, 26688, 26775, 26945,
      27199, 27538, 27664, 27720, 27991, 28050, 28392

Defoliation, decapitation, ear and root removal, effect on electron transport chain
      25189

Defoliation, decapitation, ear and root removal, effect on gas exchange
      25328, 25765, 25832, 26121, 26775, 27384, 27501, 27594, 28091

Defoliation, decapitation, ear and root removal, effect on resistance to CO_2 and
        water vapour transfer 25832, 26775

Desiccation of tissue see Water saturation deficit

Deuterium oxide, tritium oxide 25271

Development, leaf, plant see Leaf (and plant) development and ageing

Dew see Precipitation, dew ...

Dew measurement see Aerodynamic methods ...

Dew point see Humidity of air ...

Dichroisms determination (methods and results)
        25263, 25337, 25351, 25746, 25837-8, 26061, 26158, 26374, 26401, 26942,
        26992, 27088, 27128, 27193, 27340-1, 27758, 27952, 28258-9, 28414

Differentiation of tissues see Leaf (and plant) development and ageing; Ontogeny....

Diffusion, diffusion coefficient see CO_2 transfer, theory

Diffusion (diffusive) conductance see Resistance ...

Diffusion (diffusive) resistance see Resistance ...

Diurnal changes (biological clock) in algae productivity 26051, 26568, 27392, 27470,
        27611, 27838, 27959-60, 27970, 27987, 28076, 28129

Diurnal changes (biological clock) in carbon fixation pathways and parameters
        25620, 25787, 25954, 26299, 27052, 27147, 27183, 27541, 28065

Diurnal changes (biological clock) in carotenoids 28188

Diurnal changes (biological clock) in chlorophyll 25557-8, 25679, 26242, 27622,
        28076

Diurnal changes (biological clock) in chloroplast (chromatophore) 25559-60, 27497

Diurnal changes (biological clock) in ecosystem productivity and canopy functioning
        25203, 25264, 25527, 25662, 25666, 25683, 26027, 26080, 26337, 26372, 26783,
        26954, 27052, 27087, 27139, 27220, 27270, 27630, 27656, 27658, 27612, 28087,
        28158, 28283, 28350, 28397, 28408

Diurnal changes (biological clock) in gas exchange 25371, 25527, 25559, 25571,
        25596, 25666, 26027, 26046, 26075, 26085, 26121, 26166, 26265, 26303, 26368,
        26403, 26506, 26515, 26580, 26583-5, 26591, 26674, 26690, 26801, 26811, 26905,
        26909, 26912, 27095, 27107, 27139, 27180, 27267, 27271, 27359, 27574, 27638,
        27750, 27768, 27773, 27782, 27888, B27915, 28014, 28066-7, 28076, 28085,
        28087, 28152, 28193, 28230, 28342

Diurnal changes (biological clock) in photorespiration 26299

Diurnal changes (biological clock) in resistances to CO_2 and water vapour transfer
        25283, 25327, 25371, 25406, 25430, 25454, 25653, 25742, 25843, 25947, 25999,
        26015, 26237, 26265, 26303, 26612, 26739, 26909, 27139, 27269, 27271, 27356,
        27360, 27574, 27636, 27749, 27912, 28065, 28067, 28089, 28125, 28257, 28350,
        28462

Drought and carbon fixation pathways and parameters 26549

Drought and chlorophyll 25992, 26183, 26758

Drought and chloroplast (chromatophore) 26126

Drought and ecosystem productivity and canopy functioning 25970, 26685, 27325, 27658, 27710, 28241, B28482

Drought and gas exchange 25431, 25859, 25949-50, 25992, 26494, 26685, 26690, 27107, 27656

Dry-matter production, gravimetric determination 26812, 27436, 28058, 28475

E

Ear removal see Defoliation ...

Ecosystem production, primary productivity (terrestrial).
25198, 25229, 25238-9, 25253-4, 25264, 25274-5, 25299, 25303, 25312, 25331,
25336, 25357, 25412, 25435, 25472-3, 25475, 25480-1, 25518, 25523, 25538,
25576, 25581, 25593-4, 25608, 25615, 25623, 25626, B25637, 25674, 25737,
25753, 25801, 25811, 25818, 25827, 25857, 25886, 25910, 25921, 25946, 25970,
25978, 25982, 25989, 25997, 26017, 26029, 26031, 26050, B26054, 26055, 26068,
26076, 26086-7, 26092, 26096, 26106, 26112, 26124, 26132, 26145, 26151,
26182, 26189, 26202, 26204, 26256, 26263, 26267, 26284-6, 26331, 26373,
26419, 26450, 26452, 26461, 26490, 26509, 26530, 26534, 26547, 26599, 26648,
26685, 26688-9, 26719, 26724, 26735, 26742, 26778, 26784, 26787, 26803-4,
B26846, 26847, 26881, 26883, 26895, 26904, 26906, 26919-20, 26927, 26945,
26952, 26954, 26964, 27014, 27190, 27209, 27228, 27242, 27293-5, 27333, 27336,
27343, 27371, 27385, 27409, 27413, 27415, 27422, 27428, 27457, 27466, 27507,
27538, 27563, 27604, 27656, 27673, 27707, 27710, 27728, 27734-5, 27740, 27748,
27760, 27778-9, 27793, 27830, 27893-4, B27915, 27920, 27930, 27981-2, 27983,
28013, 28020, 28031, 28041, 28087, 28093, 28098-100, 28109-10, 28113, 28143,
28158, 28160, 28169, 28171, 28175, 28192, 28197, 28212, 28228, 28262, 28265,
28282, 28294, 28300, 28322, 28343, 28352, 28381, 28397, 28408, 28431, 28459-
60, 28470, 28474, 28480

Ecotypes, geographical types, and carbon fixation pathways 26116

Ecotypes, geographical types, and ecosystem productivity and canopy functioning
25660, 25807, 25811, 25909, 26017, 26881, 28470

Ecotypes, geographical types, and electron transport chain 27558

Ecotypes, geographical types, and gas exchange
26116, 26811, 26881, 26884, 27089, 27271, 27337, 27638, 28110, 28146, 28146,
28368, 28391, 28474

Ecotypes, geographical types, and photorespiration 26116

Ecotypes, geographical types, and resistances to CO_2 and water vapour transfer
26116, 26532

Ecotypes, geographical types, and respiration 26116, 27337, 27638

Ecotypes, geographical types, carotenoids in 25731-2, 27558, 28073

Ecotypes, geographical types, chlorophyll in 26116, 27124, 27558, 28073

Efficiency, photosynthetic see Irradiance and gas exchange, analysis of light
curves

Electron paramagnetic resonance see EPR, NMR

Electron spin resonance see EPR, NMR

Electron transport chain activity
        25217, 25321, 25750, 25892, 26366, 26554, 26762, 27857, 28225, 28285

Electron transport chain components see Cytochromes; Ferredoxin ...; Ferredoxin-
        NADP reductase; NADP ...; O_2 evolution ...; Photosystems; Plastocyanin; Quino-
        nes

Electron transport chain localization in thylakoid
        25182, 25228, 25241, 25244, 25331, 25339, 25421, 25626, 25762, 25789, 25884,
        25907, 26193, 26259, 26268, 26343, 26423, 26445, 26538, 26609, 26767, 26859,
        26942, 27137, 27140, 27305-6, 27399, 27617, 27904, 27966, 27979, 28137,
        28163-4, 28237, 28303, 28334, 28409-10

Electron transport chain model see Model ...

Electron transport chain, serological analysis 25891, 26711, 27138, 27745, 28281,
        28315

Emerson effect, Blinks effect 25244, 25362, B26220, 26861, 27950, 28286, 28376,
        28395-6

Energy balance, leaf 25724

Energy utilization, plant and ecosystem
        25331, 25357, 25772, 25903, 25946, 25978, 26189, 26267, 26426, 26719, 26869,
        26881, 27128, 27180, 27333, 27635, 27688, 27779, 27786, 27894, B27915, 28192,
        28247, 28470

Enzymes and chlorophyll 27027

Enzymes and chloroplast (chromatophore) 25277, 25398, 26893, 27003, 27723, 27816

Enzymes and electron transport chain 25362, 25427, 26554, 27019, 27603, 27618,
        27763, 27943-4, 28203

Enzymes and gas exchange 26067, 27382, 27618

Enzymes of carbon fixation pathways other than RuBPC, PEPC and malic enzyme
        25211-3, 25221, 25226, 25278, 25325, 25338, 25431, 25516, 25528, 25583-4,
        25671-2, 25730, 25768-9, 25786-7, 25833, 25867, 25944, 26003-4, 26059, 26165,
        26188, 26203, 26249, 26280, 26314, 26317, 26525-6, 26597, 26622-4, 26685,
        26755, 26786, 26806, 26858, 26866, 26872, 26921, 26940, 27007-8, 27053,
        27091, 27147, 27183, 27263, 27297, 27315, 27329, 27364, 27382, 27422, 27431,
        27473, 27509, 27542, 27569, 27621, 27639, 27755, 27759, 27772, 27793, 27906,
        27909, 28040, 28151, 28183, 28220, 28285-6, 28303, 28342, 28355, 28357, 28403,
        28419, 28430, 28490, 28492

Enzymes of carbon fixation pathways other than RuBPC, PEPC, malic enzyme, malate de-
        hydrogenase, methods 25235, 25656, 26356, 26978, 28492

Enzymes of carotenoids synthesis and degradation see Carotenoids, enzymes ...

Enzymes of chlorophyll synthesis and degradation see Chlorophyll, enzymes ...;
        Chlorophyllase

Enzymes of photorespiration see Photorespiration enzymes

Epidermis see Leaf epidermis ...

EPR, NMR (methods and results)
        25276, 25331, 25477, 25488, 25517, 25563, 25629, 25736, 25773, 25844, 25893,
        25963, 25968, 26176, B26220, 26394-5, 26457, 26559, 26561, 26609, 26616,
        26657, 26679-80, 26751, 26830, 26942-3, 26974, 27033, 27088, 27099-100,
        27192-3, 27277, 27282, 27308-9, 27326, 27452, 27528, 27706, 27868, 27875,
        28037, 28060, 28131, 28135-6, 28155, 28234, 28258, 28332, 28334-5, 28376,
        28420, 28437-8

Ethylene see Gases, organic ...

Evolution see Phylogeny

Excitation resistance see Resistance, carboxylation and excitation

Exhaust gases see Pollution of air ...

Exposure chamber see Assimilation chamber

Extension growth, leaf dimensions
        25170, 25209, 25230, 25250, 25318, 25319, 25481, 25521, 25526, 25538, 25605,
        25666, 25725, 25754, B25775, 25785, 25811, 25826, 25829, 25943, 25955, 26003,
        26017, 26073, 26093, 26197, 26213, 26284, 26400, 26461, 26466, 26643, 26798,
        26904, 26906, 26927, 26931, 26954, 27011, 27108, 27142, 27172, 27190, 27229,
        27270, 27404, 27457, 27488-9, 27540, 27579, 27625, 27701, 27725, 27899,
        27922, 27981, 28041, 28134, 28207, 28336, 28400, 28454, 28456

Extraction of pigments see Pigments ...

Exudation of photosynthates see Photosynthate translocation ...

F

Fatty acids see Lipids, fatty acids ...

Ferredoxin, ferredoxin-NADP reductase, methods 26066, 26654, 27310, 27628

Ferredoxin, flavoproteins, rubredoxin
        25211, 25347, 25470, 25505, 25511, 25565, 25583, 25585, 25626-9, 25635,
        25939-40, 25942, 25963, 25968, 26173, B26220, 26254, 26266, 26305-6, 26310,
        26442, 26636-7, 26653-4, 26671, 26677, 26695, 26901, 26942, 27019-20, 27048,
        27076, 27088, 27230, 27310, 27516, 27519, 27668, 27772, 27870, 27963, 27979,
        28080, 28103, 28105, 28218-9, 28306-7, 28409

Ferredoxin-NADP reductase, pteridines
        26066, B26220, 26310, 26677, 27020, 27603, 27628-9, 27963, 28105, 28218

Flashes of light see Irradiation, flash

Flavoproteins see Ferredoxin ...

Flooding and chlorophyll 25992, 26749

Flooding and ecosystem productivity and canopy functioning
        25435, 25566, 25784, 26499, 26729, 27240, 27248, 27625, B28482

Flooding and gas exchange 25780, 25992, 26749, 27248

Fluorine see Pollution of air ...

Foliage see Canopy ...

Fraction I protein see Ribulose 1,5-bisphosphate carboxylase

Frost (hardiness) see Temperature, low ...

Fungus diseases see Phytopathological effects ...

Fusicoccin see Growth regulators ...

G

Gas exchange in algae
25163, 25215, 25272, 25280, 25317, 25335, 25381-2, 25395, 25426, 25459,
25501-3, 25536, 25543, 25548-9, 25559, 25598, 25601, 25635, 25651, 25706,
25752, 25806, 25824-5, 25840, 25866-7, 25869, 25894, 25952, 25996, 26002,
26023, 26035, 26089, 26108, 26148, 26167, 26170, 26211, 26214, 26313, 26367,
26386, 26420, 26472, 26478, 26529, 26583-5, 26608, 26642-4, 26655, 26661,
26698, 26781, 26794, 26807, 26822, 26829, 26896, 27156, 27177-8, 27185, 27227,
27247, 27257-8, 27279, 27319-21, 27349, 27367-8, 27377, 27406, 27427, 27429,
27438, 27463, 27490, 27511, 27513, 27534, 27548, 27567, 27613, 27668, 27681,
27700, 27703, 27775, 27818, 27823, 27825, 27827, 27898, 27988, 27993-4,
28008, 28076-7, 28088, 28124, 28128, 28140, 28152, 28170, 28174, 28223,
28278, 28325, 28331, 28349, 28372, 28395, 28421, 28446

Gas exchange in isolated chloroplasts see Chloroplast, isolated, gas exchange by

Gas exchange in photosynthetic bacteria see Photosynthetic bacteria, gas exchange
in

Gas exchange, model see Model ...

Gas exchange of organs other than leaf 25967, 26046, 26109, 27344, 28086

Gases, organic, and carbon fixation pathways 26733

Gases, organic, and chlorophyll 25233

Gases, organic, and chloroplast (chromatophore) 26733

Gasometric methods, generally
25178, 25823, B26120, 26464, 26474, 26592, 26635, 26745, 26805, 26837

Gasometric system, conditioning of air 25666, 25911, 26635, 27369, 28366

Gasometric system, open
25666, 26044, B26120, 26164, 26227, 26370, 26375, 26384, 26593, 26661, 26745,
27658, 28102, 28380, 28475

Gasometric system, semiclosed and closed
25388, 25549, 25911, 26044, B26120, 26577, 26635, 26745, 27156, 27461, 28475

Genetics cf. also Mutagens ...; Mutants ...

Genetics and ecosystem productivity and canopy functioning
25546, 25593, B25637, 25989, 26030, 26188, 26202, 26263, 26460, 27210, 27293,
27295, 27982, 28316

Genetics of carbon fixation pathways
25259-60, 25437, 25692, 26181, 26188, 26314, 26430, 26809, 27896, 28385

Genetics of carotenoids 26125, 26186, 27165, 28455

Genetics of chlorophyll
        25259-61, 25265, 25872, 25889, 26125, 26205, 26885, 27543, 27559, 27561,
        27622, 27903, 27945, 28319, 28373, 28455

Genetics of chloroplast (chromatophore)
        25354, 25429, 25524, 25689, 25935, 26260, 26431, 26853, 27383, 27499, 27694-
        7, 27891, 28138, 28252, 28373

Genetics of electron transport chain 26424, 27161, 27365, 27559, 27996, 28385

Genetics of gas exchange
        25406-7, 25593, 25608, B25637, 26188, 26373, 26460, 26883, 26987-8, 27210,
        27278, 27421, 27638, 27896, 28168, 28316, 28475, 28478

Genetics of photorespiration 26188, 26987-8, 28477-9

Genetics of resistances to CO_2 and water vapour transfer 25261

Genetics of respiration 25406, 26988, 27638

Glycolate metabolism see Photorespiration ...

Glyoxysome see Peroxisome ...

Granum see Thylakoid ...

Gravimetric determination of photosynthesis see Dry-matter production ...

Gross photosynthetic rate
        25195, 25307, 25592, 25654, 25666, 25801, 25810, 26037, 26046, 26420, 26487,
        26642, 26884, 27460, 28086-7, 28180, 28325, 28408, 28413

Growth analysis, methods
        25469, 25475, 25531, 25569, 25575, 25707, 25910, 25921, 25972, 26534, 26789,
        27437, 27892, 28358, 28475

Growth analysis, net assimilation rate, leaf area ratio, relative growth rate
        25205, 25230, 25232, 25357, 25412, 25437, 25566, 25571, 25613, 25618, 25666,
        25681, 25729, 25758, 25771-2, 25774, B25775, 25860, 25909, 25955, 25965,
        25970, 25989, 26017, 26020, 26073, 26086-7, 26134, 26184, 26197, 26213, 26308,
        26390, 26461, 26474, 26484, 26648, 26652, 26729, 26776, 26787, 26803, 26857,
        26869, 26884, 26906, 26954, 27011, 27052, 27108, 27133, 27145, 27188, 27195,
        27200, 27228, 27234, 27242, 27283, 27333, 27336, 27382, 27385-6, 27415, 27437,
        27457, 27502, 27507, 27563, 27664, 27686, 27710, 27719-20, 27734, 27749,
        27779, 27810, 27847, 27893, 27930, 27980, 27991, 28068-9, 28086, 28098, 28169,
        28175, 28241, 28323, 28343, 28392

Growth analysis, specific leaf area, leaf area index, leaf area duration
        25167, 25169, 25189, 25203, 25264, 25303, 25336, 25412, 25454, 25482-3,
        25526, 25528, 25540, 25546, 25566, 25572, 25607, 25618, 25642, 25659-61,
        25673, 25681, 25686, 25772, 25774, B25775, 25792, 25909, 25943, 25948, 25965,
        25967, 25989, 26027, 26030, 26055, 26068, 26079, 26085-7, 26096, 26133-4,
        26164, 26197, 26202, 26204, 26213, 26263, 26308, 26337, 26372-3, 26390, 26409,
        26461, 26466, 26474, 26501, 26506, 26515, 26652, 26723, 26741, 26782, 26787,
        26798, 26803, 26828, 26878, 26903-4, 26906, 26912, 26928, 26944-5, 26954,
        26964, 27011, 27052, 27094, 27133, 27139, 27175, 27188, 27195, 27210, 27216,
        27228, 27234-5, 27242, 27269, 27294-5, 27325, 27333, 27336, 27382, 27385-6,
        27389, 27410, 27437, 27457, 27466, 27502, 27507, 27563, 27587, 27625, 27630,
        27664, 27686, 27701, 27710, 27720, 27734, 27779, 27810, 27819, 27882, 27887,
        27896, B27915, 27920, 27980, 28022, 28068-70, 28087, 28089, 28092, 28097,
        28123, 28195, 28197, 28231, 28247, 28265, 28294, 28323, 28392, 28413, 28461

Growth regulators and carbon fixation pathways 25338, 25377, 25516, 26446, 27101,
        27408, 27651

Growth regulators and carotenoids
    25432, 25877, 25905-6, 26291, 26634, 27101, 27165, 27372-3, 27443-4, 27527

Growth regulators and chlorophyll
    25236, 25377, 25432-3, 25498, 25510, 25652, 25686, 25905-6, 26291, 26494,
    26581, 26586-7, 26602, 26634, 26701, 26814, 26918, 27101, 27205, 27311,
    27372-3, 27380, 27443-4, 27446, 27497, 27527, 27560, 27727, 27854, 28243

Growth regulators and chloroplast (chromatophore)
    25187, 25333, 25377, 25551, 25995, 27160, 27373, 27443, 27497, 27667, 27716,
    27727, 28243, 28364

Growth regulators and ecosystem productivity and canopy functioning
    25209, 25256, 25510, 25686, 26782, 27444, 27538, 27821, 27990, 28207, 28246,
    28339

Growth regulators and electron transport chain
    25338, 26291, 26766, 26902, 27408, 27442, 27532, 27553, 27562, 28246

Growth regulators and gas exchange
    25256, 25338, 25516, 25642, 26237, 26536, 26824, 27101, 27311, 27380, 27408,
    27560, 27761,

Growth regulators and photorespiration 27101

Growth regulators and resistances to CO_2 and water vapour transfer 26237, 26775,
    27356, 27574, 27575, 28321, B28340

Growth regulators and respiration 26536, 27101

"Growth" respiration see Respiration, "growth" ...

H

H_2 evolution, photoreduction
    25370, 25915-6, 26171, 26544, 26640, 26659, 26677, 26805, 27151-3, 27252,
    27357, 27440, 27571, 28444

H_2 isotopes see Deuterium ...

H^+ transport in chloroplast see Chemiosmotic hypothesis

Halobacterium photosynthesis
    25191, 25296, 25298, 25337, 25351-2, 25369, 25393, 25461, 25547, 25561,
    25803, 25839, 25927, 26180, 26283, 26347, 26365, 26379, 26439, 26611, 26636-7,
    26728, 26819-21, 26850-1, 26963, 27305-7, 27441, 27545, 27599, 27626, 27828-9,
    27836, 27890, 27904-5, 28042-3, 28150, 28293, 28298-9, 28378

Hatch-Slack cycle see C_4 pathway ...

Herbicides see Pesticides, herbicides ...

Heterogeneity of leaf blade (organ) and carbon fixation pathways
    25768, 25987, 26057, 26165, 26441, 26445, 26598, 27008, 27091, 27202, 27361,
    27596, 28351

Heterogeneity of leaf blade (organ) and chlorophyll
    25204, 25310, 25877, 25913, 26099, 26444, 26666-7, 26880, 26994, 27526,
    27581-2, 27622, 28205

Heterogeneity of leaf blade (organ) and chloroplast (chromatophore)
        25428, 25431, 25512, 25530, 26665, 27106, 27157, 27164, 27351, 27621, 28357

Heterogeneity of leaf blade (organ) and ecosystem productivity and canopy functi-
        .oning 26489

Heterogeneity of leaf blade (organ) and electron transport chain
        25768, 25913, 26442-3, 26872, 27552, 27580-1, 27921, 28420

Heterogeneity of leaf blade (organ) and gas exchange 25821, 25987, 26078, 26112,
        26316, 27007, 27389, 28205

Heterogeneity of leaf blade (organ) and photorespiration 25711

Heterogeneity of leaf blade (organ) and resistances to CO_2 and water vapour
        transfer 26993

Heterogeneity of leaf blade (organ) and respiration 27389

Heterotrophy see Carbon metabolism types ...

Hill reaction see Photosystem 2 activity ...

Hill reaction, methods see Photosystem 2 activity, methods

Humidity of air and chlorophyll 25898, 27040

Humidity of air and ecosystem productivity and canopy functioning
        B25775, 25785, 25929, 26920, 27133, 27235, 27656, 28134, 28427

Humidity of air and gas exchange
        25167, 25415, 26227, 26397, 26580, 26646-7, 26905, 27768, 27888, 28133,
        28134, 28143

Humidity of air and resistances to CO_2 and water vapour transfer
        25251, 25604, 25742, 25999, 26015, 26265, 26397, 26532, 26612, 26646, 26865,
        26947, 27133, 27432, 28089, 28125, 28143

Humidity of air, methods (cf. also infra-red gas analyser for water vapour)
        26264, 26577, 26592, 26635, 26708, 27826, 28245

Hydration level of leaf and carbon fixation pathways
        25520, 25772, 26446, 26689, B26846, 26864, 27360, 27363, 27793

Hydration level of leaf and carotenoids 25716, 25905-6, 27855

Hydration level of leaf and chlorophyll
        25173, 25507, 25716, 25905-6, 25991, 26338, 26437, 27163, 27496, 27855,
        28266
Hydration level of leaf and chloroplast (chromatophore) 25906, 26090, 26128, 26141,
        27454, 28266

Hydration level of leaf and ecosystem productivity and canopy functioning
        25173, 25238, 25520-1, 25860, 25885, 25992, 26409, 26437-8, 26484, 26537,
        26759, 26970, 27749, 28456

Hydration level of leaf and electron transport chain 25520, 26007, 26335, 27163,
        27274, 27496

Hydration level of leaf and gas exchange
        25173, 25258, 25283, 25407, 25519, 25520-1, 25549, 25592, B25637, 25642,
        25653, 25755, 25772, 25774, B25775, 25828, 25860, 25902, 25928, 25991, 25993,
        26121, 26296, 26300, 26348, 26409, 26437-8, 26484, 26494, 26497, 26537,
        26580, 26685, 26689, 26759, B26846, 26863, 26970-1, 27133, 27139, 27163,

Hydration level of leaf and gas exchange (continued)
    27302, 27348, 27363, 27409, 27411, 27768, 27782, 27793, 27998, 28064, 28066-
    -7, 28342

Hydration level of leaf and photorespiration 25520, 25592, 25755, 26689, 26863

Hydration level of leaf and resistances to CO_2 and water vapour transfer
    25173, 25283, 25327, 25454, 25572, 25592, 25604, 25742, 25843, 25999, 26114,
    26265, 26438, 26532, 26843, B26846, 26865, 26970-1, 27109, 27289, 27363,
    27433, 27749, 27882, 27897, 27998, 28064, 28067, 28123, 28321, 28341, 28367

Hydration level of leaf and respiration
    25173, 25520-1, 25549, 25592, 25991, 26437, 26863, 26971, 27163

Hydrogen see H_2

Hydrogenase see O_2 evolution mechanism and kinetics; H_2 evolution ...

Hygrometer see Humidity of air, determination; Infra-red gas analyser for water
    vapour

I

Ideotype see Model ...

Immobilization of chloroplasts and photosynthetic systems see Photosystems stabili-
    zation ...

Induction phenomena see Transient phenomena...

Infra-red gas analyser for CO_2
    25666, 25828, 26027, 26335, 26577, 26592, 26635, 26708, 26743, 26745, 27369,
    27421, 27658, 27798, 27807, 27886, 28063, 28245, 28475

Infra-red gas analyser for water vapour 27798

Infra-red radiation, effect on photosynthetic parameters see Irradiance, spectral
    composition ...; Temperature, high ...

Inhibitors of electron transport chain (cf. also Pesticides ...; Antibiotics ...)
    25177, 25183, 25192, 25197, 25206, 25208, 25211, 25222, 25240, 25296, 25302,
    25315-6, 25321-2, 25339-40, 25353, 25362, 25365, 25370, 25375, 25395, 25401,
    25403, 25417, 25425, 25434, 25461, 25470, 25493-4, 25503, 25508, 25513, 25521,
    25589, 25591, 25616, 25621, 25651, 25657-8, 25671-2, 25695, 25699, 25701,
    25712, 25719, 25750, 25765, 25782, 25833, 25836, 25880, 25884, 25892, 25912-
    -3, 25918, 25951, 25964, 26009, 26084, 26099, 26101, 26137, 26152, 26162,
    26167, 26175, 26193-4, 26196, 26210, 26222, 26224, 26230-1, 26254, 26272,
    26322, 26334, 26350, 26361, 26396, 26399, 26411-2, 26422, 26424-5, 26441-3,
    26472-3, 26478, 26507, 26517, 26559, 26561, 26566, 26616, 26686, 26703-4,
    26709, 26718, 26753, 26773, 26780-1, 26792-4, 26859, 26861, 26946, 26962,
    26968, 27031, 27055, 27070-1, 27075, 27086, 27100, 27111, 27125, 27127, 27182,
    27184, 27191, 27249, 27252, 27257, 27264, 27275, 27279, 27349-50, 27353, 27355,
    27367, 27387, 27395, 27403, 27430, 27439-40, 27453, 27466, 27497, 27548, 27550,
    27551, 27571, 27588-9, 27597, 27603, 27610, 27612-3, 27618, 27627, 27637,
    27641, 27655, 27680, 27685, 27703, 27717, 27724, 27746, 27764-5, 27786, 27788,
    27827, 27832, 27840, 27842, 27856, 27862, 27878, 27880, 27907, 27917, 27921,
    27924, 27934-5, 27937, 27944, 27992, 27996, 28027, 28029, 28037, 28080-1,
    28084, 28119, 28121, 28124, 28163, 28206, 28211, 28213, 28218, 28223, 28225,
    28235-6, 28239-40, 28244, 28260, 28333-4, 28349, 28386-7, 28411, 28418, 28420,
    28446, 28449, 28457-8, 28468, 28477, 28490-1

Insertion see Ontogeny ...

Intercellular spaces, CO_2 concentration inside 25261, 25437, 25611, 25780, 26265,
    26600, 27180, 27370, 27533

Intracellular resistance see Resistance, intracellular (mesophyll)

Ionizing radiation (gamma, X, cosmic, *etc.*) and carotenoids 25297, 27885

Ionizing radiation (gamma, X, cosmic, *etc.*) and chlorophyll 25297, 25715, 25864,
    27201, 27429, 27477, 27885, 27902-3

Ionizing radiation (gamma, X, cosmic, *etc.*) and chloroplast (chromatophore)
    27054, 28149

Ionizing radiation (gamma, X, cosmic, *etc.*) and ecosystem productivity and canopy
    functioning 25239, 27821

Ionizing radiation (gamma, X, cosmic, *etc.*) and gas exchange 27429, 27477, 28487

Irradiance, compensation see Compensation irradiance

Irradiance (PhAR) and algae productivity
    25210, 25282, 25426, 25525, 25647, 25694, 25739, 25894, 26051, 26058, 26217,
    26407, 26420, 26516, 26568, 26825, 26849, 26882, 27231, 27281, 27400, 27470,
    27534, 27567, 27648, 27938, 27988, 28017, 28147, 28421, 28472

Irradiance (PhAR) and biliproteins 25867, 25869, 26168, 26608, 27568, 27901

Irradiance (PhAR) and carbon fixation pathways
    25213, 25221, 25261, 25431, 25529, 25672, 25875, 25913, 26251, 26526, 26598,
    26624, 26760, 26858, 26888, 26940, 27182, 27247, 27272, 27524, 27709, 27823,
    27901, 27911, 27928, 28116, 28405

Irradiance (PhAR) and carotenoids
    25553, 25696, 25969, 26168, 26707, 26740, 26975, 27066, 27388, 27511, 27586,
    27677, 27711, 27805, 27823, 27850, 27864, 27946, 28217, 28239-40, 28357

Irradiance (PhAR) and chlorophyll
    25186, 25237, 25248, 25268-9, 25287, 25427, 25440, 25577, 25596, 25696,
    25748, 25764, 25852, 25867, 25869, 25887, 25898, 25900, 26070-1, 26094,
    26099, 26168, 26192, 26215, 26310, 26338, 26399, 26481, 26486, 26608, 26617,
    26706-7, 26727, 26737, 26842, 26890, 26910, 26930, 26937, 26940, 26976,
    27077, 27216, 27272, 27287, 27291, 27368, 27388, 27407, 27497, 27500, 27511,
    27520, 27536, 27567-8, 27622, 27659, 27677-8, 27703, 27718, 27795, 27805,
    27823, 27850, 27854, 27857, 27864, 27885, 27901, 27997, 28017, 28156, 28166,
    28173, 28217, 28221-2, 28239-40, 28357

Irradiance (PhAR) and chloroplast (chromatophore)
    25237, 25427, 25529-30, 25802, 26103, 26279, 26323, 26435, 26715, 26769,
    27105, 27176, 27351, 27407, 27497, 27576, 27729, 27756-7, 27784, 28166,
    28217, 28221, 28274

Irradiance (PhAR) and ecosystem productivity and canopy functioning
    25190, 25198, 25209, 25231, 25264, 25291, 25484, 25659, 25681, 25772, 25781,
    25785, 25857, 26213, 26267, 26274, 26308, 26384-5, 26398, 26719, 26769,
    26798, 26857, 26906, 27216, 27325, 27360, 27540, 27687, 27701, 27812, 28031,
    28053, 28138, 28177, 28337-8, 28350, 28470

Irradiance (PhAR) and electron transport chain
    25183, 25212, 25335, 25399, 25417, 25494, 25528, 25589, 25632, 25668, 25696,
    25700, 25918, 26146, 26310, 26335, 26399, 26424, 26486, 26703, 26766, 26872,
    26910, 26925, 27093, 27104, 27126, 27184, 27279, 27332, 27354-5, 27403, 27496,
    27504, 27590, 27659, 27699, 27856-7, 27921, 27977, 28002, 28221, 28297, 28332

Irradiance (PhAR) and gas exchange
        25167-8, 25210, 25215, 25221, 25248, 25255, 25272, 25283, 25291, 25367, 25381-
        -2, 25414-5, 25419, 25426, 25431, 25437, 25476, 25521, 25530, 25536, 25549,
        25559, 25578, 25596, 25609-10, 25615, 25618, 25620, 25653, 25666, 25674,
        25684, 25772, 25801, 25806, 25824-5, 25869, 25873, 25875, 25894, 25902, 25922,
        25924, 25954, 25996, 26002, 26009, 26037, 26045-6, 26065, 26078-9, 26085-6,
        26108, 26154, 26164, 26188, 26227, 26240, 26300, 26348, 26368, 26370, 26384-5,
        26389, 26397, 26402, 26420, 26426, 26472, 26475, 26506, 26515-6, 26529, 26533,
        26544, 26593, 26646, 26662, 26698, 26708, 26719, 26746, 26775, 26785, 26796,
        26798, 26835, 26872, 26888, 26905, 26941, 26947, 26970-2, 27035-6, 27095,
        27107, 27139, 27149, 27156, 27163, 27180, 27182, 27185, 27187, 27216, 27242,
        27271-2, 27302, 27320, 27348, 27357, 27370, 27382, 27388, 27414, 27438, 27458,
        27463, 27511, 27521, 27534, 27567, 27595, 27602, 27635, 27643, 27672, 27681,
        27688, 27700, 27750, 27768, 27775, 27782, 27792, 27805, 27818, 27823, 27833,
        27887, 27911, B27915, 27930, 27988, 27995, 28024, 28067, 28076, 28085, 28087,
        28090, 28093, 28102, 28106, 28128, 28140, 28193-5, 28197, 28206, 28221, 28230,
        28300, 28303, 28317, 28350, 28372, 28380, 28413, 28421, 28426, 28462, 28478

Irradiance (PhAR) and gas exchange, analysis of light curves
        25186, 25195, 25224, 25294, 25367, 25476, 25618, 25642, 25674, 26001, 26078,
        26085, 26116, 26164, 26300, 26370, 26426, 26476, 26661, 26718, 26720, 26972,
        27035, 27069, 27139, 27180, 27210, 27227, 27270, 27317, 27348, 27399, 27422,
        27469, 27490, 27630, 27728, 27768, 27920, 27980, 27995, 28077, 28087, 28102,
        28106-7, 28110, 28192-3, 28229-30, 28247, 28287, 28300, 28337, 28413

Irradiance (PhAR) and photorespiration 25653, 26037, 26785, 27493, 27754, 27901,
        27929, 28353

Irradiance (PhAR) and resistances to CO_2 and water vapour transfer
        25327, 25430, 25604, 25620, 25653, 25843, 25947, 25999, 26065, 26237, 26240,
        26389, 26514, 26843, 26865, 26947, 26970-1, 27180, 27187, 27329, 27360, 27432,
        27595, 27939, 28067, 28125, 28341, 28350, 28367, 28462

Irradiance (PhAR) and respiration   25653, 26046, 26299, 26578, 26744, 26746, 27687,
        27873

Irradiance (PhAR, total) measurement
        25552, 25676-7, 26000, 26177, 26564, 27080, 27096, 27770, 28051-2, 28188, 28208,
        28210, 28233

Irradiance, spectral composition and algae productivity 25390, 25534, 25849, 25894,
        26217, 27144, 28147

Irradiance, spectral composition and biliproteins   25867, 26168, 27177-8, 27568

Irradiance, spectral composition and carbon fixation pathways 25515, 26077, 28009,
        28285

Irradiance, spectral composition and carotenoids 25778, 26168, 26491, 26779, 26855,
        27239, 27513, 27848

Irradiance, spectral composition and chlorophyll
        25208, 25220, 25268-9, 25688, 25829, 25867, 25892, 26168, 26285-6, 26415-6,
        26459, 26520, 26588, 26603-4, 26779, 26813, 26855, 26935-6, 27041, 27064,
        27078, 27177-8, 27238-9, 27260, 27499, 27513, 27568, 27848, 27854, 27899,
        27947, 27997, 28218, 28476

Irradiance, spectral composition and chloroplast (chromatophore)
        25289, 25373, 25530, 26088, 26321-3, 26346, 26520, 27211, 27499, 27530, 27657,
        27757, 27848, 28029, 28285

Irradiance, spectral composition and ecosystem productivity and canopy functioning
        25792, 26034, 26284-6, 26389, 26466, 26515, 27229, 28228, 28343-4

Irradiance, spectral composition and electron transport chain
        25202, 25335, 25379, 25401, 25533, 25589, 25782, 25829-30, 25869, 25892-3,
        26255, 26258, 26411-2, 26424, 26480, 26703, 26720, 26756, 26921, 26955,
        27163, 27275, 27279, 27324, 27659-60, 27724, 27764, 27809, 27901, 27916,
        27950, 27955, 28285, 28329, 28334, 28395

Irradiance, spectral composition and gas exchange
        25329, 25332, 25789, 25806, 25825, 25829, 26284, 26389, 26620, 26718, 26796,
        26955, 27041, 27178, 27229, 27475, 27513, 27848, 27899, 28228, 28286-7,
        28349, 28449

Irradiance, spectral composition and photorespiration 25778, 26620

Irradiance, spectral composition and resistances to CO_2 and water vapour transfer
        26389

Irradiance, spectral composition and respiration 26756, 27873, 27889

Irradiation, flash, and algae productivity 28172

Irradiation, flash, and carotenoids 28029, 27071

Irradiation, flash, and chlorophyll
        25544, 25630, 25645, 25840, 25861, 25988, 26459, 26481-2, 26545, 27071,
        27452, 27867, 28028-9, 28172-3, 28258

Irradiation, flash, and chloroplast (chromatophore) 25988, 26324, 27070-1, 28027,
        28189

Irradiation, flash, and electron transport chain
        25240, 25836, 25840, 26056, 26196, 26211, 26476, 26482, 26861, 27070-1, 27173,
        27447, 27518, 27529, 28028-9, 28083, 28189, 28332-3, 28409, 28473

Irradiation, flash, and gas exchange 25294, 25915, 26476, 26744, 26746, 27399, 28115,
        28171, 28334

Irradiation, flash, and photorespiration 26746

Irradiation, flash, and respiration 26744, 26746

Irradiation, illumination equipment and systems 28475

Irrigation and chlorophyll 25473-4, 25948, 27579, 28312

Irrigation and ecosystem productivity and canopy functioning
        25473, 25725, B25775, 25827, 25948, 26030, 26263, 26723, 26827, 26847, 27579,
        27710, 27748, 27793, B27915, 28099-100, 28169, 28196

Irrigation and gas exchange 25419, 27768, 28167, 28169

Irrigation and resistances to CO_2 and water vapour transfer 25971, 26282

K

Kok effect 26116

L

Laboratory for photosynthesis studies, mobile (field laboratory) 25828

Leaf anatomy (*cf.* also Leaf thickness)
        25264, 25431, 25445, 25512, B25603, 25612, 25754, 25767, 25813, 25878, 25924,
        25986, 26078, 26086, 26116, 26243, 26317, 26319, 26428, 26446, 26550, 26628,
        26776, 26816-7, 26872, 26951, 27012, 27060-1, 27141, 27158, 27270, 27361,
        27364, 27431, 27433, 27506, 27585, 27928, 27930, 28114, 28179-80, 28181,
        28195, 28205, 28288-9, 28346, 28400, B28482

Leaf and plant development and ageing, morphology (*cf.* also Ontogeny ...)
        25737, 25832, 25857, 25955, 25986, 26213, 27108, 27213, 27981

Leaf area duration see Growth analysis, specific leaf area ...

Leaf area index see Growth analysis, specific leaf area ...

Leaf area measurement
        25250, 25643, 25702, 26133, 26245, 26318, 26625, B26846, 27102, 27296, 27799,
        28044

Leaf area ratio see Growth analysis, net assimilation rate ...

Leaf chamber see Assimilation chamber ...

Leaf dimensions see Extension growth, leaf dimensions

Leaf epidermis, anatomy
        25194, 25251, 25406, 25430, 25572, 25653, 25755, 25881, 25924, 25947, 25971,
        25998, 26075, 26090-1, 26229, 26336, 26400, 27012, 27158, 27271, 27433, 27574,
        27594-5, 27609, 27897, 28093, 28123

Leaf epidermis, stomata (*cf.* also Amphistomatous leaf, gas exchange in) 25205, 28182

Leaf life span, plastochron index 25754, 25832, 25955, 26073, 26904, 26906, 26931,
        28134, 28400

Leaf morphology
        25189, 25206, 25238, 25250, 25481, 25521, 25530, 25546, 25643, 25654, 25754,
        25955, 25986, 26050, 26055, 26073, 26086, 26093, 26123, 26197, 26318, 26361,
        26373, 26465, 26512, 26533, B26551, 26688, 26788, 26812, 26904, 26906, 26927,
        27003, 27107, 27142, 27172, 27207, 27216, 27229, 27272, 27386, 27430, 27595,
        27896, 28134, 28144, 28169, 28181, 28199, 28400, 28413, 28426

Leaf movements 27820, 28051

Leaf optical properties (*cf.* also Carotenoids absorption spectra *in vivo*; Chlorophyll
        absorption spectra *in vivo*)
        25176, 25329, 25560, 25922, 25924, 26117, 26285-6, 26476, 26483, 26515, 26898,
        27036, 27370, 27388, 27460, 27630, 27728, 27981, 28141, 28188, 28197

Leaf resistance see Resistance for water vapour ...; Resistance, stomatal ...

Leaf, sun- and shade leaf see Leaf anatomy

Leaf temperature (methods and results)
        25326, 25367, 25450, 25885, 26114, 26265, 26375, 26397, 26474, 26708, 26854,
        26898, 26941, 26970, 27107, 27154, 27180, 27322-3, 27826, 28085, 28087

Leaf temperature measurement see Leaf temperature ...

Leaf thickness 25404, 25406, 25654, 25902, 27012, 27270, 27386, 27410, 27488-9

Mutants, chlorophyll in
      25184-5, 25378, 25410, 25427, 25456, 25504, 25890, 25715, 25734, 25749,
      25791, 25838, 25864, 25900, 26053, 26094, 26362, 26395, 26456, 26562, 26597,
      26632-3, 26700, 26755-6, 26768, 26774, 26829-30, 26899, 26990, 27137, 27211,
      27246, 27272, 27346, 27505, 27690, 27718, 27751, 27809, 27834-5, 27879,
      27884, 27902-3, 27924, 27973-4, 28008, 28148, 28216, 28218, 28255, 28260-1

Mutants, chloroplast (chromatophore) in
      25184, 25378, 25427, 25463, 25497, 25504, 25726, 25749, 26094, 26144, 26260,
      26362, 26597, 26649, 26699-700, 26768, 26774, 26831, 26990, 26995, 27211,
      27246, 27268, 27351-2, 27467, 27576, 27614, 27668, 27694, 27884, 27924,
      27962, 27974, 28373

Mutants, ecosystem productivity and canopy functioning of 25827, 27991

Mutants, electron transport chain in
      25184, 25335, 25370, 25456, 26146, 26362, 26456, 26557, 26632-3, 26773-4,
      26793, 26830, 26887, 26990, 27086, 27132, 27137, 27246, 27668, 27681, 27809,
      27834, 27924, 28218, 28469

Mutants, gas exchange in
      25184-5, 25335, 26362, 26597, 26632-3, 26700, 26829, 27132, 27187, 27210,
      27247, 27272, 27357, 27924

Mutants, photorespiration in 28477

Mutants, photosynthetic, isolation and selection 25378
Mutants, resistances to CO_2 and water vapour transfer in 27187

N

N_2, anaerobic atmosphere, and carbon fixation pathways 26866

N_2, anaerobic atmosphere, and chlorophyll 26736

NAD see NADP, NAD

NADP, NAD 25184, 25211, 25320, 25506, 25564, 25675, 25862, 25892, 25940, 25942,
      25983-4, 26291-2, 26380, 26486, 26574, 26677, 26718, 26722, 26765, 26767,
      26872, 26900, 26902, 26921, 27020, 27064, 27104, 27203, 27297, 27345, 27550,
      27580-1, 27603, 27610, 27628, 27681, 27772, 27776, 27841, 27856-7, 27901,
      27979, 28105, 28184, 28218, 28286, 28297, 28304, 28306-7, 28410

NADP, NAD, methods 27841

Net assimilation rate see Growth analysis, net assimilation rate ...

Net photosynthetic rate see Gas exchange ...

Nitrogen see N_2 ...; Mineral elements (N, P, K) ...

NMR see EPR, NMR

Nuclear magnetic resonance see EPR, NMR

Nucleic acids see Proteins, amino acids, nucleic acids ...

O

O_2 and carbon fixation pathways
25216, 25708, 25711, 25787, 25868, 25912, 26251, 26619, 26668, 26959, 27026, 27182, 27247, 27494, 27639, 27760, 27928, 28268-9

O_2 and carotenoids 25798, 27711, 28455

O_2 and chlorophyll 25621, 25798, 26727, 26737-8, 27494, 28240, 28379, 28455, 28476

O_2 and chloroplast (chromatophore) 25621

O_2 and ecosystem productivity and canopy functioning
25568, 26155, 26403, 26648, 26784, 27290, 27413, 27494, 27502, 27539-40

O_2 and electron transport chain 25242, 25401, 26146, 26333, 27064

O_2 and gas exchange
25249, 25346, 25395, 25448, 25457, 25568, 25592, 25684, 25708-12, 25767, 25776, 25883, 25913, 25923, 26091, 26110, 26155, 26188, 26214, 26403, 26566-7, 26619, 26775, 26785, 26835, 26858, 26863, 26880, 27156, 27182, 27289, 27370, 27424-5, 27494, 27502, 27548, 27760, 27951, 28092, 28205, 28229-30, 28268-9, 28379, 28393, 28477

O_2 and photorespiration
25448, 25776, 26427-8, 26626, 26629, 26880, 27493, 27754, 27760, 27929, 28269, 28479

O_2 and resistances to CO_2 and water vapour transfer 25604, 26091, 26403, 27187

O_2 and respiration 25710, 25883, B28482

O_2 determination in water reservoirs see Algae, primary productivity, methods

O_2 determination (other than O_2 electrode) 25480, 25823, 26064, 27123, 28205, 28415, 28475

O_2 electrode 26009, 26381, 26544, 26577, 26661, 26924, 27117, 27146, 27773, 28205, 28349, 28475

O_2 evolution mechanism and kinetic
25180, 25189, 25242-4, 25276, 25322, 25368, 25379, 25384, 25432, 25466, 25616, 25626-7, 25695, 25836, 25840, 26056, 26267, 26325, 26480, 26487, 26659, 26718, 26720, 26765, 26767, 26859-61, 26942-3, 27122, 27332, 27375, 27512, 27548, 27556, 27603, 27617, 27655, 27763, 27876, 28027, 28029, 28083, 28213, 28330, 28334, 28437-9, 28468

O_2 exchange see Gas exchange...

O_2 isotopes, use in photosynthesis measurement 26135, 27122, 27991

Ontogeny of algae and algae productivity 27652

Ontogeny of algae and biliproteins 27177, 27257

Ontogeny of algae and carbon fixation pathways 25280

Ontogeny of algae and carotenoids 25648, 26855

Ontogeny of algae and chlorophyll
25621, 25648, 25853, 26097, 26794-5, 26807, 26855, 26957, 27077, 27177, 27257, 27652, 27796

Osmotically active substances and gas exchange 26227, 26344-5, 26524, 26669, 27036

Osmotically active substances and photorespiration 26227

Osmotically active substances and resistances to CO_2 and water vapour transfer
          26524

Oxygen see O_2

Ozone see Pollution of air ...

P

P680   25276, 26156, 26422, 26559, 26694, 27100, 27571, 27863, 28334

P700, P750, P890, *etc.*
          25166, 25180, 25184, 25186, 25202, 25217, 25281, 25293, 25295, 25315, 25323,
          25347-8, 25379, 25385-6, 25470, 25488, 25517, 25570, 25606, 25629, 25668-9,
          25773, 25819, 25846, 25884, 25888, 25918, 25957, 25963, 25968, 26041-2, 26048,
          26084, 26172-3, 26223, 26257-8, 26325, 26380, 26423, 26457, 26468, 26610,
          26616, 26632-3, 26687, 26713, 26718, 26720, 26731, 26734, 26761, 26794,
          26830, 26942-3, 26974, 26990, 27031, 27033, 27050, 27072, 27086, 27099, 27173,
          27203, 27222, 27254, 27273-4, 27340, 27449, 27518, 27525, 27536, 27581, 27653,
          27668, 27681, 27724, 27804, 27843, 27856, 27863, 27866-71, 27868, 27917,
          27979, 28021, 28028, 28036, 28105, 28127, 28131, 28135-6, 28237, 28258-9,
          28332-5, 28334, 28409, 28420

Paramagnetic oxygen analyser see O_2 determination...

Paramagnetic resonance see EPR, NMR

PEP carboxylase (PEPC) see Phosphoenolpyruvate carboxylase

Peroxisome, glyoxysome, microbody
          25443, 25512, 25542, 25878, 25940, 26218, 26574, 27121, 27754, 27759, 27975-6,
          28154, 28267, 28422

Pesticides see also Inhibitors of electron transport chain

Pesticides, herbicides and carbon fixation pathways 26513, 26889

Pesticides, herbicides and carotenoids 25852, 26601, 26962, 27184, 27387, 27677,
          28012, 28217, 28240

Pesticides, herbicides and chlorophyll
          25162, 25852, 26123, 26167, 26601, 26962, 27159, 27387, 27397, 27497, 27637,
          27677, 27762, 28012, 28159, 28217-8, 28240, 28326

Pesticides, herbicides and chloroplast (chromatophore) 25545, 25852, 26601, 26613,
          27184, 27497, 27690

Pesticides, herbicides and ecosystem productivity and canopy functioning 25190, 25489,
          26123, 27145, 27338, 27397, 27918, 28254

Pesticides, herbicides and electron transport chain
          25459, 26137, 26350, 26889, 27132, 27184, 27338, 27387, 27397, 27549, 27996,
          28254

Pesticides, herbicides and gas exchange
          25190, 25349, 25395, 25418, 25459, 25476, 25489, 25658, 25744, 25752, 26013-4,
          26035, 26123, 26167, 26281, 26399, 26513, 26824, 26889, 26962, 27141, 27349,

Photoperiod and chlorophyll 25596, 26377, 27810

Photoperiod and chloroplast (chromatophore) 27714

Photoperiod and ecosystem productivity and canopy functioning
        25231, 25596, 26426, 27283, 27428, 27673, 27719, 27810, 28041, 28145, 28352

Photoperiod and gas exchange 25332, 26386, 26426

Photophosphorylation, cyclic
        25222, 25240, 25365, 25399, 25425, 25493, 25508, 25520, 25846, 25891, 26148,
        26193, B26220, 26268, 26335, 26442-3, 26472, 26556, 26566, 26711, 26718,
        26767, 26852, 26872, 26889, 26967, 27060, 27114, 27125, 27365, 27403, 27485,
        27535, 27550, 27552-3, 27580, 27588-91, 27603, 27637, 27668, 27744, 27809,
        27856, 27901, 28002, 28112, 28164, 28203, 28225, 28235, 28416, 28447, 28452,
        28457-8

Photophosphorylation in photosynthetic bacteria see Photosynthetic bacteria, pho-
        tophosphorylation

Photophosphorylation mechanism see Chemiosmotic hypothesis ...

Photophosphorylation, methods 27093, 27125, 27454-5, 28002, 28108, 28452

Photophosphorylation, model see Model ...

Photophosphorylation, non-cyclic
        25177, 25182-3, 25197, 25362, 25365-6, 25375, 25377, 25387, 25399, 25401,
        25417, 25425, 25487, 25493-4, 25508, 25520, 25706, 25789, 25829, 25846, 25891,
        25984-5, 26148, 26152, 26175, 26193, 26210, B26220, 26230, 26268, 26366,
        26418, 26423, 26473, 26496, 26500, 26507, 26566, 26615, 26711, 26718, 26720,
        26726, 26730, 26767, 26780, 26790, 26852, 26862, 26872, 26902, 26967-8, 26980,
        26984, 27010, 27017-8, 27047, 27125, 27131, 27184, 27194, 27204, 27243-4,
        27335, 27354-5, 27442, 27471, 27485-6, 27491, 27523, 27532, 27550, 27552,
        27559, 27562, 27580, 27590-1, 27640, 27661, 27676, 27744-5, 27747, 27786-7,
        27796, 27815-6, 27856-7, 27901, 27910, 27934, 27992, 27996, 28105, 28112,
        28164-5, 28211, 28227, 28235, 28246, 28306-7, 28376, 28410-1, 28416, 28445,
        28452, 28458

Photophosphorylation, pseudo-cyclic 25493, B26220, 26268, 26443, 26566, 26780

Photoreduction see H_2 evolution ...

Photorespiration enzymes
        25542, 26004, 26162, 26218-9, 26402, 26478, 26784, 26786, 26858, 27359, 27364,
        27416, 27569, 27754, 27778, 28151-4, 28183, 28220, 28229, 28267, 28342, 28478

Photorespiration, metabolic cycles
        25249, 25262, 25340, 25542, 25573, 25602, 25619, 25640, 25711-2, 25767-8,
        25940, 26038, 26162, 26218, 26272, 26314, 26402, 26574, 26595, 26620, 26623,
        26659, 26668, 26689, 26691, 26722, 26767, 26810, 26858, 26872, 26959, 26996,
        27247, 27416, 27462, 27493-4, 27524, 27754, 27778, 27797, 27989, 28154, 28230,
        28422, 28477-80

Photorespiration rate
        25167-8, 25366, 25448, 25476, 25609, 25640, 25858, 25883, 26055, 26155,
        26214, 26317, 26626, 26988, 27106, 27313, 27414, 27424, 27469, 27521, 27524,
        27760, 27837, 27989, 28112, 28153, 28268, 28386-7, 28393, 28477

Photosynthate translocation and distribution
        25170-1, 25190, 25193, 25206-7, 25226, 25229, 25256, 25299, 25324, 25328,
        25397, 25405, 25411, 25489, 25519, 25527, 25568, 25593, 25642, 25666, 25682,
        25686, 25722-3, 25727, 25740, 25765, 25774, B25775, 25799, 25801, 25806,
        25856, 25901, 25929, 25946, 25965-7, 26006, 26010, 26032, 26039, 26093, 26109,

Photosynthate translocation and distribution (continued)
        26121-2, 26135-6, 26150, 26161, 26169, 26190, 26221, 26320, 26349, 26369,
        26384-6, 26389-90, 26429, 26433-4, 26451-2, 26489, 26493, 26498, 26522,
        26569, 26596, 26607, 26623, 26742, 26778, 26784, B26815, 26816-7, 26845,
        26847, 26856, 26876, 26883, 26926, 26954, 26965, 26979, 26983-4, 27011, 27016,
        27055, 27098, 27199, 27210, 27215, 27240, 27255, 27290, 27331, 27338, 27343-
        -4, 27358, 27361, 27384, 27386, 27389, 27393, 27412, 27430, 27438, 27445,
        27477-8, 27502, 27514-5, 27540, 27573, 27585, 27593, 27600, 27624, 27687,
        27688, 27722, 27725, 27752-3, 27812, 27814, 27821, 27887, 27918, 27930, 27932,
        27942, 27990-1, 27995, 27999, 28005-6, B28015, 28018, 28020, 28023, 28053,
        28062, 28130, 28177, 28246, 28254, 28269, 28296, 28301-2, 28304, 28322, 28336-
        -9, B28340, 28343, 28374, 28382, 28394, 28425, 28435, B28482, 28495

Photosynthates and intermediates of carbon fixation pathways
        25188, 25249, 25280, 25331, 25338, 25358, 25373-4, 25449, 25515, 25584,
        25619, 25642, 25678, 25682, 25766, 25768, 25796, 25831, 25863, 25866-9,
        25912, 25954, 26005-6, 26018, 26052, 26181, 26206, 26244, 26333, 26343,
        26398, 26402, 26492, 26504, 26549-50, B26551, 26574-5, 26595-7, 26623, 26627-
        -8, 26668, 26727, 26760, 26767, 26770-2, 26864, 26866, 26888, 26911, 26996,
        27007, 27010, 27021, 27026, 27236, 27247, 27255, 27279, 27292, 27361, 27384,
        27399, 27414, 27530, 27621, 27639, 27705, 27797, 27862, 27901, 27980, 28116,
        28130, 28269, 28303, 28351, 28357

Photosynthates and intermediates of carbon fixation pathways and gas exchange
        25258, 26021, 26121, 27171, 27215, 27908, 28003

Photosynthetic bacteria carbon fixation pathways   25478, 25582

Photosynthetic bacteria carotenoids   see   Xanthophylls of photosynthetic bacteria

Photosynthetic bacteria chlorophylls   see   Bacteriochlorophyll; Chlorophylls, *Chloro-
        bium*

Photosynthetic bacteria chromatophores   see   Chloroplast and chromatophore ...; Chro-
        matophores ...

Photosynthetic bacteria electron transport chain
        25302, 25656, 25669, 25746, 25773, 25855, 25884, 25953, 25964, 26423, 26456,
        26486, 26695-6, 26720, 26808, 26902, 26942-3, 27068, 27282, 27326, 27340,
        27447-9, 27516-9, 27536, 27615, 27653, 27731, 27841, 27945, 28103, 28136,
        28163, 28237, 28259, 28376, 28469

Photosynthetic bacteria gas exchange   25249, 26727, 28484

Photosynthetic bacteria photophosphorylation
        25645, 25721, 25804, 25914, 26146, 26222, 26233, 26266, 26439, 26486, 26631,
        26969, 27374, 27447-9, 27455, 27504, 27535, 27537, 27627, 27806, 27916,
        28163, 28451

Photosynthetic bacteria reaction centres   see *P*700 ...

Photosynthetic (chlorophyll) unit
        25184-6, 25240, 25263, 25315, 25335, 25353, 25616, 25631, 25751, 25846, 26060,
        B26220, 26257, 26353-4, 26655, 26718, 26943, 26990, 27128, 27378, 27399,
        27511, 27581, 27953, 28055-6, 28488

Photosystem 1
        25166, 25174, 25180, 25182-4, 25186, 25196, 25202, 25214, 25217, 25240-1,
        25244-6, 25293, 25295, 25304, 25315, 25320, 25339, 25344, 25347-8, 25362,
        25370, 25375, 25379, 25385-7, 25401, 25403, 25425, 25438, 25456, 25470, 25477,
        25490, 25504, 25589, 25635-6, 25762, 25782, 25819, 25853, 25879, 25888, 25893,
        25900, 25918, 25942, 25963, 25968, 26040, 26056, 26084, 26148, 26172-4,
        B26220, 26223-4, 26253, 26255, 26257-8, 26268, 26288, 26294, 26298, 26310,
        26334, 26362, 26367, 26378, 26411-2, 26423-4, 26457, 26544, 26556-7, 26575,

Quantum yield and requirement (continued)
      25741, 25751, 25836, 25923, 26001, 26040, 26211, B26220, 26224, 26267, 26294,
      26332, 26423, 26472, 26476, 26655, 26718, 26720, 26780, 26794, 26859, 26861,
      27045, 27075, 27163, 27173, 27537, 27617, 27663, 27796, 27970, 27977, 28008,
      28056, 28173, 28328, 28376, 28395, 28473

Quantum yield and requirement, methods see Quantum yield and requirement

Quinones in photosynthesis
      25293, 25321, 25339, 25362, 25370, 25387, 25392, 25401, 25416, 25470, 25616,
      25773, 25846, 25953, 25962, 25964, 26147, 26196, B26220, 26231, 26257-8,
      26262, 26376, 26410, 26442-3, 26459, 26556, 26566, 26610, 26615, 26751,
      26766, 26790, 26887, 26942-3, 27070, 27100, 27309, 27326, 27328, 27345, 27447,
      27447-9, 27529, 27580, 27617, 27653, 27668, 27809, 27825, 27868, 27878, 28155,
      28376, 28409, 28423

Quinones, methods 25392

R

Radiation in canopy see Canopy, radiation ...

Radiation, light see Irradiance ...

Rain, precipitation, methods see Aerodynamic methods ...

Reaction centres see $P680$; $P700$ ....

Recycling of CO_2 inside the cell and leaf 25619, 26618, 27402, 28474

Relative growth rate see Growth analysis, net assimilation rate ...

Relative water content see Water saturation deficit

Resistance, carboxylation and excitation   25620, 25684, 27899

Resistance, cuticular 25521, 25998, 26541, 28367

Resistance, intracellular (mesophyll)
      25419, 25620, 25654, 25873, 25881, 26065, 26091, 26154, 26163-4, 26199,
      26227, 26403, 26438, 26515, 26524, 26532, 26646, 26775-6, 26785, 26863,
      26971, B27046, 27171, 27215, 27270, 27363, 27422, 27595, 27700, 27749, 28102,
      28194, B28340, 28368, 28391, 28428, 28463

Resistance, leaf boundary layer
      25283, 25419, 25451, 25620, 26015, 26114, 26154, 26199-200, 26239, 26307,
      26397, 26515, 26532, 26541, 26843, 26865, 26947, B27046, 27269, 27401, 27609,
      27728, 28087, 28209, 28257, 28366, 28369-70, 28428, 28463

Resistance, stomatal (and intercellular) (*cf*. also Resistances for water vapour ...)
      25167, 25173, 25195, 25251, 25261, 25271, 25283, 25309, 25326-7, 25332,
      25364, 25407, 25419, 25430, 25454, 25514, 25521, 25537, 25571-2, 25592,
      25620, 25642, 25654, 25742, 25755, 25813, 25832, 25842-3, 25873, 25881, 25947,
      26015, 26065, 26091, 26116, 26124, 26154, 26164, 26199, 26204, 26227, 26235,
      26239, 26264, 26282, 26336, 26389, 26392, 26403, 26421, 26436, 26438, 26514-
      5, 26532, 26535, 26579, 26612, 26646, 26684, 26739, 26775-6, 26785, 26843,
      26848, 26863, 26865, 26941, 26947, 26971, 26993, B27046, 27073, 27109, 27133,
      27148, 27171, 27180, 27187, 27269-70, 27278, 27280, 27356, 27360, 27363,
      27401, 27410, 27573-4, 27578, 27594-5, 27630, 27636, 27692, 27749, 27839,
      27846-7, 27865, 27882, 27897, 27899, 28064-5, 28067, 28089, 28123, 28143-4,

Ribulose 1,5-bisphosphate carboxylase (continued)
        27797, 27800, 27823, 27825, 27896, 27906, 27911, 27928, 27948, 27966, 27995,
        28009, 28032-3, 28071-2, 28079, 28083, 28102, 28112, 28121, 28154, 28170,
        28183, 28220, 28229, 28248, 28268, 28285, 28303, 28305, 28357, 28386-7,
        28398-9, 28430, 28477-8, 28480

Ribulose 1,5-bisphosphate carboxylase, methods
        25496, 25934, 26208-9, 26405, 26546, 26560, 26665, 27263, 27503, 27531, 27669,
        27966, 28009, 28032, 28071-2

Ribulose 1,5-bisphosphate oxygenase see Ribulose 1,5-bisphosphate carboxylase ...;
        Photorespiration enzymes

Root removal see Defoliation ...

Root, underground part, and chlorophyll 27162

Root, underground part, and ecosystem productivity and canopy functioning
        26017, 26349, 26498, 26723, 27707, 27814

Root, underground part, and gas exchange 26498

Root, underground part, and resistances to CO_2 and water vapour transfer 26421

Root, underground part, and respiration 25666, 27011

Rooted leaves, gas exchange in 27331

RuBP carboxylase, RuBPC see Ribulose 1,5-bisphosphate carboxylase

Rubredoxin see Ferredoxin ...

S

Saccharides and carbon fixation pathways
        25188, 25374, 25449, 25512, 25620, 25685, 25727, 26206, 26244, 26251, 26343,
        26360, 26513, 26525, 26770, 26777, 26907, 27091, 27101, 27284, 27509, 27589,
        27790, 28116, 28477

Saccharides and carotenoids 28240

Saccharides and chlorophyll 26292, 28240

Saccharides and chloroplast (chromatophore) 25621, 26095, 26141, 26276, 26795,
        27268, 27742

Saccharides and ecosystem productivity and canopy functioning 27430, 27918

Saccharides and electron transport chain 26292, 27284, 27589-90

Saccharides and gas exchange 26513, 26642, 27258, 27641, 27840

Saccharides and photorespiration 27284, 28477

Salinity of soil and algae productivity 26344

Salinity of soil and carbon fixation pathways 26446, 26550, B26551, 27255, 28404

Salinity of soil and chlorophyll 25236, 25510, 26550, B26551, 26939, 28268

Salinity of soil and chloroplast (chromatophore) 28266

Salinity of soil and ecosystem productivity and canopy functioning
    25299, 25510, 25526, 26437, 26452, 26685, 26939

Salinity of soil and electron transport chain 27255

Salinity of soil and gas exchange
    26344, 26669, 26685, 26939, 27198, 27321, 28078, 28128, 28200

Salinity of soil and resistances to CO_2 and water vapour transfer 26282

Salinity of soil and respiration 26669

Samples for pigment determination see Pigments ...

Seasonal changes see also Ontogeny of plant ...

Seasonal changes in algae productivity
    25210, 25218, 25270, 25286, 25390, 25426, 25455, 25483, 25633, 25647, 25694,
    25739, 25743, 25841, 25847, 25849, 26002, 26058, 26089, 26108, 26179, 26270,
    26330, 26371, 26407, 26417, 26420, 26458, 26540, 26568, 26570, 26663, 26741,
    26752, 26823, 26849, 26933, 27103, 27143, 27219, 27225-7, 27231, 27250, 27266,
    27281, 27298, 27459, 27470, 27476, 27642, 27662, 27785, 27964, 27970, 27987,
    28017, 28088, 28128, 28129, 28162, 28174, 28190, 28394, 28412, 28464-5, 28472

Seasonal changes in biliproteins 27177-8

Seasonal changes in carbon fixation pathways
    25685, 26068, 26078, 26991, 26996, 27053, 28049, 28059, 28064-6, 28102

Seasonal changes in carotenoids
    25218, 25423, 25871, 25897, 26131, 26707, 26800, 27021, 27066, 27388, 27405,
    27926, 28073, 28188, 28464

Seasonal changes in chlorophyll
    25210, 25218, 25270, 25390, 25422-3, 25507, 25644, 25679, 25685, 25743, 25822,
    25871-2, 25877, 25896, 25898, 25948, 25956, 26131, 26469-70, 26529, 26552,
    26617, 26707, 26800, 26849, 26857, 26918, 26998, 27021, 27068, 27124, 27162,
    27177-8, 27216, 27227, 27265, 27388, 27405, 27484, 27488, 27642, 27666, 27733,
    27766, 27885, 27926, 27971, 27980, 28017, 28073, 28117, 28156, 28174, 28185,
    28318, 28322, 28464

Seasonal changes in chloroplast (chromatophore) 25821

Seasonal changes in ecosystem productivity and canopy functioning
    25203, 25230, 25253-4, 25264, 25274-5, 25358, 25412, 25481, 25518, 25618,
    25737, 25771, 25792, 25826, 25886, 25909, 25946, 25948, 25967, 25997, 26017,
    26020, 26068, 26085, 26096, 26150, 26189, 26197, 26216, 26256, 26348, 26372,
    26501-2, 26506, 26652, 26724. 26747, 26783, 26787, 26803-4, 26841, 26857,
    26874, 26904, 26906, 26912, 27049, 27087, 27098, 27108, 27139, 27150, 27162,
    27216, 27220, 27228, 27336, 27384, 27404, 27604, 27625, 27630, 27656, 27710,
    27720, 27736, 27740, 27748, B27915, 28087, 28100, 28158, 28175, 28197, 28265,
    28283, 28300, 28322, 28350, 28413

Seasonal changes in electron transport chain 25618, 26001, 27161, 27559

Seasonal changes in gas exchange
    25303, 25332, 25480, 25523, 25618, 25642, 25825, 25954, 26001, 26045-6, 26055,
    26096, 26112, 26189, 26313, 26370, 26506, 26571, 26651, 26662, 26754, 26798-
    -9, 26801, 26883, 26991, 27009, 27129, 27150, 27177-9, 27185, 27216, 27301,
    27388, 27425, 27438, 27463, 27656, 27768, 27774, 27782, B27915, 27926, 27935,
    28019-20, 28024, 28064, 28086, 28102, 28133, 28140, 28174, 28197, 28300,
    28413, 28426

Seasonal changes in photorespiration 27425

Seasonal changes in resistances to CO_2 and water vapour transfer
25971, 25999, 26015, 27049, 27269, 27432, 27575, 27609, 28102, 28350

Seasonal changes in respiration
25303, 25480, 26046, 26370, 26572, 27774, 28086-7, 28102

Simulation see Model ...

Sink and source of photosynthates, CO_2, *etc.*
25193, 25207, 25229, 25318, 25576, 25723, 25765, 25774, B25775, 25965, 25967,
26028, 26032, 26121-2, 26243, 26320, 26433, 26489, 26537, 26688, 26775, 27199,
27361, 27384, 27412, 27514-5, 27573, 27594, 27722, 27725, 27991, 27995, 27999,
28053, 28098, 28296, 28336, 28339, B28340

Soil moisture and carbon fixation pathways 27360

Soil moisture and carotenoids 25930, 27398, 27855, 27925

Soil moisture and chlorophyll 25930, 26338, 26392, 27398, 27855, 27925, 28312

Soil moisture and ecosystem productivity and canopy functioning
25170, 25238, 25253, 25319, 25481, 25572, 25607, 25613, 25737, B25775, 25930,
26124, 26216, 26337, 26501-2, 26509, 26537, 26569, 26605, 26723, 26729,
26782, 26798, 26841, 26920, 27094, 27108, 27133, 27200, 27404, 27413, 27433,
27457, 27656, 27686, 27710, 27984, 28098, 28196, 28241, B28482, 28495

Soil moisture and gas exchange
25419, 25613, 25642, 25930, 25951, 26096, 26124, 26392, 26464, 26729, 27009,
27139, 27289, 27398, 27656, 28066, 28196, 28462

Soil moisture and photorespiration 26241

Soil moisture and resistances to CO_2 and water vapour exchange
25537, 25572, 25742, 25999, 26015, 26392, 26532, 27133, 27271, 27433, 28089,
28125, 28462

Soil moisture and respiration 26241, 27398

Solar radiation and canopy see Canopy, radiation distribution ...; Canopy, radiation
profile ...

Specific leaf area see Growth analysis, specific leaf area ...

Spectral methods in photosynthesis research 25491, 25907, 26301, 26381, 26556, 28028,
28030, 28328-9

Stabilization of photosystems see Photosystems stabilization ...

Stand see Canopy ...; Ecosystem ...

Steady state and non-steady state see Oscillations ...

Stem, petiole, morphology, structure and physiological activity in 26931

Stomata morphology and anatomy (number, dimensions, types, development, structure,
etc.) (*cf.*also Leaf epidermis, stomata)
25194, 25251, 25364, 25404, 25406, 25445, 25604, 25653, 25663, 25810, 25881,
25971, 26074-5, 26336, 26421, 26477, B26551, 26853, 26931, 27012, 27036,
B27046, 27073, 27109, 27158, 27433, 27467, 27555, 27585, 28179, 28182, 28367,
28399, 28456, 28463

Stomata physiology (mechanism of action, reactivity, *etc.*) (*cf.*also Leaf epidermis,
stomata)
25194, 25251, 25271, 25604, 25928, 26074-5, 26238, 26265, 26436, 26612, 26908,
B26982, B27046, 27047, 27101, 27109, 27280, 27356, 27410, 27553, 27555-6,
27573, 27578, 27846, 27936, 28182, 28209, 28244, 28275, 28463

Temperature, physiological, and gas exchange (continued)
      26618-9, 26646-7, 26708, 26719, 26754, 26776, 26785, 26811, 26835, 26854,
      26883, 26905, 26919, 26941, 26961, 26970, 26987-8, 27089, 27107, 27139, 27179-
      -80, 27198, 27270-1, 27301-2, 27311, 27320, 27348, 27401, 27414, 27422, 27458,
      27460, 27574, 27602, 27638, 27654, 27749, 27750, 27768, 27782, 27887, 27912,
      27930, 27935, 28024, 28067, 28085, 28087, 28092, 28106-7, 28110, 28128, 28133-
      -4, 28342, 28391, 28408, 28432

Temperature, physiological, and photorespiration
      25721, 25858, 26037, 26785, 26941, 26988, 27460, 27754, 27929, 28342

Temperature, physiological, and resistances to CO_2 and water vapour transfer
      25327, 25604, 25653, 25873, 25999, 26240, 26246, 26265, 26282, 26514, 26524,
      26865, 26909, 26941, 27049, 27271, 27360, 27574, 27912, 28067, 28341-2,
      28367, 28391

Temperature, physiological, and respiration
      25283, 25291, 25858, 25873, 26037, 26046, 26073, 26085, 26112, 26299-300,
      26370, 26386, 26461, 26506, 26578, 26988, 27301, 27478, 27521, 27638, 27687,
      27935, 28092, 28106

Thylakoid, granum (cf. also Chloroplast ultrastructure; Electron transport chain lo-
      calization in thylakoid)
      25180, 25184, 25196, 25217, 25228, 25240, 25246, 25313, 25315, 25337, 25386,
      25495, 25498, 25500, 25513, 25528, 25624, 25627, 25665, 25700, 25745, 25770,
      25802, 25819, 25874, 25878, 25906, 25979, 25988, 26007, 26071, 26090, 26094-
      -5, 26101, 26103-5, 26126, 26141, 26206, B26220, 26234, 26335, 26355, 26359,
      26366, 26446-8, 26527-8, 26557, 26609, 26649, 26664-5, 26667, 26674, 26699,
      26715, 26791, 26795, 26831, 26872, 26893-4, 26917, 26942, 26948, 27061,
      27081, 27135-8, 27167-8, 27170, 27186, 27211, 27249, 27261, 27305-6, 27332,
      27351, 27373, 27448, 27451, 27481-2, 27491, 27546, 27562, 27581, 27603, 27691,
      27715, 27745, 27767, 27776, 27788, 27816, 27818, 27851, 27863, 27874, 27923,
      27933, 27966, 27968-9, 27977-8, 28027, 28029, 28037, 28045-6, 28074-5,
      28111, 28119, 28149, 28166, 28217, 28221-2, 28249, 28364, 28401, 28433, 28488

Tissue cultures, carbon fixation pathways in 26628

Tissue cultures, carotenoids in 25798, 27165

Tissue cultures, chlorophyll in 25265, 25498-9, 25798, 26629, 28477

Tissue cultures, chloroplast (chromatophore) in 25486, 27002

Tissue cultures, electron transport chain in 27471

Tissue cultures, gas exchange in 26629, 27471, 28477

Tissue cultures, photorespiration in 26628-9

Transient phenomena in carbon fixation pathways 28303

Transient phenomena in gas exchange 25597, 25787, 25916, B26220, 26835, 26863,
      28029, 28193, 28244

Transpiration and photosynthesis
      25309, 25419, 25484, 25540, 25613, 26065, 26227, 26265, 26282, 26329, 26373,
      26446, 26535, 26541, 26646-7, 26684, 26690, 26708, 26785, 26947, 27271,
      27289, 27409-11, 27578, 27595, 28428

Tritium oxide see Deuterium oxide, tritium oxide ...

U

Ubiquinones see Quinones ...

Ultraviolet radiation see Irradiance, spectral composition ...

Uncouplers of electron transport chain (*cf.* also Antibiotics and electron transport
        chain)
        25177, 25222, 25320, 25322, 25385-7, 25675, 25914, 25980, 26210, 26231,
        26376, 26411, 26473, 26523, 26544, 26556, 26718, 26790, 26792, 26902, 27093,
        27184, 27264, 27276-7, 27335, 27354-5, 27453, 27465, 27486, 27641, 27655,
        27872, 27907, 27916-7, 27996, 28203, 28235, 28452, 28483

V

Virus diseases see Phytopathological effects ...

Vitamin K_3 see Quinones ...

Volume changes in leaf and other organs 25550, 25785, 25999, 26276, 26612, 26909,
        27109, 27432, 27636, 28489

Volume changes of chloroplasts (chromatophores) see Chloroplast and chromatophore
        volume changes

W

Warburg effect see O_2 and gas exchange

Water, heavy see Deuterium oxide, tritium oxide ...

Water saturation deficit
        25170, 25236, 25251, 25407, 25604, 25828, 25885, 25902, 25943, 26096, 26123,
        26204, 26265, 26296, 26392, 26437, 26494, 26828, 26947, 26993, 27139, 27280,
        27289, 27302, 27339, 27496, 27579, 27625, 27749, 27847, 28300, 28341

Water splitting mechanism see O_2 evolution mechanism and kinetics

Wind (air-flow rate) and ecosystem productivity and canopy functioning 25659, 26199,
        28197

Wind (air-flow rate) and gas exchange 26199, 26708, 26745, 27139, 28087

Wind (air-flow rate) and resistances to CO_2 and water vapour transfer 25451, 25661,
        26114, 26199-200, 27749, 28257

Wind mesurement see Aerodynamic methods ...

X

Xanthophylls 25376, 25424, 25838, 25897, 26225-6, 26270, 26400, 26666-7, 26707,
        26740, 26830, 26975, 27060, 27066, 27110, 27118-9, 27165, 27237, 27288,
        27405, 27527, 27816, 27863-4, 27874, 28060, 28356-7

The plant index presents a selection of plant genera and types interesting as experimental material for physiological, ecological and agricultural studies. In general, mainly those plant names have been included which are given in the title of the respective papers or in abstracts. Latin scientific names of plant genera are the main items which present the reference numbers.

A

Abies   B26099, 26348, 26515, 27130, 27432, 27683, 28114, 28353

Acacia   25521, 25592, 27769, 28139

Acer   25186, 25206-7, 25431, 25511, 25540, 25710, 25742, 25896, 25989, 25998-9, B26099, 26534, 26952, 27021, 27035, 27242, 27362, 27730, 28308, 28352, 28470

Acetabularia   25228, 25598, 26340, 26583-5, 27023, 27155, 27233, 27646, 28453

Aesculus   B26099, 28362

Agave   25445-6, 25926, 26303, 27271

Alder   see *Alnus*

Alfalfa   see *Medicago*

Algae (*cf.* also *Acetabularia,* A. blue-green, A. brown, A. green, A. red, *Anabaena, Anacystis, Ankistrodesmus, Chlamydomonas, Chlorella, Chrysophyta,* Diatoms, *Dinoflagellates, Dunaliella, Euglena, Nostoc, Porphyridium, Scenedesmus, Ulva*)   25165, 25210, 25218, 25286, 25331, 25341-3, 25390, 25485, 25495, 25543, 25626, 25633, 25655, 25724, 25738-9, 25743, 25763, 25770, 25805-6, 25815, 25847, 25849-51, 25894-5, 25952, 26089, 26108, 26129, 26163, 26179, 26186, 26212, 26214, 26262, 26267, 26271, 26273, 26287, 26367, 26371, 26386- -7, 26404, 26407-8, 26427, 26437, 26453, 26458, 26518, 26521, 26530, 26563, 26568, 26641, 26655, 26669, 26698, 26705, 26720, 26741, 26823, 26849, 26873, 26875, 26882, 26888, 26915, 26919, 26933, 26953, 26966, 26985, 26992, 27086, 27112, 27185, 27208, 27225, 27266, 27299, 27367-8, 27376-7, 27392, 27400, 27406, 27459, 27463, 27470, 27476, 27544, 27557, 27591, 27593, 27650, 27652, 27662, 27732, 27785, 27811, 27838, 27913, 27932, 27956, 27960-1, 27964, 27970, 27987-8, 27993, 28016, 28025, 28054, 28140, 28152, 28159, 28162, 28174, 28204, 28349, 28354, 28372, 28377, 28464-5, 28471-2, B28494

Algae, blue-green (*cf.* also *Anabaena, Anacystis, Nostoc*)   25163, 25165, 25248, 25270, 25280, 25286, 25345, 25361, 25391, 25394, 25402, 25425, 25439, 25447, 25495, 25509, 25548, 25585, 25587, 25601, 25624, 25635, 25651, 25665, 25699, 25743, 25759-60, 25770, 25847, 25849, 25875, 25883, 25958, 25963, 25996, 26006, 26016, 26048, 26058, 26072, 26130, 26148, 26158-60, 26171, 26179, 26186, 26203, 26207, 26219, B26220, 26242, 26250, 26254-5, 26268, 26271, 26288, 26305-6, 26340, 26356, 26380-1, 26407, 26457, 26478, 26520, 26528, 26570-1, 26576, 26697, 26718-9, 26725, 26805, 26823, 26916, 26978, 26994, 27033, 27086, 27100, 27104, 27122, 27151-2, 27206, 27223, 27226, 27252, 27281, 27297, 27299-300, 27312, 27393, 27418, 27426-7, 27440, 27453, 27490, 27534, 27548, 27566, 27572, 27616, 27642, 27708, 27763, 27766-7, 27823, 27825, 27827, 27858, 27898, 27911, 27940, 27943-4, 27958-60, 27964, 27986, 28004, 28017, 28026, 28072, 28079, 28094, 28111, 28126, 28142, 28151, 28159, 28173-4, 28184, 28190, 28198, 28234, 28267, 28331, 28354, 28412

Algae, brown   25179, 25317, 25447-8, 25452, 25471, 25549, 25602, 25676, 25778, 25794, 25883, 25975-6, 26025, 26038, 26063, 26186, 26203, 26313, 26521, 26549-50, 26642, 26644-5, 26661-2, 26950, 27114, 27279, 27321, 27418, 27476, 27752, 27763, 27921, 28017, 28129, 28142, 28147, 28151, 28325

Algae, green (*cf.* also *Acetabularia, Ankistrodesmus, Chlamydomonas, Chlorella, Dunaliella, Scenedesmus, Ulva*)   25179, 25286, 25306, 25317, 25333, 25345-6, 25391, 25447-8, 25459, 25471, 25486, 25490, 25500-3, 25553, 25598, 25602, 25624-5, 25633, 25676, 25743, 25770, 25794, 25847, 25849, 25876, 25882-3, 25933, 25936, 25960, 26035, 26038, 26051, 26063, 26111, 26119, 26130, 26162, 26179, 26186, 26203, 26206, B26220, 26271, 26313, 26321-4, 26344, 26346, 26356, 26407, 26429, 26435, 26437, 26504, 26518, 26549-50, B26551, 26570,

Algae, green (continued)
26643, 26645, 26662, 26683, 26752, 26771, 26797, 26807, 26823, 26826, 26838, 26873, 26889, 26896, 26962, 26989, 26994, 27051, 27057-8, 27086, 27206,. 27222-3, 27226-7, 27281, 27310, 27312, 27316, 27321, 27349-50, 27357, 27367, 27377, 27399, 27418, 27438-9, 27476, 27534, 27567-9, 27588-92, 27642, 27652, 27700, 27727, 27743, 27756-7, 27763, 27766, 27858, 27889, 27921, 27947, 27959, 27963, 27975-6, 27986, 28004, 28016-7, 28026, 28094, 28119, 28126-7, 28142, 28147, 28151, 28159, 28170, 28190, 28204, 28226, 28234, 28317, 28354, 28377, 28421

Algae, red (cf. also *Porphyridium*)
25179, 25394, 25442, 25447-8, 25471, 25587, 25602, 25665, 25676, 25770, 25794, 25824-5, 25869, 25883, 25958, 26038, 26048, 26158-9, 26170, 26186, 26203, 26313, 26382, 26558, 26645, 26662, 26698, 26718, 26720, 26770, 26772, 26916, 26994, 27086, 27177-8, 27206, 27297, 27299, 27320-1, 27438, 27450, 27476, 27567-8, 27763, 27818, 27901, 27959, 28111, 28142, 28147, 28151, 28234, 28267, 28295, 28376

Allium 25511, 25930, B26099, 26151, 26476, 28075, 28102, 28139, 28267

Alnus 25569, 25592, 25742

Alpine plants 25659-61, 26854, 27107, 28358

Amaranthus 25408, 25420, 25431, 26315, 26360, 26398, 26403. 26469-70, 26475, 26539, 27414, 27506, 27530, 27585, 28005, 28368

Anabaena 25286, 25508, 25579, 25629, 26084, 26179, 26472-3, 26527, 26544, 26575, 26608, 26866, 27258, 27440, 27572, 28072, 28105, 28354, 28446

Anacystis 25197, 25493-4, 25532, 25586, 25866-8, 25980-1, 26168, 26575, 26682, 26948, 26957, 27048, 27156, 27257, 27259, 28008, 28072, 28075, 28213

Ananas 25327, 25446, 25797, 27182, 27506, 28308

Ankistrodesmus 25286, 26147, 27591

Antirrhinum 25891, 26260, 26700, 27114, 28139

Apium 27673

Apple see *Malus*

Aquatic macrophytes (cf. also *Elodea, Phragmites, Typha*)
25120, 25253, 25282, 25374, 25436, 25518, 25523, 25826, 25878, 26038, B26099, 26417, 26549, 26724, 26767, 26824, 26888, 27013, 27025, 27060-1, 27310, 27371, 27439, 27458, 27470, 27611, 27671, 27710, 27741, 27773-4, 27989, 28229-30, 28303, 28359

Arabidopsis 25265, 27268, 28139

Arachis 25184, 25404-7, 25445, 25965, 26093, 26476, 26750, 28038, 28096, 28228, 28347, 28392

Arbor vitae see *Thuja*

Armoracia 28267

Ash see *Fraxinus*

Asparagus 27110

Aspen see *Populus*

Atriplex 25360, 25431, 25437-8, 25457, 25521, 25538, 25732, 25807, 25835, 25923, 25944-5, 26155, 26402, 26437, 26539, 26719, 26872, 27179, 27359, 27929, 28288, 28489

Avena 25340, 25431, 25516, 25545, 25624, 25697, 25729, 25745, 25874, 25877, 25961, 26003, 26073, 26096, 26277, 26279-80, 26476, 26536, 26539, 26605, 26688, 26919, 26947, 26991, 27239, 27287, 27439, 27530, 27885, 27931, 27997, 28032, 28086, 28116, 28139, 28157, 28228, 28235, 28267, 28360-1, 28406

B

Bacteria, photosynthetic (*cf.* also *Chlorobium, Chromatium, Halobacterium, Rhodo-*
        pseudomonas, Rhodospirillum)
        25263, 25380, 25478, 25488, 25490, 25495, 25554, 25582, 25585, 25587, 25606,
        25614, 25628, 25631, 25669, 25733-4, 25770, 25773, 25816, 25835, 25855,
        25888, 25957, B26220, 26247, 26394, 26404, 26410, 26504, 26539, 26609, 26687,
        26727, 26805, 26915, 26926, 26942-3, 26974, 26978, 26985, 26994, 27068,
        27151-2, 27173, 27297, 27340-2, 27374, 27419, 27441, 27449, 27509, 27519,
        27531, 27535, 27537, 27615, 27627, 27631, 27822, 27841-2, 27867, 27945,
        28071, 28163, 28178, 28335, 28409, 28451, 28466, 28484

Bamboo .see *Bambussa*

Bambussa 27506

Banana see *Musa*

Barley see *Hordeum*

Bean see *Phaseolus*

Beech see *Fagus*

Bermuda grass see *Cynodon*

Beta 25176, 25226, 25331, 25360, 25431, 25457, 25521, 25568, 25571, 25585, 25656,
        25693, 25709-10, 25714, 25723, 25772, 25872, 25877, 25935, 25940-1, 25948,
        26039, 26063, 26121-2, 26151, 26209, 26235, 26237, 26260, 26267, 26281,
        26360, 26419, 26471, 26476, 26539, 26605, 26733, 26767, 26783, 26816-7,
        26954, 26999, 27010, 27133, 27180, 27327, 27466, 27478-9, 27592, 27673,
        27692, 27715, 27748, 27779, 27918, 27998-9, 28039, 28112-3, 28139, 28252,
        28315, 28432, 28478, 28489

Betula 25710, 25998-9, B26099, 27021, 27060, 27484, 28160, 28470

Birch see *Betula*

Blackberry see *Rubus*

Blueberry see *Vaccinium*

Bluegrass see *Poa*

Brassica 25511, 25772, 25812, 25935, 25982, 26151, 26267, 26355, 26360, 26476,
        26814, 27016, 27024, 27041, 27222, 27265, 27300, 27312, 27439, 27561-2,
        27673, 28005, 28228, 28315, B28482, 28487

Brinjal see *Solanum*

Bristlegrass see *Setaria*

Broadbean see *Vicia*

Broccoli see *Brassica*

Bromegrass see *Bromus*

Bromus 26145, 28406

Brussels sprouts see.*Brassica*

Bryophyllum 26299, 26578, 27948, 28146, 28308, 28388, 28489

Buckwheat see *Fagopyrum*

C

Cabbage see *Brassica*

Cacti 25360, 25445, 25787, 25862, 28049, 28066, 28143

Cajanus 26274, 27389

Calluna see Heath plants and communities

CAM plants (*cf.* also *Agave, Ananas, Bryophyllum,* Cacti, Succulents)

CAM plants (continued)
        25379, 25787, 25862, 25926, 26690-2, 27073, 27198, 27271, 27359, 27542,
        27769, 27948, 28065-6, 28143, 28389-90, 28403-5, 28407

Canarygrass see *Phalaris*

Cannabis 25511, B27915

Capsicum 25168, 25225, 25521, 25947, 26186, 26232, 26290, 26311-2, 26775, 26956,
        27142, 27673, 27799, 27883, 27995, 28139

Carex 25254, 25887, 26490, 26532, 26854, 27013-4, 27139, 27604, 27814

Carpinus 25896, 25898, 26490, 26787-8, 27362, 28160

Carrot see *Daucus*

Carthamus 25990

Carya 25821, 26534

Cassava see *Manihot*

Castor bean see *Ricinus*

Cat's tail see *Typha*

Cattail flag see *Typha*

Cauliflower see *Brassica*

Cedar see *Cedrus*

Cedrus 25186, 26348

Celery see *Apium*

Cereals see *Avena, Hordeum, Oryza, Panicum, Secale, Sorgum, Triticum, Zea*

Chenopodium 26360, 27196, 27230, 27404, 28297, 28315

Chick pea see *Cicer*

Chinese cabbage see *Brassica*

Chlamydomonas 25286, 25370, 25378-9, 25395, 25429, 25463, 25504, 25524, 25570, 25616,
        25690, 25693, 25726, 25849, 25882, 25925, 25935, 26024, 26051, 26144, 26171,
        26295, 26340, 26371, 26414-6, 26430-1, 26557, 26573, 26575, 26659, 26774,
        26823, 26829-30, 26994, 27077, 27206, 27246-7, 27451, 27591, 27603, 27690,
        27695-7, 27766, 27858, 27891, 27962, 28074-5, 28127, 28174, 28223, 28267,
        28452

Chlorella 25163, 25174, 25179, 25208, 25243, 25331, 25335, 25345, 25366-7, 25381-3,
        25511, 25527, 25533-6, 25630-2, 25675, 25730, 25752, 25759, 25782, 25836,
        25840, 25861, 25868, 25875, 25882, 25915, 25935, 26018-9, 26021, 26024, 26063,
        26094, 26097, 26111, 26171, 26211, B26220, 26288, 26293-4, 26298, 26302,
        26340, 26437, 26454, 26504, 26517, 26538-9, 26549-50, 26639, 26683, 26693,
        26718, 26720, 26755-6, 26780-1, 26823, 26859, 26861, 26873, 26943, 26985,
        27026, 27084, 27086, 27122, 27206, 27222, 27227, 27297, 27314-5, 27399, 27439,
        27566, 27591-2, 27612-3, 27704, 27718, 27766, 27790-1, 27796, 27843, 27858-9,
        27950, 27963, 27994, 28056, 28072, 28075, 28115, 28120, 28124, 28128, 28159,
        28171-2, 28204, 28223, 28234, 28278, 28319, 28328-9, 28334, 28354, 28395, 28418

Chlorobium 25478, 25562-3, 25584, 26630, 26695, 26926, 28232

Chromatium 25249, 26677, 26702, 26959, 27262, 27326, 27517, 27868-9, 28135-6, 28238

Chrysophyta 25215, 25247, 25286, 25346, 26130, 26179, 26437, 26867, 27120-1, 27406,
        27592, 28126, 28129

Cicer 25965, 27237, 27389

Citrullus 27880, 28267

Citrus 25233, 25250, 25330, 25376, 25551, 25634, 26476, 26511, 26646-7, 26909, 27049,
        27085, 27191, 27472, 27506, 28318

Clover see *Trifolium*

Cockafoot see *Dactylis*

Cocoa see *Theobroma*

Coconut palm see *Cocos*

Cocos   27069, 27220, 27506, 28384

Coffea 27289, 27601, 27951

Coffee tree see *Coffea*

Coniferous plants (*cf.* also *Abies, Cedrus, Cupressus, Juniperus, Larix, Metasequoia,
     Picea, Pinus, Pseudotsuga, Taxus, Thuja, Tsuga*)
     B25637, 26515, 26283, 27260, 28062, 28139, 28341

Corchorus 26245

Corn see *Zea*

Cornelian cherry see *Cornus*

Cornus 25822, 25897, 26534, 27362, 28283

Cotton see *Gossypium*

Cottonwood see *Populus*

Cowberry see *Vaccinium*

Cowpea see *Vigna*

Crabgrass see *Digitaria*

Crataegus 27362, 28283

Cucumber see *Cucumis*

Cucumis 25176, 25258, 25345, 25368, 25666, 25747, 25877, 26073, 26079, 26103, 26125,
     26476, 26599, 26769, 26779, 27142, 27291, 27430, 27439, 27667, 27922, 27927,
     28182, 28239-40, 28369

Cucurbita 25511, 25614, B26099, 26151, 26186, 26476, 26718, 26720, 26961, 27162,
     27439, 27500, 27566, 27730, 28117, 28139, 28245, 28315

Cupressus 25186

Currant see *Ribes*

Cyanobacteria see Algae, blue-green

Cynodon 25679, 27895, 28100

Cyperus 26539, 26923, 27325, 27506

Cypress see *Cupressus*

D

Dactylis 25817-8, 26065, 26569, 27296, 27333, 27396, 27480, 27819, 28005, 28397

Dallis grass see *Paspalum*

Date palm see *Phoenix*

Daucus 25306, 25479, 25486, 25935, 26151, 26157, 26267, 27165, 27478, 27730, 28117

Deciduous trees and shrubs (*cf.* also *Acer, Aesculus, Alnus, Betula, Carpinus, Carya,
     Citrus, Cornus, Crataegus, Eucalyptus, Fagus, Fraxinus, Hevea, Hibiscus,
     Juglans, Malus, Morus, Olea, Persica, Pirus, Platanus, Populus, Prunus, Quer-
     cus, Ribes, Robinia, Rubus, Salix, Sambucus, Sorbus, Syringa, Tamarix, Tilia,
     Ulmus, Vitis*)
     25186, 25283, 25423-4, 25511, 25521, 25592, 25619-20, B25637, 25693, 25822,
     26533-4, 26674, 26977, 27062, 27222, 27506, 27730, 27939, 28255, 28283,
     28294, 28433, 28470

Forage crops (*cf.* also *Brassica*, Grasses, Leguminous plants, *Lupinus*, *Medicago*, *Tri-folium*, *Vicia*, *Vigna*, etc.)
     27931, 28381

Forest (including undergrowth) plants and ecosystems (*cf.* also Coniferous plants,
     Deciduous trees and shrubs, Ferns, *Fragaria*, Grasses, Heath plants and com-
     munities, Lichens, Liverworts, Medicinal plants, Mosses, *Vaccinium*, etc.)
     25195, 25331, 25423-4, 25472, 25539, B25637, 25693, 25742, 25811, 25887,
     25896, 25972, 26026, 26034, 26133, 26348, 26490, 26504, 26534, 26719, 26870,
     26883, 26895, 27797, 27894, 28068, 28102, 28294, 28470

Fountain-grass see *Pennisetum*

Foxtail millet see *Setaria*

Fragaria 26145, 26183, 26476

Fraxinus 25540, 25592, 25901, 26095, 26499, 26534, 26801, 27062, 27248

Fruit plants and trees (*cf.* also *Ananas*, *Citrullus*, *Citrus*, *Cocos*, *Cucumis*, *Cucurbita*,
     Ficus, *Fragaria*, *Malus*, *Musa*, *Persica*, *Phoenix*, *Pirus*, *Prunus*, *Ribes*, *Rubus*,
     Sorbus, *Vaccinium*, *Vitis*)
     25608

Fungi (parasitic) 25271, 25514, 25705, 26107, 26234, 26369, 26777, 26993, 27042,
     27059, 27675, 27751, 27877, 27941-2, 28296, 28374

G

Garlic see *Allium*

Ginger see *Zingiber*

Glycine 25173, 25186, 25223, 25308-9, 25331, 25336, 25407, 25521, 25576, 25646,
     25693, 25721, 25727, 25877, 25881, 25917, 25965, 25990, 26093, 26106, 26124,
     26134, 26190, 26318, 26337, 26360, 26373, 26440, 26450, 26463, 26474, 26476,
     26618-9, 26684, 26729, 26749, 26784, 26898, 27003, 27163, 27215, 27237,
     27272, 27287, 27336, 27397, 27415, 27421, 27501-2, 27530, 27540, 27707,
     27719, 27730, 27800, 27821, 27849, 27887, 27897, 27929, 27935, 28041, 28130,
     28144, 28182, 28228, 28267, 28309, 28311-3, 28342-4, 28347-8, 28426-8, 28478,
     B28482, 28495

Gossypium 25331, 25411-2, 25511, 25521, 25572, 25772, 25786, 25835, 25971, 26086,
     26093, 26260, 26263, 26267, 26651, 26777, 27036, 27059, 27102, 27212-3,
     27216, 27290, 27318, 27415, 27559, 27625, 27675, 27845-7, 27981, 28078,
     28123, 28139, 28196, 28245, 28431

Gourd see *Cucurbita*

Gram chick pea see *Vigna*

Grape fruit see *Citrus*

Grape vine see *Vitis*

Grasses (*cf.* also *Avena*, *Bromus*, *Carex*, *Cynodon*, *Cyperus*, *Dactylis*, *Digitaria*, *Festu-ca*, *Hordeum*, *Lolium*, *Oryza*, *Panicum*, *Paspalum*, *Pennisetum*, *Phalaris*, *Phleum*,
     Poa, *Saccharum*, *Secale*, *Setaria*, *Sorgum*, *Triticum*, *Zea*) .
     25331, 25422, 25431, 25473-4, 25531, 25538, 25596, 25613, 25660, 25674, 25711,
     25753, 25772, 25873, 25877, 25910, 25912-3, 25946, 25978, 26065, 26074-5,
     26145, 26164, 26181, 26189, 26216, 26267, 26308, 26316, 26320, 26390, 26442,
     26451-2, 26501, 26532-3, 26539, 26625, 26803-4, 26854, 26919, 26947, 26973,
     27002, 27098, 27139, 27242, 27322, 27421, 27428, 27439, 27480, 27493, 27530,
     27580-5, 27597, 27630, 27656, 27686, 27769, 27814, 27819, 27830, 27893-4,
     28013, 28078, 28099-100, 28158, 28160, 28187-8, 28228, 28280, 28323, 28343,
     28397, 28406, 28454, 28489

Groundnut see *Arachis*

H

Halobacterium 25191, 25296, 25298, 25337, 25351-2, 25369, 25461, 25547, 25561,
    25803-4, 25927, 26180, 26283, 26365, 26379, 26439, 26611, 26636-7, 26728,
    26819-21, 26850-1, 26963, 27305-7, 27545, 27626, 27828-9, 27836, 27890,
    27905, 28042-3, 28150, 28293, 28298-9, 28378

Halophilous plants (*cf.* also Salt marsh and strand plants)
    25510, 26033, 28071, 28138, 28151, 28288-9

Hawthorn   see *Crataegus*

Heath plants and communities 26739

Hedera 25685, 27060-1, 27109

Helianthus 25251, 25331, 25445, 25511, 25521, 25640, 25833, 25877, 25954, 25982,
    26007, 26037, 26048, B26099, 26132, 26237, 26264, 26267, 26360, 26449, 26579,
    26747, 26779, 26844, 26852, 26991, 27163, 27287, 27468-9, 27524, 27546,
    27592, 27595, 27728, 27755, 27820, 27897, 27902-3, 27980, 28303, 28425,
    28478, 28489

Hemlock   see *Tsuga*

Hemp   see *Cannabis*

Hevea   25627, 27506

Hibiscus 25170

Hickory   see *Carya*

Hop   see *Humulus*

Hordeum 25185, 25199-200, 25224, 25232, 25236, 25256, 25413, 25430-4, 25445, 25456,
    25496-7, 25521, 25564, 25585, 25693, 25715, 25748, 25772, 25838, 25858, 25912-
    -3, 25918, 25988, 26053, 26063, 26068, 26076, 26085, 26092, B26099, 26113,
    26181, 26191, 26202, 26205, 26215, 26234, 26256, 26267, 26331, 26360, 26432,
    26444, 26474, 26476, 26531, 26539, 26545, 26562, 26572, 26575, 26586, 26594,
    26626, 26699, 26758, 26802, 26925, 26927-8, 26930-1, 26940, 26947, 26980,
    26995, 27022, 27027, 27043, 27082, 27137, 27234, 27238, 27276-7, 27284, 27287,
    27292, 27346, 27439, 27443, 27488-9, 27503, 27538, 27584, 27592, 27614, 27634,
    27687-8, 27690, 27729, 27730, 27751, 27771, 27784, 27851-2, 27854, 27922-4,
    27997, 28032, 28085-6, 28139, 28189, 28200, 28205-6, 28220, 28267, 28281,
    28285, 28406, 28408, 28463, 28487

Hornbeam   see *Carpinus*

Horse chestnut   see *Aesculus*

Horseradish   see *Armoracia*

Horsetail   see *Equisetum*

Humulus 25312, 28139

I

Ipomoea 25810, 26433-4, 26476, 26607, 27404, 28245, 28355

Ivy   see *Hedera*

J

Jerusalem artichoke   see *Helianthus*

Jointgrass   see *Paspalum*

Juglans 25989, 26534, 28145

Juniper   see *Juniperus*

Juniperus 25507, 25592, 25807, 26857, 27348

Maize   see *Zea*

Malus   25162, 25328, 25683, 26013-4, 26246, 26269, 26723, 26840, 26843, 26869, 26996,
         27235, 27514-5, 27520-1, 27693, 27733, 28095, 28101, 28156, 28367

Mangold   see *Beta*

Manihot   26267, 27011, 27431, 27506

Manioc   see *Manihot*

Maple   see *Acer*

Marrow   see *Cucurbita*

Meadowgrass   see *Poa*

Medicago   25331, 25445, 25450-1, 25511, 25546, 25687, 25693, 25714, 25737, 25772,
         25852, 26055, 26145, 26267, 26305, 26409, 26546, 27108, 27284, 27398, 27461-
         -2, 27464, 28078, 28139

Medicinal plants (*cf.* also *Cynodon, Hibiscus, Papaver, Ricinus, etc.*)
         26853, 28069.

Melon   see *Colocynthis, Cucumis*

Metasequoia   25186

Millet   see *Panicum*

Morus   26534, 26775, 27207, 27725, 27730, 28315

Mosses   25252, 25700-1, 25858-60, 26088, 26166, 26186, 26300, 26370, 26601, 26952,
         26994, 27160, 27241, 27301-2, 27782, 28158, 28221-2, 28267, 28435

Mulberry   see *Morus*

Mung bean   see *Vigna*

Musa   26225-6

Musk-melon   see *Cucumis*

Mustard   see *Sinapis*

N

Napier grass   see *Pennisetum*

Nicotiana   25181, 25186, 25204, 25429, 25445, 25457, 25521, 25692-3, 25710, 25713,
         25891, 25912, 25935, 25982, 26037, B26099, 26187, 26208-9, 26297, 26340,
         26349, 26360, 26398, 26402, 26444, 26505, 26507, 26539, 26575, 26602, 26716,
         26719, 26767, 26809-10, 26814, 26886, 26899, 26997-8, 27063, 27111, 27142,
         27164, 27190, 27242, 27560, 27571, 27584, 27592, 27595, 27840, 27935, 27937,
         27967, 28003, 28032-3, 28039, 28139, 28146, 28216, 28224, 28260, 28267,
         28274, 28303, 28385, 28387, 28398, 28416-7, 28477-8, 28480

Nostoc   26179, 26306, 28072, 28354

O

Oak   see *Quercus*

Oat   see *Avena*

Oil palm   see *Elaeis*

Olea   26095, 26801

Olive   see *Olea*

Onion   see *Allium*

Orange   see *Citrus*

Orchardgrass   see *Dactylis*

Orchids 25802, 26368, 27114, 27744

Ornamental plants (*cf.* also *Agave, Antirrhinum, Asparagus,* Coniferous plants, *Cyperus,*
        Deciduous trees and shrubs, *Eucalyptus, Euphorbia, Ficus, Hedera, Hibiscus,*
        Lupinus, Pelargonium, Perilla, Rosa, Tradescantia, etc.)
        25360, 25376, 25429, 25496, 25511, 25531, 25590, 25652, 25659, 25693, 25814,
        25822, 25935, 25962, 25977, B26099, 26157, 26186, 26260, 26362, 26405, 26461,
        26476, 26534, 26683, 26719, 26853, 26918-9, 26993, 27118-9, 27404, 27457,
        27467, 27506, 27573, 27592, 27711, 28133, 28139, 28198, 28292, 28401

Oryza   25331, 25410, 25435, 25457, 25686, 25772, 25809, 25967, 26267, 26284-6,
        26307, 26474, 26476, 26539, 26549-50, 26581, 26652, 26736-8, 26806, 26904-6,
        27157-8, 27210, 27221, 27311, 27325, 27384-6, 27413, 27506, 27592, 27709,
        27721-2, 27835, 27971, 28087, 28092-3, 28096, 28109, 28130, 28139, 28195-7,
        28228, 28326, 28347, 28461-2

P

Paddy see *Oryza*

Palms see *Cocos, Elaeis*

Panicum 25164, 25431, 25445, 25573, 25913, 26052, 26181, 26249, 26308, 26315-6,
        26360, 26474, 26618-9, 26786, 26919, 26971-3, 26979, 27008, 27175, 27539,
        27582, 27584, 27596-8, 27929, 28005, 28100, 28130, 28199, 28228, 28257,
        28400

Papaver 25581

Paprika see *Capsicum*

Para-rubber tree see *Hevea*

Parasitic plants 26009-10, 26339, 26493, 27147, 27450

Parsley see *Petroselinum*

Parsnip see *Pastinaca*

Paspalum 25274, 26308, 27106, 27506, 28099

Pastinaca 25511

Pasture plants see Forage plants

Pea see *Pisum*

Peach see *Persica*

Peanut see *Arachis*

Pear see *Pirus*

Pecan see *Carya*

Pelargonium 25521, 26260, 26362, 26649, 26699, 27694, 27771, 28139, 28308

Pennisetum 25445, 25521, 28099

Pepper see *Capsicum, Piper*

Perilla 25453

Persica   25329, 26183, 26421, 26476, 27445

Petroselinum 26360, 27673

Phalaris 27720, 28099

Phaseolus 25189, 25221, 25237, 25293, 25295, 25326, 25431, 25476, 25489, 25496,
        25513, 25521, 25528-30, 25624, 25654, 25748, 25796, 25829-30, 25899, 25929,
        25935, 25965, 26011, 26073, 26091, B26099, 26109, 26121, 26205, 26221, 26264,
        26282, 26289, 26310, 25351, 26360-1, 26397, 26399, 26462, 26476, 26513,
        26524, 26535, 26560, 26587, 26620, 26675, 26683, 26744-6, 26753, 26779,
        26784, 26818, 26842, 26852, 26911, 26935-8, 26964-5, 26994, 27036, 27043,
        27078, 27106, 27109, 27168, 27170-1, 27201, 27287, 27300, 27329, 27363,

Q

Quercus   25186, 25239, 25445, 25578, 25664, 25807, 25897-8, 25989, 25998, B26099,
26369, 26490, 26499, 26534, 26787-8, 26903, 27242, 27362, 27885, 28052,
28068, 28125, 28160, 28383, 28460

R

Radish   see *Raphanus*

Rape   see *Brassica*

Raphanus   25710, 25835, 26151, 26476, 26587, 26994, 27024, 27439, 27831-2, 27991-2,
28050, 28130

Raspberry   see *Rubus*

Redwood   see *Metasequoia*

Reed   see *Phragmites*

Rheum   27673

Rhodopseudomonas   25478, 25555-6, 25645, 25668, 25741, 25746, 25889, 25964, 26205,
26395, 26406, 26456, 26610, 26631, 26660, 26681, 26696, 26702, 26714, 26731,
26808, 27174, 27254, 27273, 27447-8, 27504-5, 27517-8, 27525, 27653, 27726,
27871, 28103, 28259, 28455, 28469

Rhodospirillum   25281, 25302, 25323, 25717-8, 25756-7, 25791, 25884, 25914, 25953,
26146, 26222, 26233, 26328, 26357, 26395, 26486, 26575, 26686, 26969, 27050,
27262, 27274, 27303-4, 27455, 27517, 27536, 27806, 27869, 27879, 27916,
27965, 28048, 28072, 28131, 28202, 28237

Rhubarbe   see *Rheum*

Ribes   25569

Rice   see *Oryza*

Ricinus   25373, 25431, 25511, 26142, 27506, 28267, 28303

Robinia   B26099, 27062

Rosa   25414-5, 25569, 27362, 27753, 28436

Rose   see *Rosa*

Rubber tree   see *Hevea*

Rubus   26814, 26952

Rye   see *Secale*

Ryegrass   see *Lolium*

S

Saccharum   25190, 25198, 25331, 25431, 25445, 25457, 25482, 25626, 25769, 25913,
26267, 26539, 26550, 26719, 26806, 26872, 27055, 27242, 27506, 28005, 28247,
28267, 28301

Safflower   see *Carthamus*

Salix   25569, 25581, 25813, 25822, 27139, 27236, 28158

Salt marsh and strand plants (*cf.* also Halophilous plants)
25253, 25262, 25543, 25549, 25806, 25886, 26261, 26663, 27267, 27700, 28192

Sambucus   25511, B26099, 26186, 27284, 27566, 27737, 28315

Scenedesmus   25163, 25286, 25310, 25339, 25370, 25511, 25633, 25882, 26023-4, 26111,
B26220, 26270, 26288, 26305, 26350, 26356, 26371, 26402, 26479, 26504,
26659, 26793-7, 26807, 26822-3, 27086, 27100, 27222, 27227, 27261, 27312,
27377, 27534, 27548, 27591, 27668, 27681, 27683, 27690, 27762-3, 27795,
27858, 27875, 27974, 28009, 28074, 28174, 28223

Secale   25310, 26003-4, 26063, 26455, 26506, 27239, 27547, 27651, 27678, 27690,
       27739, 27853, 28086, 28139, 28200, 28228, 28252

Sedge   see *Carex*

Sempervirent plants (*cf.* also *Coffea*, Coniferous plants, *Hedera*, etc.)
       25685, 26095, 26799-801

Service-tree   see *Sorbus*

Sesamum   25511, 27769, 27794

Setaria   25913, 26308, 27551, 28099-100

Sinapis   25235, 25707, 25776, 26077, 26603-4, 27439, 27586, 27623, 27759

Sisal   see *Agave*

Solanum (*cf.* also *Lycopersicon*)
       25511, 25571, 25693, 25740, 25758, 25772, 25779, 25871, 25935, 25947, 26116,
       26138, 26182, 26267, 26336, 26476, 26733, 26742, 26785, 26853, 27145, 27236,
       27269, 27477-8, 27538, 27564-6, 27592, 27701, 27716, 27740, 27799, 27805,
       28097, 28139, 28175, 28199, 28315

Sorbus   25569, 27362

Sorghum   see *Sorgum*

Sorgum   25169, 25238, 25331, 25431, 25445, 25454, 25521, 25615, 25705, 25725, 25772,
       25835, 25912-3, 26027-8, 26080, 26141, 26165, 26337, 26372, 26437, 26484,
       26539, 26778, 26872, 26932, 27047, 27199, 27242, 27278, 27382, 27411, 27439,
       27506, 27582, 27789, 27896, 27983, 28098, 28123, 28228, 28231, 28257, 28264,
       28337-8, 28343-4

Soybean   see *Glycine*

Spinach   see *Spinacia*

Spinacia   25179-80, 25188, 25201-2, 25217, 25222, 25242-3, 25276, 25284, 25290, 25292,
       25301, 25304, 25313, 25315-6, 25320-2, 25331, 25347, 25353, 25370, 25372,
       25386-7, 25392, 25398-9, 25403, 25416-7, 25421, 25445, 25470, 25487, 25496,
       25506, 25511, 25544, 25565, 25570, 25585, 25591, 25636, 25638, 25671-2, 25693,
       25712, 25719-20, 25731, 25750, 25762, 25770, 25798, 25820, 25833, 25836,
       25861-2, 25879-80, 25890, 25932, 25935, 25942, 25959, 25968, 26036, 26048,
       26056, 26059, 26066, 26084, 26100, 26103-5, 26149, 26156, 26172-4, 26181,
       26192-4, 26196, 26198, 26210, B26220, 26224, 26231, 26257-8, 26268, 26275,
       26288, 26294, 26301, 26305, 26325, 26333-4, 26340, 26342-3, 26352, 26354,
       26356, 26366, 26374, 26376, 26378, 26380, 26396, 26418, 26422, 26467-8,
       26476, 26485, 26487-8, 26522-3, 26525, 26539, 26553-4, 26559, 26565-6, 26575,
       26615, 26622, 26624, 26664, 26668, 26683, 26703-4, 26718, 26726, 26730,
       26733, 26760, 26767, 26790-2, 26807, 26809, 26826, 26872, 26888, 26900-2,
       26907, 26917, 26946, 26955, 26958-9, 26967-8, 26994, 27004-6, 27031, 27039,
       27072, 27093, 27099, 27113, 27131, 27135-6, 27138, 27140, 27186, 27194,
       27204, 27222, 27242, 27245, 27263-4, 27297, 27300, 27312, 27324, 27328,
       27334-5, 27354-5, 27399, 27420, 27423, 27468-9, 27491-2, 27495, 27497-9,
       27528-9, 27546, 27566, 27571, 27617-20, 27639-41, 27655, 27657, 27665, 27680,
       27691, 27723-4, 27730, 27742, 27745, 27758, 27772, 27776, 27788, 27816,
       27875, 27881, 27901, 27906-9, 27934, 27943, 27946, 27966, 27968-9, 27977-8,
       28006-7, 28032, 28034, 28072,28075, 28080-2, 28084, 28104, 28132, 28149, 28165,
       28170, 28203, 28225, 28227, 28229, 28235-6, 28258, 28292, 28303, 28305-7,
       28315, 28327, 28332-4, 28371, 28386-7, 28396, 28402, 28409, 28411, 28419,
       28422, 28437, 28445, 28447-8, 28457, 28466, 28488-92

Spinach beet   see *Beta*

Spirodela   see *Lemnaceae*

Spruce   see *Picea*

Squash   see *Cucurbita*

Strawberry   see *Fragaria*

V

Vaccinium 25659, 26895, 26952

Vegetables (*cf.* also *Allium, Asparagus, Bambussa, Beta, Brassica, Capsicum, Cucumis,*
    Cucurbita, Daucus, Lactuca, Lycopersicon, Pastinaca, Petroselinum, Phaseolus,
    Pisum, Portulaca, Raphanus, Solanum, Spinacia)
    26218, 26360, 26809, 27024, 27478, 27673, 27799, 28127, 28130, 28227-8,
    28315, 28487

Vetch see *Vicia*

Vicia 25193, 25311, 25413, 25511, 25618, 25710, 25770, 25920, 25935, 25943, 25965,
    26145, 26252-3, 26398, 26448, 26476, 26561, 26722, 26802, 26813, 26893,
    26994, 27047, 27090, 27109, 27195, 27237, 27288, 27364, 27477, 27566, 27573,
    28037, 28045-6, 28130, 28158, 28199, 28209, 28268, 28285, 28315, 28321,
    28399, 28483

Vigna 25171, 25299, 25431, 25482, 25511, 26465, 27389, 27941-2

Vine see *Vitis*

Vitis 25203, 25521, 25642-3, 25857, 25947, 26392, 26476, 26498, 26775-6, 27009,
    27129, 27134, 27141, 27388, 28018-20

W

Walnut see *Juglans*

Watermelon see *Citrullus*

Weeds (*cf.* also *Amaranthus, Atriplex, Avena, Bromus, Chenopodium, Digitaria, Setaria,*
    etc.)
    25978, 26065, 27022, 27047, 27336, 28092

Wheat see *Triticum*

Whortleberry see *Vaccinium*

Willow see *Salix*

Wolffia see *Lemnaceae*

Y

Yew see *Taxus*

Yucca 28065-6

Z

Zea 25166, 25221, 25297, 25318-9, 25331, 25354, 25397, 25408, 25431, 25440, 25445,
    25449-50, 25457, 25482, 25484, 25511-2, 25521, 25529, 25594, 25662, 25673,
    25678, 25711, 25744, 25770, 25772, 25827, 25835, 25837, 25852, 25858, 25877,
    25900, 25912-3, 25919-20, 25935, 25947, 25967, 25982, 25987, 26020, 26057,
    26063, 26073, 26076, 26091, B26099, 26101, 26117, 26181, 26235, 26237, 26267,
    26315-6, 26318, 26327, 26403, 26437, 26444, 26447, 26474, 26476, 26559, 26575,
    26598, 26626-7, 26665-7, 26706, 26715, 26744-6, 26779, 26789, 26806, 26812,
    26833, 26852, 26862, 26872, 26887, 26890, 26893-4, 26911, 26921, 26932, 26949,
    26994, 27007-8, 27036, 27047, 27106, 27131-2, 27167-9, 27176, 27187, 27211,
    27228, 27240, 27242, 27255, 27275, 27292, 27296, 27300, 27351-2, 27365,
    27402, 27421, 27434-5, 27444, 27468-9, 27506-7, 27532-3, 27546, 27573, 27576-
    -7, 27582, 27592, 27605, 27682, 27687, 27709, 27730, 27792, 27837, 27854,
    27862, 27888, 27896, 27920-1, 27925, 27972-3, 27996, 28001, 28003, 28005,
    28010-1, 28022, 28058, 28078, 28096, 28109, 28130, 28139, 28146, 28167-9,
    28177, 28183, 28196-7, 28201, 28219, 28228, 28265, 28285, 28309-10, 28351,
    28353, 28357, 28393, 28420, 28432, B28482, 28489

Zebrina see *Tradescantia*

Zingiber 28267